Medicinal Plants

Medicinal Plants

Chemistry, Pharmacology, and Therapeutic Applications

Edited by
Mallappa Kumara Swamy
Jayanta Kumar Patra
Gudepalya Renukaiah Rudramurthy

CRC Press
Taylor & Francis Group
Boca Raton London New York

CRC Press is an imprint of the
Taylor & Francis Group, an **informa** business

CRC Press
Taylor & Francis Group
6000 Broken Sound Parkway NW, Suite 300
Boca Raton, FL 33487-2742

© 2019 by Taylor & Francis Group, LLC
CRC Press is an imprint of Taylor & Francis Group, an Informa business

---

Library of Congress Cataloging-in-Publication Data

Names: Swamy, Mallappa Kumara, editor. | Patra, Jayanta Kumar, editor. |
Rudramurthy, Gudepalya Renukaiah, editor.
Title: Medicinal plants : chemistry, pharmacology, and therapeutic
applications / editors, Mallappa Kumara Swamy, Jayanta Kumar Patra,
Gudepalya Renukaiah Rudramurthy.
Description: Boca Raton : Taylor & Francis, 2019. | "A CRC title, part of the
Taylor & Francis imprint, a member of the Taylor & Francis Group, the
academic division of T&F Informa plc." | Includes bibliographical
references and index.
Identifiers: LCCN 2019010552| ISBN 9780367111724 (hardback : alk. paper) |
ISBN 9780429259968 (ebook)
Subjects: LCSH: Materia medica, Vegetable. | Medicinal plants.
Classification: LCC RS164 .M3539 2019 | DDC 615.3/21--dc23
LC record available at https://lccn.loc.gov/2019010552

Visit the Taylor & Francis Web site at
http://www.taylorandfrancis.com

and the CRC Press Web site at
http://www.crcpress.com

Contents

Section I   Medicinal Plants

Section II   Plant Metabolites and Bioactive Compounds

Section III   Bioactive Potential of Medicinal Plants and Treatment against Diseases

Foreword

Medicinal plants are considered a rich source of natural bioactive compounds of medicinal importance. Since ancient times, plants with therapeutic properties have secured an important place in treatment of diseases. More than 30% of the entire plant species, at one time or other, was used for medicinal purposes. Over three-quarters of the world population relies mainly on plants and plant extracts for health care. The herbal drugs from these plants are highly preferred over the synthetic drugs due to safety in their uses, low cost and easy availability. Various technologies have been adopted for enhancing bioactive molecules in medicinal plants. Biotechnological tools are important for the multiplication and genetic enhancement of the medicinal plants by adopting techniques such as *in vitro* regeneration and genetic transformation. Application of biotechnological tools and techniques in medicinal plant research are moving into innovative areas and the technology is transforming itself with faster, economic and expectable options, and hence the present publication is very important.

There are 16 chapters in this book, all contributed by experts in the field of medicinal plants. The selection of chapters in this volume are diverse, providing readers an overview of current applications and potentials related to medicinal plant research and development of noble natural drugs to treat dreadful diseases.

It is fascinating because the chapters focus on various aspects including botany, phytochemistry, extraction methods, molecular mechanisms of actions of plant compounds, their delivery systems, formulation challenges, safety aspects, bioavailability studies, and production using biotechnological tools.

I congratulate the authors and editors (Dr. MK Swamy, Dr. JK Patra and Dr. GR Rudramurthy) of the book *Medicinal Plants: Chemistry, Pharmacology, and Therapeutic Applications* for their contribution through this volume, especially when discussions continue to raise on basic interventions for multifaceted problems and the need to re-visit some fundamentals of medicinal plant research and natural drug discovery to understand their advanced applications and technologies. I hope that this book will be useful to the people who are interested in medicinal plant research.

H. N. Thatoi
Professor
Department of Biotechnology
North Orissa University
Baripada, Odisha, India

Preface

Naturally occurring medicinal plants and their bioactive compounds are being widely used in various formulations as preventive and curative medicines against several acute and chronic diseases, such as atherosclerosis, cancer, neurological diseases, diabetes, arthritis, and aging. Hence, there is a high demand for these plants in traditional medicinal practices, including Ayurveda, Kampo, Traditional Chinese Medicine, and Unani. Further, herbal medicines are highly preferred over synthetic drugs due to safety, low cost and easy availability. As a result, the demand for plant-based products has increased significantly. As bioactive compounds occur in low quantities in nature, their production is now scaled up through biotechnological methods. Major bioactive compounds isolated from medicinal plants include podophyllotoxins, camptothecin, rosmarinic acid, curcumin, resveratrol, catechin, etc. However, there are number of medicinal plants that are yet to be explored for the presence of bioactive compounds and their medicinal properties. Therefore, a thorough knowledge on these plant sources, phytochemistry, and therapeutic activities of phytocompounds will assist in developing novel drug molecules.

Medicinal Plants: Chemistry, Pharmacology, and Therapeutic Applications is a timely effort in this direction. This book, with 16 chapters, provides up-to-date and comprehensive information on the botany, phytochemistry and pharmacological activities of medicinal plants. Further, it emphasizes on various aspects including botany, phytochemistry, extraction methods, molecular mechanisms of actions of plant compounds, their delivery systems, formulation challenges, safety aspects, bioavailability studies, and production using biotechnological tools. We hope that this compiled book on medicinal plants will be helpful for students, educators, researchers and industrial persons who are working in the fields of medicinal plant research, phytochemistry, drug development and discovery, pharmacology and health care sectors.

We sincerely thank and appreciate all the contributors who have readily accepted our invitation to write the chapters to finally produce this interesting book. We also thank the team of CRC Press, USA for their generous cooperation at every stage during the publication of this book.

Mallappa Kumara Swamy
Karnataka, India

Jayanta Kumar Patra
Goyangsi, South Korea

Gudepalya Renukaiah Rudramurthy
Karnataka, India

Editors

Mallappa Kumara Swamy, PhD, is working as a professor, Department of Biotechnology at East West First Grade College of Science, Bengaluru, India. He had worked as a postdoctoral researcher at the Department of Crop Science, Faculty of Agriculture, Universiti Putra Malaysia (UPM), Serdang, Selangor, Malaysia. Previously, he had worked as an associate professor and head, Department of Biotechnology, Padmashree Institute of Management and Sciences, Bangalore University, Bengaluru, India. He earned his PhD (Biotechnology) from Acharya Nagarjuna University, Guntur, India in 2013. He has more than 15 years of teaching and research experience in the fields of plant biotechnology, secondary metabolites production, phytochemistry and bioactive studies. To his credit, he has published more than 80 research publications in peer-reviewed journals and 16 book chapters by reputed book publishers. Recently, he has edited 4 books, published by Springer Nature Singapore Pvt Ltd., Singapore in 2018 and 3 books are in preparation to be published by Springer Nature Singapore Pvt Ltd., Singapore. He is also serving as the editorial board member and reviewer for a few high-impact international journals. Presently, he is working in the area of natural products research, plant cell and tissue culture technology for bioactive compounds production and evaluation of their bioactivities. Also, his research is focused on nanobiotechnology for medical applications.

Jayanta Kumar Patra, MSc, PhD, PDF, is currently working as assistant professor at Dongguk University, Gyeonggi-do, Republic of Korea. He has approximately 12 years of research and teaching experience in the field of food, pharmacological and nanobiotechnology. Dr. Patra completed his PhD (Life Sciences) from North Orissa University, India in pharmacological application of mangrove plant bioactive compounds, and PDF (Biotechnology) from Yeungnam University, South Korea. To his credit, he has published approximately 100 papers in various national and international peer-reviewed journals from various publishers such as Elsevier, Springer, Taylor & Francis Group, Wiley, and so on, and more than 20 book chapters in different edited books. Dr. Patra has also authored five books, *Industrial and Environmental Biotechnology* by STUDIUM Press (India); *Natural Products in Food, Prospects and Applications* by STUDIUM Press LLC USA; *Microbial Biotechnology: Applications in Agriculture and Environment*, and *Microbial Biotechnology: Applications in Food and Pharmacology*, both by Springer Nature publisher, and *Advances in Microbial Biotechnology: Current Trends and Future Prospects* with Apple Academic Press, Inc., Canada, CRC Press. Additionally, a number of book projects are in progress with Apple Academic Press, Inc., Canada, CRC Press, Cambridge Scholars Publishing, Springer Nature and Elsevier publisher etc. He is also serving as editorial board member on several international journals and one science magazine.

Gudepalya Renukaiah Rudramurthy, PhD, is working as a professor and department head, Department of Biotechnology at East West First Grade College of Science, Bengaluru, India. He has earned his PhD (Biotechnology) from Jawaharlal Nehru Technological University, Hyderabad (JNTUH) Telangana, India in 2016. He has more than 17 years of teaching and research experience in the fields of molecular diagnostics, natural product and nanoparticles research. To his credit, he has published more than 30 research and review publications in peer-reviewed international and national journals.

Contributors

Sevcan Adiguzel
Department of Food Engineering
Faculty of Engineering
Istanbul Okan University
Istanbul, Turkey

Manijeh Khorsandi Aghaei
Department of Horticulture
Gorgan University of Agricultural Sciences and Natural
    Resources
Gorgan, Iran

Frank Arfuso
Stem Cell and Cancer Biology Laboratory
School of Biomedical Sciences
Curtin Health Innovation Research Institute
Curtin University
Perth, Western Australia, Australia

Somorita Baishya
Department of Life Science and Bioinformatics
Assam University
Silchar, India

Satheeswaran Balasubramanian
Molecular Toxicology Laboratory
Department of Biotechnology
Bharathiar University
Coimbatore, India

Gausiya Bashri
Ranjan Plant Physiology and Biochemistry Laboratory
Department of Botany
University of Allahabad
Allahabad, India

Manabendra Dutta Choudhury
Department of Life Science and Bioinformatics
Assam University
Silchar, India

Laura A. Contreras-Angulo
Centro de Investigación en Alimentación y Desarrollo A.C.
Functional Foods and Nutraceuticals Laboratory
Culiacan, México

Subrata Das
Department of Life Science and Bioinformatics
Assam University
Silchar, India

Alexis Emus-Medina
Centro de Investigación en Alimentación y Desarrollo A.C.
Functional Foods and Nutraceuticals Laboratory
Culiacan, México

Azim Ghasemnezhad
Department of Horticulture
Gorgan University of Agricultural Sciences and Natural
    Resources
Gorgan, Iran

Mansour Ghorbanpour
Department of Medicinal Plants
Faculty of Agriculture and Natural Resources
Arak University
Arak, Iran

Sandra Gonçalves
Faculty of Sciences and Technology, MeditBio
University of Algarve
Faro, Portugal

Erick P. Gutiérrez-Grijalva
Centro de Investigación en Alimentación y Desarrollo A.C.
Functional Foods and Nutraceuticals Laboratory
Culiacan, México

J. Basilio Heredia
Centro de Investigación en Alimentación y Desarrollo A.C.
Functional Foods and Nutraceuticals Laboratory
Culiacan, México

Gnanasekeran Karthikeyan
Muthayammal Centre for Advanced Research
Muthayammal College for Arts and Science
and
Department of Microbiology
Muthayammal College for Arts and Science
Rasipuram, India

H.L. Koh
Department of Pharmacy
Faculty of Science
National University of Singapore
Singapore

Kallevettankuzhy Krishnannair Sabu
Biotechnology and Bioinformatics Division
Jawaharlal Nehru Tropical Botanic Garden and Research
  Institute
Thiruvananthapuram, India

Vipin Kumar
National Innovation Foundation India
An Autonomous Body of Department of Science and
  Technology
Government of India, Grambharti
Gandhinagar, India

Sankar Malayandi
Department of Biotechnology
Mepco Schlenk Engineering College
Sivakasi, India

Deepa Nath
Department of Botany
Karimganj College
Karimganj, India

Rajat Nath
Department of Life Science and Bioinformatics
Assam University
Silchar, India

S.Y. Neo
Department of Pharmacy
Faculty of Science
National University of Singapore
Singapore

Tugba Ozdal
Department of Food Engineering
Faculty of Engineering
Istanbul Okan University
Istanbul, Turkey

Sasikumar Arunachalam Palaniyandi
Department of Biotechnology
Mepco Schlenk Engineering College
Sivakasi, India

Rabinarayan Parhi
GITAM Institute of Pharmacy
GITAM
Visakhapatnam, India

Ekambaram Perumal
Molecular Toxicology Laboratory
Department of Biotechnology
Bharathiar University
Coimbatore, India

Sheo Mohan Prasad
Ranjan Plant Physiology and Biochemistry Laboratory
Department of Botany
University of Allahabad
Allahabad, India

Kasturi Poddar
Department of Biotechnology and Medical Engineering
National Institute of Technology Rourkela
Rourkela, India

Azhwar Raghunath
Molecular Toxicology Laboratory
Department of Biotechnology
Bharathiar University
Coimbatore, India

Karthikeyan Rajendran
Department of Biotechnology
Mepco Schlenk Engineering College
Sivakasi, India

Madheshwar RajhaViknesh
Muthayammal Centre for Advanced Research
Muthayammal College for Arts and Science
and
Department of Microbiology
Muthayammal College for Arts and Science
Rasipuram, India

Pooja Rawat
National Innovation Foundation India
An Autonomous Body of Department of Science and
  Technology
Government of India, Grambharti
Gandhinagar, India

Anabela Romano
Faculty of Sciences and Technology, MeditBio
University of Algarve
Faro, Portugal

Priyanka Saha
Department of Life Science and Bioinformatics
Assam University
Silchar, India

Angana Sarkar
Department of Biotechnology and Medical Engineering
National Institute of Technology Rourkela
Rourkela, India

Debapriya Sarkar
Department of Biotechnology and Medical Engineering
National Institute of Technology Rourkela
Rourkela, India

Jisha Satheesan
Biotechnology and Bioinformatics Division
Jawaharlal Nehru Tropical Botanic Garden and Research
 Institute
Thiruvananthapuram, India

Gautam Sethi
Department of Pharmacology
Yong Loo Lin School of Medicine
National University of Singapore
Singapore

Rajendran Shurya
Department of Biotechnology
Muthayammal College for Arts and Science
Rasipuram, India

Y.Y. Siew
Department of Pharmacy
Faculty of Science
National University of Singapore
Singapore

D. Singh
Department of Pharmacy
Faculty of Science
National University of Singapore
Singapore

Pawan Kumar Singh
National Innovation Foundation India
An Autonomous Body of Department of Science and
 Technology
Government of India, Grambharti
Gandhinagar, India

Shikha Singh
Ranjan Plant Physiology and Biochemistry Laboratory
Department of Botany
University of Allahabad
Allahabad, India

Natesan Sudhakar
Muthayammal Centre for Advanced Research
Muthayammal College for Arts and Science
and
Department of Microbiology
Muthayammal College for Arts and Science
Rasipuram, India

Mallappa Kumara Swamy
Department of Biotechnology
East West First Grade College of Science
Bengaluru, India

Anupam Das Talukdar
Department of Life Science and Bioinformatics
Assam University
Silchar, India

Gabriela Vázquez-Olivo
Centro de Investigación en Alimentación y Desarrollo A.C.
Functional Foods and Nutraceuticals Laboratory
Culiacan, México

Priyanka Velankanni
Molecular Toxicology Laboratory
Department of Biotechnology
Bharathiar University
Coimbatore, India

Phurpa Wangchuk
Centre for Biodiscovery and Molecular Development of
 Therapeutics
Australian Institute of Tropical Health and Medicine
James Cook University
Cairns, Queensland, Australia

Seung Hwan Yang
Department of Biotechnology
Chonnam National University
Yeosu, Republic of Korea

H.C. Yew
Department of Pharmacy
Faculty of Science
National University of Singapore
Singapore

About the Book

This book is a comprehensive collection of data on various aspects of medicinal plants, including phytochemistry, biological activities and therapeutic potential. Further, it emphasizes major isolated phytocompounds and their immense pharmacological significances. Herbal cures against various human health problems, such as cancer, diabetes, cardiovascular diseases, neurological diseases, microbial infections, skin diseases, aging, and so on, are being discussed in this book. In addition, extraction/isolation of pure compounds from plants, molecular mechanisms of actions of certain plant compounds, safety aspects, and production using biotechnological approaches are highlighted with the latest scientific findings. This comprehensive data will certainly benefit the scientific community to further validate and appraise medicinal benefits of several plant species and simplify the drug discovery process. Further, the content of this book could be useful to the scientific community who are engaged in the field of drug discovery and the development of new drug formulations. In addition, it can be used as a source of information to academic persons.

Section I

Medicinal Plants

1

Gloriosa superba, *a Source of the Bioactive Alkaloid Colchicine: Chemistry, Biosynthesis and Commercial Production*

Sankar Malayandi, Karthikeyan Rajendran, Sasikumar Arunachalam Palaniyandi, and Seung Hwan Yang

CONTENTS

1.1 Introduction

Gloriosa superba (G. superba) is a semi-woody, perennial climber, tuberous and monocot plant that belongs to the *Colchicaceae* family of the order *Liliales*. It is a medicinal plant, which grows in regions with an elevation of more than 1500 meters above sea level. It is native to the tropical and southern part of Africa. The name *Gloriosa* originated from the Latin words glorious and superba from the word superb. *Gloriosa superba* grows naturally in several parts of south Asia, which include India, Burma, Malaysia and Srilanka. In India, it has been predominantly found in northeast state of Assam, western state of Maharashtra, and in southern states such as Goa, Karnataka, Kerala, and Tamil Nadu.

G. superba has been recognized as the national flower of Zimbabwe (where it is a protected plant) as well as the state flower of the Indian state of Tamil Nadu (Arumugam and Gopinath, 2012) (Figure 1.1). In India, the plant has been used for various medicinal purposes by the tribal people. The common name of *G. superba* is glory lily, although it has been known by several other vernacular names in several languages.

G. superba is an herbaceous and climbing perennial plant, which grows between 3.5 and 6 m in length. The vines of the plant are tall, semi-solid, and has tuberous roots that contain high amount of colchicine. Leaves are sessile and alternate. Its flowers look light and solitary and appears greenish at the start and afterward becomes yellow and lastly evolves into scarlet. The plant generates fruit capsules containing several seeds, which are also rich in colchicine. *G. superba* is frequently cultivated during the period of August to March of the year in India.

Hot and humid regions are well suited for the cultivation of *G. superba*. Rainfall of about 400 cm is required for the growth of the plant. It does not resist permanent moisture tension and needs regular irrigation up to flowering in the dry period. Continuous rainfall in the month of December spoils the crop and lowers the yield. Fungal diseases, like leaf blight caused by continuous cloudy weather conditions, affect the crop cultivation.

FIGURE 1.1 Mature flower of *Gloriosa superba*.

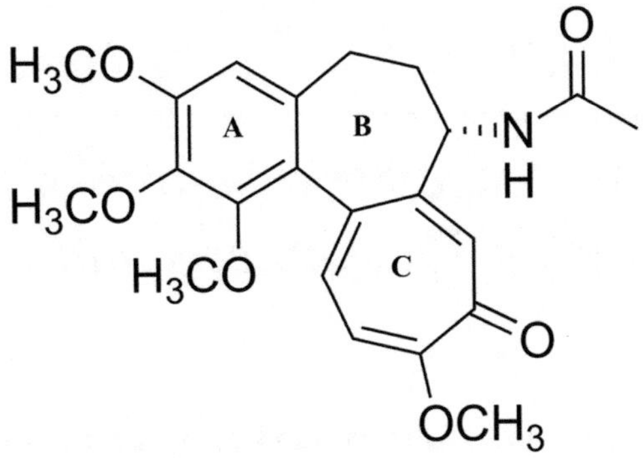

FIGURE 1.2 Structure of colchicine.

G. superba grows in sandy loam red soil with a pH of about 5.5 to 7.0. It grows well in nutrient deprived soil. Proper drainage is required for the growth of the plants to avoid water logging, which affects the rhizome of the plant.

Gloriosa have various amounts of alkaloids, flavonoids, saponins, and resins. Alkaloid is found to be largely present in the *Gloriosa* tubers. Major alkaloids present in *G. superba* are colchicine and gloriosine. In India, *G. superba* is cultivated for its high content of colchicine (Rehana and Nagarajan, 2012).

1.2 Colchicine Chemistry

Colchicine is one of the alkaloids extracted from *Colchicum autumnale* and *G. superba*. Colchicine is a tricyclic alkaloid molecule, consisting of: a trimethoxyphenyl ring (A-ring), a seven-member ring with acetamide at seventh position (B-ring) and a tropolonic ring (C-ring) (Figure 1.2).

1.2.1 Colchicine Structure-Activity Relationship

Colchicine is an anti-mitotic drug that inhibits mitotic cell division in metaphase. Colchicine binds with high affinity to tubulin and prevents the elongation of the microtubule at low concentration and microtubule depolymerization at high concentration (Leung et al., 2015). The high affinity

of colchicine towards tubulin is due to the structure of trimethoxyphenyl ring and tropolonic ring (Lee, 1999a, Bhattacharyya, 2007). Tubulin binds to colchicine between the C- and the A-ring (Boye O, 1992), Isocolchicine, an analogue of colchicine, could not bind with tubulin due to the difference in the positions of methoxy and carbonyl groups in ring C (Lee, 1999a, Bhattacharyya, 2007).

Anti-tubulin activity was retained with replacement of acetamide ring with another alkyl amides. In spite of the anti-tubulin activity retainment by alkyl amides, free amine group reduces the binding activity with tubulin. Replacement of methoxy group with hydroxyl group in the tropolonic ring inhibits the binding capacity of colchicine (Boye, 1992).

1.2.1.1 Trimethoxy Phenyl Ring (Ring A)

A number of synthetic analogues of colchicine were used to study the effect of methoxy substitutions on trimethoxy phenyl ring and its influence on anti-tubulin activity, inhibition of microtubule assembly, and induction of GTPase activity. Removal of methoxy groups at 2, 3, and 4 deteriorated all the three activities. Inhibition of microtubule assembly was strongly dependent on the nature of tropolonic ring. The 4-methoxy group of trimethoxyphenyl ring has been proposed to aid in immobilization of the drug on the protein (Andreu, 1998) (see Figure 1.2).

1.2.1.2 Seven-Member Ring with Acetamide (Ring B)

Colchicine analogues such as 2-Methoxy-5-(2,3,4-trimethoxy-phenyl) tropone and desacetamidocolchicine (Figure 1.3) are found to bind with tubulin in the presence of Taxol. Colchicine binding sites are exposed in the presence of Taxol (Choudhury et al., 1986). Analogues having trimethoxy benzene and methoxytropone combined into a single molecular entity has been shown to possess colchicine-like activity (Detrich, 1982).

Analogues of colchicine bind slowly with tubulin at 4°C and very quickly at 37°C (Ray, 1981, Bane, 1984). B-ring plays an essential role in the binding of colchicine to tubulin. Reversible

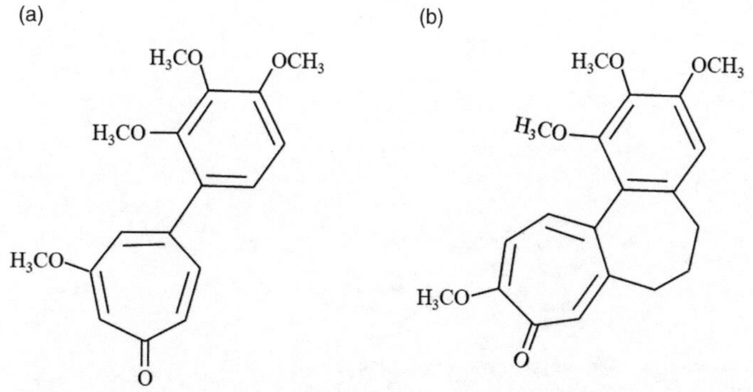

FIGURE 1.3   (a) 2-Methoxy-5-(2,3,4-trimethoxyphenyl) tropone and (b) desacetamidocolchicine. (Adapted from Detrich, H.W. et al., *Biochemistry*, 21, 9, 1982.)

binding of analogues, Colcemid and 2-methoxy-5-(2′,3′,4′-trimethoxyphenyl) tropone, without the B-ring indicate that the B-ring does not have any role in reversible binding (Bane, 1984).

1.2.1.3 Tropolonic Ring (Ring C)

A series of iminonitroso agents were added and reacted with colchicine to form C-ring modified colchicine analogues. Various products were formed by the cycloaddition reaction of colchicine with iminonitroso agents. The analogues of colchicine with C-ring modification were found to be cytotoxic in nature. C-ring modified colchicine also has anti-tubulin activity with changes in the interaction of tubulin (Yang, 2010).

1.3 Biosynthesis of Colchicine

The aromatic amino acids Phenylalanine and tyrosine are the precursors of colchicine. Phenylalanine ammonia lyase (PAL) catalyze the formation of cinnamic acid from phenylalanine (Sivakumar et al., 2004), which is reduced to cinnamaldehyde and subsequently to dihydrocinnamaldehyde (Aroud, 2005). Dihydrocinnamaldehyde is hydroxylated at 4th position to 3-(4-hydroxyphenylpropanal) (Sivakumar, 2018) and L-tyrosine is converted to dopamine by the sequential action of tyrosine hydroxylase and L-dopa decarboxylase (Aroud, 2005). Coumaric acid and dopamine are coupled to form trihydroxylated phenethylisoquinoline, which is converted to autumnaline by methylation involving S-adenosylmethionine (SAM). Autumnaline undergoes intramolecular para,para′-oxidative phenolic coupling involving cytochrome P-450 oxidase (Nasreen et al., 1996) to form isoandrocymbine (Maier and Zenk, 1997), which is converted to O-methylandrocymbine by isoandrocymbine O-methyltransferase (Funayama and Cordell,

2014). O-methylandrocymbine is oxidized by cytochrome P-450 oxidase to formyldemecolcin, which in turn is converted to demecolcine by the action of N-formyldemecolcine deformylase (Rueffer and Zenk, 1998). Demecolcine is converted to deacytylcolchicine, which is converted to colchicine by the action of deacytylcolchicine acetyltransferase (Rueffer and Zenk, 1998). The complete biosynthetic pathway of colchicine is presented Figure 1.4.

1.3.1 Colchicine Derivatives

Several derivatives of colchicine were synthesized and evaluated for their bioactivity such as anticancer and antimicrobial activities. Colchicine derivatives with low toxicity has application in the treatment of cancer (De Vincenzo, 1999). Several derivatives exhibited a high-binding affinity towards tubulin (Uppuluri et al., 1993, Lowe et al., 2001) and low toxicity when compared to native colchicine (Muzaffar et al., 1990, Kurek, 2014). Substitution of the C-3 methyl ester moiety with an amide residue displayed reduced neurotoxicity (Lee, 1999b). N-acylated analogs of colchicine and demecolcine were effective against P388 leukemia cells with lower toxicity than colchicine (Roesner et al., 1981, De Vincenzo, 1999). Derivatives with bis(2-methoxyethyl) amine showed antimicrobial activity against methicillin-resistant *Staphylococcus aureus* (MRSA) (De Vincenzo, 1999). Sulfur-containing derivatives exhibited strong affinity towards tubulin and toxicity against P388 cancer cells (Kerkes et al., 1985) and were also effective against multi-drug-resistant tumor cells (De Vincenzo, 1999).

In an interesting study, a combination of colchicine and paclitaxel was synthesized by coupling the two agents with a glutarate linker. The resulting molecule is called colchitaxel. Colchitaxel exhibited same activity on microtubules as that of a combination of colchicine and paclitaxel (Bombuwala et al., 2006). A more detailed review of colchicine derivatives and their bioactivities can be found in (Dubey et al., 2017).

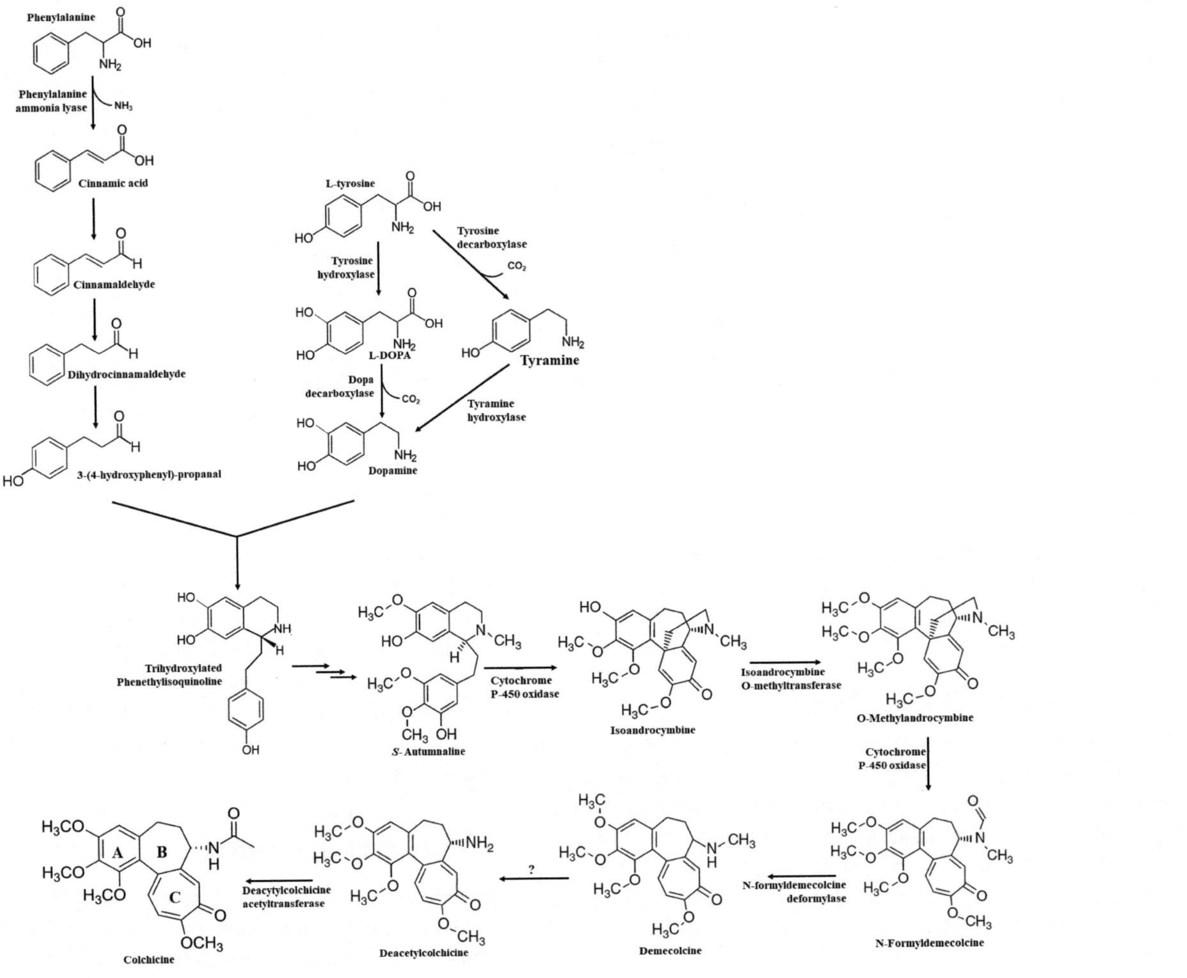

FIGURE 1.4 Biosynthetic pathway of Colchicine.

1.4 Commercial Production of Colchicine

1.4.1 Cultivation of *Gloriosa superba*

Commercial production of colchicine depends on field grown *Gloriosa superba*, which is one of the successful sources of colchicine. Rhizomes are used for the propagation of *Gloriosa superba* in the field. Field grown *G. superba* accumulate a high amount of colchicine in its seeds (~0.9%) and rhizomes (~0.3%) (Sivakumar, 2013). *G. superba* produces bi-forked rhizomes, and each of these forks have only one growing bud. In India, the tubers start sprouting in the month of May and planting the crops starts in the months of July and August. During these planting seasons the condition would favor better plant growth and tuber yield of about 2.5 to 3 ton/ha.

1.4.2 *In Vitro* Propagation of *Gloriosa* and Colchicine Production

The content of colchicine in field grown plants is low and obtaining it is limited to certain growing seasons and conditions. *In vitro* propagation presents an alternative way of producing

colchicine throughout the year. *G. superba* rhizomes were used as explants for induction of callus with IAA and 2,4-D at 0.5 and 1 mg/L, respectively. Multiple shoots were induced with BAP in half-strength MS medium (Arumugam and Gopinath, 2012). A combination of BAP (1.5 mg/L) and NAA (0.5 mg/L) induced maximum shoot formation. Addition of 10% coconut water in MS medium also influenced increased formation of shoots by 10% to 20% (Akter et al., 2014).

Elicitation of *in vitro* plant cultures with biotic and abiotic elicitors is an important strategy for the production of secondary metabolites from plants. The effect of elicitors on colchicine production by *in vitro* root cultures of *G. superba* (Ghosh et al., 2006). Four biotic elicitors such as methyl jasmonate (MJ), salicylic acid (SA), casein hydrolysate (CH), yeast extract (YE), and four abiotic elicitors such as $CdCl_2$, $AlCl_3$, $CaCl_2$, and $AgNO_3$ were tested for induction of colchicine in root cultures (Ghosh et al., 2006). A 50-fold increase in intracellular colchicine levels was observed when root cultures were treated with 5 mM methyljasmonate. Other biotic elicitors were not as effective as methyljasmonate in inducing synthesis of colchicine. A 63-fold increase in colchicine content was observed in root cultures of *G. superba* when treated with $AlCl_3$, which was higher than that of the other abiotic elicitors (Ghosh et al., 2006). In another study (Mahendran et al., 2018) callus culture of *G. superba* was treated with various elicitors such as SA, YE, CH, and AgNO$_3$, which showed that CH induced a maximum level of colchicine over a period of 15 days (Mahendran et al., 2018).

Another strategy for maximal colchicine production *in vitro* is to supply the plant cultures with precursors involved in the biosynthesis of colchicine. *G. superba* calluses were supplied with the precursor aminoacids pheylalanine and tyrosine, which resulted in high colchicine production in case of tyrosine feeding compared to phenylalanine feeding (Sivakumar et al., 2015). Another study explored the effect of precursor feeding in root cultures, which used phenylalanine, tyramine, and *p*-coumaric acid (Ghosh et al., 2002). All these precursors enhanced colchicine production without increasing root growth (Ghosh et al., 2002).

1.4.3 Colchicine Biomanufacturing

Production of colchicine through biorhizomes is a recently developed alternative approach for its commercial production. Biorhizomes are asexually developed rootstocks grown *in vitro*, which are capable of producing new shoots, adventitious roots, and daughter biorhizomes that can serve as reproductive as well as storage organs (Sivakumar et al., 2017). These *in vitro* developed organs are a unique source of manufacturing phytochemical from rhizomatous plants (Sivakumar, 2018). Biorhizomes could continuously synthesize colchicine compared to root cultures and field grown rhizomes, which has low concentration of colchicine (Sivakumar, 2013). Biosynthesis of colchicine is upregulated in the biorhizome compared to that in adventitious root cultures (Sivakumar et al., 2017). The content of colchicine in biorhizome was also high (>0.5%) compared to <0.1% in leaves and stems (Sivakumar et al., 2017). In bioreactor, the roots-derived biorhizomes continuously produced colchicine, whereas shoots-derived biorhizomes lose this ability and induce new daughter biorhizomes (Sivakumar et al., 2017).

1.4.4 Extraction of Colchicine

Colchicine extraction from *G. superba* primarily involves the use of aqueous organic solvents such as methanol (Kannan et al., 2007, Joshi et al., 2010) and ethanol (Ellington et al., 2003b). Colchicine has been extracted from dried and powdered seeds of *G. superba* by optimizing the ratio of methanol and water. Maximum yield was obtained with 100% methanol (Kannan et al., 2006). Colchicine is extracted from the *Gloriosa* seeds with methanol using soxhlet apparatus and sonicator. Extraction using Soxhlet apparatus yielded high colchicine levels compared to sonicator (Kumar et al., 2016).

Statistical optimization was performed for the maximum extraction of colchicine from *G. superba*, which analyzed several extraction parameters such as temperature, time, particle size, ratio of solvent-solids, solvent composition, pH, and number of extraction steps using Box-Behnken Design. The optimal conditions for the extraction of the colchicine are found to be single-step extraction at 35°C and pH 7 for 70 minutes with a solvent-solid ratio of 50:1 and mean particle size of 0.5 mm using 70% ethanol (Pandey and Banik, 2012).

Supercritical CO$_2$ extraction was performed in corms of *Colchicum autumnale* for the extraction of colchicine. A pressure range of 20–40 Mpa was tested at 40°C with methanol, ethanol or acetone as pre-soakers to increase mass transfer. Optimum extraction of colchicine was obtained with ethanol pre-soaked corms at 35 MPa and 40°C (Ruibin and Shihong, 1999). Optimal extraction of colchicine from seeds of *C. autumnale* was observed at a CO$_2$ density of 0.90 g/mL, carbon dioxide flux of 1.5 mL/min with 3% methanol as modifier at 35°C for 30 minutes (Ellington et al., 2003a).

1.5 Medicinal Uses of Colchicine

1.5.1 Gout Treatment

Gout is a familiar form of arthritis, which is mainly due to increased uric acid levels in the joints of the human body. Gout has been described with extreme joint pain, inflamed joints, low range of motion, and sudden attack of pain. Uric acid is not excreted due to genetic disorder in the urate transporter proteins and dietary habits. The reduced excretion of uric acid leads to hyperuricemia and results in formation of uric acid crystals in kidneys (Tausche et al., 2009). Colchicine has been prescribed for the treatment of gout. Clinical studies on colchicine for gout treatment demonstrated the superiority of colchicine over placebo in 185 patients randomized to a high dose or low dose of colchicine. A high dose of colchicine causes gastrointestinal side effects; hence, use of a low dose of colchicine is recommended for treatment of gout (Leung et al., 2015).

1.5.2 Cancer Treatment

The primary mechanism of colchicine's anti-cancer activity is by its high affinity binding to tubulin and arresting cell division. Colchicine also inhibits cancer cell migration and metastasis, cell blebbing, inhibition of angiogenesis, limit ATP influx into mitochondria and release of caspases and cytochrome-c that leads to apoptosis (Leung et al., 2015).

Patients with gout are more prone to develop cancer than the patients without gout. In a cohort study, the effects of colchicine on cancer development in male patients with gout were analyzed. Male patients undergoing colchicine treatment are found to have less incidence of cancer compared to patients not taking colchicine. Gout patients without colchicine uptake have a high incidence of all cancers (Kuo et al., 2015). Colchicine at low concentration has been found to show anti-cancer activity in gastric cell lines. *In vivo* studies show that there is a significant decrease in the tumor in mice after 14 days (Lin, 2016).

1.6 Colchicine Toxicity

In spite of the medicinal properties, consumption of *Gloriosa* results in many side effects such as hair loss, gastroenteritis, diarrhea, vomiting, and fever (Bains et al., 2016). The entire *G. superba* plant is toxic due to high content of colchicine alkaloid. Toxic effect is mainly due to the anti-mitotic activity of colchicine. Bone marrow cells, intestinal epithelium, and hair follicles were easily affected by the *G. superba* poisoning. Sixty milligrams of *G. superba* extract is toxic to adults and incurable periods are about 12–72 hours. The symptoms of colchicine poisoning by consuming the *Gloriosa* plant parts were gastrointestinal, alopecia, vomiting, numbness, dermatitis, gastroenteritis, acute renal failure, cardiotoxicity, and hematological abnormalities.

Colchicine is also cardiotoxic, affecting cardiac impulse generation and conduction. It was the first time to report the cardiotoxic nature of the colchicine with electrocardiographic images (Mendis, 1989). Consumption of *G. superba* plant leads to a delayed onset of the toxic encephalopathy due to the binding nature of colchicine with tubulin and disruption of tubulin polymerization, which leads to the neuronal cell death (Gooneratne et al., 2014).

Alopecia was reported individuals who ingested *Gloriosa* tubers. Alopecia was due to the alkaloids gloriosine and colchicine (Gooneratne, 1966). Consumption of *Gloriosa* tubers leads to *Anagen effluvium*, an abnormal loss of hair in a short period of time. The hair loss depends upon the quantity of tubers consumed.

1.7 Future Perspectives

Gloriosa superba is a very promising source for the commercial production of colchicine. The plant has been over-exploited for colchicine production and has been extensively harvested from its natural habitat. However, cultivation measures have been taken as a countermeasure to stop the overreliance on naturally grown *Gloriosa* for commercial production. Biomanufacturing technologies must be developed for its sustainable commercial production, which is still in its infancy. Novel ecofriendly methods for the extraction of colchicine from *Gloriosa* are urgently needed in order to avoid environmental pollution from manufacturing facilities due to the release of organic solvents used in the extraction process.

REFERENCES

Akter, S., Roy, P. K., Mamun, A. N. K., Islam, M. R., Kabir, M. H. & Jahan, M. T. 2014. *In vitro* regeneration of *Gloriosa superba* L.—an overexploited medicinal plant in Bangladesh. *Nuclear Science and Applications*, 23, 55–59.

Andreu, J. M., Perez-Ramirez, B., Gorbunoff, M. J., Ayala, D., Timasheff, S. N. 1998. Role of the colchicine ring A and its methoxy groups in the binding to tubulin and microtubule inhibition. *Biochemistry*, 37, 3.

Aroud, G. 2005. Production of colchicine by plant cell culture. PhD, Dublin City University.

Arumugam, A. & Gopinath, K. 2012. *In vitro* micropropagation using corm bud explants—an endangered medicinal plant of *Gloriosa superba*. *Asian Journal of Biotechnology*, 4, 120–128.

Bains, A., Verma, G. K., Vedant, D. & Negi, A. 2016. Anagen effluvium secondary to Gloriosa superba ingestion. *Indian Journal of Dermatology, Venereology, and Leprology*, 82, 677–680.

Bane, S., Puett, D., Macdonald, T. L., Williams, R. C. 1984. Binding to tubulin of the colchicine analog 2-methoxy-5-(2′,3′,4′-trimethoxyphenyl) tropone. Thermodynamic and kinetic aspects. *Journal of Biological Chemistry*, 259, 8.

Bhattacharyya, B., Panda, D., GUPTA, S., Banerjee, M. 2007. Antimitotic activity of colchicine and the structural basis for its interaction with tubulin. *Medicinal Research Reviews*, 28, 29.

Bombuwala, K., Kinstle, T., Popik, V., Uppal, S. O., Olesen, J. B., Vina, J. & Heckman, C. A. 2006. Colchitaxel, a coupled compound made from microtubule inhibitors colchicine and paclitaxel. *Beilstein Journal of Organic Chemistry*, 2, 13.

Boye O, B. A. 1992. *The Alkaloids: Chemistry and Pharmacology*, In: Brossi A, Cordell GA, editors. Academic Press, San Diego, CA, 41, 51.

Choudhury, G. G., S. Maity, B. Bhattacharyya, & Biswas. B. B. 1986. B-ring of colchicine and its role in taxol-induced tubulin polymerization. *FEBS Letters*, 197, 4.

De Vincenzo, R., Ferlini, C., Distefano, M., Gaggini, C., Riva, A., Bombardelli, E., Morazzoni, P. et al. 1999. Biological evaluation on different human cancer cell lines of novel colchicine analogs. *Oncology Research Featuring Preclinical and Clinical Cancer Therapeutics* 11, 8.

Detrich, H. W., Williams, R. C., & Wilson, L. 1982. Effect of colchicine binding on the reversible dissociation of the tubulin dimer. *Biochemistry*, 21, 9.

Dubey, K. K., Kumar, P., Labrou, N. E. & Shukla, P. 2017. Biotherapeutic potential and mechanisms of action of colchicine. *Critical Reviews in Biotechnology*, 37, 1038–1047.

Ellington, E., Bastida, J., Viladomat, F. & Codina, C. 2003a. Supercritical carbon dioxide extraction of colchicine and related alkaloids from seeds of Colchicum autumnale L. *Phytochemical Analysis*, 14, 164–169.

Ellington, E., Bastida, J., Viladomat, F., Simanek, V. & Codina, C. 2003b. Occurrence of colchicine derivatives in plants of the genus Androcymbium. *Biochemical Systematics and Ecology*, 31, 715–722.

Funayama, S. & Cordell, G. A. 2014. *Alkaloids Derived from Phenylalanine and Tyrosine*. Elsevier, Burlington, UK.

Ghosh, B., Mukherjee, S., Jha, T. B. & Jha, S. 2002. Enhanced colchicine production in root cultures of *Gloriosa superba* by direct and indirect precursors of the biosynthetic pathway. *Biotechnology Letters*, 24, 231–234.

Ghosh, S., Ghosh, B. & Jha, S. 2006. Aluminium chloride enhances colchicine production in root cultures of Gloriosa superba. *Biotechnology Letters*, 28, 497–503.

Gooneratne, B. W. 1966. Massive generalized alopecia after poisoning by Gloriosa superba. *British Medical Journal*, 1, 1023–1024.

Gooneratne, I. K., Weeratunga, P., Caldera, M. & Gamage, R. 2014. Toxic encephalopathy due to colchicine--Gloriosa superba poisoning. *Practical Neurology*, 14, 357–359.

Joshi, C. S., Priya, E. S. & Mathela, C. S. 2010. Isolation and anti-inflammatory activity of colchicinoids from Gloriosa superba seeds. *Pharmaceutical Biology*, 48, 206–209.

Kannan, S., Wesley, S. D., Ruba, A., Rajalakshmi, A. R. & Kumaragurubaran, K. 2007. Optimization of solvents for effective isolation of colchicines from Gloriosa superba L. seeds. *Natural Product Research*, 21, 469–472.

Kannan, S., Wesley, S. D., Ruba, A., Rajalakshmi. A. R., Kumaragurubaran, K. 2006. Optimization of solvents for effective isolation of colchicines from Gloriosa superba L. seeds. *Natural Product Research*, 21, 4.

Kerkes, P., Sharma, P. N., Brossi, A., Chignell, C. F. & Quinn, F. R. 1985. Synthesis and biological effects of novel thiocolchicines. 3. evaluation of N-acyldeacetylthiocolchicines, N-(alkoxycarbonyl) deacetylthiocolchicines, and O-ethyldemethylthiocolchicines.

New synthesis of thiodemecolcine and antileukemic effects of 2-demethyl- and 3-demethylthiocolchicine. *Journal of Medicinal Chemistry*, 28, 1204–1208.

Kumar, V., Gupta, G. & Rane, A. D. 2016. Standardization of drying and extraction techniques for better colchicine recovery from *Gloriosa superb*. *Der Pharmacia Lettre*, 8, 4.

Kuo, M. C., Chang, S. J. & Hsieh, M. C. 2015. Colchicine significantly reduces incident cancer in gout male patients a 12-year cohort study. *Medicine* 94, 6.

Kurek, J., Boczon, W., Myszkowski, K., Murias, M., Borowiak, T., Wolska, I. 2014. Synthesis of sulfur containing colchicine derivatives and their biological evaluation as cytotoxic agents. *Letters in Drug Design and Discovery*, 11, 11.

Lee, K. H. 1999b. Anticancer drug design based on plant-derived natural products. *Journal of Biomedical Science*, 6, 236–250.

Lee, K.-H. 1999a. Novel antitumor agents from higher plants. *Medcinal Research Reviews*, 19, 28.

Leung, Y. Y., Hui, Y. L. L. & Kraus, V. B. 2015. Colchicine—Update on mechanisms of action and therapeutic uses. *Seminars in Arthritis and Rheumatism*, 45, 341–350.

Lin, Z. Y., Kuo, C. H., Wu, D. C. & Chuang, W. L. 2016. Anticancer effects of clinically acceptable colchicine concentrations on human gastric cancer cell lines. *The Kaohsiung Journal of Medical Sciences*, 32.

Lowe, J., Li, H., Downing, K. H. & Nogales, E. 2001. Refined structure of alpha beta-tubulin at 3.5 A resolution. *Journal of Molecular Biology*, 313, 1045–1057.

Mahendran, D., Kishor, P. B. K., Sreeramanan, S. & Venkatachalam, P. 2018. Enhanced biosynthesis of colchicine and thiocolchicoside contents in cell suspension cultures of *Gloriosa superba* L. exposed to ethylene inhibitor and elicitors. *Industrial Crops and Products*, 120, 123–130.

Maier, U. H. & Zenk, M. H. 1997. Colchicine is formed by para-para phenol coupling from autumnaline. *Tetrahedron Letters*, 38, 7357–7360.

Mendis, S. 1989. Colchicine cardiotoxicity following ingestion of Gloriosa superba tubers. *Postgraduate Medical Journal*, 65, 752–755.

Muzaffar, A., Brossi, A., Lin, C. M. & Hamel, E. 1990. Antitubulin effects of derivatives of 3-demethylthiocolchicine, methylthio ethers of natural colchicinoids, and thioketones derived from thiocolchicine. Comparison with colchicinoids. *Journal of Medicinal Chemistry*, 33, 567–571.

Nasreen, A., Rueffer, M. & Zenk, M. H. 1996. Cytochrome P-450-dependent formation of isoandrocymbine from autumnaline in colchicine biosynthesis. *Tetrahedron Letters*, 37, 8161–8164.

Pandey, D. K. & Banik, R. M. 2012. Optimization of extraction conditions for colchicine from *Gloriosa superba* tubers using response surface methodology. *Journal of Agricultural Technology*, 8, 1301–1315.

Ray, K., Bhattacharyya, B. & Biswas, B. B. 1981. Role of B-ring of Colchicine in Its Binding to Tubulin. *Journal of Biological Chemistry*, 259, 4.

Rehana, B. & Nagarajan, N. 2012. Phytochemical screening for active compounds in Gloriosa superba leaves and tubers. *International Journal of Pharmacognosy and Phytochemical Research*, 4, 4.

Roesner, M., Capraro, H. G., Jacobson, A. E., Atwell, L., Brossi, A., Iorio, M. A., Williams, T. H., Sik, R. H. & Chignell, C. F. 1981. Biological effects of modified colchicines. Improved preparation of 2-demethylcolchicine, 3-demethylcolchicine, and (+)-colchicine and reassignment of the position of the double bond in dehydro-7-deacetamidocolchicines. *Journal of Medicinal Chemistry*, 24, 257–261.

Rueffer, M. & Zenk, M. H. 1998. Microsome-mediated transformation of O-methylandrocymbine to demecolcine and colchicine. *FEBS Letters*, 438, 111–113.

Ruibin, F. & Shihong, Z. 1999. Supercritical fluid CO_2 extraction of Colchicine from *Colchicine autumnele liliaceae*. *Chinese Journal of Chromatography*, 17, 249–252.

Sivakumar, G. 2013. Colchicine semisynthetics: Chemotherapeutics for cancer? *Current Medicinal Chemistry*, 20, 892–898.

Sivakumar, G. 2018. Upstream biomanufacturing of pharmaceutical colchicine. *Critical Reviews in Biotechnology*, 38, 83–92.

Sivakumar, G., Alba, K. & Phillips, G. C. 2017. Biorhizome: A biosynthetic platform for colchicine biomanufacturing. *Frontiers in Plant Science*, 8, 1137.

Sivakumar, G., Krishnamurthy, K. V., Hahn, E. J. & Paek, K. Y. 2015. Enhanced in vitro production of colchicine in Gloriosa superba L.—an emerging industrial medicinal crop in South India. *The Journal of Horticultural Science and Biotechnology*, 79, 602–605.

Sivakumar, G., Krishnamurthy, K. V., Hao, J. & Paek, K. Y. 2004. Colchicine production in *Gloriosa superba* calluses by feeding precursors. *Chemistry of Natural Compounds*, 40, 499–502.

Tausche, A. K., Jansen, T. L., Schröder, H. E., Bornstein, S. R., Aringer, M., Müller-Ladner, U. 2009. Gout—Current diagnosis and treatment. *Deutsches Ärzteblatt International*, 106, 549–555.

Uppuluri, S., Knipling, L., Sackett, D. L. & Wolff, J. 1993. Localization of the colchicine-binding site of tubulin. *Proceedings of the National Academy of Sciences*, 90, 11598.

Yang, B., Zhu, Z. C., Goodson, H. V., MILLER, M. J. 2010. Syntheses and biological evaluation of ring-C modified colchicine analogs. *Bioorganic & Medicinal Chemistry Letters*, 20, 3.

2

Botany, Phytochemistry and Pharmacological Activities of Leea Species

D. Singh, Y.Y. Siew, H.C. Yew, S.Y. Neo, and H.L. Koh

CONTENTS

2.1 Introduction

The genus *Leea* belongs to the family Vitaceae and comprises about 36 species (The Plant List, 2013). *Leea* species are distributed in the tropical and subtropical regions of Asia, Africa, and Madagascar (Ridsdale, 1974, 1976; Op de Beck et al., 2003; Molina et al., 2013). Most of the *Leea* species are grown in lowland evergreen forests up to an altitude of 2500 m (Wen, 2007). The plants are erect herbs, shrubs, and trees with terminal inflorescences and characteristically large stipules (Molina et al., 2013). Leaves comprise 1–4 pinnate with globular pearl glands, leaflets opposite, cymose inflorescence, hermaphrodite flowers in a floral disc capped by connate stamens, discoidal ovary and fruits as berries are the characteristics of the genus *Leea* (Ridsdale, 1976; Molina et al., 2013; Cabelin et al., 2015).

In the past 5 years, research interests in the genus *Leea* have expanded in the areas of systematics, isolation of secondary metabolites, analytical chemistry and pharmacological activities, especially for *L. macrophylla* and *L. indica* (Wong et al., 2012a, 2012b; Molina et al., 2013; Sen et al., 2013; Lakornwong et al., 2014; Cabelin et al., 2015; Joshi et al., 2016a, 2016b). Although there are 36 species in the genus *Leea*, only a few species have been investigated so far. Hence in this review, we discuss the findings for *L. asiatica*, *L. guineensis*, *L. indica*, *L. macrophylla* and *L. rubra* which have been investigated for both phytochemistry and pharmacological activity; *L. philippines* and *L. tetramera* which have been studied for pharmacological activity; *L. thorelii* which has been examined for phytochemistry. The botanical characteristics of these *Leea* species are also presented here.

The aim of this review is to provide updated and comprehensive information on botany, ethnomedicinal uses, phytochemistry, pharmacological activities, and safety considerations of the genus *Leea*.

2.2 Botanical Characteristics and Geographical Distribution

Leea species are widely distributed in tropical and subtropical Asian countries such as India, China, Malaysia, Vietnam, the Philippines, Singapore, Thailand, tropical Africa and Australia (Ridsdale, 1974, 1976; Op de Beck et al., 1998; Wen, 2007; Lok et al., 2011; Molina et al., 2013). Dating analyses indicated that the genus *Leea* originated from Indo-China in Late Cretaceous and subsequently spread to other places (Molina et al., 2013). The *Leea* species are small trees, shrubs or herbs with a woody base and globular or stellate "pearl" glands on the leaf base. The flowers of the genus *Leea* are characteristically present in a tube within the whorl of the stamens. On the basis of color of the flowers, taxonomist Clarke divided genus *Leea* into two series named *Rubriflorae* (red flowers) and *Viridiflorae* (green flowers) (Ridsdale, 1974).

Botanists have different opinions about assigning the family to this genus. Some considered the genus *Leea* as its own monogeneric family Leeaceae, characterized by erect trees or shrubs lacking tendrils and flowers with a clear floral tube (Ridsdale, 1974, 1976; Wen, 2007; Molina et al., 2013). Others consider the genus *Leea* to be under the family Vitaceae (Angiosperm Phylogenic Group [APG] 1998; Ingrouille et al., 2002; APG II, 2003; APG III, 2009;

Cabelin et al., 2015). APG treats *Leea* as the only genus in the subfamily Leeoideae Burmeister of Vitaceae. The genus *Leea* and Vitaceae are placed under the order Vitales (Molina et al., 2013). The inclusion of *Leea* in Vitaceae is based on the characteristics of "pearl" glands, shared corolla-stamen primordia, raphides, phloem plastids, and ruminate seeds with oily endosperm (APG III, 2009). The exact number of species in genus *Leea* is currently inconclusive. Various studies have proposed varying number of species in the genus *Leea*: 34 species (Ridsdale, 1974, 1976; Wen, 2007); 36 species (The Plant List, 2013), ~70 species (Heywood et al., 2007); 80 species (Cronquist, 1988); and 153 species (Li, 1998). Clearly, more detailed phylogenetic analyses are necessary.

In our evaluation of the published literature, we found that Op de Beck and coworkers worked on the same plant collected from Bella territory (Cameroon) and deposited under voucher specimen number 3115 at National Herbarium, Museum of Yaounde. However, they named it differently, i.e. *L. guinensis* G. Don, *L. guineensis* G. Don and *L. guineense* G. Don in their three research papers (Op de Beck, 1998, 2000, 2003). In our review, we considered the species they worked on to be *L. guineensis*. For reports discussing the pharmacological activity of *L. crispa* (Singh et al., 2002; Marles and Fransworth, 1995), we consider *L. crispa* to be a synonym of *L. asiatica* in agreement with The Plant List.

This review is the first report on 36 accepted *Leea* species (The Plant List, 2013), as shown in Table 2.1 along with their synonyms and geographical distributions. The botanical characteristics of 8 *Leea* species used for phytochemical investigation and pharmacological studies are presented below.

2.2.1 *L. asiatica* (L.) Ridsdale

L. asiatica is commonly known as Banchalita in India. It is an erect shrub with swollen nodes. Leaves are pinnately compound. Leaflets are 3–5, laterals opposite, ovate, serrate with sharp tip. Petioles and peduncles are usually narrow. Flowers are greenish white in color with 5–6 mm in diameter and cymes at the end of branches. Calyx is united, cup-like, 5 toothed, and obscure. Petals are five in number, 2–3 mm long, connate, ovate, and acute (Flowers of India, 2016).

2.2.2 *L. guineensis* G. Don

L. guineensis is commonly known as *Red Tree Vine* and marketed as *L. coccinea* (Wen, 2007). It is an evergreen shrub or small tree of 1–10 m in height. Leaves are 1 to 4 pinnate, light green at early stage and glossy dark green after maturation. Inflorescences are 3–40 cm long. Flowers are reddish orange in color and 5-merous. Stamina tube is red to citrus white in color. Flower blooms abundantly for about 3 months from August to October (Ridsdale, 1976; Molina, 2009). Fruits are red, subglobose berries and about 5–15 mm in diameter. Seeds are usually six in number and endosperms simply ruminate. It is found in Northern and Eastern Australia, New Guinea, South and Southeast Asia, Africa including Nigeria, and at 2250 m in the Himalayan range (Ridsdale, 1976; Molina, 2009).

2.2.3 *L. indica* Merr

L. indica, commonly known as bandicoot berry (English), Memali (Malay), yan tuo (Chinese), katangbai (Thai) and

TABLE 2.1

List of Reported *Leea* Species

No.	Species	Synonyms	Distributions
1	*L. aculeata* Bl. ex Spreng	*L. aculeata* var. *moluccana* Miq.; *L. sandakanensis* Ridl.; *L. serrulata* Miq.	Philippines, Java, Indonesia
2	*L. acuminatissima* Merr.	Not available	Philippines
3	*L. aequata* L.	*L. ancolona* Miq.; *L. hirsuta* Blume ex Spreng.; *L. hirta* Roxb. ex Hornem.; *L. hispida* Gagnep.; *L. Kurzii* C.B. Clarke; *L. hirsuta* Blume ex Spreng.; *L. hirta* Roxb. ex Hornem.; *L. hispida* Gagnep.; *L. kurzii* C.B. Clarke	India, Thailand, Malaysia, Vietnam, Cambodia
4	*L. alata* Edgeworth	Not available	India, Nepal, Bhutan
5	*L. amabilis* Veitch ex Mast.	*L. amabilis* var. *spledens* Linden & Rodigas	Borneo
6	*L. angulata* Korth. ex Miq.	*L. horrida* T. & B.; *L. sambucina* var. *intermedia* Ridl.	India, Thailand, Malaysia
7	*L. asiatica* (L.) Ridsdale	*L. crispa* L.; *L. edgeworthii* Santapau; *L. herbacea* Buch-Ham.; *L. pinnata* Andrews; *L. pumila* Kurz; *Phytolacca asiatica*	India
8	*L. compactiflora* Kurtz	*L. bracteata* C.B. Clarke; *L. trifoliata* M.A. Lawson	India, China, Bhutan, Vietnam
9	*L. congesta* Elmer	*L. capitata* Merr.	Philippines
10	*L. coryphantha* Lauterb.	Not available	New Guinea
11	*L. curtisii* King.	*L. stipulosa* Gagnep.	Vietnam, Malaysia
12	*L. glabra* C. L. Li	Not available	Not available
13	*L. gonioptera* Lauterb.	Not available	New Guinea
14	*L. grandifolia* Kurtz	Not available	India
15	*L. guineense* G. Don	Not available	Not available
16	*L. guineensis* G. Don	*L. acuminata* Wallich ex Clarke; *L. arborea* Sieber ex Bojer; *L. arborea* Telf. Ex Wight & Arn.; *L. aurantiaca* Zoll. & Moritzi; *L. bipinnata* Boivin; *L. bulusanensis* Elmer; *L. coccinea* Planch.; *L. cumingii* C.B. Clarke; *L. cuspidifera* Baker; *L. dentata* W.G. Craib; *L. euphlebia* Merr.; *L. guineensis* f. comoriensis Desc.; *L. guineensis* var. cuspidefera (Baker) Desc.; *L guineensis* f. *longifoliata* Desc.; *L. guineensis* f. *monticola* Desc.; *L. guineensis* f. *orientalis* Desc.; *L. guineensis* f. *spiculata* Desc.; *L. guineensis* f. *truncata* Desc.; *L. laeta* Wall. ex Kurz.; *L. luzonensis* Elmer; *L. maculata* Desf.; *L. manillensis* Walp.; *L. negrosense* Elmer; *L. palawanensis* Elmer; *L. papillosa* Merr.; *L. parva* var. Elmer; *L. parvifolia* Merr.; *L. punctata* Desf. Ex Planch.; *L. sambucina* var. *arborea* (Sieber ex Bojer) Miq.; *L. sanguinea* Wall. ex Kurz; *L. schomburgkii* W.G. Craib; *L. speciosus* Siebold ex Miq.; *L. wightii* C.B. Clarke	Africa, India, Taiwan, Thailand, Java, Sumatra, Malaysia
17	*L. indica* (Burm. f.) Merr.	*Aqilicia otilis* Gaertn.; *L. biserrata* Miq; *L. celebica* C.B. Clarke; *L. expansa* W.G. Craib; *L. fuliginosa* Miq.; *L. gigantea* Griff.; *L. gracilis* Lauterb.; *L. naumannii* Engl.; *L. novoguineensis* Val.; *L. ottilis* (Gaertn.) DC.; *L. palambanica* Miq.; *L. pubescens* Zipp. Ex Miq.; *L. ramosii* Merr.; *L. roehrsiana* Sanders ex Masters; *L. sumatrana* Miq.; *L. sundaica* Miq.; *L. sundaica* var. *fuliginosa* (Miq.) Miq.; *L. sundaica* var. *pilosiuscula* Span. Ex Miq.; *L. sundaica* var. *subsessilis* Miq.; *L. sambucina* var. *biserrata* (Miq.) Miq.; *L. divaricata* T. & B.; *L. sambucina* var. *heterophylla* Zipp ex Miq.; *L. sambucina* var. *occidentalis* C.B. Clarke; *L. sambucina* var. *robusta* Miq.; *L. sambucina* var. *roehrsiana* (Sanders ex Masters) Chitt.; *L. sambucina* var. *simplex* Miq.; *L. sambucina* var. *sumatrana* Miq.; *L. umbraculifera* C.B. Clarke; *L. viridiflora* Planch; *Staphylea indica* Burm. F.	Ceylon, India, China, Vietnam, Fiji, Myanmar, Thailand, Cambodia, Laos
18	*L. krukoffiana* Ridsdale	Not available	New Guinea
19	*L. longifolia* Merr.	Not available	Not available
20	*L. macrophylla* Roxb. ex Hornem.	*L. angustifolia* P. Lawson; *L. aspera* Wall. Ex G. Don; *L. cinarea* P. Lawson; *L. coriacea* P. Lawson; *L. diffusa* P. Lawson; *L. integrifolia* Roxb.; *L. latifolia* Wall. Ex Kurz; *L. macrophylla* var. *oxyphylla* Kurz; *L. pallida* W.G. Craib; *L. parallela* Wallich ex Lawson; *L. parallela* var. *angustifolia* (P. Lawson) Kurz; *L. parllela* var. *puberula* W.G. Craib; *L. robusta* Roxb.; *L. talbotii* King ex Talbot; *L. venkobarowii* Gamble	India, Nepal, Thailand, Bhutan, Cambodia, Bangladesh, Laos
21	*L. macropus* K. Schum. & Laut.	Not available	New Guinea
22	*L. magnifolia* Merr.	*L. banahaensis* Elmer; *L. cataduanensis* Quisumb; *L. pycnantha* Quisumb. & Merr. *L. mastersii* C.B. Clarke	Philippines
23	*L. papuana* Merr. & L.M. Perry	Not available	New Guinea
24	*L. philippinensis* Merr.	*L. nitida* Merr.; *L. philippinensis* var. *pauciflora* Merr.	Philippines, Taiwan

(Continued)

TABLE 2.1 (*Continued*)

List of Reported *Leea* Species

No.	Species	Synonyms	Distributions
25	*L. quadrifida* Merr.	*L. agusanensis* Elmer; *L. platyphylla* Merr.	Philippines
26	*L. rubra* Bl. ex Spreng.	*L. brunoninan* C.B. Clarke; *L. rubra* var. *apiifolia* Zipp. Ex Miq.; *L. linearifolia* C.B. Clarke; *L. rubra* f. *celebica* Koord; var. *polyphylla* Miq.	India, Malaysia, North Australia, Vietnam, Cambodia, Thailand, Bangladesh
27	*L. saxatilis* Ridl.	Not available	Malaysia
28	*L. setuligera* C.B. Clarke	*L. Mastersii* C.B. Clarke; *L. Mastersii* var. *siamensis* C.B. Clarke; *L. tenuifoia* W.G. Craib	India, China, Thailand
29	*L. simplicifolia* Zoll. & Moritzi	*L. forbesii* Baker f.; *L. pauciflora* King; *L. pauciflora* var. *ferruginea* W.G. craib	Thailand, Malaysia, Indonesia, Java
30	*L. smithii* Koorders	Not available	Indonesia
31	*L. spinea* Desc.	Not available	Madagascar
32	*L. tetramera* Burtt.	*L. solomonensis* Merr. & L.M. Perry; *L. suaveolens* Merr. & L.M. Perry	Solomon Islands
33	*L. thorelii* Gagnep.	*L. tetrasperma* Gagnep.	Thailand, Cambodia, Vietnam, Laos,
34	*L. tinctoria* Baker	Not available	Gulf of Guinea
35	*L. unifoliata* Merr.	*L. longipetiolata* Merr.	Philippines
36	*L. zippeliana* Miq.	*L. micholitzii* Sanders; *L. monophylla* Lauterb.; *L. zippeliana* var. *ornata* Lauterb.	New Guinea

Source: The Plant List, Version 1.1. http://www.theplantlist.org/.2013, Accessed on 5 May 2017, 2013.

kukurjihvaa or karkatajihvaa in Indian Ayurveda (Wong et al. 2012a, 2012b). It is an evergreen perennial shrub or a small tree growing up to 2–16 m in height with soft wooded and glabrous stems (Ridsdale, 1974, 1976; Khare, 2007; Rahman et al., 2013a). Leaves are 1–3 pinnate with 7 leaflets and 6–35 cm long petioles. Stipules are 2–5 cm long and obovate. Leaflets are ovate, glabrous to hairy with crenate to serrate margins. Pearl glands are angular to globose. Inflorescence is 5–40 cm long, condensed and glabrous to pubescent. Flowers are 5-merous, greenish-white in color, trichotomus, divaricated cymes on short peduncles, and blooms abundantly for about three months from June to August (Molina, 2009). Fruits are 5–10 mm in diameter that turn purple black when ripe and usually bear six seeds. Figure 2.1 shows photographs of *L. indica*. They are distributed in the rain forest of

FIGURE 2.1 Photographs of *Leea indica* (a) Foliage, (b) Stipule, (c) Flowers, (d) Berries and (e) Prop roots. (Fig. b, c and e, Photograph by: Alvin Francis S.L. Lok and reprinted with permission from Hugh T.W. Tan, *Leea* L. (Vitaceae) of Singapore. *Nature in Singapore* 4, (2011), 59–60.)

countries such as India, China, Vietnam, Malaysia, Bangladesh, Thailand, and North Australia with an altitude up to 1700 m in wet areas to ridges and ascending to 2500 m in Himalayas (Ridsdale, 1974, 1976).

2.2.4 *L. macrophylla* Roxb. Ex. Hornem

In India, *L. macrophylla* is commonly known as Hastikarna palasa due to the morphology of the leaves, which look like elephants' ears (Khare, 2007; Nizami et al., 2012). It is an erect perrenial shrub or herb of about 90 cm or more in height. Leaves are simple, broadly ovate, and large up to 60 cm, nearly as broad as long. Leaflets are unifoliate, broadly ovate, serrate margin, acuminate apex, and cordate base. Inflorescences are 12–45 cm long, terminal, corymbose cymes and peduncle up to 25 cm long. Flowers are 5-merous and greenish white in color. Berries are globose, 10–15 mm in diameter, black, with six seeds and ruminate endosperm (Ridsdale, 1974; Islam et al., 2013). Roots are tuberous. It is distributed in India, Nepal, Bhutan, Myanmar, Bangladesh, Thailand, Cambodia, Siam, and Laos (Ridsdale, 1974; Singh and Singh, 1981).

2.2.5 *L. philippinensis* Merr

The *L. philippinensis* Merr. tree can grow up to 10 m tall. Leaves are 1 or rarely 2–3 pinnate with 5–15 leaflets. Pearl glands are stellate and globose. Inflorescences are 3–25 cm long. Flowers are 4-merous and cream in color. Fruits are 10–15 mm in diameter and orange brown in color. Seeds are dark brown and four in number (Ridsdale, 1976). It is distributed in Philippines, Malaysia, and Taiwan.

2.2.6 *L. rubra* ex Spreng

L. rubra is commonly known as Hawaiian holly, Red *Leea* or West Indian holly (Fayaz, 2011). The semi-herbaceous shrub can grow up to 2.5–3 m tall. Leaves are 2–4 pinnate. Leaflets are numerous, ovate to ovate oblong, 3–15 cm long, and 1.5–5 cm wide. Inflorescences are 5–16 cm long, rusty pubescent, usually compact, bracts deltoid-triangular, and peduncle is 4–16 cm long. Flowers are 5-merous, 0.3–0.5 cm across, glabrous and bright pinkish-red. Fruits are 8–10 mm in diameter with dark red berries that turn black on ripening. *L. rubra* is found in dry monsoon forest, savannah and secondary vegetation 500 m (Figure 2.2). It is distributed throughout Malaysia, New Guinea, and North Australia (Ridsdale, 1974, 1976; Lok et al., 2011).

2.2.7 *L. tetramera* Burtt

The *L. tetramera* tree can grow up to 15 m tall. Leaves are 1-2 pinnate. Leaflets are 7–15 usually glabrous, pearl glands globose and black in color. Inflorescence is 15–35 cm long, fulvous pubescent with a 4–10 cm peduncle. Flowers are 4 or 5-merous and creamy white in color (Ridsdale, 1976). Fruits are red-orange in color, 30 mm in diameter with usually six seeds. It is distributed in Papua New Guinea (Ridsdale, 1976; Khan et al., 2003).

2.2.8 *L. thorelii* Gagnep

L. thorelii is herb or semi-herbaceous shrub which can grow up to 1 m height. Leaves are trifoliate to 1–3 pinnate, petioles 3–10 cm long with exceedingly variable leaflets on one plant. Petioles are 1–3 cm long and finely pubescent. Inflorescences

FIGURE 2.2 Photographs of *Leea rubra* (a) Plant, (b) Foliage, (c) Berries and (d) Flowers.

are up to 8 cm long, pubescent, condensed, bracts deltoid to linear up to 3 mm long, and peduncle 1–5 cm. Flowers are 5-merous and white in color. Fruits are 5–8 mm in diameter, blackish purple with four to six seeds. It is found in Thailand, Cambodia, Laos, and Vietnam (Ridsdale, 1974).

2.3 Ethnomedicinal Uses

The genus *Leea* has been traditionally used for the treatment of various conditions. The ethnomedicinal importance of *L. macrophylla* has been documented in Ayurvedic system of medicine (Joshi et al., 2016a). In ancient Indian texts, *Samhitas* leaves of *L. macrophylla* (Hastikarnapalasasya) have been reported to be anti-inflammatory and analgesic, whereas in *Garuda Purana*, they are claimed to have "*Rasayana*" (rejuvenating) properties (Singh and Singh, 1981). In most of the studies, the leaves and roots of the *Leea* species are used. An overview of different plant parts, preparation, and their traditional uses are presented below and summarized in Table 2.2.

2.3.1 Leaves

In Guinea, the leaves of *L. guineensis* are used to treat cancer and arthritis, and the fresh leaf juice is used as an enema as well as in the treatment of enlarged spleen in Nigerian children (Graham et al., 2000; Molina, 2009; Falodun et al., 2007; Cabelin et al., 2015). Traditionally, the leaves of *L. indica* are consumed for the treatment of cancer, diabetes, diarrhoea, dysentery, spasm, cold, headache, injury, rheumatoid arthritis, and skin diseases (Chatterjee and Pakrashi, 1994). In Ayurveda, *L. indica* is reported to possess rasa, kashaya, tikta, guna, seeta, lakhu, and virya properties (Mishra et al., 2016). In India, *L. indica* leaves are consumed in different ways for various therapeutic purposes: raw or concoction brewed from fresh leaves; juice by women as a remedy during pregnancy, delivery, and for birth control; decoction to treat obstetric diseases and body pain as well as herbal tea for general health (Bourdy and Walter, 1992). Ointment prepared from roasted leaves of *L. indica* relieves vertigo (Chatterjee and Pakrashi, 1994). In the Philippines, the leaves and roots of *L. indica* are used to treat skin problems and relief from dizziness (Cabelin, 2015). Leaves of *L. macrophylla* are traditionally

TABLE 2.2
Ethnomedicinal Uses of *Leea* Species

Species	Plant Part/Preparation Used	Ethnomedicinal Uses	Place	References
L. aequata L.	Roots (decoction taken orally)	Stomachache, backache, cough, leucorrhea, and liver disease	Northeast Cambodia	Chassagne et al. (2016)
L. asiatica (L.) Ridsdale	Bulbs (paste)	Joint disease	Madhya Pradesh (India)	Wagh and Jain (2015)
	Roots (pounded and applied)	Boils and blisters	Andaman Islands (India)	Prasad et al. (2008)
	Roots (paste) and leaves (juice, decoction, and kept overnight in water, taken orally)	Worm infection, liver disorders, heart disorders, diabetes	Tripura (India)	Sen et al. (2011)
	Whole plant (paste)	Bone fracture	Karnataka (India)	Bhandary et al. (1995)
	Roots along with the barks of *Boswellia serrata* (pounded and taken orally)	Antidote to snake bite	Uttar Pradesh (India)	Singh et al. (2002)
L. guineensis G. Don	Ground barks (topical)	Skin regrowth in deep wounds (regeneration)	Maroantsetra (Madagascar)	Quansah (1988)
	Leaves (not specified)	Enlarged spleen in children	Nigeria	Falodun et al. (2007)
	Leaves (decoction)	Pregnancy, gastroenteritis, antiscorbutic, strengthens the system, rheumatism, abdominal pains, and malaria.	Nigiria	Borokini and Omotayo (2012)
	Leaves (not specified)	Pelvic abscess and pelvic inflammatory diseases in women	Cameroon (South Africa)	Njamen et al. (2013)
L. indica (Burm. f.) Merr.	Stems and roots (decoction)	Hemorroid, tonic	Yuan (Thailand)	Inta et al. (2013)
	Roots and stems (decoction as potion)	Diarrhoea, hemorrhoid, and gastric ulcer	Northern Thailand	Tangjitman et al. (2015)
	Stem barks (paste)	Hydrocele	Assam (India)	Sonowal (2013)
	Leaves (decoction taken orally)	Diabetes	Kedah (Malaysia)	Mohammad et al. (2012a and 2012b)
	Leaves (burned and applied on chest)	Asthma	Kedah (Malaysia)	Mohammad et al. (2012b)
	Roots (extract)	Dysentery	Tamil Nadu (India)	Shanmugam et al. (2011)
	Roots extract with honey	Expectorant	Assam (India)	Jain and Borthakur (1980)
	Roots (decoction)	Stomachache and diarrhoea	Tripura	Sen et al. (2011)

(Continued)

TABLE 2.2 (*Continued*)

Ethnomedicinal Uses of *Leea* Species

Species	Plant Part/Preparation Used	Ethnomedicinal Uses	Place	References
	Inflorescence (extract)	Chest pain	Rajasthan (India)	Swarnkar and Katewa (2008)
	Tubers (paste)	Allergy from obnoxious weed	Rajasthan (India)	Meena and Rao (2010); Jain et al. (2005); Swarnkar and Katewa (2008)
	Not specified	Body ache	Rajasthan (India)	Choudhary et al. (2008)
	Roots (decoction)	Muscular pain	Yuan (Thailand)	Inta et al. (2013)
	Roots (decoction)	Dysmenorrhoea	Hmong (Thailand)	Srithi et al. (2012)
	Roots (decoction)	Diarrhoea	Northern Thailand	Junsongduang et al. (2014)
	Leaves and leaf buds (drink, squeezed juice of roasted hot leaves)	Facilitate birth	Vanuatu	Bourdy and Walter (1992)
	Leaves and roots (oral, bath, massage, cataplasm)	Headache, cold, injury, and rheumatoid arthritis	Mt. Yinggeling Hainan Island (China)	Zheng and Xing (2009)
L. macrophylla Roxb. Ex. Hornem.	Bulbs (poultice)	Skin disorder	Madhya Pradesh (India)	Wagh and Jain (2015)
	Seeds (wrapped in cloth and tied around the neck of children) and chewed	Stomachache and viral fever	Sikkim (India)	Pradhan and Badola (2008)
	Not specified	Fracture and sprain in cattle	Satkhera district (Bangladesh)	Mollik et al. (2009)
	Leaves (powdered leaves with honey)	Cancer	India	Swarnkar and Katewa (2008)
	Tubers (decoction)	Dysentery in animals	India	Swarnkar and Katewa (2008)
	Tubers (powder)	Sexual debility in male	Rajasthan (India)	Swarnkar and Katewa (2008); Choudhary et al. (2008)
	Leaves (paste)	Anti-inflammatory, boils, arthritis, gout, and rheumatism	Bangladesh	Yusuf (2007)
	Roots	Bone fracture, body pain, sprain, hemostatic, vermic, and wounds	Rajasthan (India)	Jain et al. (2005)
L. rubra Blume ex Spreng.	Roots (decoction taken orally)	Hemorrhoids and stomachache	Northeast Cambodia	Chassagne et al. (2016)
	Roots	Treatment of rheumatism	Vietnam	Tran et al. (2015)
	Not specified	Laxative	Vietnam	Dung and Loi (1991)
L. thorelli Gagnep.	Roots (decoction taken orally)	Hemorrhoids, stomachache, backache, cough, leucorrhea, and liver disease	Northeast Cambodia	Chassagne et al. (2016)
	Roots (decoction)	Tonic	Thailand	Lakornwong et al. (2014)

used in the treatment of urinary problems, goiter, gastric tumor, boils, lipoma, arthritis, gout, rheumatism, tetanus, and in preparation of seasonal tonic modaka by Ayurvedic physicians (Yusuf et al., 2007; Singh and Singh, 1981; Islam et al., 2013; Nizami et al., 2012). In addition, leaves of *L. macrophylla* are used as a first-line therapy for the treatment of arthritic pain and urinary problems by local tribes in Bangladesh (Nizami et al., 2012).

2.3.2 Roots

The roots and leaves of *L. asiatica* are used as a remedy to treat worm infection, liver disorder, and diabetes (Sen et al., 2011). The roots of *L. indica* are consumed as decoction or drinks for the treatment of colic, diarrhoea, cardiac disorders, skin disease, and dysentery (Chatterjee and Pakrashi, 1994; Sen et al., 2011). In Thailand, *L. indica* is used in the preparation of postpartum

herbal bath recipes (Panyaphu et al., 2012). *L. thorelli* is known as "Katang bai tia" in Thai, and water decoction of its roots is used as a tonic. Moreover, the water decoction of roots of *L. thorelii* and *L. rubra* is used as hallucinatory traditional medicine in Thailand (Chuakul et al., 2002; Lakornwong et al., 2014). The mixture of dried powdered roots of *L. macrophylla* along with butter is taken in the morning as an "age sustainer," whereas its topical application with coconut oil is used to treat wounds and sores (Jadhao and Wadekar, 2010; Islam et al., 2013; Joshi et al., 2016a). The local tribes of Uttar Pradesh (India) consume root tubers of *L. macrophylla* orally for treating wounds and apply externally to allay pain and are alexipharmic (Kirtikar and Basu, 1975). In the Philippines, pounded roots of *L. macrophylla* are used for healing wounds (Cabelin, 2015). In Bangladesh, the root tubers of *L. macrophylla* are used externally to relieve pain and stop the effusion of blood (Yusuf et al., 2007).

2.3.3 Whole Plant and Others

Leea species have also been used as a whole plant or the distinct plant part used has not specified (Dung and Loi, 1991; Choudhary et al., 2008; Mollik et al., 2009). In Karnataka (India), the paste of the whole plant of *L. asiatica* is reported to heal bone fracture in 15 days (Bhandary et al., 1995). *L. guineensis* is native to temperate zone of tropical Africa and used to treat enlarged spleen in children, pregnancy detection, purgative, toothache, gonorrhoea, general weakness, skin lesions, ulcer, diarrhoea, dysentery, pain killer, paralysis, epileptic fits, stomach troubles, herpes, and boils (Ajiboye et al., 2014). *L. guineensis* is also used for its cardiac and analgesic properties in traditional African medicine (Op de Beck et al., 2000). The whole plant of *L. indica* is used as a remedy for relieving body pain, headache, and skin problems (Burkil), and the inflorescence extract is useful to cure chest pain in children (Chatterjee and Pakrashi, 1994; Wong and Kadir, 2012b). *L. macrophylla* is traditionally used as to treat worm and ringworm infections, sore, wounds (Kirtikar and Basu, 1975; Bhavamishra, 2010), and cancer (Choudhary et al., 2008). *L. tetramera* is used to treat fever, stomachache, worm and tapeworm infections, dysentery, headache, skin infections, boils, and wounds (Khan et al., 2003). *L. thorelii* Gagnep. is used in traditional Thai medicine to treat fever and inflammation (Kaewkrud et al., 2007).

2.4 Patents

L. asiatica, *L. compactiflora*, *L. guineensis* and *L. macrophylla* have been patented in China and Japan for their pharmaceutical, healthcare and cosmeceutical properties (Ren, 2008; Miyagoshi et al., 2009a, 2009b; Kong et al., 2015; Zhang, 2015; Xu, 2016). Information on these patents is presented in Table 2.3.

2.5 Phytochemistry

The genus *Leea* has not been extensively investigated chemically. Chemical analyses on crude extracts of *L. guineensis*, *L. indica*, *L. macrophylla* and *L. thorelii* showed the presence of alkaloids, terpenoids, flavonoids, tannins, saponins, carbohydrates, and proteins (Khan et al., 2003; Falodun et al., 2007; Islam et al., 2013; Avin et al., 2014; Joshi et al., 2016a). The reported yields are shown in brackets: *L. macrophylla* fresh whole plant (5.5%), dried leaves (8.5%–20%), and dried root tubers (22.0%); *L. indica* dried leaves (~10%) and *L. guineensis* dried leaves (10.86%) (Falodun et al., 2007; Nizami et al., 2012; Reddy et al., 2012; Raihan et al., 2012; Dewanjee et al., 2013; Faruq et al., 2014; Akhter et al., 2015; Joshi et al., 2016a). Secondary metabolites

TABLE 2.3

List of Patents on *Leea* Species

Species	Composition (preparation)	Pharmacological properties	Applications	Patent
L. asiatica	Macerated methanol extract of dried whole plant (LA-D) containing kaempferol, kaempferol-3-*O*-α-L-rhamnoside, quercetin, isoquercitrin, quercetin-3-*O*-α-L-rhamnopyranoside, myricetin, myricetin-3-*O*-α-L-rhamnopyranoside, europetin, europetin-3-*O*-α-L-rhamnopyranoside, (-)-catechin and maslinic acid	Anticancer activity	Pharmaceutical	Yang et al. (2017). *Leea asiatica* anticancer active extract LA-D, its pharmaceutical composition, and method for separation and identification of chemical component thereof. CN106632197A20170510
L. compactiflora	Plant part not mentioned (Instant tea with other plants)	Promoting sleep and improving health	Health promotion	Zhang (2015). Instant tea for promoting sleep and improving health, and its preparation method. CN 104255990A20150107
L. guineensis	Fresh leaf juice (medicinal drop)	Treating rhinitis	Pharmaceutical	Ren (2008). A method for preparing Chinese medicinal drop for treating rhinitis. CN 101104073A20080116
L. macrophylla	Plant part not mentioned (simmered with other plants for 2–2.5 h)	Oral lichen planus	Pharmaceutical	Kong et al. (2015). Traditional Chinese medicine formulation for treating oral lichen planus. CN104784275A20150722
	Leaves and roots powder	Ulcerative colitis	Pharmaceutical	Li (2014). A kind of Chinese medicine for the treatment of ulcerative colitis. CN 103599266A 20140226
	Plant part not mentioned (extract)	Skin-lightening agent	Cosmeceutical	Miyagoshi et al. (2009a). The skin-lightening agent, skin applications, and the food and drink. JP 2009298711A20091224
	Plant part not mentioned (extract)	Antioxidant, anti-inflammatory, anti-aging, hair restorer, skin applications and food and drink	Pharmaceutical, cosmeceutical, general health, food and beverage	Miyagoshi et al. (2009b). The antioxidant, the anti-inflammatory drug, the anti aging agent, the hair restorer, skin applications, and the food and drink. JP 2009298712A20091224.

of different chemical classes, e.g. phenolics, flavonoids and their glycosides, terpenoids, bergenin derivatives, coumarin, phthalic acid derivatives, and steroids, have been reported. The structures of some of these compounds are shown in Figure 2.3 and Table 2.4 presents selected compounds reported in various *Leea* species and the plant parts. The following section elaborates on the components according to chemical classes.

2.5.1 Phenolics and Flavonoids

Phenolics and flavonoids are the major classes of compounds reported in the genus Leea. Phenolic compounds including chlorogenic acid (**7**), ethyl gallate (**12**), gallic acid (**15**), *p*-hydroxybenzoic acid (**17**), and syringic acid (**44**) have been reported from different *Leea* species (Siv and Paris, 1972; Op de Beck et al., 2003; Srinivasan et al., 2008; Joshi et al., 2016). Three novel hydrophilic sulphated flavonoids, namely quercitrin 3′-sulphate (**38**), quercetin 3,3′-disulphate (**39**), and quercetin 3,3′,4′-trisulphate (**40**), were isolated from the leaves of *L. guineensis* (Op de Beck et al., 1998, 2003). 0.04% w/w gallic acid (**15**) and 0.13% w/w of quercetin (**36**) were detected in the ethyl acetate fraction of *L. indica* leaves by High Performance Thin Layer Chromatography (Patel et al., 2017). The High-Performance Liquid Chromatographic analysis of methanolic

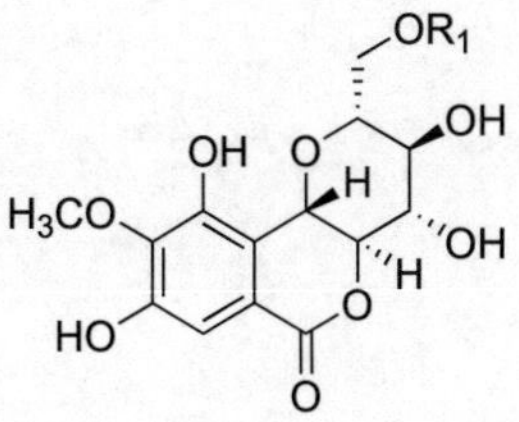

1 $R_1 = COCH_3$
4 $R_1 = H$

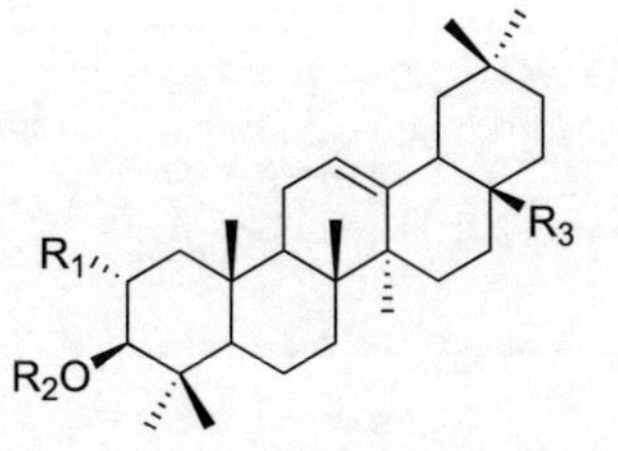

2 $R_1 = R_2 = R_3 = H$
16 $R_1 = R_3 = H, R_2 = -C_{16}H_{31}O$
22 $R_1 = OH, R_2 = H, R_3 = COOH$
33 $R_1 = R_2 = H, R_3 = COOH$

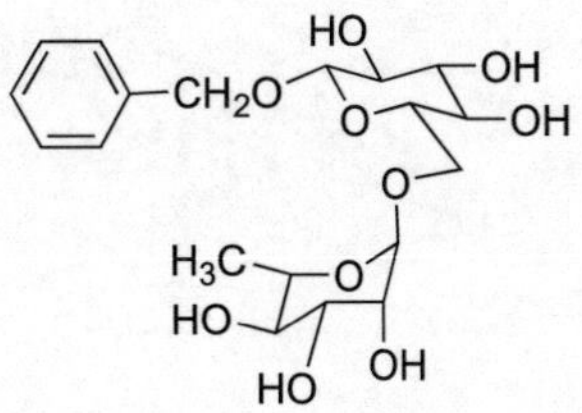

3

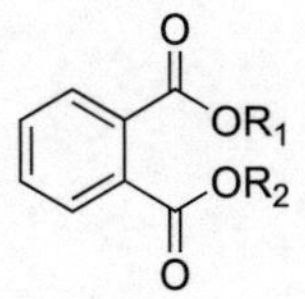

5 $R_1 = R_2 = -C_4H_9$
32 $R_1 = R_2 -C_8H_{17}$

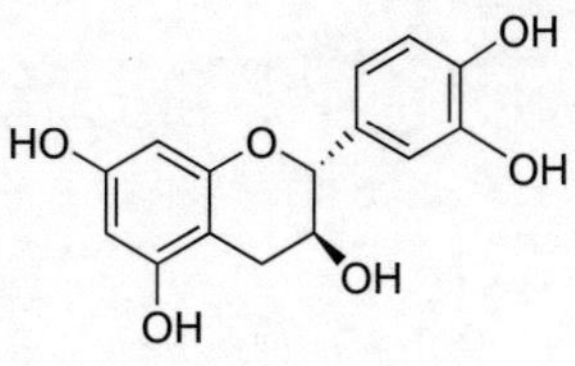

6

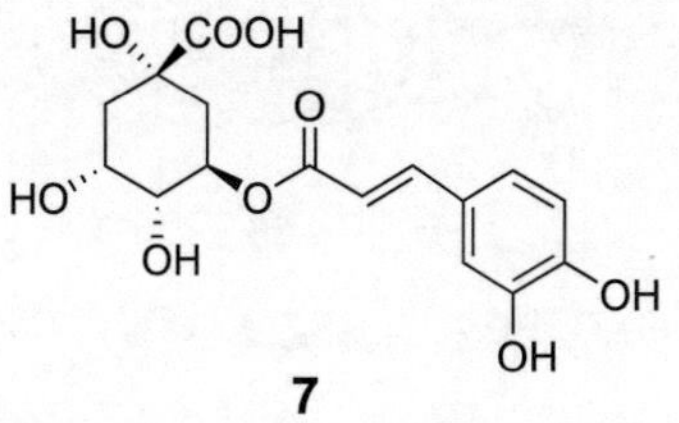

7

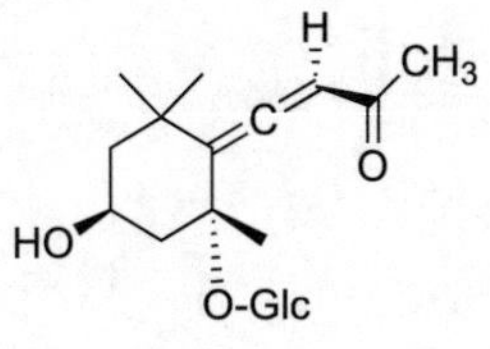

8

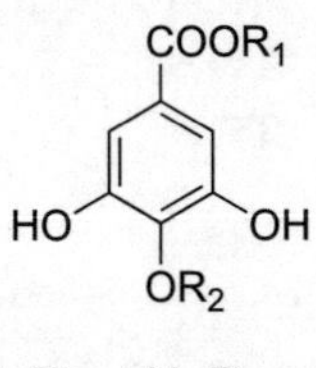

9 $R_1 = H, R_2 = CH_3$
12 $R_1 = -C_2H_5, R_2 = H$
15 $R_1 = R_2 = H$

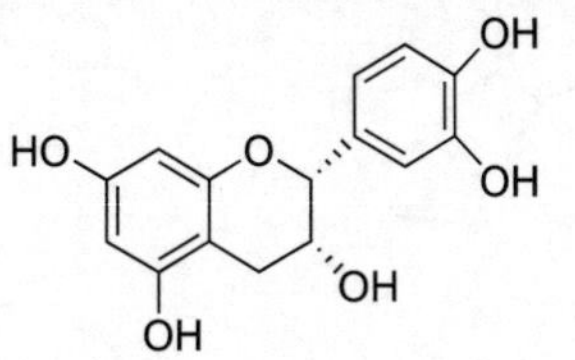

10

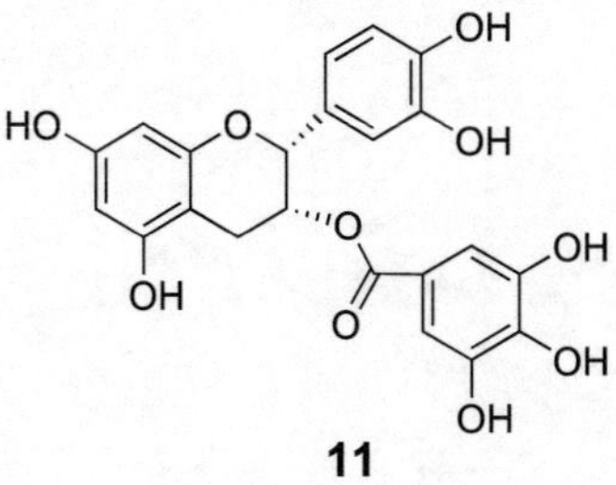

11

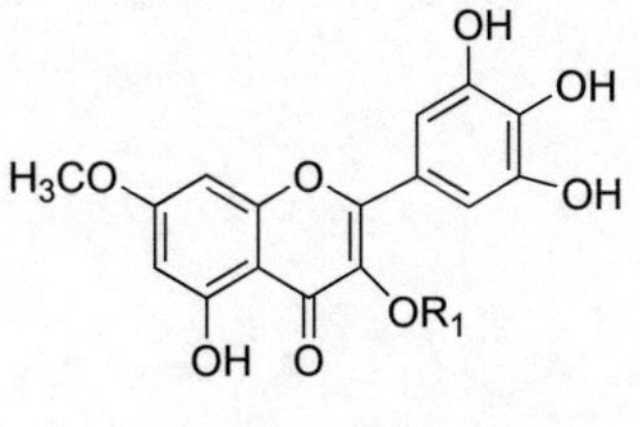

13 $R_1 = H$
14 $R_1 = Rha$

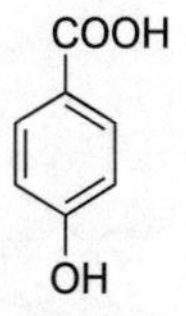

17

FIGURE 2.3 Chemical structures of selected compounds in *Leea* species. *(Continued)*

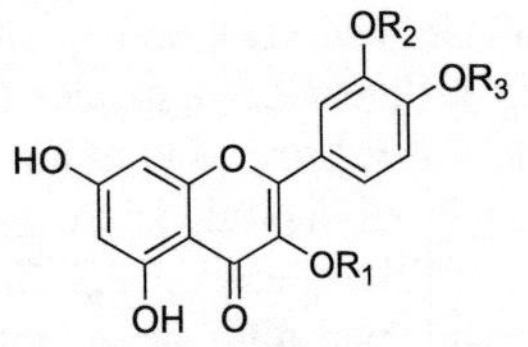

18 R₁ = Glc; R₂ = R₃ = H
36 R₁ = R₂ = R₃ = H
37 R₁ = Rha; R₂ = R₃ = H
38 R₁ = Rha; R₂ = SO₃⁻, R₃ = H
39 R₁ = R₂ = SO₃⁻, R₃ = H
40 R₁ = R₂ = R₃ = SO₃⁻

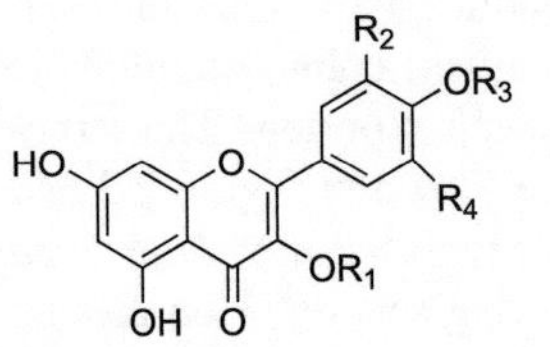

19 R₁ = R₂ = R₃ = R₄ = H
20 R₁ = Rha, R₂ = R₃ = R₄ = H
23 R₁ = Rha, R₂ = OH, R₃ = CH₃, R₄ = OH
30 R₁ = H, R₂ = OH, R₃ = H, R₄ = OH
31 R₁ = Rha, R₂ = OH, R₃ = H, R₄ = OH

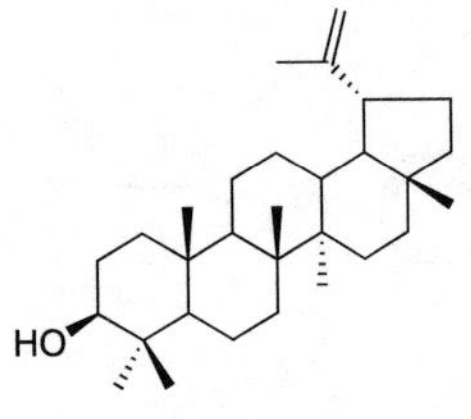

21 **24** **25**

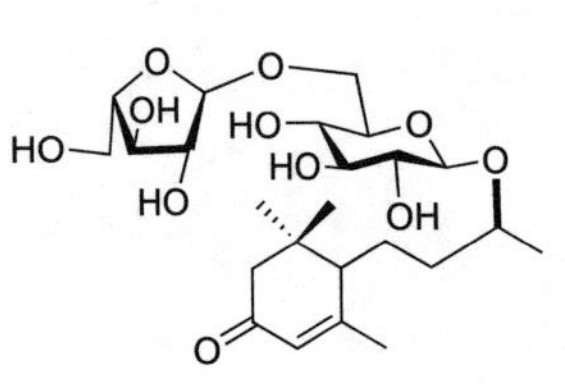

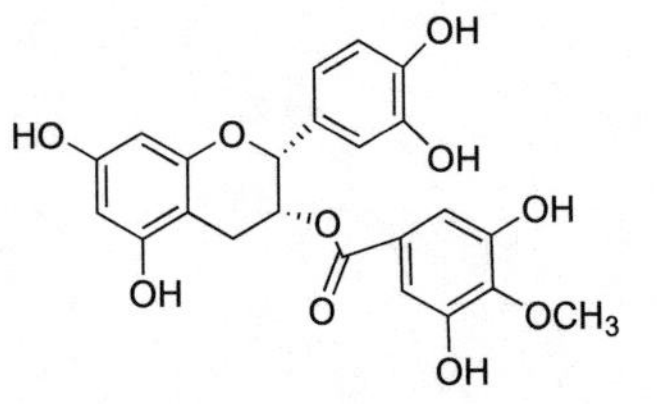

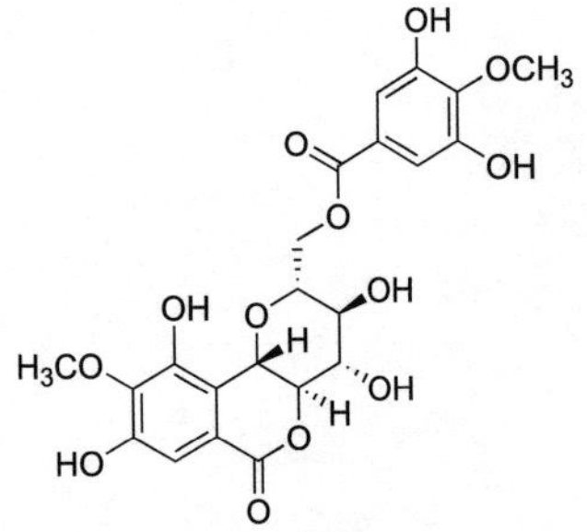

26 **27** **28** R = Ara **29** R = Xyl

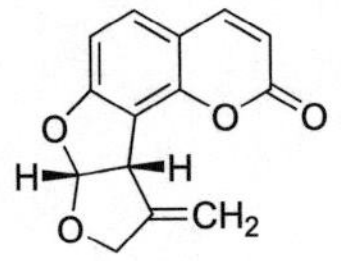

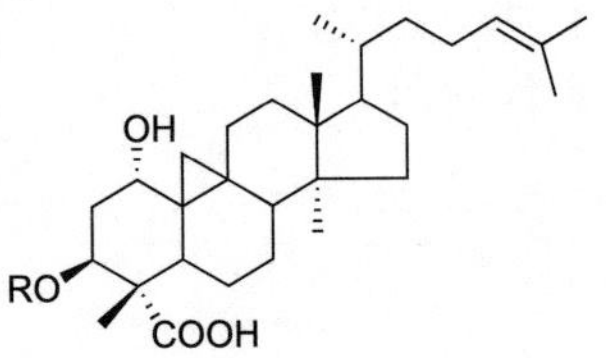

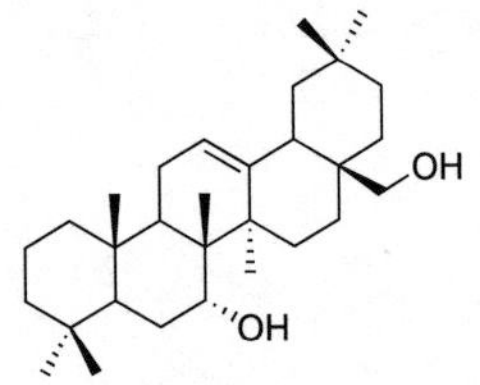

34 **35** **41** R = H **42** R = Glc

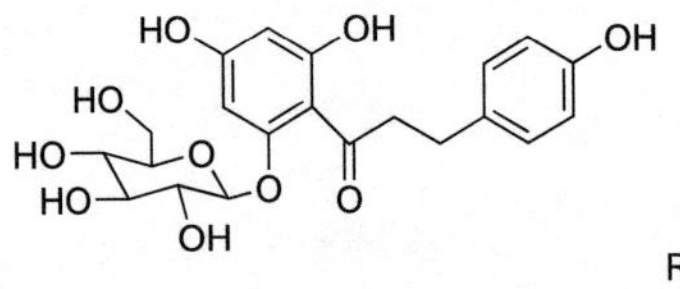

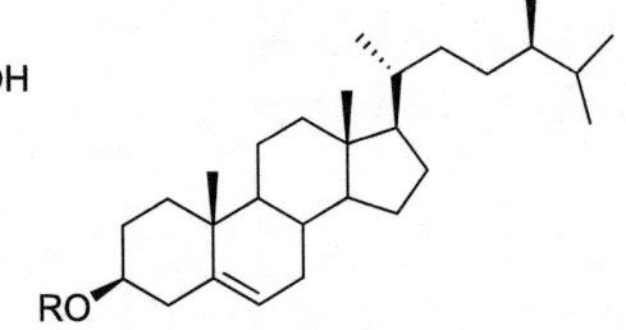

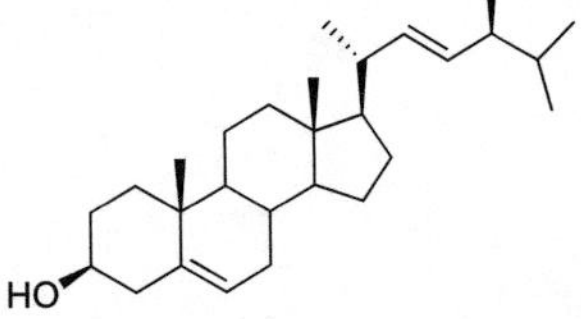

43 **44** **46**

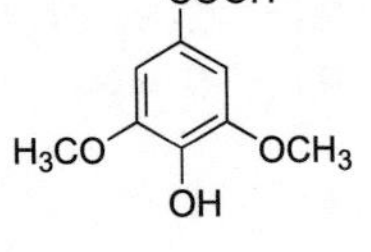

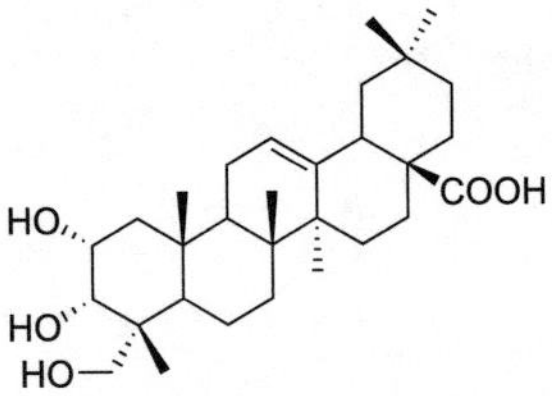

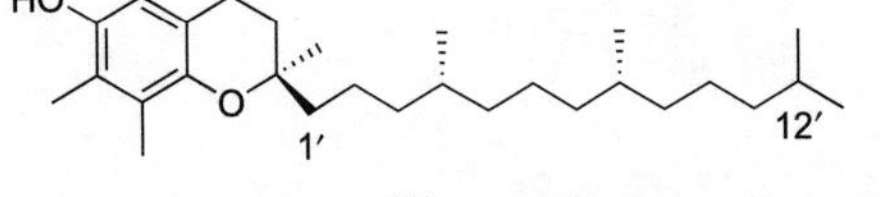

45 **47**

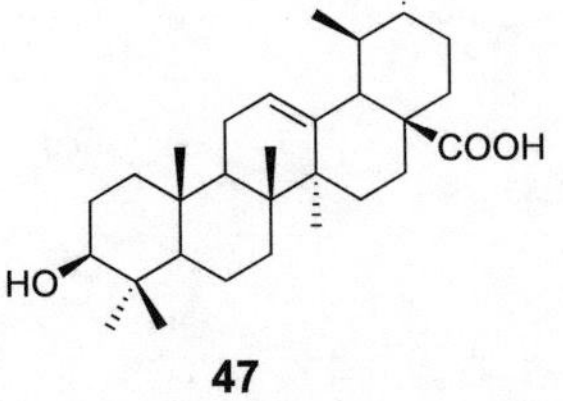

FIGURE 2.3 (Continued) Chemical structures of selected compounds in *Leea* species.

TABLE 2.4

Selected Compounds Reported in *Leea* Species

No.	Compounds	Chemical Class	Species	Plant Parts	References
1	11-*O*-Acetyl bergenin	Tannin	*L. thorelii*	Roots	Lakornwong et al. (2014)
2	*β*-Amyrin	Triterpenoid	*L. indica*	Roots	Joshi et al. (2013)
3	Benzyl-*O*-*α*-L-rhamno pyranosyl-(1→6)-*β*-D-glucopyranoside	Benzyl alcohol glycoside	*L. thorelii*	Leaves	Kaewkrud et al. (2007)
4	Bergenin	Tannin	*L. thorelii*	Roots	Lakornwong et al. (2014)
5	di-n-Butyl phthalate	Phthalic acid ester	*L. indica*	Roots	Joshi et al. (2013)
6	Catechin	Flavonoid	*L. asiatica*	Whole plant	Yang et al. (2017)
7	Chlorogenic acid	Phenolic acid	*L. macrophylla*	Roots tubers	Joshi et al. (2016b)
8	Citroside A	Megastigmane	*L. thorelii*	Leaves	Kaewkrud et al. (2007)
9	3,5-Dihydroxy-4-methoxybenzoic acid	Phenolic acid	*L. thorelii*	Roots	Lakornwong et al. (2014)
10	(-)-Epicatechin	Flavonoid	*L. thorelii*	Roots	Lakornwong et al. (2014)
11	(-)-Epicatechin gallate	Flavonoid	*L. thorelii*	Roots	Lakornwong et al. (2014)
12	Ethyl gallate	Phenolic acid	*L. guineense*	Leaves	Op de Beck et al. (2003)
13	Europetin	Flavonoid	*L. asiatica*	Whole plant	Yang et al. (2017)
14	Europetin-3-O-*α*-L-rhamnoside	Flavonoid	*L. asiatica*	Whole plant	Yang et al. (2017)
15	Gallic acid	Phenolic acid	*L. indica*	Leaves	Srinivasan et al. (2008)
				Roots	Joshi et al. (2013)
			L. guineensis	Leaves	Op de Beck et al. (2003)
			L. rubra	Not specified	Siv and Paris (1972)
16	*O*-Hexadecanoyl-*β*-amyrin	Triterpenoid	*L. indica*	Not specified	Saha et al. (2007)
17	*p*-Hydroxybenzoic acid	Phenolic acid	*L. rubra*	Not specified	Siv and Paris (1972)
18	Isoquercitrin	Flavonoid	*L. asiatica*	Whole plant	Yang et al. (2017)
19	Kaempferol	Flavonoid	*L. guineensis*	Leaves	Op de Beck et al. (2003)
20	Kaempferol-3-O-*α*-rhamnoside	Flavonoid	*L. asiatica*	Whole plant	Yang et al. (2017)
21	Lupeol	Triterpenoid	*L. indica*	Leaves	Srinivasan et al. (2008)
				Roots	Joshi et al. (2013)
22	Maslinic acid	Triterpenoid	*L. asiatica*	Whole plant	Yang et al. (2017)
23	Mearnsitrin	Flavonoid	*L. guineensis*	Leaves	Op de Beck et al. (2003)
24	(6R,9S)-Megastigman-3-on-4-ene-9-ol 9-*α*-L-arabino furanosyl-(1→6)-*β*-D-gluco pyranoside (Leeaoside)	Megastigmane	*L. thorelii*	Leaves	Kaewkrud et al. (2007)
25	4″-*O*-Methyl-(-)-epicatechin gallate	Flavonoid	*L. thorelii*	Roots	Lakornwong et al. (2014)
26	11-*O*-(4′-*O*-Methylgalloyl) bergenin	Tannin	*L. thorelii*	Roots	Lakornwong et al. (2014)
27	Microminutinin	Coumarin	*L. thorelii*	Roots	Lakornwong et al. (2014)
28	Mollic acid arabinoside	Triterpenoid	*L. indica*	Leaves	Wong et al. (2012a)
29	Mollic acid xyloside	Triterpenoid	*L. indica*	Leaves	Wong et al. (2012a)
30	Myricetin	Flavonoid	*L. asiatica*	Whole plant	Yang et al. (2017)
31	Myricetin-3-*O*-*α*-L-rhamnopyranoside (Myricitrin)	Flavonoid	*L. thorelii*	Leaves	Kaewkrud et al. (2007)
32	di-n-Octyl phthalate	Phthalic acid ester	*L. indica*	Roots	Joshi et al. (2013)
33	Oleanolic acid	Triterpenoid	*L. macrophylla*	Leaves,	Dewanjee et al. (2013)
				Roots	Mahmud et al. (2017)
34	7α, 28-Olean diol	Triterpenoid	*L. macrophylla*	Roots	Mahmud et al. (2017)
35	Phloridzin	Dihydrochalcone	*L. indica*	Not specified	Saha et al. (2007)
36	Quercetin	Flavonoid	*L. guineensis*	Leaves	Op de Beck et al. (2003)
			L. indica	Whole plant	Patel et al. (2017)
37	Quercetin 3-*O*-*α*-L-rhamno pyranoside (Quercitrin)	Flavonoid	*L. guineensis*	Leaves	Op de Beck et al. (2003)
			L. indica	Roots	Joshi et al. (2013)
			L. thorelii	Leaves	Kaewkrud et al. (2007)
38	Quercetin 3-*O*-*α*-L-rhamno pyranoside-3′-sulphate (Quercitrin 3′-sulphate)	Flavonoid	*L. guineensis*	Leaves	Op de Beck et al. (1998)
39	Quercetin 3,3′-disulphate	Flavonoid	*L. guineensis*	Leaves	Op de Beck et al. (2003)

(Continued)

TABLE 2.4 (*Continued*)

Selected Compounds Reported in *Leea* Species

No.	Compounds	Chemical Class	Species	Plant Parts	References
40	Quercetin 3,3′,4′-trisulphate	Flavonoid	*L. guineensis*	Leaves	Op de Beck et al. (2003)
41	β-Sitosterol	Steroid	*L. indica*	Roots	Joshi et al. (2013)
42	β-Sitosterol-3-O-β-D-glucopyranoside	Steroid	*L. indica*	Roots	Saha et al. (2005)
43	Stigmasterol	Steroid	*L. thorelii*	Roots	Lakornwong et al. (2014)
			L. macrophylla	Roots	Mahmud et al. (2017)
44	Syringic acid	Phenolic acid	*L. rubra*	Not specified	Siv and Paris (1972)
45	α-Tocopherol	Phenolic	*L. indica*	Roots	Joshi et al. (2013)
46	2α,3α,23-Trihydroxy-12-oleanen-28-oic acid	Triterpenoid	*L. indica*	Not specified	Saha et al. (2007)
47	Ursolic acid	Triterpenoid	*L. indica*	Leaves	Srinivasan et al. (2008)

extract of *L. macrophylla* leaves showed the presence of oleanolic acid (**33**), ursolic acid (**47**), and ascorbic acid (Dewanjee et al., 2013), while the ethanolic extract of root tubers contained 9.01% *w/w* of chlorogenic acid (**7**) (Joshi et al., 2016b).

A chalcone, phloridzin (**35**), was isolated from *L. indica* (Saha et al., 2007). Flavonoids such as kaempferol (**19**), myricitrin (**31**), quercetin (**36**), and quercitrin (**37**) were also reported in *L. guineensis, L. indica, L. rubra* and *L. thorelii* (Op de Beck et al., 2003; Kaewkrud et al., 2007; Singh et al., 2019).

2.5.2 Terpenoids

A megastigmane citroside A (**8**), and a megastigmane diglycoside (6R,9S)-megastigman-3-on-4-ene-9-ol 9-α-L-arabinofuranosyl-(1→6)-β-D-glucopyranoside (leeaoside) (**24**) were isolated from the leaves of *L. thorelii* (Kaewkrud et al., 2007). This is the first report of megastigmanes from the genus *Leea*. Different classes of triterpenoids, e.g., lupane (lupeol), oleane (β-amyrin, O-hexadecanoyl-β-amyrin, maslinic acid, oleanolic acid and 2α,3α,23-trihydroxy-12-oleanen-28-oic acid), ursane (ursolic acid) and cycloartane (mollic acid arabinoside and mollic acid xyloside), have been reported from three *Leea* species, namely, *L. asiatica, L. indica,* and *L. macrophylla* (Yang et al., 2017; Saha et al., 2007; Wong et al., 2012a; Joshi et al., 2013; Dewanjee et al., 2013).

2.5.3 Steroids

β-Sitosterol (**41**) and its glycoside β-sitosterol-3-O-β-D-glucopyranoside (**42**) were isolated from *L. indica* (Saha et al., 2005; Srinivasan et al., 2008), while stigmasterol (**43**) from *L. thorelii* and *L. macrophylla* (Lakornwong et al., 2014; Mahmud et al., 2017).

2.5.4 Essential Oils

The flowers of *L. indica*, leaves and wood of *L. guineensis*, as well as leaves of *L. longifolia* have been analyzed by gas chromatography-mass spectrometry (GC-MS) (Srinivasan et al., 2009; Op de Beck et al., 2000; He-Ping et al., 2006). A total of 28 compounds were identified from the essential oil of *L. longifolia* leaves. The major constituents in *L. longifolia* oil are phenols (15.72%), n-hexadecanoic acid (11.25%), phenylethyl alcohol (8.25%), octadec-9-enoic acid (7.96%),

2,3-dihydro-benzofuran (6.20%), and pentatriacontane (5.90%) (He-Ping et al., 2006). The composition and yield of essential oils can differ with the geographical distribution and plant parts. Srinivasan et al. (2009) investigated the yield and composition of essential oils of *L. indica* flowers from two different regions of Kerala (India). The essential oil from Dhoni forest yielded 0.18% *v/w* and predominantly contained di-isobutylphthalate (79.00%), di-n-butylphthalate (7.18%), n-butylisobutylphthalate (6.11%), and butylisohexylphthalate (3.67%). Di-isobutylphthalate (75.64%), di-n-butylphthalate (7.48%), n-butylisobutylphthalate (7.87%), and butylisohexylphthalate (3.71%) were predominant compounds found in the essential oil (yield 0.13%) from Calicut University. Thus, phthalic acid esters were the main components that accounted >95% of the total essential oil composition of *L. indica*. These results will be helpful to establish the quality standards of *L. indica* essential oil.

The hydrodistilled volatile oils of *L. guineensis* leaves and wood from Cameroon were identified by GC-MS (Op de Beck et al., 2000). A total of 69 components were identified from the volatile oil of *L. guineensis*, with abundance of fatty acids in wood and terpenoids in leaves. The leaves contained 0.3% of volatile oil, whereas the wood contained 0.4%. The major constituents of the volatile oil from the wood were fatty acids and their esters (37.3%), terpenoids (17.6%), aldehydes (14.2%), and phenylpropanoids (8.6%), while the leaves predominantly contained terpenoids (46.7%), fatty acids/esters (28.7%), aldehydes (5.5%), fatty compounds (3.0%), and ketones (2.0%). The rare compounds with 13 carbons, namely trimethyl dihydronaphthalene (0.3%), and vitispirane (1.2%) were also found in the volatile oil of *L. guineensis* leaves.

2.5.5 Other Secondary Metabolites

A coumarin named microminutinin (**27**) was isolated from the roots of *L. thorelii* (Lakornwonga et al., 2014). Bergenin, a *C*-glucoside of 4-*O*-methyl gallic acid (**4**) together with its derivatives 11-*O*-acetyl bergenin (**1**) and 11-*O*-(4-*O*-methylgalloyl) bergenin (**26**) were isolated from the roots of *L. thorelii* (Lakornwonga et al., 2014). Petroleum ether fractions of *L. indica* leaves were found to contain phthalic acid and its esters, lycopersene and n-heptadecane as major chemical constituents followed by n-tetratriacontane, n-tetratetracontane, heptacosane, n-tetracosane, 1-eicosanol, solanesol, farnesol, and others as minor constituents (Srinivasan et al., 2008).

Two phthalic acid derivatives, namely dibutyl phthalate (**5**) and di-n-octyl phthalate (**32**), and α-tocopherol (**45**) were isolated from *L. indica* roots (Joshi et al., 2013). The only chemical study available on *L. rubra* revealed the presence of myrecitol 3-rhamnoside, p-hydroxybenzoic acid, syringic acid, and gallic acid (Siv and Paris, 1972). An alcohol with a molecular formula of $CH_3(CH_2)_{50}CH_2OH$ was the first and only isolated compound of *L. asiatica* leaves (Sen et al., 2014).

2.5.6 Nutritional Contents

The nutritional composition of *L. guineensis* leaves was analyzed by using proximate and mineral analysis (Fagbohun et al., 2012). The leaves contained ash (7.43%), moisture (5.69%), crude protein (19.3%), fat (7.28%), crude fiber (9.61%) and carbohydrate (50.7%). The minerals identified in *L. guineensis* leaves were calcium (36.29 mg/100 g), phosphorus (35.53 mg/100 g), sodium (31.51 mg/100 g), potassium (31.21 mg/100 g), zinc (30.18 mg/100 g), magnesium (28.68 mg/100 g) and iron (5.08 mg/100 g). The seeds of *L. guineensis* contained vitamin A (1264.4 ± 0.12 μg/100 g), vitamin C (8.29 ± 0.15 μg/100 g), vitamin D (9.72 ± 0.05 μg/100 g) and vitamin E (16.54 ± 0.12 μg/100 g) (Ajiboye et al., 2014). In terms of minerals, the most prevalent was manganese (156.87 ± 0.01 mg/kg), followed by zinc (48.67 ± 0.01 mg/kg), selenium (2.63 ± 0.12 μg/kg), calcium (0.16 ± 1.20%) and sodium (0.08 ± 0.01%).

The whole plant parts of *L. macrophylla* including leaves, stems and roots were evaluated for their mineral contents. The leaves, roots, and stems of *L. macrophylla* are a rich source of potassium and sulphur. The leaves and roots of *L. macrophylla* contained 3550 and 2700 mg/100 g of potassium and 3375 and 2574 mg/100 g of sulphur, respectively (Jadhao et al., 2009).

2.6 Pharmacological Activities

Leea species possess many different pharmacological activities such as analgesic, antiangiogenesis, antiatherothrombotic, anticollagenase, antielastase, anthelmintic, antihypertensive, anti-inflammatory, antimicrobial, antioxidant, antiproliferative, antipyretic, antityrosinase, anxiolytic, diuretic hepatoprotective, nephroprotective, phosphodiesterase inhibitory activity, and wound healing. The reported pharmacological activities of 7 *Leea* species are shown in Table 2.5, listed according to the species. Details of the extraction method, plant part used, dose and key results are provided wherever available. In the following sections, the reported pharmacological activities of various *Leea* species are listed in alphabetical order.

2.6.1 Analgesic Activity

Leea species have been traditionally used to relieve body pain and headache (Bourdy and Walter, 1992; Kirtikar and Basu, 1975; Chatterjee and Pakrashi, 1994; Wong et al., 2012a). However, only *L. indica* and *L. macrophylla* leaves have been assessed for their analgesic effects (Emran, 2012; Emran et al., 2012; Dewanjee et al., 2013). Both plants were found to exhibit central and peripheral analgesic effects in mice. The ethanolic extract (200 mg/kg, *p.o.*) of *L. indica* leaves significantly

($p < 0.05$) inhibited the writhing response in acetic acid induced writhing test. The drug diclofenac sodium (40 mg/kg, *i.p.*) was used as a positive control (Emran et al., 2012). The ethanolic extract (200 mg/kg, *p.o.*) also suppressed the pain response (8.18%) in formalin induced licking test compared to diclofenac sodium (66.45%; 0.5 mg/kg, *i.p.*). Based on these results, it was concluded that the ethanolic extract exhibited anti-nociceptive activity, supporting its traditional uses.

In a separate study, the methanolic extract of *L. macrophylla* leaves was shown to exhibit central ($p < 0.01$) and periphery ($p < 0.05$–0.01) analgesic effects in mice (Dewanjee et al., 2013). This extract reduced the acetic acid-induced writhing at 100 and 200 mg/kg, *p.o.* doses by 40.8% and 56.2%, respectively, compared to the drug paracetamol (50 mg/kg, *p.o.*), which showed 65.7% inhibition of writhing ($p < 0.01$). Moreover, the methanolic extract significantly increased the hot plate reaction time in the hot plate test. The drug pentazocine (10 mg/kg, *p.o.*) was used as the positive control. The triterpenoids (oleanolic acid, ursolic acid, and lupeol) may have contributed to the analgesic activity (Emran et al., 2012; Dewanjee et al., 2013). Further studies are required to establish the exact mechanism of action contributing to the analgesic effect and the active chemical constituents responsible.

2.6.2 Anthelmintic Activity

The ethyl acetate and methanolic fractions of *L. asiatica* leaves showed anthelminthic activity against Indian adult earthworm (*Pheretima posthuma*) (Sen et al., 2012) in terms of reducing the time taken for the worm to be paralyzed and to die, albeit at higher concentrations than the standard drug piperazine citrate. The results support the traditional use of *L. asiatica* leaves in the treatment of worm infections (Sen et al., 2011). However, the active chemical constituents and mechanism of action are unknown.

2.6.3 Antiangiogenic and Antiatherothrombotic Activities

The ethanolic extract of *L. indica* leaves was evaluated using for preliminary angiogenesis assays (e.g., recombinant vascular endothelial growth factor ($rVEGF_{165}$) induced *in vivo* chorioallantoic membrane, rat corneal micropocket and tumor induced peritoneal angiogenesis assays) (Avin et al., 2014). The ethanolic extract (50 mg/kg) treated CAM showed a decrease of ~25% of the total vessels, which was similar to the results in the rat corneal pocket assay. Moreover, the ethanolic extract showed potent inhibition of tumor induced angiogenesis with a decrease of ~12% of the total vessel length in the peritoneum. Avin and coworkers measured mRNA from Ehrlich Ascites Carcinoma (EAC) cells treated with ethanolic extract using reverse transcription-PCR analysis and found that the extract inhibited VEGF secretion up to four fold compared to the control. The authors suggested the antiangiogenesis activity of *L. indica* might be due to the presence of triterpenoids. The ethanolic leaf extract of *L. macrophylla* at a dose of 5 μg/μL showed 20.61% clot lysis after 90 min of incubation, compared to the positive control streptokinase of 81.53% clot lysis, suggesting the plant has weak antiatherothrombotic activity (Faruq et al., 2014).

TABLE 2.5

Pharmacological Activities of Extracts, Fractions and Isolated Compounds of *Leea* Species

Species	Type of Extract /Fraction (Plant Part)/Compound	Pharmacological Activity	Doses	Results	References
L. aequata	Ethanol (leaves)	Anticonvulsant activity	2.5 mg/mL	Showed smooth muscle relaxant effect $105.42 \pm 2.91\%$ compared to atropine sulfate $113.97 \pm 4.58\%$ at 6.95×10^{-3} mg/mL against acetylcholine induced contraction in isolated guinea pigs ileum.	Ginting et al. (2017)
L. asiatica	Petroleum ether, ethyl acetate, and methanol fractions (leaves)	Anthelmintic activity	10, 20 and 50 mg/mL	At 50 mg/mL, both the ethyl acetate and methanolic fractions reduced the time to paralyse *Pheretima posthuma* and to kill the earthworm.	Sen et al. (2012)
		Antioxidant activity	32, 80 and 120 μg/mL	Ethyl acetate fraction showed potent antioxidant activity with IC_{50} values 9.5, 13.0 and 57.0 μg/mL in DPPH, NO radical scavenging, and lipid peroxidation assays, respectively.	Sen et al. (2012)
	80% Methanol (leaves)	Antioxidant activity	—	Ferric reducing power 42.76 ± 28.54 mg of Trolox Equivalent/g dry extract and 2,2'-azino-bis(3-ethylbenzothiazoline-6-sulphonic acid (ABTS) radical scavenging activity with $IC_{50} = 11.90$ μg/mL using standards trolox ($IC_{50} = 4.90$ μg/mL), and ascorbic acid ($IC_{50} = 4.10$ μg/mL), respectively.	Evans et al. (2015)
	80% Methanol (roots)	Antioxidant activity	—	Ferric reducing power 37.60 ± 28.36 mg of Trolox Equivalent/g dry extract and ABTS radical scavenging activity with $IC_{50} = 5.0$ μg/mL using standards trolox ($IC_{50} = 4.90$ μg/mL), and ascorbic acid ($IC_{50} = 4.10$ μg/mL), respectively.	Evans et al. (2015)
	Methanol extract and fractions (petroleum ether and ethyl acetate), (leaves)	Nephroprotective effect	Extract (150 and 300 mg/kg) and fractions (75 and 150 mg/kg)	Ethyl acetate fraction (150 mg/kg) reduced the BUN (29.11 ± 1.88 mg/dL), serum creatinine (0.93 ± 0.20), uric acid (5.01 ± 0.39 mg/dL) and MDA levels (2.11 ± 0.25 nM/min/mg protein), and increased total protein (7.32 ± 0.68 g/dL) and albumin level (3.14 ± 0.21 g/dL) in blood serum against cisplatin-induced toxicity in mice.	Sen et al. (2013)
	80% Aqueous acetone (fruits)	Antioxidant activity	—	Free radical scavenging activity of DPPH (3990.10 ± 18.10 mg ascorbic acid equivalents/100 g fruit weight), ABTS (2989.0 ± 154.50 mg butylated hydroxyanisole equivalent/100g fruit weight), superoxide (52500.71 ± 1345.0 mg catechin equivalent/100 g fruit weight), linoleate hydroperoxide (10019.26 ± 395.30 mg butylated hydroxyanisole equivalent/100 g fruit weight), ferric reducing (6366.60 ± 409.0 mg ascorbic acid equivalents/100 g fruit weight) and ferrous metal chelating activity (15.37 ± 4.80 mg EDTA equivalent/100 g fruit weight).	Singh et al. (2015)
		Antielastase and anticollagenase activity	—	Significant ($p < 0.05$) inhibitory activity against elastase (1426.72 ± 5.21 mg EGCG equivalents/100 g fruit weight) and collagenase (8560.90 ± 77.42 mg EGCG equivalents/100 g fruit weight) enzymes.	Singh et al. (2015)
		Antityrosinase activity	—	Significant ($p < 0.05$) antityrosinase activity 30661.60 ± 1849.0 mg kojic acid equivalents/100 g fruit weight.	Singh et al. (2015)
	Aqueous (aerial parts)	Antidiabetic activity	—	Normal, active (details are not available).	Marles and Farnsworth (1995)

(Continued)

TABLE 2.5 (Continued)

Pharmacological Activities of Extracts, Fractions and Isolated Compounds of *Leea* Species

Species	Type of Extract/Fraction (Plant Part)/Compound	Pharmacological Activity	Doses	Results	References
L. aqeuata	95% Methanol (leaves)	Antimicrobial activity	—	Showed 13.33 ± 0.58 mm zone inhibition against *Vibrio cholerae* using gentamycin as positive control (17.33 ± 0.58 mm) in agar well diffusion method.	Chander and Vijayachari et al. (2017)
L. guineensis	Kaempferol (**19**)	Antioxidant activity	—	Free radical scavenging activity of DPPH ($IC_{50} = 120$ µM).	Op de Beck et al. (2003)
	Quercetin (**36**)	Antioxidant activity	—	Free radical scavenging activity of DPPH ($IC_{50} = 49$ µM).	Op de Beck et al. (2003)
	Quercitrin (**37**)	Antioxidant activity	—	Free radical scavenging activity of DPPH ($IC_{50} = 88$ µM).	Op de Beck et al. (2003)
	Quercitrin 3′-sulphate (**38**)	Antioxidant activity	—	Free radical scavenging activity of DPPH ($IC_{50} >> 200$ µM).	Op de Beck et al. (2003)
	Quercetin 3,3′-disulphate (**39**)	Antioxidant activity	—	Free radical scavenging activity of DPPH ($IC_{50} >> 200$ µM).	Op de Beck et al. (2003)
	Quercetin 3,3′,4′-trisulphate (**40**)	Antioxidant activity	—	Free radical scavenging activity of DPPH ($IC_{50} = 150$ µM).	Op de Beck et al. (2003)
	Gallic acid (**15**)	Antioxidant activity	—	Free radical scavenging activity of DPPH ($IC_{50} = 35$ µM).	Op de Beck et al. (2003)
	Ethyl gallate (**12**)	Antioxidant activity	—	Free radical scavenging activity of DPPH ($IC_{50} = 67$ µM).	Op de Beck et al. (2003)
	Aqueous, ethanol and acetone (leaves)	Diuretic	0.33 mg/mL	Ethanolic extract inhibited angiotensin converting enzyme (90%–100%).	Adsersen and Adsersen (1997)
	Aqueous (leaves)	Anti-inflammatory activity	400 mg/kg	Significantly reduced the oedema level at 400 mg/kg dose (73%), compared to indomethacin (46% at 10 mg/kg) in carrageenan-induced rat paw oedema model.	Falodun et al. (2007)
	90% Ethanol (leaves, root barks and stem barks)	Antiparasitic activity	500 to 0.07 µg/mL	Weak antitrypanosomal activity against *Trypanosoma brucei rhodesiense* (leaves, root barks, and stem barks extracts, $IC_{50} = 13$, 20 and 8.0 µg/mL, respectively) and antiplasmodial activity against *Plasmodium falciparum* (leaves, root barks, and stem barks extracts, $IC_{50} = >5$ µg/mL).	Atindehou et al. (2004)
	Aqueous (seeds)	Hepatoprotective activity	200 and 400 mg/kg	At 400 mg/kg, decreased the level of MDA in serum (0.21 ± 0.58 nmol/mL) and liver (3.27 ± 0.17 nmol/mL) compared to dichlorvos fed rats as negative control 2.21 ± 0.06 and 6.23 ± 0.35 nmol/mL in serum and blood respectively. Increased the activities of CAT, SOD and GPx against in DDVP-induced toxicity in Wistar rats.	Ajiboye et al. (2014)
L. indica	Mollic acid arabinoside (**28**)	Antiproliferative activity	—	Inhibited the growth of Ca Ski cells with $IC_{50} = 11.60 \pm 0.29$ µg/mL compared to camptothecin $IC_{50} = 2.51 \pm 0.33$ µg/mL.	Wong et al. (2012a)
	Mollic acid xyloside (**29**)	Antiproliferative activity	—	Inhibited the growth of Ca Ski cells with $IC_{50} = 20.13 \pm 0.21$ µg/mL compared to camptothecin $IC_{50} = 2.51 \pm 0.33$ µg/mL.	Wong et al. (2012a)

(Continued)

TABLE 2.5 (Continued)
Pharmacological Activities of Extracts, Fractions and Isolated Compounds of *Leea* Species

Species	Type of Extract/Fraction (Plant Part)/Compound	Pharmacological Activity	Doses	Results	References
	Ethanol (leaves)	Sedative, analgesic and anxiolytic activity	200 mg/kg	Suppressed motor activity, exploratory behavior and prolongation of thiopental induced sleeping time. Exhibited analgesic activity by reducing the writhing (20.25 min) and licking response (92.78 min) compared to diclofenac Na (19.75 min at 40 mg/kg) and (1.02 min in late phase) in acetic acid-induced writhing test and formalin-induced licking test, respectively, in Swiss-webstar strain.	Emran et al. (2012); Emran (2012)
		Antioxidant activity	—	Significant ($p < 0.05$) DPPH scavenging effect with IC_{50} values (139.83 ± 1.40 μg/mL) compared to ascorbic acid (AA) 1.46 ± 0.06 μg/mL; $FeCl_3$ reduction (16.48 ± 0.64 μg/mL) compared to AA 14.04 ± 1.20 μg/mL; superoxide scavenging (676.08 ± 5.80 μg/mL) compared to curcumin 60.48 ± 0.53 μg/mL and iron chelating (519.33 ± 16.96 μg/mL) compared to AA 8.81 ± 0.90 μg/mL.	Rahman et al. (2013a)
		Cytotoxic activity	20–1000 μg/mL	LC_{50} = 2.4771 μg/mL with 95% of confidence limit (2.25–2.69 μg/mL) in brine shrimp lethality assay.	Paul and Saha (2012)
		Cytotoxic activity	10 mg/mL	LC_{50} = 2.65 ± 0.16 μg/mL, compared to reference vincristine sulfate (LC_{50} = 0.76 ± 0.05 μg/mL) in brine shrimp lethality assay.	Rahman et al. (2013a); Rahman et al. (2013b)
		Antibacterial activity	1, 2 and 3 mg/disc	Showed significant ($p < 0.05$) zone inhibition against Gram positive bacteria *B. subtilis* (12 ± 1.0 mm), *B. megaterium* (19 ± 1.0 mm) and *S. aureus* (11 ± 1.0 mm), and Gram negative bacteria *S. typhi* (10 ± 1.0 mm), *S. paratyphi* (11 ± 1.5 mm), *P. aeroginosa* (10 ± 0.0 mm), *S. dysenteriae* (11 ± 2.0 mm) and *V. cholerae* (10 ± 0.25 mm) at 3.0 mg/disk, compared to positive control tetracycline and ampicillin (16 and 20 mm) at 30 μg/disc using disc diffusion assay.	Rahman et al. (2013a)
		Antifungal activity	1 and 10 mg	At 10 mg/disc showed 38.09 ± 0.5, 22.58 ± 2.2 and 61.82 ± 2.7% growth inhibition against *A. flavus*, *C. albicans* and *F. equisetii* respectively, compared to standard drug fluconazole 67.01 ± 1.8, 40.00 ± 2.5 and 72.32 ± 2.3% at 100 μg/disc.	Rahman et al. (2013a)
		Thrombotic activity	10 mg/mL	Significant ($p < 0.0001$) clot lysis 39.30 ± 0.96% compared to reference streptokinase (75.00 ± 3.04%).	Rahman et al. (2013b)
		Antiangiogenic activity	50 mg/kg	Inhibited the sprouting blood vessels both in non-tumorigenic and tumorigenic conditions. ~25% decrease in total vessel, inhibited ~72% of the $rVEGF_{165}$ induced sprouting vessels in cornea and ~78% of growing blood vessels in the peritoneum using $rVEGF_{165}$ induced Swiss albino mice.	Avin et al. (2014)

(Continued)

TABLE 2.5 (Continued)

Pharmacological Activities of Extracts, Fractions and Isolated Compounds of *Leea* Species

Species	Type of Extract/Fraction (Plant Part)/Compound	Pharmacological Activity	Doses	Results	References
	80% Ethanol (leaves)	Cytotoxic activity	0.001–0.1 mg/mL	Inactive against Vero and HeLa cell lines using microtitration cytotoxicity assay.	Ali et al. (1996)
		Antiviral activity	0.001–0.1 mg/mL	Selectively inhibited HSV-1 with minimum inhibitory concentration (MIC = 0.05 mg/mL) in Herpes simplex virus type-1 (HSV-1) and vesicular stomatitis virus (VSV) using simplified plaque reduction assay.	Ali et al. (1996)
	Ethanol extract and fractions (hexane, ethyl acetate and water) (leaves)	Antioxidant activity	—	Water fraction exhibited potent reducing power (2.70 ± 0.02 compared to ascorbic acid 2.73 ± 0.03 at 0.8 mg/mL) and strong DPPH radical scavenging activity (EC_{50} = 48 μg/mL compared to ascorbic acid EC_{50} = 15 μg/mL).	Reddy et al. (2012)
		Cytotoxic activity	—	The extract and all three fractions did not show cytotoxic effects (IC_{50} ≥ 100 μg/mL) against colon cancer cell lines (HT-29, HCT-15 and HCT-116). Cisplatin was used as standard drug using MTT assay.	Reddy et al. (2012)
	Aqueous (leaves)	Antidiabetic	—	Normal, active (details are not available).	Marles et al. (1995)
	Ethanol and ethanol:water (3:1) (leaves)	Antihyperglycemic and hypolipidemic activity	200 and 400 mg/kg (oral)	Reduced the serum glucose, triglycerides, cholesterol, low-density lipoprotein, alanine transaminase, aspartate aminotransferase, and elevated HDL level. At a dose of 400 mg/kg, hydroalcoholic extract (110.12 ± 3.23 mg/dL) reduced serum glucose level more effectively than alcoholic extract (117.83 ± 2.41 mg/dL) compared to glibenclamide (108.66 ± 4.35 mg/dL, 10 mg/kg) in alloxan induced diabetic rats.	Dalu et al. (2014)
	Ethanol extract and fractions (hexane, ethyl acetate and water) (leaves)	Antiproliferative activity	10–200 μg/mL	Antiproliferative activity against Ca Ski, MCF-7, MDA-MB-435, KB, HEP G2 and WRL68 cell lines. Ethyl acetate fraction showed maximum growth inhibitory activity against Ca Ski (IC_{50} = 85.83 ± 6.01 μg/mL) followed by ethanol extract, hexane, and water fractions with IC_{50} values 188.03 ± 2.87, >200 and >200 μg/mL, respectively.	Wong and Kadir (2011)
	Methanol (leaves)	Antilipase activity	300 μg/mL	Lipase inhibitory activity (48.5%) compared to orlistat (0.05 μg/mL) using porcine pancreatic lipase assay.	Ado et al. (2013)
		Sedative and anxiolytic effects	200 and 400 mg/kg	Prolonged the duration of sleeping time in thiopental sodium-induced sleeping test, and suppressed locomotor activity in hole cross and open field tests for sedative activity.	Raihan et al. (2011)
		Antitumor activity	40 mg/kg	Reduction of tumor weight (7.90 g) and inhibition of cell growth (77.29%) at a dose of 40 mg/kg (*i.p*) compared to drug bleomycin 7.05 g and 92.02% at a dose of 0.3 mg/kg respectively against Ehrlich Ascites Carcinoma bearing mice.	Raihan et al. (2012)

(Continued)

TABLE 2.5 (Continued)

Pharmacological Activities of Extracts, Fractions and Isolated Compounds of *Leea* Species

Species	Type of Extract/Fraction (Plant Part)/Compound	Pharmacological Activity	Doses	Results	References
	Methanol and ethanol (leaves)	Anticancer activity	3.125–200 µg/mL	Methanolic and ethanolic extract showed cytotoxic effect with IC_{50} values 529.44 ± 42.07 µg/mL and 677.11 ± 37.01 µg/mL respectively against DU-145 cell line, and 547.55 ± 33.52 µg/mL and 631.99 ± 50.24 µg/mL, respectively, against PC-3 cell line. Paclitaxel ($IC_{50} = 0.3$ µM) was used as standard drug.	Ghagane et al. (2017)
	Methanol (leaves)	Antioxidant activity	100–500 µl	DPPH free radical scavenging activity with percentage inhibition of 57.11 ± 0.43, 64.15 ± 0.49, 73.24 ± 0.29, 78.16 ± 0.15 and 82.86 ± 0.25% compared to standard ascorbic acid 62.45 ± 0.17, 66.96 ± 0.25, 75.03 ± 0.19, 82.15 ± 0.14 and 90.78 ± 0.12% at 100, 200, 300, 400 and 500 µL, respectively.	Ghagane et al. (2017)
	Methanol (leaves)	Antimicrobial activity	1 mg/disc	Inactive against *B. cereus*, *B. subtulis*, *E. coli*, *P. aeruginosa*, and *S. aureus* using gentamicin (10 µg) and nystatin (20 µg) as positive control in disc diffusion method.	Wiart et al. (2004)
	Methanol:water (4:1) (leaves)	Antibacterial activity	—	Exhibited 12 and 10 mm zones of inhibition against *E. coli* and *Shigella flexneri* compared to gentamicin with 26 and 24 mm zones of inhibition, respectively, at 10 µg.	Panda et al. (2016)
	MeOH extract (leaves, stems and barks)	Antiproliferative activity	1–100 µg/mL	Inactive against breast cancer cell lines MCF-7 and T47D ($IC_{50} > 100$ µg/mL) using sulforhodamine B (SRB) assay.	Nurhanan et al. (2008)
	Ethanol (stem barks)	Hepatoprotective effect	200 and 400 mg/kg	Pretreatment of extract at 400 mg/kg dose significantly reduced the level of serum glutamic pyruvic transaminase (SGPT) 85.62 ± 0.601 IU/L; serum glutamic oxaloacetate transaminase (SGOT) 130.57 ± 0.67 IU/L; serum alkaline phosphatase (SALP) 81.00 ± 3.12 IU/L; total bilirubin (0.65 ± 0.04 mg/dL) and triglyceride (158.70 ± 1.301 mg/dL) compared to standard silymarin at 100 mg/kg (SGPT 74.64 ± 0.90; SGOT 121.13 ± 0.60; SALP 79.66 ± 1.18; bilirubin total 0.64 ± 0.04 and triglyceride 145.7 ± 0.88) in paracetamol-induced hepatotoxicity in Wistar albino rats.	Mishra et al. (2014)
	Ethanol (roots)	Phosphodiesterase inhibitory activity	0.1 mg/mL	Complete inhibitory effect against phosphodiesterases ($IC_{50} = 2.62 ± 0.25$ µg/mL), compared to standard inhibitor 3-isobutyl-1-methylxanthine ($IC_{50} = 0.68 ± 0.13$ µg/mL) using scintillation proximity radioassay.	Temkitthawon et al. (2008)
	Dichloromethane: Methanol (1:1) and water (seeds)	Cytotoxic activity	1 mg/mL	$LC_{50} > 1000$ µg/mL for both organic and aqueous extracts using cycloheximide ($LC_{50} = 40.0$ µg/mL) as positive control in brine shrimp (*Artemia salina*).	Cantrell et al. (2003)
	Methanol (whole plant)	Antioxidant activity	—	Strong DPPH free radical scavenging activity with $IC_{50} = 25$ µg/mL compared to IC_{50} of standards vitamin C, quercetin and BHT 9.0, 11.0, 12.5 µg/mL, respectively.	Saha et al. (2004)

(Continued)

TABLE 2.5 (*Continued*)

Pharmacological Activities of Extracts, Fractions and Isolated Compounds of *Leea* Species

Species	Type of Extract /Fraction (Plant Part)/Compound	Pharmacological Activity	Doses	Results	References
		Anti-inflammatory activity	—	Percentage of NO inhibition 83.63, 80.42 and 74.91% at concentrations of 250, 125.5 and 62.5 µg/mL, respectively, using Griess assay for NO inhibition in lipopolysaccharide (LPS) and interferon-γ (IFN-γ) induced RAW264.7 cells.	Saha et al. (2004)
L. macrophylla	Ethanol (leaves)	Anti-inflammatory activity	500 µg/mL	Inhibited protein denaturation by 47.4% compared to positive control aspirin (52.35 ± 0.00% at 0.1% conc.).	Faruq et al. (2014)
		Membrane stabilizing activity	500 µg/mL	Inhibited hemolysis of human RBCs by 57.63 ± 0.00% compared to standard aspirin 89.83 ± 0.00% using albumin denaturation method.	Faruq et al. (2014)
		Antithrombosis activity	5 µg/µL	Exhibited 20.61 ± 1.76% clot lysis compared to positive control streptokinase (81.53 ± 0.39%).	Faruq et al. (2014)
		Antibacterial activity	500 µg/disc	Showed moderate antibacterial activity against Gram positive (*Bacillus cereus, B. megaterium, B. subtilis,* and *S. aureus*) and Gram negative (*E. coli, P. aeruginosa, Salmonella paratyphi, S. typhi, Shigella dysentriae, S. sonnei,* and *V. cholera*) strains with zone of inhibition 9–12 mm.	Faruq et al. (2014)
		Antifungal activity	500 µg/disc	Showed strong zone of inhibition against antifungal strains *A. niger* (16 ± 2.65 mm), *Blastomyces dermatidis* (25 ± 1 mm), *C. albicans* (26 ± 1 mm), *Cryptococcus neoformans* (26 ± 1 mm), *Microsporum sp.* (26 ± 1 mm), *Pityrosporum ovale* (31 ± 1 mm), and *Trichophyton sp.* (28 ± 1 mm).	Faruq et al. (2014)
	Methanol (leaves)	Anti-inflammatory activity	100 µg/mL	Significantly reduced IL-1β production (8.05 ± 0.78 pg/mg) in lipopolysaccharide stimulated macrophages.	Deewanjee et al. (2013)
		Anti-inflammatory activity	100 and 200 mg/kg (oral)	Inhibited carrageenan induced inflammation by 15.7% and 17.4%, and reduction of granuloma tissue formation by 27.5% and 38.4% at 200 and 400 mg/kg, respectively, using paw adema and cotton pellet granuloma assays in rats. Indomethacin (10 mg/kg) was used as standard drug.	Deewanjee et al. (2013)
		Analgesic activity	100 and 200 mg/kg (oral)	Significantly decreased the acetic acid-induced writhing) in mice by 40.8% ($p < 0.05$) and 56.2% ($p < 0.01$) at 100 and 200 mg/kg doses respectively, compared to paracetamol 65.7% ($p < 0.01$) at 50 mg/kg.	Deewanjee et al. (2013)
	Methanol extract and fractions (chloroform and ethyl acetate) (leaves)	Hepatoprotective effect	100 and 200 mg/kg	Ethyl acetate (100 mg/kg) and chloroform (100 mg/kg) decreased the high level of low-density lipoprotein (LDL) to 37.00 ± 2.00 and 34.67 ± 1.15 mg/dL. Methanol extract at 200 mg/kg dose normalized the increased triglyceride level to 95.00 ± 2.00 mg/dL. Ethyl acetate (200 mg/kg) and chloroform fractions (100 mg/kg) significantly restored the serum CK-MB level in hepatic damage in CCl_4-induced liver injury in Wistar rats.	Akhter et al. (2015)

(Continued)

TABLE 2.5 (Continued)

Pharmacological Activities of Extracts, Fractions and Isolated Compounds of *Leea* Species

Species	Type of Extract/Fraction (Plant Part)/Compound	Pharmacological Activity	Doses	Results	References
	Ethanol-water (root tubers)	Wound healing activity	250, 500 and 750 mg/kg	Oral treatment (500 mg/kg) of ethanol extract increased wound breaking strength by 23.41%, while topical application produced complete wound contraction in 20 days with significant ($p < 0.05$) increase in antioxidant GPx, SOD and CAT. Ethanolic extract reduced the level of proinflammatory cytokines like interleukin-1β (IL-1β), IL-6 and TNF-α along with the promoting cell proliferation during wound healing.	Joshi et al. (2016b)
	Ethanolic extract and fractions (hexane, ethyl acetate, chloroform, butanol and water) (root tubers)	Antibacterial activity	50 mg/mL for extract and 100 mg/mL for fractions	Ethanolic extract was found effective against *S. aureus*, *S. flexneri*, and *S. boydii* with MIC value ranging from 0.195 to 3.125 mg/mL.	Joshi et al. (2016a)
		Antioxidant activity	—	Total antioxidant activity of ethanolic extract > fractions (butanol > aqueous > ethyl acetate > chloroform > hexane) using DPPH, NO scavenging, H_2O_2 and hydroxyl radical assay.	Joshi et al. (2016a)
	Ethanol extract (whole plant)	Antilithiatic effect	500 mg/kg	Exhibited therapeutic antiurolithiatic effects on urinary parameters by decreasing calcium (0.97 ± 0.05 mg), phosphate (9.32 ± 0.26 mg) and oxalate (3.88 ± 0.25 mg), and increasing magnesium (0.26 ± 0.01 mg) and creatinine (6.34 ± 0.18 mg) concentrations in urine on 28 days experimental period in rats, compared to ethylene glycol fed rats (calcium, 1.78 ± 0.08 mg; phosphate, 14.36 ± 0.51 mg; oxalate, 12.68 ± 0.58 mg; magnesium 0.18 ± 0.01 mg and creatinine 5.61 ± 0.20 mg).	Nizami et al. (2012)
	n-Hexane, chloroform, ethyl acetate, and methanol fractions (seeds)	Antimicrobial activity	500 µg/disc	Showed antibacterial activity against Gram positive bacteria (*S. aureus*) with zone of inhibition (7–19 mm) compared to standard drug kanamycin with zone of inhibition 28 mm at 30 µg/disc and inactive against Gram negative bacteria (*E. coli*, *P. aeruginosa* and *S. tiphy*). All fractions except hexane exhibited antifungal activity against *C. albicans* with zone of inhibition of 10–15 mm using kanamycin standard with zone of inhibition 25 mm, at 30 µg/disc.	Islam et al. (2013)
L. rubra	Ethanol (aerial parts)	Inhibition of angiotensin-converting enzyme	0.1 mg/mL	Inhibited $57.0 \pm 12.5\%$ of angiotensin converting enzyme using colorimetric assay. Captopril was used as positive control ($IC_{50} = 21.2$ nmol/L).	Braga et al. (2007)
	Ethanol (leaves)	Inhibition of angiotensin-converting enzyme	100 µg/mL	Inhibited 87.1% and 57.0% of angiotensin converting enzyme using HPLC and colorimetric assays, respectively. Captopril was used as standard drug with $IC_{50} = 14.1$ nmol/L in colorimetric assay.	Serra et al. (2005)

(Continued)

TABLE 2.5 (*Continued*)

Pharmacological Activities of Extracts, Fractions and Isolated Compounds of *Leea* Species

Species	Type of Extract /Fraction (Plant Part)/Compound	Pharmacological Activity	Doses	Results	References
L. tetramera	Methanol extract and fractions (petroleum ether, dichloromethane, ethyl acetate and butanol) (leaves, stems and root barks)	Antibacterial activity	4 mg/disc	Butanolic fraction of root barks showed good antibacterial activity against Gram positive (*Bacillus cereus, B. coagulans, B. megaterium, B. subtilis, Lactobacillus casei, Micrococcus luteus, M. roseus, Staphylococcus albus, S. aureus, S. epidermidis, Streptococcus faecalis, St. pneumeniae,* and *St. mutans*) and Gram negative (*Agrobacterium tumefaciens, Citrobacter freundii, Enterobacter aerogenes, E. coli, Klebsiella pneumonia, Neisseria gonorrhoeae, Proteus mirabilis, P. vulgaris, Pseudomonas aeruginosa, Salmonella typhi, Sa. typhymurium,* and *Serratia marcescens*) strains with zone of inhibition (16–18 mm) using disc diffusion method. Chloramphenicol was used as standard drug (10 µg/disc).	Khan et al. (2003)
		Antifungal activity	4 mg/disc	Methanolic extract and all the fractions were inactive against fungal strain (*Aspergilus niger, A. rubrum, A. versicolor, A. vitis, Candida albicans, C. tropicalis, Cladosporium cladosporiods, Penicillium notatum, Trychophyton mentagrophytes,* and *T. tronsurum*).	Khan et al. (2003)
		Antiprotozoal activity	4 mg/disc	Butanolic fraction of root barks showed maximum zone of inhibition (20 mm) against *Trichomonas vaginalis* compared to standard drug chloramphenicol with zone of inhibition (16 mm at 10 µg/disc).	Khan et al. (2003)

2.6.4 Antielastase, Anticollagenase and Antityrosinase Activities

The aqueous acetone extract of *L. asiatica* fruits showed significant ($p < 0.05$) antielastase (1426.72 ± 5.21 mg EGCG equivalents/100 g fruit weight) and anticollagenase (8560.90 ± 77.42 mg EGCG equivalents/100 g fruit weight) activity. Also, the fruit extract exhibited significant ($p < 0.05$) antityrosinase (30661.60 ± 1849.00 mg kojic acid equivalents/100 g fruit weight) activity compared to the positive control kojic acid (0.5 mg/mL) using the mushroom tyrosinase model (Singh et al., 2015).

Collagenase, elastase, and tyrosinase are metalloproteinase enzymes and their inhibition significantly improves mechanical strength, elasticity and whitening of the skin, respectively (Taofiq et al., 2016). These enzymatic activities suggest that the edible fruit berries of *L. asiatica* have good potential in cosmeceutical applications for skin anti-aging and whitening. Gallic acid, known to be present in *Leea* species (see Table 2.5), has been reported to have antityrosinase activity (Taofiq et al., 2016). However, no chemical study has been conducted on *L. asiatica*, and the identities of the components responsible for the antielastase, antityrosinase and anticollagenase activities are as of yet unknown. Further chemical investigations of the extract as well as biological studies are warranted.

2.6.5 Antihypertensive and Diuretic Activity

Traditionally, a decoction of *L. asiatica* is used to treat heart disorders (Sen et al., 2011) while *L. guineensis* is used in folk medicine for its cardiac activities and diuretic effects (Adsersen and Adsersen, 1997; Op de Beck et al., 2003). The potential antihypertensive activity of an ethanolic extract of *L. rubra* aerial parts was evaluated *in vitro* by its ability to inhibit the angiotensin-converting enzyme (ACE) (Braga et al., 2007). The % ACE inhibition of extract tested (0.10 mg/mL) using colorimetric assay was reported as 57.0 ± 12.5 and was considered to be active (>50% ACE inhibition, with the positive control captopril giving an IC_{50} of 21.2 nmol/L). The % inhibition of ACE was 87.1% according to the HPLC assay method (Serra et al., 2005). Furthermore, the ethanolic and acetone extracts of *L. guineensis* leaves showed more than 60% ACE inhibition (Adsersen and Adsersen, 1997). Flavonoids present in these species are hypothesized to be responsible for the ACE inhibition (Braga et al., 2007), and more in-depth studies are recommended given their medicinal uses in heart disorders.

2.6.6 Anti-inflammatory Activity

Traditionally, the leaves of *L. macrophylla* and *L. guineensis* have been used to treat inflammatory diseases (Njamen et al., 2013; Yusuf et al., 2007). The methanolic extract of *L. macrophylla* leaves (at 100 µg/mL) significantly ($p < 0.05$) inhibited prostaglandin E_2 (PGE_2) production, interleukin-6 (IL-6) and IL-1β expression along with the reduction of tumor necrosis factor-α (TNF-α) expression in *in vitro* lipopolysaccharide stimulated murine peritoneal macrophages (Dewanjee et al., 2013). In addition, the methanolic extract reduced carrageenan induced inflammation by 15.7% (100 mg/kg, *p.o.*) and 17.4% (200 mg/kg, *p.o.*), likely due to inhibition of the release of inflammatory mediators like PGE_2. Similarly, the methanolic extract reduced cotton pellet granulomatous tissue formation in rats by 27.5% ($p < 0.05$) and 38.4% ($p < 0.01$) at doses 100 and 200 mg/kg, respectively. The *in vivo* anti-inflammatory results were compared with the standard drug indomethacin (10 mg/kg) which resulted in 48% inhibition. In another study, the anti-inflammatory activity of ethanolic extract of *L. macrophylla* leaves was determined by using inhibition of albumin denaturation method (Faruq et al., 2014). The ethanolic extract (500 µg/mL) resulted in $47.4 \pm 0.001\%$ inhibition of the protein denaturation compared to the positive control aspirin (0.1% w/w) of $52.35 \pm 0.0007\%$ inhibition. Additionally, the *L. macrophylla* ethanolic extract (500 µg/mL) inhibited the heat induced total hemolysis of red blood cells ($57.63 \pm 0.002\%$) compared to aspirin (89.83 ± 0.002) (Faruq et al. 2014). The anti-inflammatory activity of *L. macrophylla* may be attributed to phytoconstituents like oleanolic acid, ursolic acid, β-amyrin, stigmasterol, β-sitosterol and ascorbic acid (Dewanjee et al., 2013).

A study on the methanolic extract of *L. indica* whole plant demonstrated strong inhibitory effect of nitric oxide (NO) production in lipopolysaccharide (LPS) and interferon-γ induced mouse macrophage RAW 264.7 cells (Saha et al., 2004).

L. guineensis is traditionally used to treat inflammatory disorders in Nigeria. The aqueous extract of *L. guineensis* leaves has been evaluated for antiedematogenic activity in carrageenan-induced rat paw oedema assay (Falodun et al., 2007). Oral administration of the aqueous extract significantly ($p < 0.001$) and dose dependently reduced the oedema levels by 6%, 56%, and 73% at doses 100, 200, and 400 mg/kg, respectively, compared to control. The 200 mg/kg dose of aqueous extract showed significant ($p < 0.05$) anti-inflammatory effect, compared to indomethacin with 46% reduction (10 mg/kg, *p.o.*). High saponin content in the extract was thought to be responsible for the antiedematogenic effect of *L. guineensis* (Falodun et al., 2007). *Leea* species with anti-inflammatory activity could be further investigated to identify the anti-inflammatory components for development of new anti-inflammatory drugs.

2.6.7 Antimicrobial Activity

Different plant parts of *L. macrophylla* including roots, leaves and seeds have been investigated for their antimicrobial activity. The ethanolic extract of *L. macrophylla* root tubers demonstrated the most potent antibacterial activity, followed by its successive fractions, butanol > aqueous > ethyl acetate > chloroform > hexane against gram-negative and gram-positive bacteria using disc diffusion assay (Joshi et al., 2016a). The ethanolic extract (100 mg/mL) showed a maximum inhibitory effect against *Staphyllococcus aureus*, *Shigella flexneri* and *Shigella boydii* with the zones of inhibition (ZOI) at 18.54 ± 0.34, 12.54 ± 0.30 and 14.80 ± 0.25 mm, respectively, along with minimum inhibitory concentrations (MIC) of 0.195, 0.390 and 0.781 mg/mL, respectively. The ZOI of drug ciprofloxacin (0.5 mg/mL) was found to be 27.85 ± 0.25, 21.33 ± 0.45 and 25.81 ± 0.60 mm for *S. aureus*, *S. flexneri* and *S. boydii*, respectively. The antibacterial activity of the ethanolic extract may be attributed to the presence of phytochemicals such as chlorogenic acid, damaging the cell membrane and leakage of cellular materials (Joshi et al., 2016a), thus killing the microorganisms. In a separate study, the ethanolic extract of *L. macrophylla* leaves showed strong antifungal activity at 500 µg/disc against various fungi, with ZOI in brackets: *Candida albicans* (26 ± 1 mm), *Cryptococcus neoformans* (26 ± 1 mm), *Microsporum sp.* (26 ± 1 mm), *Pityrosporum ovale* (31 ± 1 mm), and *Trichophyton sp.* (28 ± 1 mm) with MIC value of 31.25 µg/mL, compared to the standard antifungal drug fluconazole 30 µg/disc with ZOI against the different fungi (11.5 ± 1.50, 14.5 ± 0.50, 11.5 ± 1.32, 13.0 ± 0.50 and 12.7 ± 1.26 mm) respectively. The ethanolic extract displayed moderate antibacterial activity (Faruq et al., 2014). The ethyl acetate extract of *L. macrophylla* seeds (500 µg/disc) had antimicrobial activity against *S. aureous* and *C. albicans* with 19 mm and 15 mm of ZOI, respectively compared to standard drug kanamycin 28 mm and 25 mm at 30 µg/disc (Islam et al., 2013).

The antimicrobial activity of ethanolic extract of *L. indica* leaves was examined against Gram positive bacteria (*Bacillus subtulis, Staphylococcus aureus, Bacillus cereus, Bacillus megaterium*) and Gram negative bacteria (*Salmonella typhi, Salmonella paratyphi, Pseudomonas aeruginosa, Vibrio cholerae, Shigella dysenteriae* and *Escherichia coli*) along with human pathogenic fungal strains (*Aspergillus flavus, C. albicans* and *Fusarium equisetii*) (Rahman et al., 2013a) were investigated. The ethanolic extract at 10 mg/disc showed significant ($p < 0.05$) % growth inhibition against fungal strains *A. flavus* (38.09 ± 0.59%), *C. albicans* (22.58 ± 2.2%) and *F. equisetii* (61.82 ± 2.7%) compared to standard drug fluconazole 67.01 ± 1.8%, 40.00 ± 2.5% and 72.32 ± 2.3%, respectively at 100 µg/disc. Also, the ethanolic extract showed significant ($p < 0.05$) antibacterial activity with ZOI (9.0–12.0 mm) at 1–3 mg/disc, compared to the antibiotic drugs tetracycline and ampicillin (16–20 mm) at 30 µg/disc. These results showed that the ethanol extract of *L. indica* has potent antimicrobial activity. However, in a separate study, the methanolic extract (1 mg/disc) of *L. indica* leaves had neither antibacterial activity against *B. cereus, B. subtilis, E. coli, P. aeruginosa* and *S.aureus*, nor antifungal activity against *C. albicans* (Wiart et al., 2004). This difference in antimicrobial activity results of *L. indica* leaves reported by Rahman et al., 2013a and Wiart et al., 2004, while working on similar bacterial and fungal strains, may be attributed to the different solvents and amounts of the extract tested. Further investigations are warranted. The essential oil extracted from the flowers of *L. indica* was evaluated against three Gram positive (*B. subtulis, B. cereus* and *S. aureus*) and two gram negative bacteria (*E. coli* and *Salmonella typhimurium*) along with five fungal strains (*Alternaria alternata, A. flavus, Aspergilus niger, Fusarium monelliforme* and *Pencillium notatum*). The oil exhibited good antibacterial activity against the bacterial strains *E. coli* and *S. typhimurium*, and fungus *P. notatum* with ZOI about 10, 11 and 21 mm, respectively (Srinivasan et al., 2009).

Additionally, a comprehensive study on the methanolic extract and fractions (petroleum ether, dichoromethane, ethyl acetate and butanol) of *L. tetramera* leaves, stem and root bark was conducted against 25 bacterial, 11 fungal and 1 protozoan strains (Khan et al., 2003). The butanol fraction of root bark showed antibacterial activity, while none of the fractions was active against the fungal strains tested (Table 2.5).

2.6.8 Antioxidant Activity

Various *Leea* extracts were investigated for antioxidant potential using different assays that measure free radical scavenging activities, such as 2,2-diphenyl-2-picrylhydrazyl hydrate (DPPH) radical scavenging activity, 2,2′-azino-bis-3-ethylbenzothiazoline-6-sulfonic acid (ABTS) radical scavenging activity, ferric thiocyanante (FTC), thiobarbituric acid (TBA), superoxide dismutase (SOD), superoxide anion radical scavenging activity, linoleate hydroperoxide radical scavenging activity, nitric oxide (NO) radical scavenging activity, hydrogen peroxide (H_2O_2) scavenging activity, hydroxyl (OH) radical scavenging activity, ferric reducing activity, ferrous metal chelating activity and lipid peroxidation assay. The antioxidant activity of the genus *Leea* was mainly attributed to the high total phenolic content (Reddy et al., 2012; Dhatri et al., 2015). The presence of secondary metabolites like gallic acid, epicatechin, chlorogenic acid and quercetin from *Leea* species might be responsible for their antioxidant potential (Iacopini et al., 2008; Sato et al., 2011; Badhani et al., 2015).

The methanolic, ethyl acetate, and petroleum extracts of *L. asiatica* leaves were evaluated for antioxidant effects (Sen et al., 2013). The methanolic leaf extract of *L. asiatica* exhibited the highest total phenolic content (77.75 ± 0.87 mg gallic acid equivalent (GAE)/g of dry material) and total flavonoid content (60.98 ± 0.58 mg quercetin equivalent (QE)/g of dry material) (Sen et al., 2013). Moreover, the methanolic leaf extract significantly inhibited MDA formation in liver tissues homogenates. Both methanolic leaf extract and positive control rutin displayed nearly similar effects at 120 µg/mL, whereas at 160 µg/mL, methanolic leaf extract and rutin showed 82.10% and 91.53% inhibition, respectively (Sen et al., 2013). Different fractions of the methanolic leaf extract were further evaluated for antioxidant activity, and ethyl acetate fraction was found more superior than other fractions (namely petroleum ether and methanol) in antioxidant activity, with IC_{50} values 9.5, 13.0, and 57.0 µg/mL in DPPH, NO radical scavenging and lipid peroxidation assays, respectively (Sen et al., 2012). A novel fatty acid alcohol, $CH_3(CH_2)_{50}CH_2OH$ from the leaves of *L. asiatica* at 20 µg/mL showed DPPH and NO free radical scavenging activity 39.12 ± 0.40% and 35.26 ± 0.22%, respectively (Sen et al., 2014).

The free radical scavenging activities of 80% aqueous acetone extract of *L. asiatica* fruits have been evaluated and documented to exhibit significant DPPH (3990.10 ± 18.10 mg ascorbic acid equivalent, $p < 0.05$), superoxide (52500.71 ± 1345.00 mg catechin equivalent) free radicals scavenging and ferric reducing (6366.60 ± 409.00 mg ascorbic acid equivalent, $p < 0.05$) activities (Singh et al., 2015). The high total phenolic and flavonoid contents were 1860.49 ± 64.80 mg GAE/100 g fruit weight and 1963.75 ± 134.20 mg catechin equivalent/100 g fruit weight, respectively in *L. asiatica* fruits, further supporting its free radical scavenging effect.

The *in vivo* antioxidant potential was evaluated in aqueous extract of *L. guineensis* seeds at 200 and 400 mg/kg doses (orally) against dichlorvos-induced toxicity in rats (Ajiboye et al., 2014). Application of aqueous seed extract significantly ameliorated liver toxicity in these rat models, as demonstrated by the increase in the activities of antioxidant enzymes CAT, SOD and GPx. The antioxidant effects of quercitrin 3′-sulphate, quercetin 3,3′-disulphate, quercetin 3,3′,4′-trisulphate, along with kaempferol, quercetin, quercitrin, gallic acid and ethyl gallate, that were isolated from *L. guineensis* leaves were screened for DPPH radical scavenging activity (Op de Beck et al., 2003). It was concluded that the loss of antioxidant activity of quercitrin-3'-sulphate and quercetin-3,3'-disulphate compared to quercetin is due to the substitution of the hydroxyl group at C-3 position (ring B) by a sulphate group.

The methanolic extract of *L. indica* whole plant exhibited potent antioxidant effect for DPPH radicals with IC_{50} value of 25.0 μg/mL, compared to the standards ascorbic acid (AA) (9.0 μg/mL), quercetin (11.0 μg/mL) and butylated hydroxyl toluene (12.5 μg/mL) (Saha et al., 2004). Ethanolic extract of *L. indica* leaf and its fractions (hexane, ethyl acetate and water) were analyzed for antioxidant activity using DPPH, reducing power assay and SOD activity assays (Reddy et al., 2012). Water fraction showed strongest DPPH radical scavenging activity (EC_{50} 48.0 μg/mL, compared to AA 15.0 μg/mL), significant ($p < 0.05$) high reducing power (2.70 ± 0.02 compared to AA 2.73 ± 0.03 at 0.8 mg/mL), and the strongest inhibition rate ($p < 0.05$) in SOD assay. Separately, studies of ethanolic extract of *L. indica* leaves exhibited strong antioxidant activity in different assay systems: DPPH (IC_{50} = 139.83 ± 1.40 μg/mL, compared to AA IC_{50} 1.46 ± 0.06 μg/mL), $FeCl_3$ reduction (IC_{50} = 16.48 ± 0.64 μg/mL, compared to AA IC_{50} = 14.04 ± 1.20 μg/mL) and superoxide radical scavenging effect (49.54 ± 0.51%, IC_{50} = 676.08 ± 5.80 μg/mL), significant compared to curcumin IC_{50} = 60.48 ± 0.53%) along with less potent effect for iron chelating activity (IC_{50} = 519.33 ± 16.96 μg/mL, compared to AA IC_{50} = 8.81 ± 0.90 μg/mL) (Rahman et al., 2013a).

The methanolic extract and chloroform, and ethyl acetate fractions of *L. macrophylla* leaves were evaluated for DDPH radical scavenging, $FeCl_3$ reducing effect, superoxide radical scavenging and iron chelating effect (Akhter et al., 2015). Interestingly, the chloroform fraction demonstrated significant ($p < 0.05$) radical scavenging effect against DPPH free radicals with IC_{50} value 2.64 μg/mL, which was even lower than standard AA (IC_{50} 8.47 μg/mL), and methanolic extract (IC_{50}, 55.46 μg/mL) and ethyl acetate fraction (946.24 μg/mL). The potent free radical scavenging activity on DPPH, H_2O_2, OH, and NO radicals was observed in the ethanolic extract of *L. macrophylla* root tubers, with total antioxidant capacity (365.67 ± 1.08 μg/mL) and DPPH reducing potential (1.73 ± 0.05 μg/mL) (Joshi et al., 2016a). The ethanolic extract showed potent antioxidant activity against DPPH with IC_{50}= 39.80 ± 2.05 μg/mL (compared to AA, IC_{50} = 23.67 ± 1.67 μg/mL) and OH free radicals with IC_{50} value 52.22 ± 0.97 μg/mL (compared to butylated hydroxyl anisol (BHA) IC_{50} = 17.59 ± 1.00 μg/mL). The ethanolic extract had the most potent antioxidant activity followed by the butanol, aqueous, ethyl acetate, chloroform, and hexane fractions. The authors proposed that the presence of chlorogenic acid in the ethanolic extract (9.01% *w/w*) may have contributed to the antioxidant activity of *L. macrophylla*, and further studies are needed to confirm the active components.

Different solvent extracts of *L. philippinensis* leaves were measured for total phenolic content (TPC) and total flavonoid content (TFC). The highest TPC and TFC values were observed for acetone extract (TPC = 83.48 ± 0.04 GAE mg/g, TFC = 87.13 ± 0.03 QE mg/g) and methanolic extract (TPC = 41.71 ± 0.06 GAE mg/g,

TFC = 83.28 ± 0.03 QE mg/g) (Bartolome and Santiago, 2015). In addition, when measured for DPPH radical scavenging activity, acetone extract and methanolic extract exhibited IC_{50} values of 0.66 ± 0.12 mg/mL and 1.17 ± 0.26 mg/mL respectively, versus AA IC_{50} of 0.66 ± 0.3 mg/mL. Further, the chromatographic fractions with the largest TPC and TFC showed concentration-dependent activity for DPPH, OH, NO and H_2O_2 inhibition with IC_{50} values ranging from 0.10–0.29, 0.17–0.80, 0.29–0.84 and 3.40–9.75 mg/mL, respectively. The antioxidant effects of *L. philippinensis* may be associated to the presence of appreciable phenolic and flavonoid contents. The compounds responsible for the antioxidant activity in *Leea* species have yet to be identified.

2.6.9 Antiproliferative Activity

Leea species including *L. indica*, *L. guineensis* and *L. macrophylla* have been traditionally used to treat cancer (Wong et al., 2012a; Graham et al., 2000; Choudhary et al., 2008). The antitumor activity of an methanolic extract of *L. indica* leaves (20, 30 and 40 mg/kg, *i.p.*) was evaluated against EAC cells in Swiss albino mice (Raihan et al., 2012). The methanolic extract at a dose of 40 mg/kg, *i.p.* showed antitumor activity with 77.29% inhibition of cell growth compared to the positive control bleomycin with 92.02% of cell growth inhibition (0.3 mg/kg, *i.p.*). Also, the 40 mg/kg dose of methanolic extract showed maximum reduction in tumor weight (7.90 g) along with the enhancement of the life span by 69.33%, compared to the positive control bleomycin at a dose of 0.3 mg/kg (*i.p.*) with 7.05 g of tumor weight reduction and 94.66% ($p < 0.01$) increase in life span.

A methanolic extract derived from *L. indica* leaf and stem/bark showed the antiproliferative effects in breast cancer cells MCF-7 and T47D with IC_{50} values greater than 100 μg/mL by sulforhodamine-B assay (Nurhanan et al., 2008). Wong and Kadir (2011) found that the ethyl acetate fraction of *L. indica* leaf ethanolic extract reduced cell viability of MCF-7 breast cancer cells (IC_{50} = 138.05 ± 19.16 μg/mL), KB nasopharyngeal cancer cells (IC_{50} = 146.9 ± 10.41 μg/mL), while MDA-MB-235 melanoma cells and HepG2 liver cancer cells had IC_{50} > 200 μg/mL using methyl-thiazoyl-tetrazolium (MTT) assay. Ethanolic leaf extract of *L. indica* did not show significant effect on the viability of Vero cells (Wong and Kadir, 2011) or HeLa cells (Ali et al., 1996). Studies of *L. indica* leaf ethanolic extract and fractions on cell viability of colon cancer cells HT-29, HCT-115, and HCT-116 showed IC_{50} > 100 μg/mL (Reddy et al., 2012). A recent investigation of the anticancer activity of *L. indica* methanolic, ethanolic, and aqueous extracts revealed that methanolic extract significantly inhibited human prostate cancer cell lines DU-145 and PC-3 with IC_{50} values 529.44 ± 42.07 μg/mL and 547.55 ± 33.52 μg/mL, respectively (Ghagane et al., 2017). This extract showed no cytotoxic effect on normal mice embryo fibroblast cell line (MEF-L929).

The effect of ethanolic extract and fractions (hexane, ethyl acetate and water) of *L. indica* leaves on cell viability were evaluated against human Ca Ski (cervical epidermoid carcinoma cells), MCF-7 (breast carcinoma cells), MDA-MB-435 (melanoma cells), KB (nasopharyngeal epidermoid carcinoma cells, HeLa derivative), HEP G2 (hepatocellular carcinoma cells), WRL 68 (liver embryonic cells, HeLa derivative), Vero (kidney epithelial cells), and colon cancer cell lines (HT-29, HCT-15 and HCT-116) using MTT assay (Wong and Kadir, 2011; Reddy et al., 2012). Amongst all fractions, ethyl acetate fraction reduced the viability of Ca Ski cells most compared with all the cell lines tested, (IC_{50} = 85.83 ± 6.01 µg/mL) (Wong and Kadir, 2011). The ethyl acetate fraction activated apoptosis *via* nuclear shrinkage, chromatin condensation, increase in sub-G_1 cells, DNA fragmentation, intracellular GSH depletion and caspase-3 activation (Wong and Kadir, 2011). Further, two cycloartane triterpenoid glycosides named mollic acid arabinoside (**23**) and mollic acid xyloside (**24**) were identified in the active ethyl acetate fraction (Wong et al., 2012a). Both mollic acid glycosides **23** and **24** inhibited growth of Ca Ski cells with IC_{50} of 19.21 and 33.33 µM, respectively. Compared to the human fibroblast cell line MRC5, **23** and **24** was about 8 times and 4 times more cytotoxic respectively to Ca Ski cells. The cytotoxicity of mollic acid arabinoside was associated with a decrease in proliferating cell nuclear antigen gene expression, cell cycle arrest at S and G2/M phases, as well as induction of hypodiploid cells (Wong et al., 2012a). In addition, mollic acid arabinoside induced mitochondrial-mediated apoptosis in Ca Ski cells through an increase in Bax/Bcl-2 ratio (Wong and Kadir, 2012b).

Using brine shrimp lethality assay, the ethanolic extract of *L. indica* leaves showed cytotoxicity at a lethal concentration (LC_{50}) of 2.65 ± 0.16 µg/mL, which was significantly ($p < 0.01$) different from the standard vincristine sulfate (LC_{50} of 0.76 ± 0.04 µg/mL) (Rahman et al., 2013a, 2013b). Separately, Paul and Saha (2012) assessed the toxicity of ethanolic extract of *L. indica* leaves by the same assay and found the LC_{50} to be 2.4771 µg/mL with 95% confidence limit where the lower and upper limits were 2.2561 and 2.698 µg/mL, respectively. Cantrell and coworkers (2003) investigated the cytotoxicity of dichloromethane-methanol (1:1) seed extract of *L. indica* using brine shrimp assay and found the LC_{50} > 1000 µg/mL. Taken together, these studies suggest the potential clinical importance of *Leea* species against tumor cells.

2.6.10 Antipyretic Activity

Oral administration of a methanolic extract of *L. macrophylla* leaves (100 and 200 mg/kg) significantly reduced the yeast-induced elevated body temperature in rats within 1 h ($p < 0.05$) and the effect was maintained ($p < 0.05$–0.01) up to 4 h (Dewanjee et al., 2013). This effect was compared to the standard drug paracetamol (150 mg/kg, *oral*) and the authors proposed the antipyretic activity might be due to the inhibition of TNF-α, IL-1β and PGE2.

2.6.11 Hepatoprotective Activity

The aqueous extract of *L. guineensis* seeds (200 and 400 mg/kg) was found to protect the liver against dichlorvos induced toxicity in rats (Ajiboye et al., 2014). Exposure of rats to dichlorvos significantly decreased total protein, albumin, bilirubin, and the activities of GPx, SOD, CAT, transaminases alanine aminotransferase (ALT) and aspartate aminotransferase (AST). The administration of aqueous extract (400 mg/kg) significantly ($p < 0.05$) restored these parameters to near normal levels. Further, the aqueous extract exhibited antioxidant activity as described above (in antioxidant activity section). The author concluded the hepatoprotective effect of *L. guineensis* seeds may be attributed to their antioxidant property.

The ethanolic extract of *L. indica* stem bark (200 and 400 mg/kg) was found to have protective effects against paracetamol induced hepatotoxicity in rats (Mishra et al., 2014). The pretreatment of ethanolic extract (400 mg/kg) significantly attenuated the increased level of serum glutamic pyruvic transaminase (SGPT, 85.62 ± 0.60 IU/L), serum glutamic oxaloacetate transaminase (SGOT, 130.57 ± 0.672 IU/L), serum alkaline phosphatase (SALP, 81.00 ± 3.12 IU/L), bilirubin direct (0.25 ± 0.01 mg/dL), bilirubin total (0.65 ± 0.04 mg/dL) and triglycerides (158.70 ± 1.30 mg/dL) compared to the compound silymarin (100 mg/kg) SGPT, 74.64 ± 0.90 IU/L; SGOT, 121.13 ± 0.60 IU/L; SALP, 79.66 ± 1.18 IU/L; bilirubin direct, 0.20 ±0.01 mg/dL; bilirubin total, 0.640 ± 0.04 mg/dL), and triglycerides (145.7 ± 0.88 mg/dL). Histopathological study of liver tissues showed that the paracetamol treated group had hepatic cell necrosis and deterioration of central vein, in contrast to the ethanolic extract treated group with minimal necrosis regeneration of hepatocyte, while positive control silymarin had mild diffuse granular degeneration and necrosis.

The methanolic extract as well as its chloroform and ethyl acetate fractions obtained from *L. macrophylla* leaves were assessed for hepatoprotection against carbon tetrachloride (CCl_4)-induced damage in rats (Akhter et al., 2015). Exposure of rats to CCl_4 increased the level of serum AST, ALT, alkaline phosphatase (ALP), total protein, creatinine kinase-MB (CK-MB), and lipid profiles (cholesterol, HDL and LDL). The hepatoprotective effect of *L. macrophylla* was due to the normalization of liver functions by reducing the serum AST, ALT, ALP along with the restoring lipid profiles, total protein and CK-MB (Akhter et al., 2015). Chloroform fraction (200 mg) showed the highest liver protection by repairing lobular pattern with mild fatty changes, necrosis and lymphocytes infiltration. Future work on determining the hepatoprotective mechanism and chemical identification in *L. guineensis*, *L. indica*, and *L. macrophylla* is recommended.

2.6.12 Nephroprotective Activity

The methanolic extract and ethyl acetate fraction of *L. asiatica* leaves were found to have protective effects against cisplatin-induced nephrotoxicity in mice (Sen et al., 2013). Negative control mice treated with cisplatin showed significant increase in blood urea nitrogen (BUN), serum creatinine, uric acid and malondialdehyde (MDA) levels, along with lowered total protein and albumin level ($p < 0.05$, 0.01) with respect to normal control mice. Pretreatment of mice with methanolic extract (150 and 300 mg/kg) and its ethyl acetate fraction (75 and 150 mg/kg) significantly reduced the BUN, serum creatinine, uric acid, and MDA levels as well as increased total protein and albumin level ($p < 0.05$, 0.01) in blood serum. The ethyl actetate fraction (150 mg/kg) showed the best activity and almost similar to that of rutin (20 mg/kg). The ethyl acetate fraction had potent antioxidant activities (Sen et al., 2013) and proposed that the nephroprotective effect of *L. asiatica* may be by inhibiting lipid peroxidation process. Further studies are warranted to explore the potential use of *L. asiatica* as an adjunct therapy to improve the effectiveness of drugs which may have nephrotoxic side effects.

L. macrophylla is traditionally used by the local tribes of India to treat urinary problems (Nizami et al., 2012). To evaluate its traditional use, the ethanolic extract of the fresh whole plant including roots of *L. macrophylla* was orally administered (500 mg/kg body weight/day) to ethylene glycol-induced urolithiasis model of male Wistar albino rats (Nizami et al., 2012). Treatment with ethanolic extract altered urinary excretion such as decreased calcium, phosphate, and oxalate along with increased magnesium and creatinine. The activity results suggest *L. macrophylla* has potent antiurolithiatic effect. The extract significantly reduced the growth of kidney stones or crystal deposition and improved renal impairment. The exact mechanism of the antiurolithiatic action of *L. macrophylla* extract is presently unclear.

2.6.13 Neuroprotective Activity

The neuroprotective effects of methanolic extract of *L. macrophylla* roots (100 and 200 mg/kg body weight) were investigated using open field test, hole cross test, elevated plus maze (EPM) test and thiopental sodium-induced sleeping time test on diazepam-induced memory impairment in amnesic Wistar rats (Ferdousy et al., 2016). The methanolic extract significantly ($p < 0.05$) decreased the number of movements in open field test animals (10.0 ± 2.8 and 8.0 ± 2.1 at doses of 100 and 200 mg/kg, respectively) compared to 13.0 ± 1.4 in normal control group. Similar effects were observed on number of holes crossed at 200 mg/kg extract (1.5 ± 0.7), which was similar to the standard drug diazepam (1.5 ± 0.7). The 200 mg/kg extract increased the time spent in the open arms (16 ± 0.08) compared to normal control group (10.51 ± 1.72), and also significantly ($p < 0.05$) increased the duration of sleep in thiopental sodium induced sleeping time test. In addition, the authors speculated that the neuropharmacological effects of *L. macrophylla* root extract might be mediated by the gamma-amino-butyric acid (GABA) receptor. The antioxidant effects like increase in SOD, catalase and glutathione enzymes as well as decrease in lipid peroxidation, NO, and advanced protein products showed the importance of this plant in reduction of oxidative stress-associated neurodegeneration.

2.6.14 Phosphodiesterase Inhibitory Activity

Phosphodiesterase (PDE) inhibitors are widely used for the treatment of many disorders like cardiovascular diseases, chronic obstructive pulmonary diseases, immune response, platelet aggregation, erectile dysfunction, and pulmonary hypertension (Boswell-Smith et al., 2006; Rahimi et al., 2010). Based on its traditional use as an aphrodisiac and neurotonic remedies in traditional Thai medicine, an ethanolic extract of *L. indica* roots was evaluated for its PDEs inhibitory activity using PDE [3H] cGMP (cyclic guanosine 3′,5′-monophosphate) scintillation proximity assay (Temkitthawon et al., 2008). The ethanolic extract (0.1 mg/mL) of *L. indica* roots showed >95% PDE inhibitory activity with IC_{50} value 2.62 ± 0.25 µg/mL, compared to the PDE inhibitor 3-isobutyl-1-methylxanthine (IC_{50} 0.68 ± 0.14 µg/mL). Furthermore, the ethanolic extract of *L. indica* roots showed $31.36 \pm 7.47\%$ PDE-5 inhibition (Temkitthawon et al., 2011). Compound(s) responsible for the reported PDE-5 inhibitory

activity of *L. indica* root extract is unknown and further work is warranted.

2.6.15 Sedative and Anxiolytic Activities

Traditionally, the leaves of *L. indica* have also been used to treat insomnia. The methanolic extract of *L. indica* leaves has been evaluated for its sedative and anxiolytic effects (Raihan et al., 2011; Emran et al., 2012). The methanolic extract (200 and 400 mg/kg, *p.o.*) induced sleep at an earlier stage and also prolonged the duration of thiopental induced sleeping time in mice, compared to controls (Raihan et al., 2011). Additionally, the methanolic extract significantly decreased ($p < 0.01$) the frequency and amplitude of movements in the hole cross and open field tests for locomotor activity, suggesting sedative effect of the plant extract. The methanolic extract (400 mg/kg, *p.o.*) significantly ($p < 0.01$) increased the exploration and time spent by treated mice in elevated plus-maze (EPM) test for anxiolytic activity, comparable to standard diazepam (1.0 mg/kg). Separately, a methanolic extract of *L. indica* leaves (200 mg/kg, *p.o.*) showed significant sedative, anxiolytic and analgesic effects (Emran, 2012). Components responsible for these observed sedative, anxiolytic, and analgesic effects remain to be determined.

2.6.16 Wound Healing Activity

Leea species including *L. asiatica*, *L. guineensis*, *L. indica*, *L. macrophylla*, and *L. tetramera* have been traditionally used in the treatment of skin lesions, boils, wounds, and skin regeneration (Sen et al., 2013; Cabelin et al., 2015; Ajiboye et al., 2014; Chatterjee and Pakrashi, 1994). In traditional Indian medicine, root tubers of *L. macrophylla* have been used for the treatment of ringworm, sores, and wounds (Bhavamishra, 2010). The ethanolic extract of *L. macrophylla* root tubers was evaluated *in vivo* for wound breaking strength (WBS) after oral administration (250, 500 and 750 mg/kg, *p.o.*) and topical application as a bioadhesive gel (2.5%, 5.0%, and 7.5% *w/v*) in excision and incision wound models (Joshi et al., 2016b). A significant increase in the percentage of WBS was observed in oral treatment at 500 mg/kg, *p.o.* (23.41%) and 750 mg/kg, *p.o.* (27.66%). The increase in percentage WBS observed in topical application of extract gel was 2.5% *w/v* (27.13%) compared to the increase (30.85%) using oral standard vitamin E (200 mg/kg, *p.o.*). Moreover, the topical application of extract bioadhesive gel at doses 5% *w/v* and 7.5% *w/v* resulted in an increase in WBS to 44.68% and 47.34%, compared to topical standard *Aloe vera* cream (53.73%). In both bioassay models, *L. macrophylla* showed potent wound healing effects and topical application (5% *w/v*; 100% wound contraction in 20 days) was found to have faster and more significant effect than oral treatment (500 mg/kg, *p.o.*; 100% wound contraction in 22 days). The topical application of *L. macrophylla* ethanolic extract significantly increased ($p < 0.05$) the antioxidant enzymes glutathione (GSH), SOD, catalase (CAT) along with reduced levels of lipid peroxidase (LPO), NO, and inflammatory marker myeloperoxidase (MPO). In addition, the reduced level of proinflammatory cytokines including IL-1β, IL-6, TNF-α and VEGF along with increased cellular proliferation supported the wound healing potential of *L. macrophylla*. The authors suggested chlorogenic acid might be responsible for the wound healing potential of *L. macrophylla* (Joshi et al., 2016b).

2.7 Safety Considerations

The safety of the medicinal plants, including the safe doses and effects on long-term use, has always been a great concern for medicinal practitioners and also affects the commercial value of the plants. The toxicity/safety studies on plants from the genus *Leea* is limited to only a few species like *L. macrophylla*, *L. indica* and *L. guineensis*. The alcoholic and hydroalchoholic extracts of *L. indica* leaves at the doses of 300, 2000, and 3000 mg/kg were evaluated for acute toxicity using male Wistar rats. Both types of extracts were safe up to a dose of 3000 mg/kg with no observed toxicity (Dalu et al., 2014). The oral lethal dose (LD_{50}) for *L. indica* was found to be more than 3000 mg/kg in rats (Mishra et al., 2014). The cell viability assay of *L. indica* ethanolic leaf extract on colon cancer cells showed $IC_{50} > 100$ µg/mL; this was considered as non-cytotoxic (Reddy et al., 2012). Based on the United States National Cancer Institute plant screening program, plant extracts are typically considered to have cytotoxic effects upon 48–72 h incubation and if $IC_{50} < 20$ µg/mL (Lee and Houghton, 2005).

A methanolic extract of *L. macrophylla* leaves was found to be non-toxic in both acute (2 g/kg) and sub-acute toxicity studies (oral doses of 1 and 2 g/kg for 14 days) and without having significant effects on body weight and internal organs like liver, kidney, and brain (Dewanjee et al., 2013). The intraperitoneal injection of an ethanolic extract of the whole plant of *L. macrophylla* was evaluated for acute toxicity in mice at doses of 0.75, 1.5, 2.5, and 3.5 g/kg for ten days and found to be non-toxic (Nizami et al., 2012). The toxicity studies suggest that the whole plant of *L. macrophylla* can be used as a safe herbal remedy for clinical applications against anti-inflammatory and other disorders. The aqueous extract of *L. guineensis* leaves was found to be partially non-toxic at a dose range of 1-5 g/kg in mice (Falodun et al., 2007). The acute toxicity of methanolic extract and fractions of *L. asiatica* leaves were evaluated in mice by oral administration at a dose of 2000 mg/kg (ALD_{50}) and did not exhibit any common side effect within 7 days of observation (Sen et al., 2013).

2.8 Conclusions

The genus *Leea* has long been used as traditional medicine in the treatment of wounds and bone fractures, sexual debility, boils, cancer, arthritis, gout, rheumatism, stomachache, diarrhoea, dysentery and skin disorders. The ethnomedicinal importance of *Leea* species is supported by their wide spectrum of pharmacological activities such as anti-inflammatory, antioxidant, antiproliferative, anxiolytic, hepatoprotective, sedative, and wound healing in different *in vitro* and *in vivo* models. This review summarizes the ethnomedicinal uses, phytochemistry, pharmacological activities, patents, and safety considerations of medicinal plants in the genus *Leea*. The promising pharmacological activities of *Leea* species

highlight the importance of their cultivation and conservation to prepare for further development of their potential applications in the pharmaceutical, nutraceutical, cosmeceutical, and food and beverage industries. This review also provides valuable insights into their diverse potential applications, leading to greater understanding and appreciation of the various *Leea* species. It will also serve to stimulate future research to discover novel lead compounds and useful therapeutics from *Leea* species.

2.9 Future Prospects

Traditional and recent studies indicate that certain species from the genus *Leea* are endowed with medicinally important properties, some of which have potential to be developed into new therapeutics, nutraceuticals and health supplements. However, majority of the species are either not studied or minimum information is available. The finding of anti-apoptotic triterpenoids mollic acid glycosides suggests the presence of other as of yet unidentified bioactive compounds, indicating an urgent need to deepen our understanding of the biology and chemistry of the genus *Leea* for drug discovery. We propose the following aspects to fill the current gap in knowledge of *Leea* species: (1) botanical characterization of the *Leea* species, including their species authentication, quality control, and phylogenetic analysis; (2) isolation and identification of chemical constituents from the bioactive extracts and fractions along with comprehensive qualitative and quantitative analyses; (3) extensive pharmacological investigations of different extracts/fractions of various plant parts from the various *Leea* species; (4) thorough evaluation of safety/ toxicity of *Leea* species for long-term use, including dose range, pharmacokinetics, and target organ toxicity. The pharmacological studies have supported the traditional uses of *Leea* species and further research would harness the potential benefits and helps in the development of novel therapeutics.

ACKNOWLEDGMENTS

This work is supported by research collaboration with Leeward Pacific Pte Ltd. (R-148-000-172-592) to KHL and National University of Singapore Provost Industrial PhD Programme Research Scholarship to SYY.

REFERENCES

Ado, M.A., Abas, F., Mohammed, A.S. and Ghazali, H.M., 2013. Anti- and pro-lipase activity of selected medicinal, herbal and aquatic plants, and structure elucidation of an anti-lipase compound. *Molecules* 18, pp. 14651–14669.

Adsersen, A. and Adsersen, H., 1997. Plants from Reunion Island with alleged antihypertensive and diuretics effects-an experimental and ethnobotanical experience. *J. Ethnopharmacol.* 58, pp. 189–206.

Ajiboye, B.O., Salawu, S.O., Okezie, B., Oyinloye, B.E., Ojo, A.O., Onikanni, S.A., Oso, A.O., Asoso, O.S. and Obafemi, T.O., 2014. Mitigating potential and antioxidant properties of aqueous seed extract of *Leea guineensis* against dichlorvos-induced toxicity in Wistar rats. *J. Toxicol. Environ. Health Sci.* 6, pp. 132–146.

Akhter, S., Rahman, M.A., Aklima, J., Hasan, M.R. and Chowdhury, J.M.K.M., 2015. Antioxidative role of Hatikana (*Leea macrophylla* Roxb.) partially improves the hepatic damage induced by CCl₄ in Wistar albino rats. *Biomed. Res. Int.* 1–12. doi:10.1155/2015/356729.pdf.

Ali, A.M., Mackeen, M.M., Ei-Sharkawy, S.H., Hamidi, J.A., Ismaili, N.H., Ahmadi, F.B.H. and Lajis, N.H., 1996. Antiviral and cytotoxic activities of some plants used in Malaysian indigenous medicine. *Pertanika J. Trop. Agric. Sci.* 19, pp. 129–136.

APG 1998. An ordinal classification for the families of flowering plants. *Ann. Missouri Bot. Gard.* 85, pp.531–553.

APG II. 2003. An update of the Angiosperm Phylogeny Group classification for the orders and families of flowering plants. *Bot. J. Linn. Soc.* 141, pp. 399–436.

APG III. 2009. An update of the Angiosperm Phylogeny Group classification for the orders and families plants. *Bot. J. Linn. Soc.* 161, pp. 105–121.

Atindehou, K.K., Schmid, C., Brun, R., Kone, M.W. and Traore, D., 2004. Antitrypanosomal and antiplasmodial activity of medicinal plants from Cote d'Ivoire. *J. Ethnopharmacol.* 90, pp. 221–227.

Avin, B.R.V., Thirusangu, P., Ramesh, Vighneshwaran, V., Prashanth Kumar, M.V., Mahmood, R. and Prabhakar, B.T., 2014. Screening for the modulation of neovessel formation in non-tumorigenic and tumorigenic conditions using three different plants native to Western ghats of India. *Biomed. Aging Pathol.* 4, pp. 343–348.

Badhani, B., Sharma, N. and Kakkar, R., 2015. Gallic acid: A versatile antioxidant with promising therapeutic and industrial applications. *RSC Adv.* 5, pp. 27540–27557.

Bartolome, M. and Santiago, L., 2015. Chromatographic separation of the free radical scavenging components of the leaf extracts of *Leea philippinensis* Merr. *Int. Food Res. J.* 22, pp. 1396–1403.

Bhandary, M.J., Chandrashekhar, K.R. and Kaveriappa, K.M., 1995. Medical ethnobotany of the Siddis of Uttara Kannada district, Karnataka, India. *J. Ethnopharmacol.* 47, pp. 149–158.

Bhavamishra. 2010. Bhavaprakash Nighantu (Indian Materia Medica).In: Chunekar, K.C., (Ed.), Chaukhambha Bharati Academy, Varanasi, pp. 686–687.

Borokini, T.I. and Omotayo, F.O., 2012. Phytochemical and ethnobotanical study of some selected medicinal plants from Nigeria. *J. Med. Plants Res.* 6, pp. 1106–1118.

Boswell-Smith, V., Spina, D. and Page, C.P., 2006. Phosphodiesterase inhibitors. *Br. J. Pharmacol.* 147, pp. S252–S257.

Bourdy, G. and Walter, A., 1992. Maternity and medicinal plants in Vanuatu I. The cycle of reproduction. *J. Ethnopharmacol.* 37, pp. 179–196.

Braga, F.C., Serra, C.P., Viana Junior, N.S., Oliveira, A.B., Cortes, S.F. and Lombardi, J.A., 2007. Angiotensin-converting enzyme inhibition by Brazilian plants. *Fitoterapia* 78, pp. 353–358.

Cabelin, V.L.D., Santor, P.J.S. and Alejandro, G.J.D., 2015. Evaluation of DNA barcoding efficiency of cpDNA barcodes in selected Philippine *Leea* L. (Vitaceae). *Acta. Bot. Gallica* 162, pp. 317–324.

Cantrell, C.L., Berhow, M.A., Phillips, B.S., Duval, S.M., Weisleder, D. and Vaughn, S.F., 2003. Bioactive crude plant seed extracts from the NCAUR oilseed repository. *Phytomedicine* 10, pp. 325–333.

Chander, M.P. and Vijayachari, V., 2017. Vibriocidal activity of selected medicinal plants used by Nicobarese Tribes of Andaman and Nicobar Islands, India. *JPP* 5, pp. 164–168. doi:10.17265/2328-2150/2017.03.008.

Chassagne, F., Hul, S., Deharo, E. and Bourdy, G., 2016. Natural remedies used by Bunong people in Mondulkiri province (Northeast Cambodia) with special reference to the treatment of 11 most common ailments. *J. Ethnopharmacol.* 191, pp. 41–70.

Chatterjee, A. and Pakrashi, S.C., 1994. The treatise on Indian medicinal plants. NISCAIR, New Delhi India: Publications and information Directorate, pp. 173–174.

Choudhary, K., Singh, M. and Pillai, U., 2008. Ethnobotanical survey of Rajasthan-an update. *Am-Eurasian J. Bot.* 1, pp. 38–45.

Chuakul, W., Saralamp, P. and Boonpleng, A., 2002. Medicinal plants used in the Kutchum district, Yasothon province. *Thai. J. Phytopharmacy* 9, pp. 22–49.

Cronquist, A., 1988. The Evolution and Classification of Flowering Plants, second Ed. New York Botanical Garden, New York, pp. 396–397.

Dalu, D., Duggirala, S. and Akarapu, S., 2014. Antihyperglycemic and hypolipidemic activity of *Leea indica*. *Int. J. Bioassays* 3, pp. 3155–3159.

Dewanjee, S., Dua, T.K. and Sahu, R., 2013. Potential anti-inflammatory effect of *Leea macrophylla* Roxb. leaves: A wild edible plant. *Food Chem. Toxicol.* 59, pp.514–520.

Dhatri, R., Dhanya, R., Prashith, K.T.R., Onkarappa, R., Vinayaka, K.S. and Raghavendra, H.L., 2015. Antifungal and radical scavenging activity of leaf and bark of *Leea indica* (Burm. f.) Merr. *JCPR* 7, pp. 105–110.

Dung, N.X. and Loi, D.T., 1991. Selection of traditional medicines for study. *J. Ethnopharmacol.* 32, pp. 57–70.

Emran, T.B., 2012. Sedative, anxiolytic and analgesic effects of the ethanolic extract of *Leea indica* (Burm. f.) Merr. leaf. 2nd International conference and exhibition on pharmaceutical regulatory affair, Hyderabad International Convention Centre, India. 1, 253.

Emran, T.B., Rahman, M.A., Zahid Hosen, S.M., Rahman, M.M., Islam, A.M.T., Chowdhury, M.A.U. and Uddin, M.E., 2012. Analgesic activity of *Leea indica* (Burm. f.) Merr. *Phytopharmacology* 3, pp. 150–157.

Evans, E., Witabouna, K.M., Honora, T.B.F. and Adama, B., 2015. Iron reducing and radical scavenging activities of 13 medicinal plants from Côte d'Ivoire. *Pharmacogn. J.* 7, pp. 266–270.

Fagbohun, E.D., Lawal, O.U. and Ore, M.E., 2012. The proximate, mineral and phytochemical analysis of the leaves of *Ocimum gratissimum* L., *Melanthera scandens* A. and *Leea guineensis* L. and their medicinal value. *Int. J. Appl. Biol. Pharm.* 3, pp. 15–22.

Falodun, A., Okunrobo, L.O. and Agbo, L.O., 2007. Evaluation of the anti-edematogenic activity of the aqueous extract of *Leea guineensis*. *Afr. J. Biotechnol.* 6, pp. 1151–1153.

Faruq, A.A., Ibrahim, M., Mahmood, A., Chowdhury, M.M.U., Rashid, R.B., Kuddu, M.R. and Rashid, M.A., 2014. Pharmacological and phytochemical screenings of ethanol extract of *Leea macrophylla* Roxb. *IPP* 2, pp. 321–327.

Fayaz, Z., 2011. Encyclopedia of tropical plants: The identification and cultivation of over 3,000 tropical plants. David Bateman, New Zealand, p. 346.

Ferdousy, S., Rahman, M.A., Al-Amin, M.M., Aklima, J. and Hasan Chowdhury, J.M.K., 2016. Antioxidative and neuroprotective effects of *Leea macrophylla* methanol root extracts on diazepam-induced memory impairment in amnesic Wistar albino rat. *Clinical Phytosci.* 2, p. 17.

Flowers of India, 2016. http://www.flowersofindia.net/. (Accessed on 31 May 2018).

Ghagane, S.C., Puranik, S.I., Kumbar, V.M., Nerli, R.B., Jalalpure, S.S., Hiremath, M.B., Neelagund, S. and Aladakatti, R., 2017. *In vitro* antioxidant and anticancer activity of *Leea indica* leaf extracts on human prostate cancer cell lines. *Integr. Med. Res.* 6, pp. 79–87.

Ginting, N., Suwarso, E. and Nerdy Singa, E., 2017. Anticonvulsant activity of Tetanus leaves (*Leea aequata* L.) ethanolic extract on guinea pig (*Cavia cobaya*) isolated ileum by *in vitro* method. *Int. J. Chemtech Res.* 10, pp. 38–48.

Graham, J.G., Quinn, M.L., Fabricant, D.S. and Farnsworth, N.R., 2000. Plants used against cancer—an extension of Jonathan Hartwell. *J. Ethnopharmacol.* 73, pp. 347–377.

He-Ping, B., Chang-Ri, H., Zhen-Yi, L. and Si-Cong, Y., 2006. Chemical constituents of essential oil from *Leea longifolia* Merr. *Zhiwu Ziyuan Yu Huanjing Xuebao* 15, pp. 72–73.

Heywood, V.H., Brummitt, R.K., Culham, A. and Serberg, O., 2007. Flowering Plant Families of the World. Kew Publishing, London, UK, p. 334.

Iacopini, P., Baldi, M., Storchi, P. and Sebastiani, L., 2008. Catechin, epicatechin, quercetin, rutin and resveratrol in red grape: Content, *in vitro* antioxidant activity and interactions. *J. Food Comp. Anal.* 21, pp. 589–598.

Ingrouille, M.J., Chase Fls, M.W., Fay Fls, M.F., Bowman, D., Van Der Bank, M. and Bruijn, A.D.E., 2002. Systematics of Vitaceae from the viewpoint of plastid *rbcL* DNA sequence data. *Bot. J. Linn. Soc.* 138, pp. 421–432.

Inta, A., Trisonthi, P. and Trisonthi, C., 2013. Analysis of traditional knowledge in medicinal plants used by Yuan in Thailand. *J. Ethnopharmacol.* 149, pp. 344–351.

Islam, M.B., Sarkar, M.M.H., Shafique, M.Z., Jalil, M.A., Haque, M.Z. and Amin, R., 2013. Phytochemical screening and antimicrobial activity studies on *Leea macrophylla* seed extracts. *J. Sci. Res.* 5, pp. 399–405.

Jadhao, K.D. and Wadekar, M.P., 2010. Evaluation and study of minerals from *Leea macrophylla* Roxb. (Leeaceae). *Asian J. Chem.* 22, pp. 2480–2482.

Jadhao, K.D., Wadekar, M.P. and Mahalkar, M. S., 2009. Comparative study of availability of vitamins from *Leea macrophylla* Roxb. *Biosci. Biotechnol. Res. Asia* 6, pp. 847–849.

Jain, S.K. and Borthakur, S.K., 1980. Ethnobotany of the Mikirs of India. *Econ. Bot.* 34, pp. 264–272.

Jain, A., Katewa, S.S., Galav, P.K. and Sharma, P., 2005. Medicinal plant diversity of Sitamata wildlife sanctuary, Rajasthan, India. *J. Ethnopharmacol.* 102, pp. 143–157.

Joshi, A., Prasad, S.K., Joshi, V.K. and Hemalatha, S., 2016a. Phytochemical standardization, antioxidant, and antibacterial evaluations of *Leea macrophylla*: A wild edible plant. *J. Food Drug Anal.* 24, pp. 324–331.

Joshi, A., Joshi, V.K., Pandey, D. and Hemalatha, S., 2016b. Systematic investigation of ethanolic extract from *Leea macrophylla*: Implications in wound healing. *J. Ethnopharmacol.* 191, pp. 95–106.

Joshi, A.B., Tari, P.U. and Bhobe, M., 2013. Phytochemical investigation of the roots of *Leea indica* (Burm. F.) Merr. *Int. J. Res. Pharm. Biomed. Sci.* 4, pp. 919–925.

Junsongduang, A., Balslev, H., Inta, A., Jampeetong, A. and Wangpakapattanawong, P., 2014. Karen and Lawa medicinal plants use: Uniformity or ethnic divergence? *J. Ethnopharmacol.* 151, pp. 517–527.

Kaewkrud, W., Otsuka, H., Ruchirawat, S. and Kanchanapoom, T., 2007. Leeaoside, a new megastigmane diglycoside from the leaves of *Leea thorelii* Gagnep. *J. Nat. Med.* 61, pp. 449–451.

Khan, M. R., Omoloso, A. D. and Kihara, M., 2003. Antibacteraial activity of *Alstonia scholaris* and *Leea tetramera*. *Fitoterapia* 74, pp. 736–740.

Khare, C.P., 2007. *Indian Medicinal Plants: An Illustrated Dictionary*. Springer, New York, pp. 366–367.

Kirtikar, K.R. and Basu, B.D., 1975. Indian Medicinal Plants, Dehradun, India, Nirali Prakashan, p. 617.

Kong, Z., Wang, F. and Du, Y., 2015. Traditional Chinese medicine formulation for treating oral lichen planus. Faming Zhuanli Shenqing, CN 104784275 A 20150722.

Lakornwong, W., Kanokmedhakul, K. and Kanokmedhakul, S., 2014. Chemical constituents from the roots of *Leea thorelii* Gagnep. *Nat. Prod. Res.* 28, pp. 1015–1017.

Lee, C.C. and Houghton, P., 2005. Cytotoxicity of plants from Malaysia and Thailand used traditionally to treat cancer. *J. Ethnopharmacol.* 100, pp. 237–243.

Li, C.L., 1998. *Leea. Flora Republicae Popularis Sinicae* 48, pp. 3–12.

Li, K., 2014. A kind of Chinese medicine for the treatment of ulcerative colitis. Faming Zhuanli Shenqing CN 103599266 A 20140226.

Lok, A.F.S.L., Ang, W.F., Ng, B.Y.Q., Suen, S.M., Yeo, C.K. and Tan, H.T.W., 2011. *Leea* L. (Vitaceae) of Singapore. *Nature in Singapore* 4, pp. 55–71.

Mahmud, Z.A., Bachar, S.C., Hasan, C.M., Emran, T.B., Qais, N. and Uddin, M.M.N., 2017. Phytochemical investigations and antioxidant potential of roots of *Leea macrophylla* (Roxb.). *BMC Res. Notes* 10, p. 245. doi:10.1186/s13104-017-2503-2.

Marles, R.J. and Farnsworth, N.R., 1995. Antidiabetic plants and their active constituents. *Phytomedicine* 2, pp. 137–189.

Meena, A.K. and Rao, M.M., 2010. Folk herbal medicines used by the Meena community in Rajasthan. *Asian J. Tradit. Med.* 5, pp. 19–31.

Mishra, G., Khosa, R.L., Singh, P. and Jha, K.K., 2014. Hepatoprotective activity of ethanolic extract of *Leea indica* (Burm. f.) Merr. (Leeaceae) stem bark against paracetamol induced liver toxicity in rats. *Niger J. Exp. Clin. Biosci.* 2, pp. 59–63.

Mishra, G., Khosa, R.L., Singh, P. and Tahseen, M.A., 2016. Ethnobotany and phytopharmacology of *Leea indica*: An overview. *J. Coast Life Med.* 4, pp. 69–72.

Miyagoshi, M., Shinbo, D., Mizutani, K. and Watanabe, T., 2009a. The skin-lightening agent, skin applications, and the food and drink. Japanese Kokai Tokkyo Koho JP 2009298711 A 20091224.

Miyagoshi, M., Shinbo, D., Mizutani, K. and Watanabe, T., 2009b. The antioxidant, the anti-inflammatory drug, the anti aging agent, the hair restorer, skin applications, and the food and drink. Japanese Kokai Tokkyo Koho JP 2009298712 A 20091224.

Mohammad, N.S., Milow, P. and Ong, H.C., 2012a. Traditional medicinal plants used by the Kensiu tribe of Lubuk Ulu Legong, Kedah, Malaysia. *Ethno Med.* 6, pp. 149–153.

Mohammad, N.S., Milow, P. and Ong, H.C., 2012b. Traditional knowledge on medicinal plant among the Orang Asli villagers in Kampung Lubuk Ulu Legong, Kedah, Malaysia. *Technology, Science, Social Sciences and Humanities International Conference.* doi:10.13140/2.1.1147.5847.

Molina, J., 2009. Floral biology of Philippine morphospecies of the grape relative *Leea* (Leeaceae). *Plant Spec. Biol.* 24, pp. 53–60.

Molina, J., Wen, J. and Struwe, L., 2013. Systematics and biogeography of the non-viny grape relative *Leea* (Vitaceae). *Bot. J. Linn. Soc.* 171, pp. 354–376.

Mollik, A.H., Azam, N.K., Ferdausi, D., Jahan, R. and Rahmatullah, M., 2009. A survey of medicinal plants used to treat cattle diseases in satkhira district, Bangladesh. *Planta Med.* 75, PD54.

Nizami, A.N., Rahman, M.A., Ahmed, N.U. and Islam, M.S., 2012. Whole *Leea macrophylla* ethanolic extract normalizes kidney deposits and recovers renal impairments in an ethylene glycol-induced urolithiasis model of rats. *Asian Pac. J. Trop. Med.* 5, pp. 533–538.

Njamen, D., Mvondo, M.A., Djiogue, S., Ketcha Wanda, G.J., Magne Nde, C.B. and Vollmer, G., 2013. Phytotherapy and women'sreproductive health: the Cameroonian perspective. *Planta Med.* 79, pp. 600–611.

Nurhanan, M.Y., Asiah, O., Mohd Ilham, M.A., Siti Syarifah, M.M., Norhayati, I. and Lili Sahira, H. (2008). Anti-proliferative activities of 32 Malaysian plant species in breast cancer cell lines. *J. Tro. For. Sci.* 20, pp. 77–81.

Op de Beck, P., Bessiere, J.M., Dijoux-Franca, M.G., David, B. and Mariotte, A.M., 2000. Volatile constituents from leaves and wood of *Leea guineensis* G. Don (Leeaceae) from Cameroon. *Flavour Fragr. J.* 15, pp. 182–185.

Op de Beck, P., Cartier, G., David, B., Dijoux-Franca, M.G. and Mariotte, A.M., 2003. Antioxidant flavonoids and phenolic acids from leaves of *Leea guineense* G Don (Leeaceae). *Phytother. Res.* 17, pp. 345–347.

Op de Beck, P., Dijoux, M.G., Cartier, G. and Mariotte, A.M.,1998. Quercitrin 3′-sulphate from leaves of *Leea guinensis*. *Phytochemistry* 47, pp. 1171–1173.

Panda, S.K., Mohanta, Y.K., Padhi, L., Park, Y.H., Mohanta, T.K. and Bae, H., 2016. Large scale screening of ethnomedicinal plants for identification of potential antibacterial compounds. *Molecules* 21, p. 293.

Panyaphu, K., Sirisa-ard, P., Na Ubol, P., Nathakarnkitkul, S., Chansakaow, S. and Van On, T., 2012. Phytochemical, antioxidant and antibacterial activities of medicinal plants used in Northern Thailand as postpartum herbal bath recipes by the Mien (Yao) community. *Phytopharmacology* 2, pp. 92–105.

Patel, A.A., Amin, A.A., Patwari, A.H. and Shah, M.B., 2017. Validated high performance thin layer chromatography method for simultaneous determination of quercetin and gallic acid in *Leea indica*. *Rev. Bras. Farmacogn.* 27, pp. 50–53.

Paul, S. and Saha, D., 2012. Cytotoxic activity of ethanol extract of *Leea indica* leaf. *Asian J. Res. Pharm Sci.* 2, pp. 137–139.

The Plant List. 2013. Version 1.1. http://www.theplantlist.org/.2013. (Accessed on 5 May 2017).

Pradhan, B.K. and Badola, H.K., 2008. Ethnomedicinal plant use by Lepcha tribe of Dzongu valley, bordering Khangchendzonga biosphere reserve, in North Sikkim India. *J. Ethnobiol. Ethnomed.* 4, p. 22.

Prasad, P.R.C., Reddy, C.S., Raza, S.H. and Dutt, C.B.S., 2008. Folklore medicinal plants of North Andaman Islands, India. *Fitoterapia* 79, pp. 458–464.

Quansah, N., 1988. Ethnomedicine in the Maroantsetra region of Madagascar. *Econ. Bot.* 42, pp. 370–375.

Rahimi, R., Ghiasi, S., Azimi, H., Fakhari, S. and Abdollahi, M., 2010. A review of the herbal phosphodiesterase inhibitors; future perspective of new drugs. *Cytokine* 49, pp. 123–129.

Rahman, M.A., Imran, T.B. and Islam, S., 2013a. Antioxidative, antimicrobial and cytotoxic effects of the phenolics of *Leea indica* leaf extract. *Saudi J. Biol. Sci.* 20, pp. 213–225.

Rahman, M.A., Sultana, R., Emran, T.B., Rahman, M.A., Chakma, J.S., Rashid, H. and Monirul Hasan, C.M., 2013b. Effects of organic extracts of six Bangladeshi plants on *in vitro* thrombolysis and cytotoxicity. *BMC Complement. Altern. Med.* 13, p. 25.

Raihan, M.O., Habib, M.R., Brishti, A., Rahman, M.M., Saleheen, M.M. and Manna, M., 2011. Sedative and anxiolytic effects of the methanolic extract of *Leea indica* (Burm. f.) Merr. Leaf. *Drug Discov. Ther.* 5, pp. 185–189.

Raihan, M.O., Tareq, S.M., Brishti, A., Alam, M.K., Haque, A. and Ali, M.S., 2012. Evaluation of antitumor activity of *Leea indica* (Burm. f.) Merr. extract against ehrlich ascites carcinoma (EAC) bearing mice. *Am. J. Biomed. Sci.* 4, pp. 143–152.

Reddy, N.S., Navanesan, S., Sinniah, S.K., Wahab, N.A. and Sim, K.S., 2012. Phenolic content, antioxidant effect and cytotoxic activity of *Leea indica* leaves. *BMC Complement. Altern. Med.* 12, p. 128.

Ren, A., 2008. A method for preparing Chinese medicinal drop for treating rhinitis. Faming Zhuanli Shenqing CN 101104073 A 20080116.

Ridsdale, C.E., 1976. Leeaceae. Flora Malesiana, Series I-Spermatophyta flowering plants, Noordhoff International Publishing, Leyden, the Netherlands, 7, pp. 755–782.

Ridsdale, C.E., 1974. A revision of the family Leeaceae. *Blumea-Biodiversity, Evolution and Biogeography of Plants* 22, pp. 57–100.

Saha, K., Lajis, N.H., Israf, D.A., Hamzah, A.S., Khozirah, S., Khamis, S. and Syahida, A., 2004. Evaluation of antioxidant and nitric oxide inhibitory activities of selected Malaysian medicinal plants. *J. Ethnopharmacol.* 92, pp. 263–267.

Saha, K., Lajis, N.H., Shaari, K., Hamzah, A.S. and Israf, D.A.,2005. Chemical constituents of *Leea indica* (Burm. f.) Merr. (Leeaceae). *Malays J. Sci.* 24, pp. 75–78.

Saha, K., Shaari, K. and Lajis, N.H., 2007. Phytochemical study on *Leea indica* (Burm. F.) Merr. (Leeaceae). *J. Bangladesh Chem. Soc.* 20, pp. 139–147.

Sato, Y., Itagaki, S., Kurokawa, T., Ogura J., Kobayashi, M., Hirano, T., Sugawara, M. and Iseki, K., 2011. *In vitro* and *in vivo* antioxidant properties of chlorogenic acid and caffeic acid. *Int. J. Pharm.* 403, pp. 136–138.

Sen, S., Chakraborty, R., De, B. and Devanna, N., 2011. An ethnobotanical survey of medicinal plants used by ethnic people in West and South district of Tripura, India. *J. Forest Res.* 22, pp. 417–426.

Sen, S., De, B., Devanna, N. and Chakraborty, R., 2012. Anthelmintic and *in vitro* antioxidant evaluation of fractions of methanol extract of *Leea asiatica* leaves. *Anc. Sci. Life* 31, pp. 101–106.

Sen, S., De, B., Devanna, N. and Chakraborty, R., 2013. Cisplatin-induced nephrotoxicity in mice: protective role of *Leea asiatica* leaves. *Ren. Fail.* 35P, pp. 1412–1417.

Sen, S., Devanna, N., Chakraborty, R., Choudhury, R. and De, B., 2014. A fatty alcohol with antioxidant potential isolated from *Leea asiatica*. *J. Indian Chem. Soc.* 91, pp. 1317–1319.

Serra, C.P., Cortes, S.F., Lombardi, J.A., Oliveira, A, B.de. and Braga, F.C., 2005. Validation of a colorimetric assay for the *in vitro* screening of inhibitors of angiotensin-converting enzyme (ACE) from plant extracts. *Phytomedicine* 12, pp. 424–432.

Shanmugam, S., Annadurai, M. and Rajendran, K., 2011. Ethnomedicinal plants used to cure diarrhoea and dysentery in Pachalur hills of dindigul district in Tamil nadu, southern India. *JAPS* 1, pp. 94–97.

Singh, A.K., Raghubanshi, A.S. and Singh, J.S., 2002. Medicinal ethnobotany of the tribals of Sonaghati of Sonbhadra district, Uttar Pradesh, India. *J. Ethnopharmacol.* 81, pp. 31–41.

Singh, H., Lily, M.K. and Dangwal, K., 2015. Evaluation and comparison of polyphenols and bioactivities of wild edible fruits of North-West Himalaya, India. *Asian Pac. J. Trop. Dis.* 5, pp. 888–893.

Singh, R.S. and Singh, A.N., 1981. On the identity and economico-medicinal uses of Hastikarnapalasa (*Leea macrophylla* Roxb., Family: Ampelidaceae) as evinced in the ancient (Sanskrit) texts and traditions. *Indian J. Hist. Sci.* 16, pp. 219–222.

Singh, D., Siew, Y.Y., Chong, T.I., Yew, H.C., Ho, S.S.W., Lim, C.S.E.S., Tan, W.X., Neo, S.Y. and Koh, H.L. 2019. Identification of phytoconstituents in *Leea indica* (Burm. F.) Merr. leaves by high performance liquid chromatography micro time-of-flight mass spectrometry. *Molecules* 24, doi: 10.3390/molecules24040714.

Siv, Y.Y. and Paris, R.R., 1972. Flavonoids from Cambodian plants belonging to genera *Cananga*, *Colona*, *Grewia*, *Leea*, and *Melastoma*. *Plantes Medicinales et Phytotherapie* 6, pp. 299–305.

Srinivasan, G.V., Ranjith, C. and Vijayan, K.K., 2008. Identification of chemical compounds from the leaves of *Leea indica*. *Acta Pharm.* 58, pp. 207–214.

Srinivasan, G.V., Sharanappa, P., Leela, N.K., Sadashiva, C.T. and Vijayan, K.K., 2009. Chemical composition and antimicrobial activity of the essential oil of *Leea indica* (Burm. f.) Merr. flowers. *Nat. Prod. Rad.* 8, pp. 488–493.

Sonowal, R., 2013. Indigenous knowledge on the utilization of medicinal plants by the Sonowal Kachari tribe of Dibrugarh district in Assam, North-East India. *Int. Res. J. Biological. Sci.* 2, pp. 44–50.

Srithi, K., Trisonthi, C., Wangpakapattanawong, P. and Balslev, H., 2012. Medicinal plants used in Hmong women's healthcare in northern Thailand. *J. Ethnopharmacol.* 139, pp. 119–135.

Swarnkar, S. and Katewa, S.S., 2008. Ethnobotanical observation on tuberous plants from tribal area of Rajasthan (India). *Ethnobot. Leaflets* 12, pp. 647–666.

Tangjitman, K., Wongsawad, C., Kamwong, K., Sukkho, T. and Trisonthi, C., 2015. Ethnomedicinal plants used for digestive system disorders by the Karen of northern Thailand. *J. Ethnobiol. Ethnomed.* 11, p. 27.

Taofiq, O., Gonzalez-Paramas, A.M., Martins, A., Barreiro, M.F. and Ferreira, I.C.F.R., 2016. Mushrooms extracts and compounds in cosmetics, cosmeceuticals and nutricosmetics-A review. *Ind. Crops Prod.* 90, pp. 38–48.

Temkitthawon, P., Hinds, T.R., Beavo, J.A., Viyoch, J., Suwanboriux, K., Pongamornkul, W., Sawasdee, P. and Ingkaninan, K., 2011. *Kaempferia parviflora*, a plant used in traditional medicine to enhance sexual performance contains large amounts of low affinity PDE5 inhibitors. *J. Ethnopharmacol.* 137, pp. 1437–1441.

Temkitthawon, P., Viyoch, J., Limpeanchob, N., Pongamornkul, W., Sirikul, C., Kumpila, A., Suwanborirux, K. and Ingkaninan, K., 2008. Screening for phosphodiesterase inhibitory activity of Thai medicinal plants. *J. Ethnopharmacol.* 119, pp. 214–217.

Tran, T.V., Malainer, C., Schwaiger, S. Hung, T., Atanasov, A.G., Heiss, E.H., Dirsch, V.M. and Stuppner, H., 2015. Screening of Vietnamese medicinal plants for NF-$_\kappa$B signaling inhibitors: Assessing the activity of flavonoids from the stem bark of *Oroxylum indicum*. *J. Ethnopharmacol.* 159, pp. 36–42.

Wagh, V.V. and Jain, A.K., 2015. Inventory of ethnobotanicals and other systematic procedures for regional conservation of medicinal and sacred plants. *Environ. Syst. Decis.* 35, pp. 143–156.

Wen, J., 2007. Leeaceae. *The Families and Genera of Vascular Plants-Flowering plants Eudicots.* Springer-Verlag, Berlin, Germany, Vol. 9, pp. 221–225.

Wiart, C., Mogana, S., Khalifah, S., Mahan, M., Ismail, S., Buckle, M., Narayana, A.K., Sulaiman, M., 2004. Antimicrobial screening of plants used for traditional medicine in the state of Perak, Peninsular Malaysia. *Fitoterapia* 75, pp. 68–73.

Wong, Y.H. and Kadir, H.A., 2011. *Leea indica* ethyl acetate fraction induces growth-inhibitory effect in various cancer cell lines and apoptosis in Ca Ski human cervical epidermoid carcinoma cells. *Evid. Based Complement Alternat. Med.* 2011, doi:10.1155/2011/293060.

Wong, Y.H., Kadir, H.A. and Ling, S.K. (2012a). Bioassay-guided isolation of cytotoxic cycloartane triterpenoid glycosides from the traditionally used medicinal plant *Leea indica*. *Evid. Based Complement Alternat. Med.* doi:10.1155/2012/164689.

Wong, Y.H., Kadir, H.A. (2012b). Induction of mitochondria-mediated apoptosis in Ca Ski human cervical cancer cells triggered by mollic acid arabinoside isolated from *Leea indica*. *Evid. Based Complement Alternat. Med.* 2012. doi:10.1155/2012/684740.

Xu, S., 2016. Ointment for moistening lung and arresting cough. CN 105311386 A 20160210.

Yang, J., Huang, J., Ke, L., Wang, Y., Luo, J. and Li, H., 2017. *Leea asiatica* anticancer active extract LA-D, its pharmaceutical composition, and method for separation and identification of chemical component thereof. Faming Zhuanli Shenqing CN 106632197 a 20170510.

Yusuf, M., Wahab, M.A., Yusuf, M., Chowdhury, J.U. and Begum, J., 2007. Some tribal medicinal plants of Chittagong Hill Tracts, Bangladesh. *Bangladesh J. Plant Taxon.* 14, pp. 117–128.

Zhang, Q., 2015. Instant tea for promoting sleep and improving health, and its preparation method. Faming Zhuanli Shenqing CN 104255990 A 20150107.

Zheng, X.L. and Xing, F.W., 2009. Ethnobotanical study on medicinal plants around Mt. Yinggeling, Hainan Island, China. *J. Ethnopharmacol.* 124, pp. 197–210.

3

Centella asiatica *(L.) Urb., an Endowment from Traditional Medicine*

Jisha Satheesan and Kallevettankuzhy Krishnannair Sabu

CONTENTS

3.1 Introduction

Centella asiatica (L). Urb., Gotukola is a creeper with leaves of the family Apiaceae (*Umbelliferae*) and *Calicyflorae* series. The *C. asiatica* is indigenous to most parts of Asia (Figure 3.1). The *C. asiatica* is known by different local names in different places (Jamil et al. 2007). The plant often shows different morphology when collected from different regions. Genetic variability and heritability of certain plant characters of *C. asiatica* have also been reported (Thomas et al. 2010a). The Latin word "Centum" means "hundred" is responsible for its name, as the plant is abundantly branched. The *C. asiatica* stem is with small, creeping stolons, and its colour varies from green to reddish green in colour. The stolons interconnects with each other. Leaves are elongated and kidney-shaped. Leaves are with curvy apices, long petioles and glabrous with palmate veins. *C. asiatica* grows preferably in damp swampy areas and contains extended horizontal, filiform stems with lengthy internodes. The creamish rooting starts from the nodes and grows perpendicularly down. Roots are protected with root hairs, while leaves are one to five in number from each node and are deeply cordate which differs in diameter from 1–7 cm. The shapes of the leaves vary from kidney-shaped, oval or orbicular, which starts from pericladial petioles and are glabrous on both surfaces. The flowers are enclosed partly in green bracts and are hermaphrodites minute, purple to white-green in colour. Each flower is composed of five stamens and two styles. They are arranged in umbels, which originate from the axils of the leaves near the surface of the soil. *C. asiatica* bears long and ovoid fruits with strongly thickened pericarp (Emboden, 1985; Warrier et al. 1994). The plant has thickly reticulate fruits which are harvested manually, the characteristic of *C. asiatica*, differentiating it from species of *Hydrocotyle*.

3.1.1 Taxonomic Position and Distribution

Kingdom: Plantae

Order: Apiales

Family: Apiaceae

Sub family: Mackinlayoideae

Genus: *Centella*

Species: *asiatica*

C. asiatica (L). Urb. is propagated by stolons and the flowers are sessile in simple umbels. Environmental conditions normally affect the morphological characteristics of the plant

FIGURE 3.1 *Centella asiatica* (L). Urban, (Hydrocotyle, Indian pennywort, Gotukola) under field conditions.

(Zheng and Qin, 2007). The entire ever-growing demand is met from the natural populations. Unlimited and overexploitation of the plant results in the depletion of the livestock (Thomas et al. 2010b). In order to overcome the scarcity of availability, alternative methods should be practised. A report by the Export and Import Bank of India proves *C. asiatica* is one of the most noted plant in the global market (Paramageetham, 2004).

3.2 Traditional Uses of *C. asiatica*

C. asiatica is with different neutraceutical properties and is recorded as a traditional medicine in the Indian and German Homoeopathic medicines. In Ayurveda, *C. asiatica* is used for developing memory and learning capabilities (Zheng and Qin, 2007) and has given cognitive benefits to the human brain (Tao et al. 2008). Oxidative damage as a consequence of aging can be mitigated by the use of *C. asiatica* plant extract (Subathra et al. 2005) and also stimulates neurotrophic effects (Rao et al. 2006). *C. asiatica* is also reported to be effective for autoimmune rheumatic disease like scleroderma, cirrhosis of liver (Darnis et al. 1979), asthma, bronchitis, eczemas, anxiety, mental disorders, urethritis (Bylka et al., 2014; Mathur et al. 2007; Gohil et al. 2010; Prasad et al. 2013) and in digestive disorders. Memory of mentally weak children can be increased by the use of the *C. asiatica* (Apparao et al., 1973; Kakkar, 1990). The active principles responsible for its medical potential are pentacyclic triterpenoid saponins which include asiaticoside, madecassoside, madecassic acids, asiatic acid, etc. (James and Dubery, 2009), together known as centelloids.

C. asiatica has proven to advance the blood circulation in small blood vessels and effective for the veins in lower limbs. It also helps to lower the blood pressure (Belcaro et al. 1990; Arpaia et al. 1990). Cardio-protective potential in the alcoholic extract of *C. asiatica* has been reported by Pragada et al. (2004). The plant product has recorded protective effect from cardiac damage caused by another anti-cancer drug, adriamycin in rats (Peskan-Berghöfer et al. 2004). Antimicrobial activities are reported in extracts of *C. asiatica* against a broad spectrum of microbes. Bauer et al. (1966) and Rahman and Rashid (2008) determined the antimicrobial activities in the crude extracts from *C. asiatica* by the most common disc diffusion method against microbial strains. Punturee et al. (2005) conducted their studies on cell mediated and humoral immune responses upon treating with *C. asiatica* water extracts. Peripheral blood mononuclear cells significantly increased the multiplication and synthesis of interleukin-2 and tumour necrosis factor-α. The role of *C. asiatica* extract in the alleviation of amyloid β quantities in hippocampus of animal model shows its possible potential in the treatment of Alzheimer's disease (Dhanasekharan et al. 2009). *C. asiatica* of different genotypic origin were used to assess the anxiolytic activities in rodents and no sedative effect was generated by the chief elements of the plant (Wijeweera et al. 2006). Wound healing through the production of type I collagen and mitigation of inflammatory reaction were also documented for *C. asiatica* extracts (Widgerow and Chait, 2000). Potential components responsible for the treatment of leprosy are structurally steroids from *C. asiatica* and is also beneficial in the regeneration of brain and neural cells (Brinkhause and Lindner, 2000). The triterpenes of the plant are known to be highly useful for skin care and toning and also to prevent premature wrinkling (Kumar et al. 2009).

3.3 Major Secondary Metabolites and Their Pharmacological Uses

Vascular plants are capable of synthesising a mass of organic molecules or phytochemicals, collectively known as metabolites (Harborne and Turner, 1984). The metabolites synthesized in the plant body are classified as primary and secondary metabolites. Primary metabolites are engaged in an array of tasks in the existence of plants, range from structural ones to protection. Primary metabolites are known to regulate cell and plasma membrane integrity and is photo-protective. These metabolites are also involved in transposition of neurodevelopments and attaching crucial biochemical purposes to the membrane systems (Strack and Fester, 2006). However, secondary metabolites are natural products synthesised from primary metabolites, accumulated in the plant body due to different stress conditions and are not essential components of the plants. Secondary metabolites play significant role in the improvement of human health and nutritional features. These compounds also have a biological part in balancing the plants and their ecological relations. Secondary metabolites include phytoalexins and phytoanticipins, which are defensive in function.

The major secondary metabolite such as triterpenoids, are synthesized from a biologically active isoprene. Each class of isoprenoid was received from a single parent compound (Ruzicka, 1953). Several interactions including plant-plant, plant-insect and plant-pathogen interactions can be correlated to the secondary metabolite production (Stoessl et al. 1976). The triterpenes of the plant are known to be highly useful for skin care and toning and also to prevent premature wrinkling (Kumar et al. 2009). The mechanism of action of triterpenes includes deterioration of the membranous tissues by dissolving the cell walls of the microorganisms and subsequent elimination (Dutta and Basu, 1968; Bisignano et al. 1999). The triterpenoids produced from *C. asiatica* are regarded as phytoanticipins due to their antimicrobial and defensive actions against different diseases (Dash et al. 2011). The major triterpenoids of *C. asiatica*, are asiaticoside, madecassoside, asiatic acid and madecassic acid.

3.3.1 Asiaticoside

The chief trisaccharide triterpene is asiaticoside (Maquart et al. 1999) and is with intense antibacterial and fungicidal activity (Hausen, 1993). Antibacterial functions of the plant against different strains of bacteria are reported by Oyedeji and Afolayan (2005). Asiaticoside production is tissue-specific and the maximum quantity of asiaticoside is synthesised in the leaves, which is evidenced by the use of leaf tissue for different herbal preparations (Brinkhause and Lindner, 2000). Addition of leucine during the callus culture of *C. asiatica* resulted in many-fold increase in the concentration of asiaticoside accumulation (Kiong et al. 2005). The pathway precursors like squalene or β-Amyrin can also be supplied for inducing asiaticoside production in these plants. *C. asiatica* biosynthesizes asiaticosides in a dose and age-related method under *in vitro* conditions (Mangas et al. 2006; Hernandez-vazquez et al. 2010; Prasad et al. 2012). However, the callus culture cell suspensions, transformed hairy roots or adventitious roots regenerated from leaf derived callus are reported to produce least amount of asiaticosides (Aziz et al. 2007; Kim et al. 2007; Mangas et al. 2008; Bonfill et al. 2011; Mercy et al. 2012). Undifferentiated cultures of *C. asiatica* showed less asiaticoside production; however, the whole plant cultures carried out in a growth-related mode is shown to enhance the amount more favourably (Mangas et al. 2006; Aziz et al. 2007).

Asiaticoside enhances the wound healing process because of its stimulation in tensile strength (Suguna et al. 1996). It fuels the epidermis by stimulating the pig skin cells, and also helps in keratinization under *in vitro* conditions. Morisset et al. (1987)

documented the exploitation of asiaticoside in the management of the presence of excessive amounts of collagen and in keloid disorder. It is also effective against ethanol induced mucosal lesions in the stomach (Cheng and Koo, 2000).

3.3.2 Asiatic Acid

Asiatic acid is also a biologically active five-ringed triterpenoid found in *C. asiatica*. Earlier studies report the bactericidal properties and neuroprotective activities of asiatic acid (Bonfill et al. 2006). Asiatic acid has shown potential antioxidant, anti-inflammatory activities and makes secure against glutamate/β-amyloid-related toxicity of neurons (Krishnamurthy et al. 2009). The pharmacological properties of asiatic acid in inhibiting multiple pathways of intracellular signalling molecules was also described (Junwei et al. 2018). The study was also extended to find out the pharmacological activities of asiatic acid by comparing with two natural compounds: curcumin and resveratrol. Recent study by Mavondo et al. (2016) documented the use of asiatic acid as a substitute for chloroquine treatment with associated amelioration of malarial disease. The compound was also found effective in combination therapy with chloroquine. Decreased H_2O_2-interrelated cell death and reduced intracellular free radical absorption were also observed in the presence of the pentacyclic asiatic acid (Mook-Jung et al. 1999).

3.3.3 Madecassoside

One of the most prominent *Centella* saponins is madecassoside along with asiaticoside. Similar to asiaticoside, madecassoside has been found to be organ specific with the highest content being reported in leaf tissue and lowest content being reported in the petiole portion (Madhusudanan et al. 2014; Aziz et al. 2007). The healing of wounds by the application of madecassoside from *C. asiatica* is also reported (Hou et al. 2016). Madecassoside when orally administered at concentrations of 6, 12 and 24 mg/kg of body weight has resulted in closure of the wounds in mice. The effect of madecassoside was observed in a time-dependent manner (Liu et al. 2008). The study also reports the possible role of Madecassoside in having antioxidant potential, formation of new blood vessels and synthesis of collagen.

3.3.4 Madecassic Acid

Madecassic acid is one of the major aglycones belonging to ursane-type triterpenoids and has several medicinal functions. Madecassic acid has a key role in the treatment of skin wound healing (Hong et al. 2005). It is also a chief component in *C. asiatica* to treat inflammation (Won et al. 2010) and possesses antitumor activities (Hsu et al. 2005). Madecassic acid also has proven antioxidant activities along with asiatic acid (Ramachandran and Saravanan, 2013).

In addition, other compounds like tannins, amino acids, etc., are also present in lesser amounts (Leung and Foster, 1998). The plant extract contains 19% ash, 6% water-soluble and 9.5% alcohol-soluble compounds. Several compounds like saponins (asiaticoside B, brahminoside etc), triterpenoid acids (brahmic acid, isobrahmic acid etc), glycosides (3-glucosylquercetin, 3-glucosylkaempferol etc) (Jiang et al. 2005; Zheng and Qin, 2007; Zainol et al. 2003). A new triterpene called 2α, 3β, 20, 23-tetrahydroxy urs-28-oic acid was obtained from the leaf and petiole parts of *C. asiatica* (Yu et al. 2006).

3.4 Biosynthetic Pathway Involved in Secondary Metabolite Production

Most of the microorganisms and animals contain terpenoids, similar to terpenoids from plant secondary metabolites. The triterpenes (C30) and sterols (C18–C29) are dissimilar molecules, but originates from the same molecule called squalene. The squalene is known to form widespread series of complexes with distinct branch points that expand the final compounds. Sterols like sitosterol play a significant and primary role in the membrane architecture. These molecules also act as a precursor for metabolites, specific to every plant species (Seigler, 1998). Asiaticoside, in the leaves of *C. asiatica* is synthesized through isoprenoid pathway through the interchange of triple acetyl-CoA to 3-hydroxy-3-methylglutaryl-coenzyme A (HMG-CoA) in preliminary step. Further, the reduction of HMG-CoA results in mevalonate by the enzyme 3-hydroxy-3-methylglutaryl-coenzyme A reductase (HMGR). The consecutive 1-4 condensation reactions including the action of Farnesyl diphosphate synthase (FPS) yields geranyl diphosphate (GPP). The GPP then produces farnesyl diphosphate (FPP), the condensation of two molecules of FPP by the action of Squalene synthase (SQS) results in the formation of squalene (Chappell et al. 1995).

The enzymatic activity of OxidoSqualene Cyclase (OSC) results in the cyclisation of 2,3-oxidosqualene. The cyclisation is catalysed via cationic to cyclic triterpene skeletons. α/β Amyrin Synthases (BAS) form the lupenyl cation. Ring extension and reorganizations are needed to form the corresponding products before the deprotonation of α/β amyrin. These alterations are catalyzed by triterpene synthases (OSCs) (Haralampidis et al. 2002). Plants produce different groups of triterpenoids by encoding different OSC enzymes. The structural diversity of triterpenes is influenced by the cyclization of 2,3-oxidosqualene catalysed by Cycloartenol Synthase (CYS), Lupeol Synthase (LS) and BAS (Mangas et al. 2006). The SQS is the regulatory enzyme which controls the carbon flow into sterols and triterpenes biosynthesis. There are reports regarding the obstruction of the SQS enzyme, which forwards FPP to pathways other than the sterol biosynthesis, in turn results in the production of other economically important isoprenoids (Zhi-lin et al. 2007). Cyclization of 2, 3-oxidosqualene generates lanosterol and cycloartenol through a protosteryl cation intermediate, the precursors. Lupeol and α/β Amyrin are generated with the cyclization of different cation intermediates (Jenner et al. 2005). The α/β BAS cyclize oxidosqualene followed by ring expansion and reorganization before removal of the proton ton form α Amyrin and β Amyrin respectively (Haralampidis et al. 2002). a Amyrin contains C19, C20 dimethyl and β Amyrin C20 dimethyl substitution patterns respectively.

Stimasterol and sitosterol are formed from cycloarterol by formation of 2,3-Oxido Squalene Cycloartenol Cyclase (OSCC), whereas 2,3-Oxido Squalene β Amyrin Cyclase (OSβ AC) is responsible for an enormous variety of five-ringed triterpenoid secondary metabolites in plants. The functions of both OSCC and OSBAS isolated from pea seedlings is documented, and the CYS gene is cloned and expressed in *Arabidopsis thaliana* and yeast, respectively (Corey et al. 1993).

3.5 Current Research in *C. asiatica*

3.5.1 Neuroprotective Role of *C. asiatica*

The *C. asiatica* is broadly used in the handling of various neuro-ailments. Since time immemorial *C. asiatica* has been used as a brain stimulant in India, as per the Ayurvedic literature. *C. asiatica* is also listed in the management of dementia in the ancient Indian Ayurvedic medicine, Charaka Susmita. In some parts of India, milk in which *C. asiatica* is added is used as a potential memory enhancer (Kirtikar and Basu, 1993).

Neurodegeneration is the age-related progressive loss of structure or function of neurons. Nerve degradation results in diseases including amyotrophic lateral sclerosis, which comprises group of rare neural abnormalities. Parkinson's, Alzheimer's, and Huntington's diseases are also share a common factor of neurodegeneration. The *C. asiatica* is known to possess neuroprotective and also positive effects on brain aging. The principal neuroprotective activity lies in the major compounds like triterpene saponosides; asiatic acid, madecassic acids, asiaticosides and madecassosides.

3.5.2 *C. asiatica* in the Treatment of Alzheimer's Disease

One of the most predominant form of dementia is Alzheimer's disease. Extracellular dumps of amyloid-β peptides, intracellular neurofibrillary tangles, neuronal and synaptic degeneration/loss, and neuroinflammation are the chief symptoms of Alzheimer's disease. Acetylcholine esterase (AChE) is the crucial enzyme taking a critical role in the pathogenesis of Alzheimers disease. Orhan et al. (2006) analysed the neuroactive effect of *C. asiatica* alcoholic extract *in vitro* against AChE. Patients with Alzheimer's diseases were with low concentrations of acetylcholine because of the action of AChE and drugs were developed by inhibiting the action of AChE and butyrylcholine esterase (BChE). The drugs currently used in the control of the disease have numerous side effects. In a recent study, *C. asiatica* ethanolic extracts were evaluated to determine the inhibitory activity of AChE for its quantification by Ellman's colorimetric method to quantify AChE inhibitor activity. The results were suggestive of its use in the treatment of Alzheimer's disease, due to the presence of natural antioxidants and acetylcholine esterase inhibitors in *C. asiatica*.

Senile plaques are the major characteristic of Alzheimer's disease and β amyloid is the chief component. β amyloid synthesis and deposition in Alzheimer's disease were studied using several transgenic mouse lines. This study on transgenic mice (Duff et al. 1996), has observed the significance of *C. asiatica* extract in reducing the amount of beta amyloid in hippocampus of its' animal model (Dhanasekaran et al. 2009).

3.5.3 Elicitors Reported from *C. asiatica*

"Elicitor may be defined as a material which can excite any kind of defence in the presence of stress elements when given in minute quantity to an organism, improves the biosynthesis of specific compound" (Radman et al. 2003). Elicitors can be of endogenous (secreted from plants by the action of pathogens) or exogenous (produced from the pathogen) and can be either abiotic or biotic that can enhance stress, which finally lead to the augmented production and increase of metabolites (Zhao et al. 2005). Additions of elicitors to the organism is a reliable method to enhance the synthesis of bioactive principles. Moreover, elicitors also act as signalling molecules for defence mechanism in all plants and it was documented in *C. asiatica* that elicitors augment collection of major secondary metabolites (Loc and Giang, 2012).

The *C. asiatica* is required in huge quantities in both national and international markets for the preparation of pharmaceuticals and cosmetics. The growing demand for the plant had resulted in it's over exploitation. The plant is thus marked as an endangered plant species (Singh, 1989). It is therefore extremely significant to build up more helpful and discriminating methods for the maximum revival of the wanted bioactives from the herb *C. asiatica*.

3.5.3.1 Abiotic Elicitors

All the factors which cannot be regarded as natural components of the environment of a plant cell are considered as abiotic elicitors. Elicitor of non-biological origin can also be used as one of the stress agents that increase the biogenesis of secondary metabolites in a particular tissues, organs or cells. Major abiotic stress factors can be chemicals or physical agents. Physical agents have the ability to promote plant defense mechanism and fuel the secondary metabolite production in plants. In *C. asiatica*, low temperatures (4°C–17°C) for many days and dehydration (at 28°C or 34°C for 16 h) enhance the level of triterpene glycosides, asiaticosides and madecassosides.

Numerous chemical elicitors are used to govern the metabolic flow among different pathways, mainly by regulating the action of cyclases. Biosynthesis of centellosides by methyl jasmonate (MJ) (Kim et al. 2004), and thidiazuron (Kim et al. 2004). The treatment with dimethyl sulphoxide (DMSO) resulted in permeabilization and feeding strategy in cell cultures and root/whole plant tissue. The effect of DMSO can obtained even in combination with β Amyrin (Hernandez-Vazquez et al. 2010). Mangas et al. (2006) reported the accumulation of MJ resulted in notable enhancement of the triterpenoid content and enhanced plant growth accompanied by decreased sterol content. The same pattern of results was observed in the hairy root cultures of the plant with MJ (Kim et al. 2007). The production of triterpenes in leaf cultures of *C. asiatica* was enhanced by feeding amino acids (Karppinen et al. 2007) and yeast extract (Kim et al. 2004).

3.5.3.2 Biotic Elicitors

The biotic elicitors from fungi including *Colletotrichum lindemuthianum* and *Trichoderma harzianum* have been reported to influence the *in vitro* asiaticoside production in *C. asiatica* shoot cultures (Prasad et al. 2013). The growth promotion effect of an axenically cultivable symbiotic fungus *Piriformospora indica* on *C. asiatica* has been described (Jisha et al. 2011). Recently it is recorded that the *P. indica* cell wall extract is the more potential secondary metabolite enhancing agent in *C. asiatica*, when compared to the live *P. indica* biomass (Jisha et al. 2018). *P. indica* augments nutrient uptake along with abiotic and biotic stress tolerance to *C. asiatica*.

3.6 Conclusion

The world-wide market of the active principles from *C. asiatica* is more than USD 60 billion due to its high neutraceutical properties. There is a need to grow *C. asiatica* as a field crop by the farmers to benefit from its commercialization and to conserve its natural populations and biodiversity. Secondary metabolites from *C. asiatica* offer several alternatives to modify the progress and symptoms of Alzheimer's disease. The chemical synthesis of plant secondary metabolites is either enormously in dilemma or infeasible. Thus, the major source for its production depends on the wild or cultivated plants, which is not seems to be enough for the growing demand for the medicinal compounds from *C. asiatica*. Result obtained from co-culture with *P. indica* proves that asiaticoside and other major secondary metabolites from *C. asiatica* actively participate in *C. asiatica-P. indica* interaction. Thus the *C. asiatica- P. indica* dual culture is the most significant and efficient tool for boosting secondary metabolite production.

REFERENCES

Apparao, M.V.R., Srinivasan K. and Rao K., 1973. Effect of Mandookaparni (*Centella asiatica*) on the general mental ability (Medhya) of mentally retarded children. *Journal of Research in Indian Medicine*, 8, pp. 9–16.

Arpaia, M.R., Ferrone, R. and Amitrano, M., 1990. Effects of *Centella asiatica* extract on mucopolysaccharide metabolism in subjects with varicose veins. *International Journal of Clinical Pharmacology Research*, 10, pp. 229–233.

Aziz, Z.A., Davey, M.R, Power, J.B, Anthony, P., Smith, R.M. and Lowe, K.K., 2007. Production of asiaticoside and madecassoside in *Centella asiatica in vitro* and *in vivo*. *Plant Biology*, 51, pp. 34–42. doi:10. 1007/5 10535-007.

Bauer, A.W., Kirby, W., Sherries, M.M. and Tuck, M., 1966. Antibiotic susceptibility testing by a standardized disc diffusion method. *Journal of American Clinical Pathology*, 45, pp. 493–496.

Belcaro, G.V., Grimaldi, R. and Guidi, G., 1990. Improvement of capillary permeability in patients with venous hypertension after treatment with TTFCA. *Angiology*, 4, pp. 533–540.

Bisignano, G., Tomaino, A., Lo Cascio, R., Crisafi, G., Uccella, N. and Saija A., 1999. On the *in vitro* antimicrobial activity of oleuropein and hydroxytyrosol. *Journal of Pharmaceutical Pharmacology*, 51, pp. 971–974.

Bonfill, M., Mangas, S., Moyano, E., Cusido, R.M. and Palazon J., 2011. Production of centellosides and phytosterols in cell suspension cultures of *Centella asiatica*. *Plant Cell Tissue and Organ Culture*, 104, pp. 61–67.

Bonfill, M., Mangas, S., Cusidu, R.M., Osuna, L., Pinol, M.T. and Palazun J., 2006. Identification of tripernoid compounds of *Centella asiatica* by thin-layer chromatography and mass spectrometry. *Biomedical Chromatography*, 20, pp. 151–153.

Brinkhause, B. and Lindner, M., 2000. Chemical, pharmacological and clinical profile of the East Asian medical plant *Centella asiatica*. *Phytomedicine*, 7, pp. 427–448.

Bylka, W., Awizen, Z.P., Stroka, S.E., Pazdrowska, D.A. and Brzezinska, M., 2014. *Centella* in dermatology: an overview. *Phytotherapy Research*, 28(8), pp. 1117–1124.

Chappell, J., Wolf, F., Proulx, J., Cuellar, R. and Saunders, C., 1995. Is the reaction catalyzed by 3-hydroxy-3-methyl-glutaryl Coenzyme-A reductase a rate limiting step for isoprenoid biosynthesis in plants. *Plant Physiology*, 109, pp. 1337–1343.

Cheng, C.L. and Koo, M.W., 2000. Effects of *Centella asiatica* on ethanol induced gastric mucosal lesions in rats. *Life Science*, 67, pp. 2647–2653.

Corey, E.J., Matsuda, S.P. and Bartel, B., 1993. Isolation of an Arabidopsis thaliana gene encoding Cycloartenol synthase by functional expression in a yeast mutant lacking Lanosterol synthase by the use of a chromatographic screen. *Proceedings on National Academy of Science*, 90(24), pp. 11628–11632.

Darnis, F., Orcel, L., de Saint-Maur, P.P. and Mamou, P., 1979. Use of a titrated extract of *Centella asiatica* in chronic hepatic disorders (author's trans). *Sem Hop*, 55, pp. 1749–1750.

Dash, B.K., Faraquee, H.M., Biswas, S.K., Alam, M.K., Sisir, S.M. and Prodan, U.K., 2011. Antibacterial and antifungal activities of several extracts of *Centella asiatica* L. against some human pathogenic viruses. *Life Sciences and Medicine Research*, http://astonjournals.com/lsmr.

Dhanasekaran, M., Holcomb, L.A., Hitt, A.R, Tharakan, B., Porter, J.W., Young, K.A., Manyam, B.V., 2009. *Centella asiatica* extract selectively decreases amyloid β levels in Hippocampus of Alzheimer's disease animal model. *Phytotherapy Research*, 23, pp. 14–19.

Duff, K., Eckman, C., Zehr, C., Yu, X., Prada, C.M., Perez-tur, J., Hutton, M. et al., 1996. Increased amyloid β42 (43) in brains of mice expressing mutant presenilin 1. *Nature*, 383, pp. 710–713.

Dutta, T. and Basu, U.P., 1968. Isothankunic acid-a new triterpene acid from *Centella asiatica* (URB). *Bulletin of National Institute of Science*, 37, pp. 178–184.

Emboden, W.A., 1985. The ethnopharmacology of *Centella asiatica* (L.) Urban (*Apiaceae*). *Journal of Ethnobiology*, 5(2), pp. 101–107.

Gohil, K.J., Patil, J.A. and Gajjar, A.K., 2010. Pharmacological review on *Centella asiatica*: a potential herbal cure-all. *Indian Journal of Pharmacological Science*, 72, pp. 546–556.

Haralampidis, K., Trojanowska, M. and Osbourn, A.E., 2002. Biosynthesis of triterpenoid saponins in plants. *Advanced Biochemical of Engineering and Biotechnology*, 75, pp. 31–49.

Harborne, J.B. and Turner, B.L., 1984. *Plant Chemosystematics*, Academic Press, London, UK.

Hausen, B.M., 1993. *Centella asiatica* (Indian pennywort), an effective therapeutic but a weak sensitizer. *Contact Dermatitis*, 29, pp. 175–179.

Hernandez-Vazquez, L., Bonfill, M., Moyano, E., Cusido, R.M., Navarro-Ocaña, A. and Palazon, J., 2010. Conversion of α-amyrin into centellosides by plant cell cultures of *Centella asiatica*. *Biotechnology Letters*, 32, pp. 315–319.

Hong, S.S., Kim, J.H., Li, H. and Shim, C.K., 2005. Advanced formulation and pharmacological activity of hydrogel of the titrated extract of *C. asiatica*. *Archives of Pharmacological Research*, 28, pp. 502–508.

Hou, Q., Li, M., Lu, Y.H., Liu, D.H. and Li, C.C., 2016. Burn wound healing properties of asiaticoside and madecassoside. *Experimental and Therapeutic Medicine*, 12, pp. 1269–1274.

Hsu, Y.L., Kuo, P.L., Lin, L.T. and Lin, C.C., 2005. Asiatic acid, a triterpene, induces apoptosis and cell cycle arrest through activation of extracellular signal-regulated kinase and p38 mitogen-activated protein kinase pathways in human breast cancer cells. *The Journal of Pharmacology and Experimental Therapeutics*, 313, pp. 333–344.

James, J.T. and Dubery, I.A., 2009. Pentacyclic triterpenoids from the Medicinal herb *Centella asiatica* (L) Urban. *Molecules*, 14, pp. 3922–3941.

Jamil, S.S., Nizami, Q. and Salam, M., 2007. *Centella asiatica* (Linn.) Urban a review. *Natural Product Radiance*, 6, pp. 58-170.

Jenner, L., Romby, P., Rees, B., Schulze-Briese, C., Springer, M., Ehresmann, C., Ehresmann, B., Moras, D., Yusupova, G. and Yusupov, M., 2005. Translational operator of mRNA on the ribosome. How repressor proteins exclude ribosome binding. *Science*, 308, pp. 120–123.

Jiang, Z.Y., Zhang, X.M., Zhou, J. and Chen, J., 2005. New triterpenoid glycosides from *Centella asiatica*. *Helvetica Chemica Acta*, 88, pp. 297–303.

Jisha, S., Anith, K.N. and Manjula, S., 2011. Induction of root colonization by *Piriformospora indica* leads to enhanced asiaticoside production in *Centella asiatica*. *Mycorrhiza*, 22, pp. 159–202. doi:10.1007/s00572-011-0394-y.

Jisha, S., Gouri, P.R., Anith, K.N. and Sabu, K.K., 2018. *Piriformospora indica* cell wall extract as the best elicitor for asiaticoside production in *Centella asiatica* (L.) Urban, evidenced by morphological, physiological and molecular analyses. *Plant Physiology and Biochemistry*, 125, pp. 106–115.

Junwei, L.V., Sharma, A., Zhang, T., Wu, Y. and Ding, X., 2018. Pharmacological review on Asiatic acid and its derivatives: A potential compound. *SLAS TECHNOLOGY: Translating Life Sciences Innovation*, 23(2), pp. 111–127.

Kakkar, P., 1990. Mandukaparni-medicinal uses and therapeutic efficacy. *Probe*, 29, pp. 176–182.

Kim, O.T., Bang, K.H., Shin, Y.S., Lee, M.J., Jung, S.J., Hyun, D.Y., Kim, Y.C. et al., 2007. Enhanced production of asiaticoside from hairy root cultures of *Centella asiatica* (L.) Urban elicited by methyl jasmonate. *Plant Cell Report*, 26, pp. 1941–1949.

Kim, O.T., Kim, M.Y., Hong, M.H., Ahn, J.C. and Hwang, B., 2004. Stimulation of asiaticoside production from *Centella asiatica* whole plant cultures by elicitors. *Plant Cell Report*, 23, pp. 339–344.

Kiong, A.L.P., Mahmood, M., Fadzillah, M. and Daud, S.K., 2005. Effect of precursor supplementation on the production of triterpenes by *Centella asiatica* callus cultures. *Pakisthan Journal of Biological Science*, 8, pp. 1160–1169.

Kirtikar, K.R. and, Basu, B.D., 1993. *Indian Medicinal Plants*, Periodical Experts Book Agency, New Delhi, India.

Krishnamurthy, R.G., Senut, M.C., Zemke, D, Min, J., Frenkel, M.B., Greenberg, E.J., Yu, S.W. et al., 2009. Asiatic acid, a pentacyclic triterpene from *Centella asiatica*, is neuroprotective in a mouse model of focal cerebral ischemia. *Journal of Neuroscience Research*, 87, pp. 2541–2550.

Kumar, M., Yadav, V., Tuteja, N. and Johri, A.K., 2009. Antioxidant enzyme activation in maize plants colonized with *Piriformospora indica*. *Microbiology*, 155, pp. 780–790.

Leung, A.Y. and Foster, S., 1998. *Encyclopedia of Common Natural Ingredients Used in Food, Drugs, and Cosmetics*, 2nd ed, John Wiley & Sons, New York, p. 284.

Liu, M., Dai, Y., Li Y., Luo, Y., Huang, F., Gong, Z., Meng, Q., 2008. Madecassoside isolated from *Centella asiatica* herbs facilitates burn wound healing in mice. *Planta Medica*, 74(8), pp. 809–815. doi:10.1055/s-2008-1074533

Loc, N.H. and Giang, N.T., 2012. Effects of elicitors on the enhancement of asiaticoside biosynthesis in cell cultures of Centella (*Centella asiatica* L. Urban). *Chemical Papers*, 66, pp. 642–648.

Madhusudanan, N., Neeraja, P. and Devi, P., 2014. Comparative analysis of active constituents in *Centella asiatica* varieties (Majjaposhak and Subhodak). *International Journal of Pharmaceutical and Phytopharmacological Research*, pp. 2250–1029.

Mangas, S., Bonfill, M., Osuna, L., Moyano, T., Tortoriello, J., Guido, R.M., Piñol, M.T. and Palazón, J., 2006. The effect of methyl jasmonate on triterpene and sterol metabolism of *Centella asiatica, Ruscus aculeatus* and *Galphimia glauca* cultured plants. *Phytochemistry*, 67, pp. 2041–2049.

Mangas, S., Moyano, E., Osuna, L., Cusido, R.M., Bonfill, M. and Palazón, J., 2008. Triterpenoid saponin content and the expression level of some related genes in calli of *Centella asiatica*. *Biotechnology Letters*, 30, pp. 1853–1859.

Maquart, F.X., Chastang, F., Simeon, A., Birembaut, P., Gillery, P. and Wegrowski, Y., 1999. Triterpenes from *Centella asiatica* stimulate extracellular matrix accumulation in rat experimental wounds. *European Journal of Dermatology*, 9, pp. 289–296.

Mathur, A., Mathur, A.K., Yadav, S. and Verma, P., 2007. *Centella asiatica* (L.) Urban-Status and scope for commercial cultivation. *Journal of Medicinal and Aromatic Plant Science*, 129, pp. 151–162.

Mavondo, G.A., Mkhwananzi, B.N., Mabandla, M.V. and Musabayane, C.T., 2016. Asiatic acid influences parasitaemia reduction and ameliorates malaria anaemia in *P. berghei* infected Sprague-Dawley male rats. *BMC Complementary and Alternative Medicine*, 16(1), p. 357. doi:10.1186/s12906-016-1338-z.

Mercy, S., Sangeetha, N. and Ganesh, D., 2012. *In vitro* production of adventitious roots containing asiaticoside from leaf tissues of *Centella asiatica* L. *in vitro. Cell Development and Plant Biology*, 48, pp. 200–207.

Mook-Jung, I., Shin, J.E., Yun, S.H., Huh, K., Koh, J.Y., Park, H.K., Jew, S.S., Jung, M.W., 1999. Protective effects of asiaticoside derivatives against beta-amyloid neurotoxicity. *Journal of Neuroscience Research*, 58, pp. 417–425.

Morisset, R., Cote, N.G., Panisset, J.C., Jemni, L., Camirand, P. and Brodeur, A., 1987. Evaluation of the healing activity of Hydrocotyle tincture in the treatment of wounds. *Phytotherapy Research*, 1, p. 117.

Orhan, G., Orhan, I. and Sener, B., 2006. Recent developments in natural and synthetic drug research for Alzheimer's Disease. *Letters in Drug Design and Discovery*, 3(4), pp. 268–274.

Oyedeji, O.A. and Afolayan, A.J., 2005. Chemical composition and antibacterial activity of the essential oil of *Centella asiatica*. Growing in South Africa. *Pharmaceutical Biology*, 43, pp. 249–252. doi:10.1080/13880200590928843.

Paramageetham, C., Babu, G.P. and Rao, J.S., 2004. Somatic embryogenesis in *Centella asiatica* L. An important medicinal and neutraceutical plant of India. *Plant Cell, Tissue and Organ Culture*, 79, pp. 19–24.

Peskan-Berghöfer, T., Shahollari, B., Giong, P.H., Hehl, S., Markert, C., Blanke, V., Kost, G., Varma, A. and Oelmuller, R., 2004. Association of *Piriformospora indica* with *Arabidopsis thaliana* roots represents a novel system to study beneficial plant-microbe interactions and involves early plant protein modifications in the endoplasmic reticulum and at the plasma membrane. *Plant Physiology*, 122, pp. 465–477.

Pragada, R., Veeravalli, K.K., Chowdy, K.P. and Routhu, K.V., 2004. Cardioprotective activity of *Hydrocotyle asiatica* L. in ischemia-reperfusion induced myocardial infarction in rats. *Journal of Ethnopharmacology*, 93(1), pp. 105–108.

Prasad, A., Mathur, A., Kalra, A., Gupta, M.M., Lal, R.K. and Mathur, A.K., 2013. Fungal elicitor-mediated enhancement in growth and asiaticoside content of *Centella asiatica* L. shoot cultures. *Plant Growth Regulation*, 69, pp. 265–273. doi:10.1007/s10725-012-9769-0.

Prasad, A., Mathur, A., Singh, M., Gupta, M.M., Uniyal, G.C., Lal, R.K. and Mathur, A.K., 2012. Growth and asiaticoside production in multiple shoot cultures of a medicinal herb *Centella asiatica* L. under the influence of nutrient manipulations. *Journal of Natural Medicine*, 66, pp. 383–387.

Punturee, K., Wild, C.P., Kasinrerk, W. and Vinitketkumnuen, U., 2005. Immunomodulatory activities of *Centella asiatica* and *Rhinacanthus nasutus* extracts. *Asian Pacific Journal of Cancer Prevention*, 6, pp. 396–400.

Radman, R., Saez, T., Bucke, C. and Keshavarz, T., 2003. Elicitation of plant and microbial systems. *Biotechnology and Applied Biochemistry*, 37, pp. 91–102.

Rahman, M.S. and Rashid, M.A., 2008. Antimicrobial activity and cytotoxicity of *Eclipta prostrata*. *Oriental Pharmocology and Experimental Medicine*, 8, pp. 47–52.

Ramachandran, V. and Saravanan, R., 2013. Asiatic acid prevents lipid peroxidation and improves antioxidant status in rats with streptozotocin-induced diabetes. *Journal of Functional Foods*, 5, pp. 1077–1087.

Rao, M.K.G., Rao, M.S. and Rao, G., 2006. *Centella asiatica* (L.) leaf extract treatment during the growth spurt period enhances hippocampal CA3 neuronal dendritic arborization in rats. *Evidence-Based Complementary and Alternative Medicine*, 3(3), pp. 349–357.

Ruzicka, L., 1953. The isoprene rule and the biogenesis of terpenic compounds. *Journal of Cellular and Molecular Life Sciences*, 9(10), pp. 357–367.

Seigler, D.S. (Ed.), 1998. *Plant Secondary Metabolism*, Kluwer Academic Publishers, Boston, MA, pp. 16–41.

Sharma, B.L. and Kumar, A., 1998. Biodiversity of medicinal plants of Triyugi Narain cha (Garhwal Himalaya) and their conservation. *National Conference on Recent Trends in Spices and Medicinal Plant Research*, Calcutta, WB, India. p. 78.

Singh, H.G., 1989. Himalayan herbs and drug importance and extinction threat. *Journal of Medicinal and Aromatic Plant Science*, 19, pp. 1049–1056.

Stoessl, A., Stothers, J.B. and Ward, E.W.B., 1976. Sesquiterpenoid stress compounds of the *Solanaceae*. *Phytochemistry*, 15, pp. 855–873.

Strack, D. and Fester, T., 2006. Isoprenoid metabolism and plastid reorganization in arbuscular mycorrhizal roots. *New Phytologist*, 172, pp. 22–34.

Subathra, M., Shila, S., Devi, M.A. and Panneerselvam, C., 2005. Emerging role of *Centella asiatica* in improving age-related neurological antioxidant status. *Experimental Gerontology*, 400, pp. 707–715.

Suguna, L., Sivakumar, P. and Chandrakasan, G., 1996. Effects of *Centella asiatica* extract on dermal wound healing in rats. *Indian Journal of Experimental Biology*, 34, pp. 1208–1211.

Tao, Y., Li, J.M., Li, Y.C., Pan, Y., Qun Xu and Ling-Dong, K., 2008. Antidepressant-like behavioural and neurochemical effects of the citrus-associated chemical apigenin. *Journal of Life Science*, 82, pp. 741–751.

Thomas, M.T., Kurup, R., John, A., Sreeja, P.C., Mathew, P.J, Dan, M. and Sabulal, B., 2010a. Elite genotypes/chemotypes, with high contents of madecassoside and asiaticoside, from sixty accessions of *Centella asiatica* of south India and the Andaman Islands: For cultivation and utility in cosmetic and herbal drug applications. *Industrial Crops and Products*, 32, pp. 545–550.

Thomas, M.T., Mathew, P.J., Dan, M. and Kumar, V.R., 2010b. Genetic variability, correlation and heritability of certain agronomic characters of *Centella asiatica* (L.) Urb. *Journal of Cytology and Genetics*, 11, pp. 77–81.

Warrier, P.K., Nambiar, V.P.K. and Ramankutty, C., 1994. *Indian Medicinal Plants*. Orient Longman, India. 2, pp. 52–56.

Widgerow, A.D. and Chait, L.A., 2000. New Innovations in Scar Management. *Aesthetic Plastic Surgery*, 24, pp. 227–234.

Wijeweera, P., Arnason, J.T., Koszycki, D. and Merali, Z., 2006. Evaluation of anxiolytic properties for Gotukola (*Centella asiatica*) extracts and asiaticoside in rat behavioural models. *Phytomedicine*, 13, pp. 668–678.

Won, J.H., Shin, J.S., Park, H.J., Jung, H.J., Koh, D.J., Jo, B.G., Lee, J.Y., Yun, K., Lee, K.T., 2010. Anti-inflammatory effects of madecassic acid via the suppression of NF-B pathway in LPS-induced RAW 246.7 macrophage cells. *Planta Medica*, 76, pp. 251–257.

Yu, Q.L., Duan, H.Q., Takaishi, Y. and Gao, W.Y., 2006. A novel triterpene from *Centella asiatica*. *Molecules*, 11(9), pp. 661–665. doi:10.3390/11090661.

Zainol, M.K., Abdul-Hamid, A., Yusof, S. and Muse, R., 2003. Antioxidative activity and total phenolic compounds of leaf, root and petiole of four accessions of *Centella asiatica* (L.) Urban. *Food Chemistry*, 81, pp. 575–581.

Zhao, J., Davis, L.C. and Verpoorte, R., 2005. Elicitor signal transduction leading to production of plant secondary metabolites. *Biotechnology Advances*, 23, pp. 283–333.

Zheng, C.J. and Qin, L.P., 2007. Chemical components of *Centella asiatica* and their bioactivities. *Chinese Journal of Integrative Medicine*, 5, pp. 348–351.

Zhi-lin, Y., Chuan-chao, D. and Lian-qing, C., 2007. Regulation and accumulation of secondary metabolites in plant-fungus symbiotic system. *African Journal of Biotechnology*, 6 (11), pp. 1266–1273.

4

The Importance of Iranian Borage (Echium amoenum *Fisch. Mey): A Critical Review*

Azim Ghasemnezhad, Manijeh Khorsandi Aghaei, and Mansour Ghorbanpour

CONTENTS

4.1 Introduction

Since the introduction of chemical drugs, humans relied on the therapeutic potential of herbs. Some people believe that these plants are used as food for the treatment of disease and human nutrition. Several herbs are used as traditional drugs, and nearly 50% of people in less-developed countries use medicinal herbs for their health care. There are nearly 2,000 racial groups in the world, and almost every group has accumulated knowledge about the use of traditional medicines, especially medicinal herbs (Ahvazi et al. 2012). The medicinal properties of Iranian borage (Echium amoenum Fisch. Mey) were first recognized by the Romans in the early third century BC and were introduced throughout Europe. Homer, the Greek poet and historian, described the plant as a joyous and sad word. Romans have been celebrating many years of plant leaves in wine and used it as an effective drink to calm down and fix the nervous system. Ancient Iranian therapeutic texts also mention the curative effects of this medicine. It is mentioned in the "Tamfe al-Hakim" that the plant has a warm temperament. Avicenna, a famous Iranian physician and scientist in the book "The Law in Medicine," described that the combination of this plant with vine increases the relaxation, and mixing with honey facilitates the heart palpation (Amirghofran et al. 2000). Khorasani, in the book "Makhzan Al-ADawiya," referred to the plant as a cure for cough, sore throat, shortness of breath, melancholia, and the treatment of a range of childhood fever (Sayyah et al. 2006). In traditional Turkish medicine, the roots of this plant are used for the treatment of ulcers.

4.2 Botanical Properties of Iranian Borage

This plant belongs to the order of Lamiales. The order consists of four families and close to 7,800 species (Hosseini et al. 2016). *E amoenum* (Figure 4.1). Boraginaceae is one of the most well-known plants of Iran, which is located in the north part of Iran (Javid and Hosseini 2016).

In general, climatic and soil conditions of cold and semiarid mountainous regions, which are relatively frequent during spring and summer, are suitable for planting (Ashoori et al. 2014). The best-known species of this family is the Borago officinalis (*Borago officinalis*), a native of Western Europe and the United States of America (Sayah et al. 2006). Iranian borage is a two-year-old herb that is distributed in mountainous areas of northern Iran (Rabbani et al. 2011). Its purple-blue dried petal, which is used in traditional Iranian medicine, contains yellow-lime essential oil with strong and exceptional aroma (Ghassemi et al. 2003). The seeds of plant also contains little amount of essential oil (Ghoreishi and Bataghva 2014). The flowers of the plant are purple. Its flowering begins in May and varies depending on the area cultivated until August (Lashkari et al. 2016). The flowers appear in one side of the branch. The flap is a crocheted plant that, after falling, can be seen as small and white prominences (Akbarinia et al. 2007). The seeds are 3–4 mm long and usually have a width of 2 millimeters with highlights (Akbarinia et al. 2007), which grows well on the normal soil. In addition to the seed, this plant can also be propagated with cuttings.

(a) (b)

FIGURE 4.1 Iranian borage (Echium amoneum) single plant (a) and farm (b).

Plant growth is possible at a wide range from the height of 60 to 2,200m (Abed et al. 2014).

4.3 Pharmaceutical and Health Care of Iranian Borage in Traditional Medicine

The part of the Iranian borage most used is a blue-violet dried flower, which is considered to be a forceful, sedative, mosaic, cough and healing agent for ulcers in traditional Iranian medicine (Ghassemi et al. 2003). This plant with a warm nature strengthens the body and also is used in kidney diseases, especially kidney inflammation (Naderi and Rezaei 2003). In past, too many people were prescribed to prevent various behavioral abnormalities. It is also an appropriate drug to regulate the blood pressure (Safavi and Khajehpour 2007). Iranian borage is used to treat inflammation, respiratory problems, blood purification, heart strengthening and as a diuretic (Abolhassani 2010). In the European Pharmacopoeia, flowers and leaves of the European-type of bouillon are used as anti-depressants, anti-anxiety, heart-softening, heart disease, pulmonary disorders, swelling and inflammation of the wrists, diuretics, laxatives, plasticizers and softeners (Ranjbar et al. 2006).

4.4 The Importance of Iranian Borage in Modern Medicine

Due to the improper use of chemical drugs and their side effects, in recent decades efforts have been made to increase the use of combined herbal remedies. Investigations have shown that Iranian borage flower has anti-viral properties and has a protective effect on the pancreas. Also, reductions of parasitic problems in lymph nodes, headache and nausea have also been reported from this plant (Abed et al. 2014). Antimicrobial effect of Iranian borage flower decoction on *Candida albicans* and *Pseudomonas aeruginosa* bacteria has been reported. This plant is effective in treating skin diseases (Sabour et al. 2017, Eruygur et al. 2016). The therapeutic effect of this plant methanolic extract has been demonstrated in the treatment of amoebiasis (Hamedi and Mawtani 2015). In recent studies, it has been shown that the use of this plant strengthens the immune system and, due to the presence of such substances as pyrolyzidine, alkaloid, quinone and Kinofuran, has antimicrobial and anti-infectious effects (Amiri et al. 2017). The effect of extraction of Iranian borage is similar to diazepam (Rabbani et al. 2011). Pyrolyzidine alkaloids and especially I-IV compounds (Figure 4.2) in this plant have a strong detoxification effect (Mehrabani et al. 2006).

Echimidine

Echimidine isomer

7-Angeloyl retronecine

7-Tigloyl retronecine

FIGURE 4.2 Structures of pyrrolizidine alkaloids of Echium amoenum Fisch & Mey. (From Mehrabani, M. et al., *DARU J. Pharm. Sci.*, 14, 125–126, 2006.)

Due to strong antioxidant compounds, the plant has shown to improve rat pearl (Dailami et al. 2016). The effect of the mixed extracts of *Viola odorata, E amoneum* and *Physalis alkekeng* showed good results in prostate disease symptoms (Beiraghdar et al. 2017). The effect of Iranian borage on the stimulation of white blood cell proliferation and the synthesis of tetralogestric antibodies in the body has been reported. The antibacterial effect of this plant also on the *Staphylococcus aureus* and its anticoagulant activity in rats was also reported (Gholamzadeh et al. 2007). The results of the studies indicate that the plant improves and prevents the spread of congenital skin disease of the atopic dermatitis. Fatty acids with several unsaturated bands in bouillon seedlings can play an important role in cell membranes, gene expression, prostaglandin biosynthesis, and the body's nervous system (Daneshfar et al. 2012). The anti-herpetic effects of green tea and Iranian borage extract on human Hep-2 cell, a cancer cell of human laryngeal epithelial, were examined (Farahani, 2016). Due to the variation of tannins and flavonoids compounds, iranian borage has good antiviral effects on virus replication and inhibits their activity (Farahani, 2016). The seed oil of the plant, like borage, is a rich source of unsaturated fats, such as acid gamma linolenic (GLA) and (SDA). This fatty acid has the potential to eliminate tumor cells without damaging the cells in the body. Therefore, GLA and SDA products can be used to provide anticancer drugs, and this result requires extensive research in the pharmaceutical industries (Ghoreishi and Bataghva 2014). It has been showed that the encapsulation of oil as a rich source of ω-6 and ω-3 fatty acids associate with cardiovascular disease of patients (Comunian et al. 2016). Due to the special properties, unsaturated fats improve the bio-membranes function and have a significant role in the transmission of cellular messages. These vital compounds are important in reproduction, improving immune responses, and growth and development of the brain and retinal network in neonates. In this regard, it has been shown that Iranian borage oil is very important. Shokri et al. (2011) reported the impact of GLA in improving the skin's efficiency in the elderly, reducing body fat, mastalgia, improving MS, colon and breast cancer, and treating melanoma (Shokri et al. 2001). In another study, an antimicrobial evaluation of Iranian borage extract against *Acinetobacter baumannii* (serious infectious bacteria) was carried out. *A. baumannii* has high resistance to one of the strongest antibiotics available under the name Imipenem®. This medication is prescribed for dangerous infections of the respiratory, urinary, skin and cardiovascular systems when antibiotics are not effective. The results of this study showed that the effect of Iranian borage extract in controlling the bacterium is similar to Imipenem (Sabor et al. 2015). The mean diameter of the colony of *A. baumannii* bacterium was affected by 50 ppm of the Iranian borage extract compared to 250 ppm of the antibiotic ciprofloxacin (an antibiotic used in the treatment of some infections including respiratory, urinary, prostate, gastrointestinal, genital and bone infections) was statistically significant (Sabor et al. 2015). This study also showed the inhibitory effect of Iranian borage extract at the concentration of 50 ppm against *A. baumannii* bacteria was significantly better than that of Colistin® at 10 µg/d for each disc (Sabor et al. 2015). In another study, the effect of plant extract on rat depression was evaluated, and the results showed that the use of aqueous Iranian borage extract caused changes in CSF serotonin and dopamine levels (Faryadian et al. 2014). These compounds play an important role in depression in mice and humans. The effect of Iranian borage extract on neurotransmitters demonstrates the positive role of this plant in strengthening the nervous system of human. In this study, it was found that the Iranian borage extract reduces stress by increasing the level of CSF serotonin and dopamine (Faryadian et al. 2014). It was also found that hydro-alcoholic extract of the Iranian borage had hypoglycemic effects in rats and caused positive changes in the level of lipids and plasma lipoproteins in these animals. Mahmoudi et al. (2015a and 2015b) demonstrated that the hydroalcoholic extract of *E. amoneum* had significant positive effect on the diabetic rats and leads to valuable changes in blood lipid profiles as well as lipoprotein levels. After treatment of animals and induction of diabetes, serum glucose level in diabetic specimens was significantly increased at 1% level. Also, the effects of different levels of extract in concentrations of 400 and 600 mg/kg body weight were significantly different in glucose levels. In addition to blood glucose, the triglycerides concentration of serum, total cholesterol and LDL decreased in all the groups receiving the extract. On the other hand, compared to the control group, the serum HDL levels in the diabetic group showed a significant reduction in the level of 1% (Mahmoudi et al., 2015a and 2015b). In a study, the researchers compared the effect of 10 mg/kg of methanolic extract of Iranian borage on albino mice compared to morphine. Although the extract of the plant has an analgesic effect, its effect was significantly less than that of 2.5 mg/kg morphine (Heidari et al. 2006). Table 4.1 presents information about benefits of Iranian borage in modern medicine.

4.5 The Antioxidant Activity of Iranian Borage (*E. amoneum*)

The destructive role of computers, mobile phones, food, biotech, waves, satellite and other modern technologies in a variety of stress is undeniable. Increasing the accumulation of free radicals caused by environmental stresses is one of the most important causes of the occurrence of physiological diseases, especially cancer. Therefore, considering this issue for human health has become a concern for governments. Among these, antioxidants have been able to react well to radicals to reduce the stress and limit damage to the cell and other harmful effects. Main ingredients involved in the radical scavenging capacity of the foods include polyphenols, carotenoids, and vitamins C and E. There are two types of natural and synthetic antioxidants. Vitamin C, carotenoids, and tocopherol are the most important natural antioxidants that can be used in the industry to improve the shelf life of products (Pilerood and Prakash 2014, Fathi and Mohammadi 2016). In an experiment with the aim of increasing the antioxidant capacity of Iranian borage using humic acid and Folivic acid, the results showed an increase of 38% and 33% of the total anthocyanin content of the plant petal, respectively (Amiri et al. 2017). The amount of anthocyanin is one of the important antioxidant indices of plants, especially Iranian bell peppers. It has been showed that the application of bio phosphorous and bio sulfuric fertilizers also increased the antioxidant activity by 7% and 8%, respectively (Amiri et al. 2017). The production

TABLE 4.1

Benefits of Iranian Borage (*Echium amoneum*) in the Modern Medicine

Research Title	Results	References
E. Amoenum Fisch Et Mey: Pharmacological and Medicinal Properties.	The reduction of parasitic problems in lymph nodes, headache and nausea	Abed et al. (2014)
Looking for antibacterial effect of *E. amoenum* against antibiotic resistant *Acinetobacter baumannii* strains isolated from burn infection.	A strong antimicrobial agent	Sabor et al. (2015)
Persian herbal medicines with anxiolytic properties. Journal of Medicinal Plants.	The anxiolytic effect of the flower of *Iranian borage* was shown	Rabbani et al. (2011)
Toxic pyrrolizidine alkaloids of *E. amoenum* Fisch and Mey.	The use of this plant increases the immune systems strength	Mehrabani et al. (2006)
A preliminary randomized double-blind clinical trial on the efficacy of *E. amoenum* in the treatment of mild to moderate major depression.	Aqueous extract of *E. amoenum* affects the mild and moderate depression	Sayyah et al. (2006)
Aqueous extract of *E. amoenum* elevate CSF Serotonin and Dopamine level in depression rats.	*E. amoenum* action similar to Dopamine	Faryadian et al. (2014)
Evaluation of the anxiolytic effect of *E. amoenum* petals extract, during chronic treatment in rats.	*E. amoenum* petals extract improved chronic disorders in rat	Gholamzade et al. (2007)
Effect of *E. amoenum* hydroalcoholic extract on blood glucose, lipid profiles and lipoproteins in male streptozotocin diabetic rats.	*E. amoenum* hydroalcoholic extract influenced the glucose, lipid profiles and lipoproteins of blood of male streptozotocin diabetic rats	Mahmoudi et al. (2015a and 2015b)
The use of endemic Iranian plant, *E. amoenum*, against the ethyl methanesulfonate and the.	Recovery of mutagenic effects	Uysal et al. (2015)
Antiviral effect assay of Iranian borage extract against HSV-1.	HSV-1 influenced by aqueous extract of *E. Amoenum* L	Farahani (2016)

of anthocyanins depends on the type and genetic ability of the plant. Environmental factors such as high light intensity, low temperatures, dryness, nitrogen deficiency and phosphorus, which are appropriate for increasing the content of sugar in the plant tissue, usually enhance the anthocyanins production of the tissue. On the other hand, the synthesis of anthocyanins with violet light and pathogens is stimulated. Anthocyanins influence a wide range of physiological activities, such as obesity, protect the liver from temporary anemia, inhibit lipoprotein oxidation, reduce permeability and fragility of the capillaries, and affect visual acuity (Babakhanzadeh Sajirani, 2007).

In another finding, the protective and antioxidant effects of anthocyanin-rich extract of Iranian borage flower on human vascular endothelial cells (HUVECs) under oxidative stress were investigated (Bekhradnia and Ebrahimzadeh, 2016). The results showed that antioxidants produce a protective effect against H_2O_2 oxidative stress in HUVECs (Miraj and Kiani, 2016).

4.6 Chemical Compounds of Iranian Borage

Chemical compounds or, in other words, secondary metabolites contain quaternary divisions include alkaloids, glycosides, phenolics and terpenoids compounds. Bitter materials, gum, mucilage, vitamins, pectin are other constituents of the plant (Hosseini et al. 2016). Approximately 26 compounds have been identified in Iranian borage seed oil, 20 of which represent 74.7% of essential oil (Ghasemi et al. 2003). This plant is an important source of unsaturated fatty acids such as α-linolenic acid (ALA) and γ-linoleic acid (GLA) (Eksiri et al. 2017). Eight types of palmitic, stearic, oleic, succinic (minimal), linoleic, α-linolenic (most), γ-linolenic and stearidonic in seed oil of this plant exist (Hosseinpour Azad et al. 2009). Linolenic acid, linoleic acid and oleic acid, the major fatty acids identified in the oil, with more

than 75% of the oil composition. Due to the high percentage of unsaturated fatty acids, in particular linolenic acid, Iranian borage seed oil is valuable in terms of nutrition (Daneshfar et al. 2012). The results showed that Iranian borage seed contains 25% oil, and the maximum extraction of oil is achieved by using super critical CO_2 method under ideal operating conditions. Using this technique is an acceptable method for the separation of gamma-linolenic acid (7%–8%), palmitic acid (6%–7%), stearic acid (3%–4%), oleic acid (12%–13%), linoleic acid (20%–20%), alpha-linolenic acid (40%–41%), and stearidonic acid (8%–9%) in the (Ghoreishi et al. 2011). The most important secondary metabolite of Iranian borage is divided into three major groups of alkaloids, terpenes and phenols (Amiri et al. 2017). Another important compound is rosmarinic acid, which has proven its antimicrobial properties (Mehrabani et al. 2006). More than 200 types of pyrrolizidine alkaloids are identified among the 6,000 herbaceous species, which represent 3% of the globular plants. Pyrolyzidine alkaloids are abundant in the families of Fabaceae, Asteraceae, Apocyaceae and Boraginaceae. These alkaloids are both toxic and non-toxic N-oxides. Among the different species of Boraginaceae family, the leaf extract of Iranian borage had 369.9 µgr/g of alkaloid. Also, in the flower extract of plant, 26.26 µg/g alkaloid and n-oxide was observed (Azadbakht et al. 2012). The pyrolyzidine alkaloid of *E. amoneum* include echimidine, echihumiline, echiupinine, echiumine, myoscorpine, uplandicine, intermidine, lycopsamin, retronecine and derivatives of each of these compounds (Mehrabani et al. 2006). The flower of the plant has a very small amount of essential oil and various salts such as Mn, Mg and P. This plant does not contain tannin compounds (Naderi and Rezaei 2003). Other important secondary metabolites of this plant include anthocyanin (13%) and flavonoid (15%) (Ghassemi et al. 2003). In addition to the above, the plant extract contains saponins, sterol, Rosmarinic acid and Anthocinidine (Ranjbar et al. 2006).

4.7 Economic Importance of Iranian Borage

In order to achieve food security, agricultural production needs of the community must be met. Agricultural production requires two factors in production. First, the physical factors of production, such as land, seed, water, labor, etc., are considered as prerequisites for quantitative and qualitative agricultural production. Second, non-physical factors are rooted in management and agricultural economics. In this regard, full attention is needed to the cost of product production. In this regard, full attention is needed to the cost of product production. The main objective is to generate more profits, reduce risk, fluctuate income, and have the ability to use sustainable capital. Therefore, in addition to optimal use of inputs, application of new techniques and technologies in production is important. In fact, it is necessary to consider the proper use of chemical fertilizers and irrigation in plant selection (Tafazolli et al. 2016). Iran's economy is based on oil revenues and is vulnerable to price volatility and supply of this product. One of the ways to cope with this challenge is to introduce non-oil exports. Production of medicinal plants, while improving the economic situation of the country, raises supply and demand in the market and improves the level of processing of medicinal plants, as one of the most important approaches to the economic evaluation of a medicinal plant. The existence of a huge national potential, as well as the culture of the use of medicinal plants in the position of medicinal herbs in the country in the prevalence of economic growth, employment it is possible today to grow it. In order to increase non-oil exports, it is considered as one of the main indicators in the growing of the national economy (Babakhanzadeh Sajirani, 2007). This plant is resistant to drought (a point that becomes more important with the critical water flow in the country). From the customer's point of view, the quality of the flowers grown in Alamut of Qazvin provinces is one of the best cultivars of Iranian borage (Akbarinia et al. 2007). Because wild plants are found in a wide range of geographies, collecting and accessing them is not economically feasible. On the other hand, such a massive collection of plant from the wild will destroy nature and expose plants to extinction. Therefore, it is important to consider the possibility of cultivating this plant in areas susceptible to cultivation. Borage and evening primrose seeds oil containing gammalinolic acid (GLA), extensively are used to prevent the progression of MS and Alzheimer's. Studies have shown that Iranian borage seed also contain this valuable compound (Hosseinpour Azad et al. 2009) and can be used in this case. Since the flowers of this plant are not harvestable via machinery, the plant is harvested by hand. Studies have shown that 70 to 80 workers per day are needed to harvest a hectare of flower; therefore, the product is important in terms of employment. Average production of Iranian borage per hectare is about 400 to 700 kilograms of dry flowers (Mir Amadi et al. 2012). Because the flowers of this plant produce a lot of nectar, they can aid in the production of honey with pharmaceutical value. Therefore, beekeeping can be done along with the production of Iranian borage (Mir Amadi et al. 2012). In comparison, if a hectare of vegetable is produced, there is a need for plenty of water, along with the cost of pests, diseases and weeds controlling. Due to lack of water resources, most crops are damaged or farmers are forced to use abnormal

water supplies. Meanwhile, the cultivation of Iranian borage with limited water requirements, high price of product (10–30 USA$ kg/dry flower depending on presenting time) can be extended. As mentioned earlier, the resistance of plant to soil drought and salinity, low nutritional requirements and minimum diseases and weeds controlling practice, and most importantly the high demand for these products, made this plant one of the most suitable planting options. Medicinal herbs have grown to create economic growth and jobs. Another point to consider economically is the drying of Iranian borage flower. Studies have shown that the drying method has a significant effect on the drying time, chlorophyll, flavonoids and anthocyanin content of this plant (Nadi 2017). The physiochemical variation during dehydration can strongly affect the physiological activities of this plant. Using new drying techniques play an important role in reducing the cost of production and increasing the product's yield quality. Studies have shown that the maximum and minimum drying time was observed in shade conditions and microwave, respectively (Saeedi and Asadi-Gharneh 2017). Controlling drying conditions, specifically temperature and air velocity, can lead to a high quality of product. It has been shown that the highest content of bioactive compounds, with minimum energy consumption, was obtained after drying at 60°C and air velocity of 0.86 m/s (Nadi 2017). The long drying time will increase the risk of contamination and corruption of the product to the open environment and reduce the economic value of the product. The anthocyanin compounds are sensitive to high temperatures and long drying times. It is recommended to use electric ovens as well as microwaves in choosing the appropriate alternative method for flower drying. However, it should be considered that the use of the above methods will be success if the temperature and time of drying be optimized. It was shown that the highest drying value of petals was obtained at a velocity of 3.7×10^{-6} g min^{-1} at an air temperature of 60°C and an air speed of 1 meter/second. In contrast, the minimum value was 0.8×10^{-5} g min^{-1} at 40°C and the air velocity was half a meter/second (Nadi and Abdanan 2017). Therefore, it is suggested that further studies be carried out to reduce the post-harvest losses and increase the economic income of producers. Note that the qualitative, marketable, and also phytochemical characteristics of the plant are one of the most important factors to be considered.

4.8 Oral Application of Iranian Borage

Probiotic drinks in combination with herbal medicines have a beneficial effect on human health. Today, probiotic beverages and fruit juices containing the flower of Iranian borage are commonly used (Eksiri et al. 2017). The extract of Iranian borage flower petals is used to prepare the formula of traditional medicine in different ways. It is also commonly used to color vinegar because of the colored ingredients. Different parts of the plant, especially the flower, have mucilage and are used in traditional medicine for treatment of digestive and respiratory diseases (Naderi and Rezaei 2003). Due to the use of anti-oxidant compounds it is also widely used as a natural preservative as a natural preservative to improve the shelf life and stability of food (Pillerroud et al. 2014). The flowers and leaves of this plant can be used as a seasoning in summer drinks. The flowers can be used to make jam. Flowers and leaves of this plant are used in salads. Due to the presence of vitamin C the leaves of the plant are used in the soup; for example, spinach the leaves can be used as cooking (Mir Amadi et al. 2012).

4.9 Side Effects of Iranian Borage Consumption

The consumption of medicinal herbs often does not have side effects and toxicity. It is nevertheless believed that their misuse, especially in conjunction with chemical drugs, may be harmful and even dangerous. The problem becomes more serious when most herbal medicines and herbs are consumed at the lowest standard. Therefore, the toxicological studies as well as the effects of the combined use of herbal and chemical drugs should be considered. There are many reports on the benefits of Iranian borage and its active ingredients, which indicate the usefulness and unhealthiness of this plant, and so far there is no strong evidence of its toxicity (Etebari et al. 2012). Nevertheless, in the evaluation of the effect of hydroalcoholic extract of plant at high concentrations of 25 mg/mL, DNA damage was identified, and it was found to play a fundamental role in most human diseases, such as cancer. Therefore, crop consumption should be taken with caution (Etebari et al. 2012). The side effects of this plant can be attributed to persistent and high levels of sleepiness, vomiting, dry mouth, constipation, and blurred vision, which are usually the most frequent reports (Sayyah et al. 2006). Also, the use of this plant for pregnant women is not recommended because it contains alkaloid pyrrolizidine and causes abortion (Babakhanzadeh Sajirani, 2007).

4.10 Conclusions

Since traditionally Iranian borage flower consumes a lot and is one of the most important native species of the Iran, the study of phytochemical characteristics of the flower of this plant is very important. The use of this valuable herbal medicine and its products not only reduce the side effects of chemical drugs, but also contribute to the economic prosperity and employment of the region and non-oil exports. Due to the problem of water scarcity, which annually shortens the scope for agriculture and production in the country, the cultivation of drought-tolerant plants, with low water requirements, is important and expanding. In this regard, Iranian borage is recommended as a valuable plant for cultivation in areas with low water reserves. Due to its limited distribution and inadequate understanding of pharmaceutical and agricultural capabilities, Iranian borage is less known to the world. Therefore, efforts are being made to do more research on the needs of agriculture and medical applications in order to introduce more and more Iranian borage to the world.

REFERENCES

Abed, A., Vaseghi, G., Jafari, E., Fattahian, E., Babhadiashar, N. and Abed, M., 2014. *Echium Amoenum* Fisch. Et Mey: A review on its pharmacological and medicinal properties. *Asian Journal of Medical Pharmaceutical Researches*, 4, pp. 21–23.

Abolhassani, M., 2010. Antiviral activity of borage (Echium amoenum). *Archives of Medical Science*, 6(3), pp. 366–369.

Akbarinia, A., KeramatiTarghi, M., HadiToutari, M.H., 2007. Effect of Irrigation Intervals on Flower Yield of *Echium amoenum* Mey & Fisch. *Research and Development in Natural Resources*, 76, pp. 124–123 (in Farsi).

Ahvazi, M., Khalighi-Sigaroodi, F., Charkhchiyan, M.M., Mojab, F., Mozaffarian, V.A. and Zakeri, H., 2012. Introduction of medicinal plants species with the most traditional usage in Alamut region. *Iranian Journal of Pharmaceutical Research*, 11(1), pp. 185–186.

Amirghofran Z, Azadbakt M and Keshavarzi F., 2000. Echium amoenum stimulate of lymphocyte proliferation and inhibit of humeral antibody synthesis. *Iranian Journal of Medical Science*, 25, pp. 119–124.

Ashoori, D., Noorhosseini, S.A., Safarzadeh, M.N., 2014. Effect of plant density and planting arrangement on Iranian borage. *Journal of Horticulture (Agricultural Sciences and Technology)*, 28(2), pp. 136–135 (In Farsi).

Azadbakht, M., Nematzade, GH.A., Hosseinpour Azad, N., Shokri, E., 2012. Quantitative and qualitative investigation of pyrrolyzidine alkaloids in roots, leaves, petals and seed of Iranian Echium Amoenum Mey& Fisch. *Journal of Mazandaran University of Medical Science*, 22(1), pp. 132–134.

Babakhanzadeh Sajirani, A., 2007. Iranian borage, the brilliance of Eshkahorat region of Gilan province, Publishing Asar Nafis (in Farsi).

Beiraghdar, F., Einollahi, B., Ghadyani, A., Panahi, Y., Hadjiakhoondi, A., Vazirian, M., Salarytabar, A., Darvish, B., 2017. A two-weeks, double-blind, placebo-controlled trial of *viola odorata*, *Echium amoenum* and *Physalis alkekengi* mixture in symptomatic benign prostate hyperplasia (BPH) men. *Pharmaceutical Biology*, 55, pp. 1804–1805.

Bekhradnia, S., Ebrahimzadeh, M.A. 2016. Antioxidant activity of *Echium amoenum.Revista de Chimi*, 67(2), pp. 224–225.

Comunian, T.A., Boillon, M.R.G., Thomazini, M., SayuriNogueira, M., Alves de Castro, I., S. Favaro-Trindade, C., 2016. Protection of Echium oil by microencapsulation with phenolic compounds. *Food Research International*, 88, pp. 114–115.

Dailami, K.N., Azadbakht, M., Lashgari, M., Rashidi, Z., 2016. Prevention of selenite-induced cataract genesis by hydroalcoholic extract of *Echium amoenum*: An experimental evaluation of the Iranian traditional eye medication. *Pharmaceutical and Biomedical Research*, 4, pp. 40–41.

Daneshfar, A., Ali Razaloo, K., Ahmadi Hosseini, S.M., Naghavi, M.R., Omid Beigi, R., 2012. Study of the composition of fatty acids and physicochemical properties of seed oil of a number of accessions of Echium amoenum. *Journal of Research in Iranian Herbs and Flowers*, 28(4), pp. 701–706 (in Farsi).

Eksiri, M., Nateghy, l., Rahmani, A. 2017. Production of probiotic drink using pussy willow and Echium amoenum extracts. *Applied Food Biotechnology*, 4(3), pp. 155–156.

Eruygur, N., Ylmaz, G., Kutsal, O., Yucel, G., Ustum, O., 2016. Biomassay-guided isolation of wound healing active compounds from Echium amoenum species growing in Turkey. *Journal of Ethnopharmacology*, 185, pp. 370–371.

Etebari, M., Zolfaghari, B., Jafarian-Dehkordi, A., Rakian, R., 2012. Evaluation of DNA damage of hydro-alcoholic and aqueous extract of Echium amoenum and Nardostachysjatamansi. *Journal of Research in Medical Sciences*, 17(8), pp. 783–785.

Farahani M., 2016. Antiviral effect of three Iranian herbs on the herpes simplex virus type I. *Journal of Lorestan University of Medical Sciences*, 18(1), pp. 118–117 (in Farsi).

Faryadian, S., Sydmohammadi, A., Khosravi, A., Kashiri, M., Faryadayn, P., Abasi, N., 2014. Aqueous extract of Echium amoenum elevate CSF Serotonin and Dopamine level in depression rat. *Biomedical and Pharmacology Journal*, 7(1), pp. 137–142.

Fathi, H., Mohammadi, H.R., 2016. A study the phenol and flavonoid content and evaluation of antioxidant activity of methanolic extracts of aerial part of the plant in a laboratory condition. *Complementary Scientific-Research Quarterly*, 1, pp. 1443–1442 (in Farsi).

Ghassemi, N., Sajjadi, S.E., Ghannadi, A., Shams-Ardakani, M., Mehrabani, M., 2003. Volatile constituents of a medicinal plant of Iran, *Echium amoenum* Fisch & Mey. Daru. *Journal of Pharmaceutical Sciences*, 11(1), pp. 32–33.

Gholamzadeh, S., Zare, S., Ilkhanipoor, M., 2007. Evaluation of the anxiolytic effect of Echium amoenum petals extract, during chronic treatment in rat. *Research in Pharmaceutical Sciences*, 2(2), pp. 91–92.

Ghoreishi, S.M., Bataghva, E., 2014. Supercritical extraction of essential oil from *Echium amoenum* seed, experimental, modeling and genetic algorithm parameter estimation. *Korean Journal of Chemical Engineering*, 31(9), pp. 1632–1633.

Hamedi, J., Mawtani, M., 2015. Anti-bacterial and anti-fungal effects of oil of *Oenothera biennis* L. and *Echium amoenum* Fisch. & C.A. Mey. *Recent Findings in Biological Sciences*, 2(3), 201–200.

Heidari, M.R., Moein Azad, E., Mehrabani, M., 2006. Evaluation of the analgesic effect of Echium amoenum Fisch & Mey. Extract in mice: Possible mechanism involved. *Journal of Ethnopharmacology*, 103, pp. 345–349.

Hosseinpour Azad, N., 2009. Study of genetic diversity of Echium amoenum Fisch & Mey., with RAPD molecular marker and gama linolinic fatty acid (GLA) with TLC. Master's thesis in Plant Breeding, pp. 3–4.

Hosseini N., Abdollahi M., MalekiRad, A.A., 2016. Identification and application of medicinal and aromatic herbs. Agricultural Research Publishing Center, Tehran, 2016.

Javid, S., Hosseini, S.E., 2016. Comparison of the effect of Echium amoenum with Buspirone on the Anxiety of cruciate maze in adult male rats. *Quarterly of the Horizon of Medical Sciences*, 22(1), pp. 52–53.

Lashkari, A., Rezvani Moqadam, P., Amin Ghafouri, A., 2016. Determination of minimum, optimum and maximum germination rates of *Echium amoenum* using regression models. *Iranian Journal of Agronomy Research*, 12(2), pp. 165–164 (in Farsi).

Mahmoudi, M., Shahidi, S., Golmohammadi, H., Mohammadi, S., 2015a. Effect of *Echium amoenum* hydroalcoholic extract on blood glucose, lipid profiles and lipoproteins in male streptozotocin diabetic rats. *Journal of Zanjan University of Medical Sciences*, 23, pp. 81–72 (in Farsi).

Mehrabani, M., Ghannadi, A.R., Sajjadi, E., Ghassemi, N., Shams-Ardakani, M.R., 2006. Toxic pyrrolizidine alkaloids of *Echium amoenum* Fisch&Mey. *DARU Journal of Pharmaceutical Sciences*, 14, pp. 125–126.

Miraj, S., Kiani, S., 2016. A review study of therapeutic effects of Iranian borage (EchiumamoenumFisch). *Der Pharmacia Letter*, 8(3), pp. 102–109.

Mahmoudi, M., Shahidi, S., Golmohammadi, H., Mohammadi, S., 2015b. The effect of *Echium amoenum* hydro-alcoholic extract on blood glucose level, lipid profile and lipoproteins in streptozotocin-induced diabetic male rats. *Journal of Zanjan University of Medicine*, 97, pp. 72–81.

Naderi, H.B.M., Rezaei, M., 2003. Phytochemical study of Echium amoenum. *Quarterly Journal of Medicinal Plants and Aromatic Plants of Iran*, 20(3), pp. 379–383 (in Farsi).

Nadi, F., Abdanan, S., 2017. An investigation into the effect of drying conditions on kinetic drying of medicinal plant of Echium amoenum. *International Medical Journal*, 1, pp. 87–91.

Nadi, F., 2017. Bioactive compound retention in Echium amoenum. Fisch & C.A. Mey. Petals: Effect of fluidized bed drying conditions. *International Journal of Food Properties*, 20(10), pp. 2249–2260.

Pilerood, S.A. and Prakash, J., 2014. Evaluation of nutritional composition and antioxidant activity of Borage (*Echium amoenum*) and Valerian (*Valerian officinalis*). *Journal of Food Science and Technology*, 51(5), pp. 845–854.

Rabbani, M., Vaseghi, G., Sajjadi, S.E, Amin, B., 2011. Persian herbal medicines with anxiolytic properties. *Journal of Medicinal Plants*, 39, pp. 7–8.

Ranjbar, A., Khorami, S., Safarabadi, M., Shahmoradi, A., Malekirad, A.A., Vakilian, K., Mandegary, A., Abdollahi, M.C.A., 2006. Antioxidant activity of Iranian Echium amoenum Fisch & C.A. Mey flower decoction in humans: A cross-sectional before/after clinical trial. *Evidence-Based Complementary and Alternative Medicine*, 4, pp. 469–473.

Saeedi, A., Asadi-Gharneh, H.A, 2017. Effect of different drying methods on some biochemical properties of Iranian borage (Echium amoenum Fisch & Mey). *Journal of Herbal Drugs*, 8(2), pp. 87–92.

Sabour, M., Hakemi-Vala, M., Mohammadi-Motamed, S., Eiji, M., 2017. Evaluation of the anti-bacterial effect of Echium Amoenum against multidrug resistant Acinetobacter baumannii strains isolated from burn wound infection. *Novelty in Biomedicine*, 1, pp. 38–42.

Safavi, S., Khajehpour, M.R., 2007. Effect of salinity on Na, K and Ca contents of borage (Borago officinalis L.) and Echium (*Echium amoenum* Fisch & Mey). *Research in Pharmaceutical Sciences*, 2, pp. 25–26.

Sayyah, M., Sayyah, M., Kamali-nejad, A., 2006. Preliminary randomized double blind clinical trial on the efficacy of aqueous extract of Echium amoenum in the treatment of mild to moderate major depression. *Progress in Neuro-Psychopharmacology and Biological Psychiatry*, 30(1), pp. 166–167.

Shokri, A., Kazemetibar, S.K., Z. Ali, J., Nematzadeh, G., Hosseinpoor Azad, N., Nasiri N., 2001. Cloning and characterizing the delta -6-desaturase gene of *Echium amoenum*. *Genetic and Breeding Research of Iranian Grassland and Forest Plants*, 19(2), 207–206.

Tafazolli, E., Zeraatkish, Y., Moosavi, S.N., 2016. The impacts economic analysis of deficit irrigation and integrated application of biological fertilizers on cultivation of borago in Iran. *Journal of Medicinal Plants and By-products*, 1, pp. 59–65.

Uysal, H., Kızılet, H., Ayar, A., Taheri, A., 2015. The use of endemic Iranian plant, Echium amoenum, against the ethyl methanesulfonate and the recovery of mutagenic effects. *Toxicology and Industrial Health*, 31(1), 44–51.

Section II

Plant Metabolites and Bioactive Compounds

5

Plant Metabolites as New Leads to Drug Discovery: Approaches and Challenges

Angana Sarkar, Debapriya Sarkar, and Kasturi Poddar

CONTENTS

5.1 Introduction

From the initial stage of the human civilization until now humans rely on the plants for various resources, viz, food, energy, shelter as well as medicine. Different raw plant parts, such as leaves, petals, roots, fruits, and bark of different plants, have been used by our ancestors as therapeutics for common discomforts, and that was the initial step for modern pharmacology. Based on that indigenous knowledge, modern science has discovered lots of bioactive metabolites from plants that have been successfully administered against different diseases listed in Table 5.1.

It has been estimated that over 700,000 plant species including angiosperm, gymnosperm, ferns, and mosses are present in the biosphere. Only 6% of have been screened for their biological activity and 15% is reported with potential medical benefits. The statistics are showing there are many more to explore (Fabricant and Farnsworth, 2001). Different research institutions across the globe have started huge-scale screening programs but few are achieving the success. For example, National Cancer Institute of National Institute of Health, USA studied for two

decades (1960–1980) and only screened 35,000 plant species for anticancer activity. This project reports finding only two successful anti-carcinogenic metabolites—paclitaxel and camptothecin (Katiyar et al., 2012).

This small rate of success shows there are certain limitations with the drug discovery from plants. This can be the rarity of the plant, high demand, complex structure, non-compatibility of plant extracts with high throughput screening, cultivation of medical plants, market competition, and threat to extinction. These limitations are discussed in detail later in this chapter.

Apart from these, there are other reasons such as new biodiversity conservation rule, market competition, legal issues regarding intellectual property rights (IPR) and challenges with low production rate of plant metabolites, which are discouraging the pharmaceutical giants all over the world to progress further with the screening programs for plant metabolites with therapeutic benefits. As a result the work remains somewhat restricted within the laboratories and small start-up companies that are not sufficient to disclose the opportunities hidden within the shield of this huge floral community of the biosphere.

TABLE 5.1

List of Different Plant Metabolites Used in Therapeutics

Target Disease	Molecule	Metabolite Class	Botanical Agent	Synthetic Analogues	Mode of Action	References
Anti-cancer	Catharanthus alkaloids	Alkaloids	*Catharanthus roseus*	Navelbine (vinorelbine), Oncovin (vincristine), Velban (vinblastine)	Inhibits the topoisomerase I (topo I) enzyme to prevent mitotic cell division	Alam et al. (2017); Sears and Boger (2015); Lee (2004)
	Taxol	Alkaloids	*Taxus brevifolia*	Docetaxel	Its interaction with the cellular tubulins produce microtubule assembly and therefore inhibits the cellular mitotic cell division process	Malik et al. (2011); Lee (2004); Balandrin et al. (1993)
	Camptothecin	Alkaloids	*Camptotheca acuminata*	Topotecan and Irinotecan	Inhibits the topoisomerase I (topo I) enzyme to prevent cellular mitotic cell division	Wang et al. (2016); Liu et al. (2015); Lee (2004)
	Podophyllotoxin	Lignan	*Podophyllum pelatum, Podophyllum emodi, Podophyllum pleianthum*	Teniposide	Inhibits the topoisomerase II (topo II) enzyme activity in the cells thus leading to cleavage of the DNA molecule	Kamal et al. (2015); Liu et al. (2014); Lee (2004)
Anti-cholinergic	Belladonna-type solanaceous tropane alkaloids	Alkaloids	*Atropa belladonna, Datura metel, Datura stramonium, Hyoscyamus niger; Mandragora officinarum*	Glycopyrrolate	It prevents nerve impulse transmission by binding with some specific receptor molecules present on the nerve cells and blocking its site of Acetylcholine binding	Balandrin et al. (1993); http://en.intoxication-stop.com/atropin_html_default.htm
	Physostigmine	Alkaloids	*Physostigma venenosum*	Phenserine	It interrupts the acetylcholine metabolism by inhibiting the activity of Acetylcholinesterase (Ache) and Butyrylcholinesterase (BChe) enzymes which are responsible for the nervous collapse caused due to acetylcholine interruption	Zhan et al. (2010)
	Pilocarpine	Alkaloids	*Pilocarpus jaborandi*	N-methyl derivative containing (S)-3-ethyl-4-[(4,-imidazolyl) methyl]-2-oxazolidinone	It can bind well with the muscarinic receptor molecules triggered by the acetylcholine from the parasympathetic nervous system and hence activates the secretions from exocrine glands, and dilates different smooth muscles of intestine, lungs, etc.	Sauerberg et al. (1989); https://pubchem.ncbi.nlm.nih.gov/compound/pilocarpine#section=Top
Antihypertensive, psychotropic	Reserpine	Alkaloids	*Rauvolfia serpentine*	Reserpine methonitrate	It controls the hyper-neurotransmission by increasing the rate of monoamine metabolism and simultaneously reducing the amount of released monoamine	Rawat et al. (2016); Eiden and Weihe (2011); Sreemantula et al. (2004)
Antimalarial, cardiac antiarrhythmic	Cinchona alkaloids	Alkaloids	*Cinchona sp.*	Artemisinin	It induces antimalarial effect in parasites by reducing their cellular oxygen consumption and carbohydrate metabolism and increasing the pH of their intracellular organelles	Balandrin et al. (1993); https://www.sciencedirect.com/topics/agricultural-and-biological-sciences/cinchona

(Continued)

TABLE 5.1 (Continued)

List of Different Plant Metabolites Used in Therapeutics

Target Disease	Molecule	Metabolite Class	Botanical Agent	Synthetic Analogues	Mode of Action	References
Anti-gout	Colchicine	Alkaloids	*Colchicum autumnale*	Desacetylmethyl colchicin (colcimid) and Demecolcin	Decreased uric acid production and accumulation	Kumar et al. (2016); Vane and Botting (1987); Hartung (1961)
Local anesthetic	Cocaine	Alkaloids	*Erythroxylum coca*	R-pseudococaine, Salicylmethylecgonine, Methylvanillylecgonine	It suppresses the ability of sensation by either reducing the stimulation of nerve endings or by preventing transmission of impulses by completely blocking the activity of neurons in the peripheral nervous system	McLeod, (2017); Balandrin et al. (1993)
Skeletal muscle relaxant	d-Tubocurarine	Alkaloids	*Strychnos toxifera, Chondodendron tomentosum*	—	It obstructs the intraneuronal activity and arrests the nerve impulse conduction through the poly-synapses at the spinal cord	Waldman (1994)
Analgesics, antitussive	Opium alkaloids	Alkaloids	*Papaver somniferum*	—	It works through the activation of the specific G protein-coupled receptor molecules in the central and peripheral nervous systems	https://www.sciencedirect.com/topics/neuroscience/opium
Oral contraceptives and other steroid drugs and hormones	Diosgenin, Hecogenin and Stigmasterol derived hormones	Steroids	*Dioscorea spp.*	Diosgenone	It controls the gene expression of the target cells. It selectively alters the transcription of some specific set of genes to code for their respective new proteins, which induce phenotypic changes in those set of target cells. Being soluble in lipids, steroid hormones can easily pass through the cell membrane or nuclear membrane, which enhances its ability to reach the target	http://www.vivo.colostate.edu/hbooks/pathphys/endocrine/moaction/intracell.html
Cardiotonic glycosides	*Digoxin and digitoxin* (Digitalis glycosides)	Steroids	*Digitalis purpurea, Digitalis lanata*	—	It acts through the specific inhibition of membrane transport processes of the cell, like Na^+-K^+ pump or Na^+-Ca^{2+} pump	Balandrin et al. (1993); Woods (1986)

5.2 Plant Metabolites as Drugs

With the recent advancement, we have successfully identified a number of plant metabolites with therapeutic applications. The initial work was started with some random biochemical process, but we cannot ignore the importance of ethno-pharmacology or indigenous knowledge as most of the discoveries related to medicinal plant metabolites was initiated by the lead provided by the knowledge we acquired from our ancestors; for example, *Rauvolfia sepertine* (Indian snakeroot) was mentioned in Ayurveda Shastra as having medical benefits against hypertension. After the screening process of the metabolites, researchers have found a chemical named Reserpine in that particular plant. Likewise, there are several examples of plant metabolites with their plant source and their respective therapeutic applications, which are listed in Table 5.1.

5.3 Screening Methods for Drug Isolation from Plants

The two stages of drug isolation involve the primary stage in which the particular plant species that possess medical benefits are identified. In the secondary stage, the necessary compounds in that plant species pertaining to druggability are screened out and separated.

5.3.1 Identification of Candidate Plant Species

The estimated number of total plant species present in this biosphere is 700,000. Among this high verity, the biological activity of 6% and phytochemical activity of 15% are studied. This statistics reveals that more remain unknown and left to discover (Fabricant and Farnsworth, 2001).

To select a particular plant species as a candidate for drug discovery, there are different approaches mentioned by Fabricant and Farnsworth as the random approach; ethno-pharmacology approach; traditional system of medicine approach; and zoo pharmacology approach (Fabricant and Farnsworth, 2001).

5.3.1.1 Random Approach

There are two different strategies in this approach:

a. *Screening for metabolites like alkaloids or flavonoids, and so on*: It is the simplest strategy to screen plants that are potential of drug source. In this process, we search for plants that are producing the particular metabolites and then we go for the druggability study. The chances of success are quite fair but the main problem with this approach is that this process cannot provide any idea about the biological efficacy.

b. *Screening of randomly selected plant for particular bioassay*: This is a trial and error approach. In this process, we randomly consider a plant species and study its metabolites for druggability. This process has the lowest outcome. For example, Central

Drug Research Institute has studied over 2,000 plant species but did not get a single break through. However, National Cancer Institute also tried to discover new anti-cancer drug and screen over 35,000 plant species for 20 years (1960–1980) and get only two breakthrough, viz. paclitaxel and camptothecin (Katiyar et al., 2012).

5.3.1.2 Ethno-Pharmacology Approach

This approach is based on the cumulative study of the indigenous drugs to develop better and new biologically active chemical entities (NCEs) that include analysis through observation, description and experimental inferences. It involves knowledge from various fields like chemistry, biology, biochemistry, pharmacology, and history. This ethno-medicinal approach has achieved several successful attempts like Andrographolide from Andrographispaniculata, Picroside from Picrorrhizakurroa, Morphine from Papaversomniferum, Berberine from Berberisaristata. However, several compounds, like L-Dopa and paclitaxel, have been successfully isolated from plants irrespective of their ethnomedical usefulness, such as Mucunaprurita and Taxusbrevifolia, respectively (Ganesan, 2008).

5.3.1.3 The Traditional System of Medicine Approach

The traditional approach makes the use of the huge stock of available botanical sources as medicines. Being a constant trend of medicinal approach, it is somehow different from the ethno-medicinal system in terms of:

a. It is being widely accepted and used compared to the ethnomedicinal system which is practiced by few families within a community.

b. The pharmacological process is well described and standardized unlike that of ethnomedicinal technique.

c. Though the ethnomedicine is a purely cumulative analysis of treatment, the traditional system comprises of the analytical relationship of experimental inferences with human physiological character and pharmacological concepts (Katiyar et al., 2012).

5.3.1.4 Zoo Pharmacology Approach

Close observation of cattle grazing pattern and monitoring their behavior pattern has led to the identification of plants with certain health benefits. Cattles attracted to a particular type of plant species were a clue to recognize the plant species to carry certain characteristics that were beneficial to the health and well-being of its consumers. Such cattle-grazing habits were observed in certain regions of Southern America that led to the identification of the plant species of the Solanaceae family, which are enriched in vitamin D3 derivatives (Katiyar et al., 2012).

5.3.2 Screening of Compounds with Therapeutic Use

5.3.2.1 Parallel Approach

This approach is adopted when the biological activity of the plant is well known either by is traditional use or by indigenous knowledge. The main objective of this screening process is to isolate the particular compound from the plant which is responsible for its biological activity or medical importance (Katiyar et al., 2012). The strategy is described in Figure 5.1.

5.3.2.2 Sequential Approach

This approach is the alternative of parallel screening approach. It is often used in random strategy or when the particular biological activity of the plant is unknown (Katiyar et al., 2012). The flow diagram in Figure 5.2 describes the process.

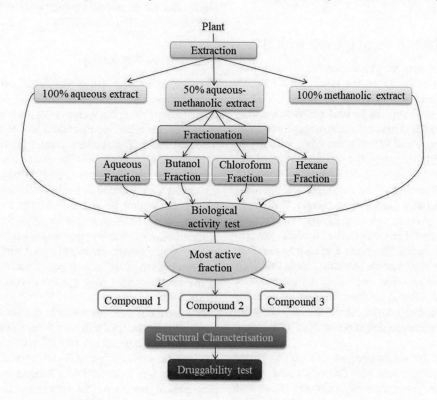

FIGURE 5.1 Schematic diagram of parallel approach.

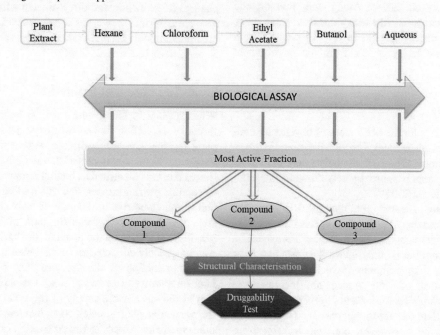

FIGURE 5.2 Schematic diagram of sequential approach.

5.4 Challenges and Overcome Strategy

As has already been discussed, global flora has provided lots of therapeutic agents to human society. What already has been discovered is only the tip of the iceberg. Many more remain to be spotted, but there are certain challenges hindering the path of discoveries of more novel drugs from plant sources as discussed in the following sections.

5.4.1 Non-compatibility of Plant Extracts with HTS

One of the major challenges for the discovery of novel drug compound is the incompatibility of the crude plant extracts with the High Throughput Screening (HTS) process. HTS is preferred as it provides us the option for both cell-free and cell-based assay systems. HTS is characterized by its high accuracy, reproducibility, robustness and efficient handling with liquid systems. To perform HTS properly, the test sample must not be decomposed, coagulated, precipitated or inert to unwanted chemical reactions. But in most of the cases, plant metabolites or plant extracts fail to fulfill any of these criteria, if not each of them. For example, plant samples are highly viscous, they tend to get coagulated and if left undisturbed tends to get precipitated out from the solution. Therefore it is hard to maintain the homogeneity in the HTS system. Another fact is that HTS results and observations are often fluorescent dependent. Now plant samples, in general, always contain fluorescent particles itself or fluorescent quenching substances which always contributes to misleading colorimetric endpoints of HTS (Atanasova et al., 2015).

Apart from this, plant extract contains lots of highly polarised compounds such as polyphenols and flavonoids and highly non-polarised compounds, like fatty acids, which can inhibit with the HTS assay type reaction. Even chlorophyll can be classified in this group (Zhou et al., 2014). Along with this organic materials, inorganic compounds like heavy metal salts can cause false positive or false negative results in HTS as plants have a tendency to accumulate heavy metals from the environment (Fernando et al., 2013).

5.4.2 High Demand

Plant-based raw materials for the production of novel drug molecules are currently in high demand. The global demand is growing at a rapid rate of almost 15%–20% annually where the Indian market of medicinal plants is approximately around US $1 billion every year (Joshi et al., 2004).

However, the rising demand for medicinal plants has increased the over-harvesting of many plant species that possess plant parts with medical benefits. For instance, Taxusbaccata has been harvested from the Himalayan wild forests at a huge quantity to extract out taxol compound which is highly effective in the treatment of ovarian cancer. Similar instances were observed with plants like *Gloriosa superba* (flame lily), *Aconitum heterophyllum, Arnebia benthamii, Dactylorhiza hatagirea, Megacarpoea polyandra*, etc. which were over-harvested and over-used for its significant demand and thus are now listed as endangered species. Certain medicinal plants that are effective against multiple diseases have been more extensively collected and over-exploited so that they are under threat to extinction from the wild forests (Kala et al., 2005). Plants like *Aegle marmelos* can be used to cure about 31 diseases. Similarly, *Hemidescus indicus* is used against 34 types of diseases. This over usage of the target plants resulted in rapid cutting down, which reduced their availability and supply from the wild. Simultaneously, the genetic diversity was interrupted, which affected the normal ecology and livelihoods of the locals in those regions (Rao et al., 2004).

5.4.3 Increasing Rarity

The continuous over use of the beneficial plant species with significant medicinal values led to the population declination of those particular target species in the geographical regions of its origin. This resulted in substantial loss of habitat of livelihood based on them. The primary reason for threat is the dilution of common social practices for protecting and preserving the natural resources (Belt et al., 2003). Weakening of the customary laws due to different powerful socio-economic forces along with several other threatening factors like, alteration of the habitat, climatic changes, invasion of non-natives, over grazing of livestock, high-density population, genetic drift, interruption in land usage, genetic drift, pathogens, and predators have limited the high availability and posed a threat to the abundance of the rare medicinal plant species in that geographical region (Kala, 2005).

In some cases there are multiple species of plants of the same genera that can be used to treat a particular disease; for example, different *Swertia sp.*, like, *S. augustifolia, S. cordata, S. chirajyta,* etc., can be used against malarial fever. Therefore there is no intense pressure on a particular *Swertia sp.* On the other hand, the species like *Rauvolfia serpentine* is feeling the intense pressure as they do not have any other load sharer. Hence the chances of extinction and threat-related with increasing rarity is a severe issue for those types of species with the metabolites that are not commonly found in other species as mentioned in case of *R. serpentine* (Kala et al., 2006).

5.4.4 The Threat of Extinction

When a plant is found with medical benefits, it faces a huge threat of extinction due to over-exploitation of that particular plant species. This was well observed previously in the incident known as "taxol supply crisis" (Cordell, 2011). In that particular time, taxol was identified as having remarkable therapeutic ability against ovarian cancer, and this finding resulted in a sudden and tremendous rise in demand of taxol. But taxol was accessible at that time only from the bark of Western Yew (*Taxus brevifolia* L.). To meet the demand, the bark of the Western Yew was intensively collected in unsustained manner. This leads to almost extinction of the Western Yew population. Although later on alternative access of taxol was identified and the problem of the sustainable supply of the taxol was met. This saves the particular plant species. This incident is quite common for other medicinal plants as the cultivation of most of these plants is quite a tedious job and non-profitable too as the generation time is very low (Kingston, 2011).

5.4.5 Cultivation of Medicinal Plants

Instead of collecting the parts of a plant of interest from the forest it would have been better if those plants can be cultivated. This can eliminate the risk of extinction and the problem with the biodiversity and forest reservation. But the information regarding the propagation of plants is very limited (only 10%) and the agro-technology has been developed successfully for only 1% of the plants (Khan and Khanum, 2000). Therefore it can be assumed that most of the plants with medical benefits cannot be cultivated properly because of lack of information. This fact implies the development of the agro-technology must be one of the largest areas of research. This development will help to cope with the increasing market demand and maintain consistency in product quality (Kala et al., 2006).

Agro-forestry could be another way out as it has also been reported that the forest environment helps to grow many medicinal plants and produce the particular metabolites with medical benefits. Introduction of agro-forestry can help us in the following ways:

a. Shade tolerant medicinal plants can be used in the lower strata in a multi-strata system, and this will give us the opportunity to produce.

b. Short-cycle medicinal plants can be cultivated as a secondary crop.

c. Larger trees with medicinal benefits can provide shades and boundary marks in agro-forestry.

d. Medicinal plants can also be inter-planted with food crops (Rao et al., 2004).

Although the investments in cultivation on medicinal plants are higher than the common food crops and the return is also lower for the same. This is the main reason why common farmers are not showing interest in the farming of medicinal plants. Some organizations are making attempts to culture different medicinal plants which are rare or endangered. They have tried to cultivate 20 endanger medicinal plant species from northern India but only succeeded to cultivate 10 of them and among them, *Rheum emobi* was the most efficient in terms of economics (Ghayur and John, 2004).

5.4.6 Bio-piracy

The overexploitation of medicinally benefitted target plants has led to restricted access to them via different legal clauses that control the access to those beneficial plants, sharing of the plant benefits and different patent issues that are controlled by the local administration of its origin. In order to preserve the genetic biodiversity of the plants, the UN Convention on Biological Diversity (CBD) was started in 1992 for the conservation of the natural biodiversity, to make sustainable use of the genetic resources and to make fair and justified distribution and sharing of the benefits obtained from the natural resources. Since the CBD regulations were implemented the countries and its representatives gained importance for the usage of the medicinally important originating plant by researchers for plant based drug discovery programs (Cragg et al., 2012). However many developing countries were still under the disbelief of destruction of their natural ecology during the search programs for therapeutic compounds from plant extract. As a result several countries started having strict regulations for obtaining permission for access of the flora. Simultaneously, obtaining positive results very rarely even after the examination and utilization of a vast range of species for a significantly long time discouraged the pharmaceutical companies from investment of time and money in this work. As for example, the US National Cancer Institute conducted a two-decade-long investigation on 12,000 species to obtain 114,000 plant extracts from which they obtained only two compounds that were eligible for being used as drug compound—taxol and camptothecin. To remove such hindrances and unfavorable situations, the Nagoya Protocol was published in , which was then agreed upon by 50 countries and became valid in September, 2014. The protocol declared fair and justified usage and sharing of genetic resources by following the guidelines to prevent destruction of the natural ecology and conservation of the biodiversity (Burton and Evans-Illidge, 2014).

In addition to the above-mentioned challenges there are more problems related to drug discovery from plants. Complex structure of these natural products in one of them. Due these complex structures they are very hard to reproduce synthetically, if not at all. This is one of the main drawbacks when working with plant metabolites as the production from the plant is low and slow. Initially, to identify the druggability, a small sample is required. But then to study characterization and other assays, the plant sample will be required in ampoule quantity. Sometimes it is very hard to obtain that much amount of the plant sample as very often the source of the plant sample and test laboratory are situated in two different continents. This problem is also encouraged by different rules regarding biopiracy and biodiversity protection and IPR. Thus it has been observed that many pharmaceutical giants have stopped their screening programs and work with plant metabolites from the early part of this century. Now field work is somewhat restricted in academics and small start-up companies (David et al., 2015).

5.5 Current Scenario

Synthetic compound obtain from comnetorial assays were expected to increase the rate of successful hits. But in reality it was found that the number of approved drug decreased at that particular time (David et al., 2015). The reason behind this is multiple. Many of the synthetic material showing positive results in HTS against the disease are not biocompatible or have side effects. Some of promising synthetic leads do not poses proper cyral centers hence has lower adaptability. Another interesting reason was in that same generation all of pharmaceutical companies has nearly same approach resulting same outcome. Because of these reasons recently the interests are again getting focused on the plants.

Another reason behind the more acceptance of natural product obtained from plants or any other biological system as drug is that natural substances are optimised by the evolutionary path. When comparison was done between natural products and synthetic substances obtained from combinatorial approaches, it was found that the major differences are the result of the modifications that has been done to increase the efficiency of the HTS.

For example, molecules with higher number of chiral centres are very hard to synthesize and assay as optical isomers are very hard to be separated and the process is also very expensive. But natural product always has higher number of chiral centres which make them adaptable for biological use. The molecules that have synthesized by combinatorial assays are expected to have lower molecular weight, shorter chains and less reactive groups so that they can easily be controlled and examined which also making them incapable. After understanding these problems the importance of plant metabolites has increased (Koehn and Carter, 2005; Klebe, 2009).

Another important approach has been currently adapted. Instead of opting purely natural or purely synthetic compounds, researchers are going for semisynthetic or mimicking the natural metabolites (Newman, 2008). Scientists have identified some special molecules with specific structures which are somewhat very common plant metabolites and termed them as privileged structures, like, benzodiazepines (Evans et al., 1988), N-acylhydrazone (Duarte et al., 2007) and indoles (Mason et al., 1999). Now research is going on around these privileged structures. They are synthesizing new molecules around these structures which are semisynthetic or completely synthetic in nature. Mimicking these particular structures also provide us with positive results to an extent. Different methods for this approaches are well explained by Newman (Melvin et al., 2005; Tsang et al., 2007).

5.6 Conclusion

At the starting of this chapter we have discussed about the approaches researchers follows to obtain plant metabolites with druggability. These approaches are very simple to perform but sometimes become a huge task if we consider the number of plants that remain to be screened for medicinal application. Therefore modified protocols can be investigated. The major problem we have discussed regarding the plant metabolites is that it is often non-compatible with modern high throughput screening technique which makes the screening of plants for medicinal application a tedious job. Apart from this problem there are also problem related to the production time, i.e. collection of the plant part from the forest and then processing it for a particular metabolite is long process. It makes it very difficult to cope the production process with the market demand. There are even issues related to biodiversity, biopiracy and threat of extinction. We have also discussed the agro-forestry and agro-cultivation approaches which is quite popular in Northern Himalayan region of India. But the problem is information related to cultivation protocol for medicinal plants is almost unavailable. Hence the extensive study in this particular field can be profitable. Due to these reasons along with some legal and trade issues in the beginning of the twenty-first century many of the pharmaceutical companies terminated there work with plant metabolites and preferred synthetic molecules obtained from combinatorial assays. But the outcome of this was unexpectedly poor as most of these synthetic compounds failed to demonstrate desired bio-adaptability. Therefore currently the importance of plant metabolites regained. We can justify these things by the recent popularity of Ayurvedic products.

REFERENCES

Alam, M.M., Naeem, M., Khan, M.M.A. & Uddin, M. 2017. Vincristine and vinblastine anticancer *Catharanthus* alkaloids: Pharmacological applications and strategies for yield improvement, In: *Catharanthus roseus*, Naeem M., Aftab T. & Khan M.M.A., editors. Springer, Cham, Switzerland, pp. 277–307.

Atanasova, A.G., Waltenbergerb, B., Pferschy-Wenzig, E.M., et al. 2015. Discovery and resupply of pharmacologically active plant-derived natural products: A review. *Biotechnology Advances*, 33, 1582–1614.

Balandrin, M.F., Kinghorn, A.D. & Farnsworth, N.R. 1993. Plant-derived natural products in drug discovery and development an overview, In: *Human Medicinal Agents from Plants*. ACS Symposium Series, 534, pp. 2–12.

Belt, J., Lengkeek, A., Zant, Jeroen van der, Yadav, A.K., Dimri, A.K. & Alam, G. 2003. *Cultivating a Healthy Enterprise: Developing a Sustainable Medicinal Plant Chain in Uttaranchal, India*. Amsterdam, the Netherlands: KIT Royal Tropical Institute, Bulletin 350.

Burton, G. & Evans-Illidge, E.A. 2014. Emerging R and D law: The Nagoya Protocol and its implications for researchers. *ACS Chemical Biology*, 9, 588–591.

Cinchona (Cinchona sp.). https://www.sciencedirect.com/topics/agricultural-and-biological-sciences/cinchona.

Cordell, G.A. 2011. Sustainable medicines and global health care. *Planta Medica*, 77, 1129–1138.

Cragg, G.M., Katz, F., Newman, D.J. & Rosenthal, Joshua. 2012. The impact of the United Nations Convention on Biological Diversity on natural products research. *Natural Product Reports*, 29, 1407–1423.

David, B., Wolfender, J.L. & Dias, D.A. 2015. The pharmaceutical industry and natural products: Historical status and new trends. *Phytochemistry Reviews*, 14, 299–315.

Duarte, C.D., Barreiro, E.J. & Fraga, C.A. 2007. Privileged structures: A useful concept for the rational design of new lead drug candidates. *Mini Reviews in Medicinal Chemistry*, 7, 1108–1119.

Eiden, L.E. & Weihe, E. 2011. VMAT2: A dynamic regulator of brain monoaminergic neuronal function interacting with drugs of abuse. *Annals of the New York Academy of Sciences*, 1216, 86–98.

Evans, B.E., Rittle, K.E., Bock, M.G., et al. 1988. Methods for drug discovery: Development of potent, selective, orally effective cholecystokinin antagonists. *Journal of Medicinal Chemistry*, 31, 2235–2246.

Fabricant, D.S. & Farnsworth, N.R. 2001. The value of plants used in traditional medicine for drug discovery. *Environmental Health Perspectives*, 109, 69–75.

Fernando, D.R., Marshall, A.T., Forster, P.I., Hoebee, S.E. & Siegele, R. 2013. Multiple metal accumulation within a manganese-specific genus. *American Journal of Botany*, 100, 690–700.

Ganesan, A. 2008. The impact of natural products upon modern drug discovery. *Current Opinion in Chemical Biology*, 12, 306–317.

Ghayur, A. & John, B. 2004. *Searching Synergy: Stakeholder Views on Developing a Sustainable Medicinal Plant Chain in Uttaranchal, India*. Amsterdam, the Netherlands: KIT Royal Tropical Institute, Bulletin 359.

Hartung, E.F. 1961. Colchicine and its analogs in gout: A brief review. *Arthritis & Rheumatism: Official Journal of the American College of Rheumatology*, 4, 18–26.

http://en.intoxication-stop.com/atropin_html_default.htm "Atropine is a vegetable alkaloid."

Joshi, K., Chavan, P., Warude, D. & Patwardhan, B. 2004. Molecular markers in herbal drug technology. *Current Science*, 87, 159–165.

Kala, C.P. 2005. Indigenous uses, population density, and conservation of threatened medicinal plants in protected areas of the Indian Himalayas. *Conservation Biology*, 19, 368–378.

Kala, C.P., Dhyani, P.P. & Sajwan, B.S. 2006. Developing the medicinal plants sector in northern India: Challenges and opportunities. *Journal of Ethnobiology and Ethnomedicine*, 2, 32.

Kala, C.P., Farooquee, N.A. & Dhar, U. 2005. Traditional uses and conservation of timur (Zanthoxylum armatum DC.) through social institutions in Uttaranchal Himalaya, India. *Conservation and Society*, 3, 224–230.

Kamal, A., Mohammed Ali Hussaini, S. & Shaheer Malik, M. 2015. Recent developments towards podophyllotoxin congeners as potential apoptosis inducers. *Anti-Cancer Agents in Medicinal Chemistry* (Formerly Current Medicinal Chemistry - Anti-Cancer Agents), 15, 565–574.

Katiyar, C., Gupta, A., Kanjilal, S. & Katiyar, S. 2012. Drug discovery from plant sources: An integrated approach. *Ayu*, 33, 10–19.

Khan, I.A. & Khanum, A. 2000. *Role of Biotechnology in Medicinal and Aromatic Plants* (Vol. III). Ukaaz Publications, Hyderabad, India.

Kingston, D.G. 2011. Modern natural products drug discovery and its relevance to biodiversity conservation. *Journal of Natural Products*, 74, 496–511.

Klebe, G. 2009. *Wirkstoffdesign: Entwurf und Wirkung von Arzneistoffen* (2nd ed.). Springer-Verlag.

Koehn, F.E. & Carter, G.T. 2005. The evolving role of natural products in drug discovery. *Natural Reviews Drug Discovery*, 4, 206–220.

Kumar, A., Singh, B., Sharma, P.R., Bharate, S.B., Saxena, A.K. & Mondhe, D.M. 2016. A novel microtubule depolymerizing colchicine analogue triggers apoptosis and autophagy in HCT-116 colon cancer cells. *Cell Biochemistry and Function*, 34, 69–81.

Lee, K.H. 2004. Current developments in the discovery and design of new drug candidates from plant natural product leads. *Journal of Natural Products*, 67, 273–283.

Liu, Y.Q., Li, W.Q., Morris-Natschke, S.L. et al. 2015. Perspectives on biologically active camptothecin derivatives. *Medicinal Research Reviews*, 35, 753–789.

Liu, Y.Q., Tian, J., Qian K. et al. 2014. Recent progress on C-4-modified podophyllotoxin analogs as potent antitumor agents. *Medicinal Research Reviews*, 35, 1–62.

Malik, S., Cusidó, R.M., Mirjalili, M.H., Moyano, E., Palazón, J. & Bonfill, M. 2011. Production of the anticancer drug taxol in Taxus baccata suspension cultures: A review. *Process Biochemistry*, 46, 23–34.

Mason, J.S., Morize, I., Menard, P.R., Cheney, D.L., Hulme, C. & Labaudiniere, R.F. 1999. New 4-point pharmacophore method for molecular similarity and diversity applications: Overview of the method and applications, including a novel approach to the design of combinatorial libraries containing privileged substructures. *Journal of Medicinal Chemistry*, 42, 3251–3264.

Mcleod, I.K. 2017. Local Anesthetics. Medscape.

Mechanism of Action: Hormones with Intracellular Receptors. http://www.vivo.colostate.edu/hbooks/pathphys/endocrine/moaction/intracell.html.

Melvin, J.Y., Bruce, A.L. & Yoshito, K. 2005. Discovery of E7389, a fully synthetic macrocyclic ketone analog of Halichondrin B, In: *Anticancer Agents from Natural Products*, Cragg, G.M., Kingston, D.G.I. & Newman, D.J., editors. Taylor & Francis Group, Boca Raton, FL, 241–265.

Newman, D.J. 2008. Natural products as leads to potential drugs: An old process or the new hope for drug discovery? *Journal of Medicinal Chemistry*, 51, 2589–2599.

Pilocarpine. https://pubchem.ncbi.nlm.nih.gov/compound/pilocarpine#section=Top.

Rao, M.R., Palada, M.C. & Becker, B.N. 2004. Medicinal and aromatic plants in agro-forestry systems, In: *New Vistas in Agroforestry*, Nair P.K.R., Rao M.R. & Buck L.E., editors. Springer, Dordrecht, the Netherlands, 61, 107–122.

Rawat, P., Singh, P.K. & Kumar, V. 2016. Anti-hypertensive medicinal plants and their mode of action. *Journal of Herbal Medicine*, 6, 107–118.

Sauerberg, P., Chen, J., Woldemussie, E. & Rapoport, H. 1989. Cyclic carbamate analogues of pilocarpine. *Journal of Medicinal Chemistry*, 32, 1322–1326.

Sears, J.E. & Boger, D.L. 2015. Total synthesis of vinblastine, related natural products, and key analogues and development of inspired methodology suitable for the systematic study of their structure–function properties. *Accounts of Chemical Research*, 48, 653–662.

Sreemantula, S., Boini, K.M. & Nammi, S. 2004. Reserpine methonitrate, a novel quaternary analogue of reserpine augments urinary excretion of VMA and 5-HIAA without affecting HVA in rats. *BMC Pharmacology*, 4, 30.

Toxicology and Human Environments. https://www.sciencedirect.com/topics/neuroscience/opium.

Tsang, C.K., Qi, H., Liu, L.F. & Zheng, X.S. 2007. Targeting mammalian target of rapamycin (mTOR) for health and diseases. *Drug Discovery Today*, 12, 112–124.

Vane, J.O.H.N. & Botting, R. 1987. Inflammation and the mechanism of action of anti-inflammatory drugs. *The FASEB Journal*, 1, 89–96.

Waldman, H.J. 1994. Centrally acting skeletal muscle relaxants and associated drugs. *Journal of Pain and Symptom Management*, 9, 434–441.

Wang, X., Tanaka, M., Krstin, S., Peixoto, H.S., De Melo Moura, C.C. & Wink, M. 2016. Cytoskeletal interference – A new mode of action for the anticancer drugs camptothecin and topotecan. *European Journal of Pharmacology*, 789, 265–274.

Woods, K.L. 1986. II. The mode of action of cardiac glycosides. *Journal of Clinical Pharmacy and Therapeutics*, 11, 11–13.

Zhan, Z.J., Bian, H.L., Wang, J.W. & Shan, W.G. 2010. Synthesis of physostigmine analogues and evaluation of their anticholinesterase activities. *Bioorganic & Medicinal Chemistry Letters*, 20, 1532–1534.

Zhou, J., Du, G. & Chen, J. 2014. Novel fermentation processes for manufacturing plant natural products. *Current Opinion in Biotechnology*, 25, 17–23.

6

Introduction to Herbs and Their Therapeutical Potential: Recent Trends

Shikha Singh, Gausiya Bashri, and Sheo Mohan Prasad

CONTENTS

6.1 Introduction

Human beings and other life forms have always been familiar with the use of plants to satisfy their need for food. Humans eventually began to identify specific plants that have properties to treat a particular disease, and a relationship was developed between plants and man. As the knowledge of humans increased regarding the cure of diseases by plant-derived products, these plants were documented in ancient Unani manuscripts, Egyptian papyrus, and Chinese writings. Similarly, Indian Vedas have also documented the use of plants for the curing of different diseases in humans. So, we can say that the use of plants and herbs as medicine was established very earlier and has been accepted and followed worldwide. For example, India has recognized herbs from ancient time and developed many traditional systems of remedy, such as Ayurveda, Unani, Siddha, homeopathy, yoga, and naturopathy (Kumar et al. 2015).

These traditional systems of remedy have gained worldwide popularity in part because of the high cost of medical treatments, the side effects of a number of man-made drugs, and the insufficient supply of drugs. These reasons have led to increased emphasis on the use of plant materials as a source of medicines for a wide variety of human diseases. Herbal medicines or medicinal plants are those plant or plant parts that are used for the therapeutic purpose for remedial of human diseases and improved health. Medicinal plants contain chemical compounds that have been produced within the plants through *secondary metabolic* or *phytochemical* pathways. Some secondary metabolites have antioxidant and defense property; therefore, they protect plants against biotic and abiotic stress. Phytochemicals also have therapeutic properties, which is why they are used in the formulation of drugs for remedy of diseases. The drugs that are formulated by medicinal plants either have no side effects or having least side effect which in part is why the traditional system of remedy is increasing and more popular worldwide.

6.2 Importance of Plants and Plant-Derived Products

Since ancient times, human society has used plant and animal products for food, clothing, shelter and for the cure of diseases (De Pasquale 1984). Nature has been a cradle of herbal medicines that have been explored and practiced all around the world. Herbal medicine includes plants and plant parts or preparations, processed and finished herbal products, and active ingredients (Sen et al. 2010; Pan et al. 2014). In diverse outmoded medicinal systems, plants are always the main basis of treatment strategy. Reports from the World Health Organization (WHO) state that the herbal medicines/products have been recognized as an essential component for primary health care and used by a large number of the population for basic health care needs (Taylor Leslie 2000). About 80% of the world's population has been using the herbal medicines for the curing of common ailments like cold, stomach disorders, toothache, and others as reported by WHO (Fabricant and Farnsworth 2001; Shakya 2016). Because of the adverse side effects of recent drugs, the hostility of micro-bodies, and also a failure of therapies being used to treat chronic disorders, a massive resurrection of herbal products has taken place in current years. In India, approximately 70% of contemporary medications and their mock equivalents have been prepared from original amalgams isolated from herbal plants. (Pan et al. 2011, 2014). At present, about 80% of anticancer drugs, antimicrobial, immunosuppressive, and cardiovascular are obtained from plant sources. Herbal plants have been adapted over millions of years for treatment against bacteria, fungi, and insects, and these plants also have unique and structurally varied secondary metabolites, therefore are proven to be a source for bioactive natural products. Various therapeutic agents like alkaloids, anthraquinone glycosides, cardiac glycosides, and lignans are the secondary metabolites that are synthesized by using biosynthetic precursor viz. amino acids, polypeptides and isoprenoids etc. (Fransworth et al. 1985; Tyler et al. 1988). In the Indian traditional medicine system, a number of plants are used. It is estimated that the Ayurveda uses 1,200 to 1,800 plants; Siddha medicine utilizes 500 to 900 plants; the Unani medicinal system includes 400 to 700 herbal plants; and the Amchi medicine practice uses nearly 300 plants. Folk healers in India use more than 7,500 plants in different treatments (Shankar and Majumdar 1997; Debnath et al. 2015; Sen and Chakraborty 2015). According to the report of the United Nations Development Project (UNDP 1994), the annual worth of medicinal plants obtained from emerging countries is about $32 billion, i.e. Rs. 32,00,000 crore.

The innovation of drugs from herbal plants continually provides new and significant leads against many pharmacological targets such as cancer, malaria, cardiovascular diseases and neurological disorders (Ramawat et al. 2009). Many herbal plants have been analyzed for the comprehensive chemical investigations that led to the exclusion of pharmacologically evaluated pure bioactive molecules. Traditionally, numbers of new remedies have been made from plant products and also from the compounds obtained from natural products. Factually, the natural products are secondary metabolites that are derived from primary metabolites, i.e. amino acids, carbohydrates, and fatty acids by secondary metabolic pathways, and these secondary metabolites are potent drugs in their natural unmodified form (Ramawat et al. 2009).

These secondary metabolites are referred to as phytochemicals (from Greek word "phyto" meaning "plant"). These secondary metabolites defend plants against microbial infections or by pest invasion. The phytochemicals which have therapeutic properties are considered for the production of valuable drugs. Consequently, the natural products obtained from medicinal plants are classified by the chemical nature of their alleged active components/chemical compositions (Table 6.1). These chemical compounds that are obtained from the herbal plants are composed of a few main phytochemical groups, i.e. alkaloids, terpenes, saponins, phenolic compounds, and so on.

6.2.1 Alkaloids

Alkaloids are wide-ranging chemicals with an unpleasant taste and are often toxic. They are found in many medicinal plants (Aniszewski and Tadeusz 2007). Structurally alkaloids are groups of heterocyclic nitrogenous compounds. About 20% of plant species contain alkaloids. In plants, alkaloids act as a defense system against insects, feeding deterrents and to the other herbivores (Harborne 1993). Morphine was the first medically useful alkaloid that isolated from *Paver somniferum* in 1805 (Fessenden and Fessenden 1982), and codeine and heroin are the derivatives of morphine. Diterpenoid alkaloids have been isolated from plants of the Ranunculaceae family, which is commonly known to have anti-microbial properties (Samy and Gopalakrishnakon 2010). With diverse action approaches (both recreational and pharmaceutical), there are several classes of alkaloids that includes atropine, scopolamine, and hyoscyamine (from the plant *Atropa Belladonna*), nicotine, caffeine, methamphetamine (ephedrine), cocaine, and opiates and so on (Kennedy and Wightman 2011). Similarly, the traditional medicine Berberine is obtained from plants *Berberis* and *Mahonia*, ephedrine (from *Ephedra*), cocaine (from *Coca*), caffeine (from *Coffea*), morphine (from opium poppy), and nicotine (from tobacco). The

TABLE 6.1

Natural Products Derived from Medicinal Plants with Their Chemical Nature and Putative Active Components/Chemical Compositions

S. No.	Phytochemicals	Chemical Structure	Examples
1.	Alkaloids	Heterocyclic ring containing nitrogen	Morphine, caffeine, berberine, codeine
2.	Anthraquinones	Derived from phenolics and glycosidic compounds	Luteolin, rhein, and salinosporamide
3.	Glycosides	Derivatives of carbohydrate and non-carbohydrate molecules	Amygdalin, gentiopicrinandrographolide, polygalin, cinnamyl acetate
4.	Polyphenols	Aromatic-aliphatic ring containing phenols	Quercetin, resveratrol, kaempferol and quercetin, caffeic acid, rutin, naringin, hesperidin, and chlorogenic, tannic acid, gallic acid, and ellagic acid
5.	Flavonoids	Aromatic ring with phenol	Flavones, iso-flavones etc.
6.	Saponins	Sugar attached to terpene or steroid aglycone	Diosgenin and hecogenin
7.	Terpenes	Long unsaturated aliphatic chains i.e. isoprene units	Artemisinin, alpha and beta-carotene, lycopene, lutein, and zeaxanthin
8.	Essential oil (monoterpene)		Alpha-pinine, beta-pinine, terpinine, carvacrol, cineole and camphor, para-cymene

alkaloid reserpine is isolated from the plant *Rauwolfia serpentine*, quinidine and quinine (from *Cinchona*), vincamine (from *Vinca minor*), and vincristine from the plant *Catharanthus roseus* (Elumalai and Eswariah 2012).

6.2.2 Terpenes

Terpenes are derived from lipids and constitute almost 30,000 lipid-soluble compounds; they contain an additional element, usually oxygen (e.g., menthol, camphor, carotenoids). These are lipid-like carbohydrates formed of one or more 5-carbon isoprene units, which are universally synthesized by all organisms via the mevalonate pathway and recently identified deoxy-*d*-xylulose pathways (Rohmer 1999). Terpenoids are classified on the basis of a five-carbon unit called isoprene (5C) as a hemiterpene; monoterpenes which incorporate 2 isoprene units i.e. 10 carbon compound (5Cx2), sesquiterpenes incorporate 3 units (5Cx3; 15 carbon compound), diterpenes comprise 4 units (5Cx4; 20 carbon compound), sesterpenes include 5 units (5Cx5; total 25 carbon), triterpenes that is 30 carbon compound via incorporating 6 units (5Cx6), and tetraterpenes which contains 8 units i.e. 40 carbon compound (5Cx8). Most of the terpenes or terpenoids have antibacterial and antifungal activity, and nearly 60% of all essential oil derivatives possess inhibitory effects upon fungi while 39% inhibited bacteria (Chaurasia and Vyas 1997; Samy and Gopalakrishnakone 2010). Several categories of terpenoids are found in a range of herbal plants, including conifers (Wiart 2014). They help repel herbivores because of their strongly aromatic nature. Along with this, their aroma makes them beneficial for the aromatherapy and rose and lavender plants are used for this perfumes (Tchen 1965; Elumalai and Eswariah 2012). Beside this, essential oil obtained from thyme (*Thymus vulgaris*) has a monoterpene (thymol) thast is used an antiseptic, antifungal, and also as a vermifuge viz. antiworm medicine. Moreover, *Ginkgo biloba* leaf extracts contain bilobalide and ginkgolides, a kind of biologically active terpenes that have been used medicinally for several millennia (Kleijnen and Knipschild 1992). *Melissa officinalis* has been in medicinal use as a mnemonic and anxiolytic psychotropic for more than two generations (Kennedy and Scholey 2006; Kennedy and Wightman 2011). Thymoquinone, a volatile oil obtained from the seeds of *Nigella sativa* (family-Ranunculaceae) shows protection against nephrotoxicity and hepatotoxicity generated by either disease or chemicals (Samy and Gopalakrishnakon 2010).

6.2.3 Glycosides

The nitrogenous protective compounds found in plants except for alkaloids are substances like cyanogenic glycosides and glucosinolates. They are not themselves toxic but are readily broken down to give off the poison. Hydrogen cyanide (HCN) is a well-known poisonous gas released by cyanogenic glycosides. The cardiac glycosides have been employed to treat patients with cardiac insufficiency and were also used in the past as poison for the tips of arrows. Cardiac glycosides slow down heartbeat and exhibit positive inotropic and bathmotropic effect along with negatively regulated chronotropic and dromotropic heart activity. They comprise digoxin and digitoxin that maintain the heart beating and also work as a diuretics. Most of the

cardiac glycosides are reported in plants of the Apocynaceae, Brassicaceae, Ranunculaceae, Asparagaceae and many more dicot families (Wink et al. 2010; Van Wyk et al. 2015a, 2015b). Furthermore, the Anthraquinone glycosides are obtained from the herbal plants such as rhubarb, cascara, *Alexandrian senna*, and *Aloe* and have been used as a laxative for millennia (Elumalai and Eswariah 2012). The steroidal glycosides and cucurbitacins (obtained from family Cucurbitaceae and few other families) inhibit tumor growth in vivo as well as in vitro while showing substantial cytotoxic activities.

6.2.4 Saponins

Saponins are the glycosides of triterpenes that include the group of cardiac glycosides and steroidal alkaloids. They are totally absent in gymnosperm (Van Wyk et al. 2015a, 2015b). The triterpene Saponins are abundant in dicot families like Ranunculaceae, Chenopodiaceae, Caryophyllaceae, Primulaceae, Phytolaccaceae, Sapotaceae, and Poaceae. Contra to this, the steroidal saponins are typically distributed in several families of monocots while less frequently in dicot families. In a few cases, the steroids, triterpenes, and saponins show structural resemblance with endogenous hormone glucocorticoids, which is anti-inflammatory in nature. A number of saponins like bidesmosidic compounds having two sugar chains are stored in vacuole, which can be changed to active form monodesmosidic compounds by the action of the enzyme β-glucosidase. Apart from this, few saponins have an additional functional group, like cardiac glycosides which contains a 5 or 6 membered cardenolide ring. The cardenolides viz. *Thevetia, Apocynum, Xysmalobium, Periploca, Strophanthus*, and *Nerium* have been obtained from the plants of the Apocynaceae family; *Cheiranthus, Erysimum* from plants of the Brassicaceae family; *Digitalis* obtained from Plantaginaceae family, formerly known as Scrophulariaceae; *Adonis* and *Helleborus* occurs in Ranunculaceae; *Bufadienolides* occur in Hyacinthaceae (Wink et al. 1998, 2010; Van Wyk et al. 2015a, 2015b).

6.2.5 Flavonoids and Anthocyanins

The characteristic feature of flavonoids and anthocyanins are the presence of two aromatic rings that have many phenolic, hydroxyl or methoxyl groups; they are the active components of many herbal plants (Harborne and Baxter 1993; Dewick 2001). They are frequently found as glycosides and are stored in vacuoles. Anthocyanins are the active antioxidants and are therefore used in phytomedicine to prevent reactive oxygen species (ROS)-related health disorders. The reddish-blue color of anthocyanins depends on many factors like hydrogen ion concentration, level of glycosylation, and the presence of some metals viz. aluminum ions present in the vacuole (Dewick 2001). In addition to this, the isoflavones that are common in legumes (subfamily Papilionoideae) exhibited similarities with the female sex hormone estradiol; hence termed as "phytoestrogens" that might play a role in women having menopausal- or bone-related disorders. The phytoestrogens are often observed as beneficial compounds that show antioxidant properties, and estrogenic exhibited a role in the prevention of cancers. Isoflavones obtained from *Trifolium pratense* and *Glycine max* are used as a nutraceutical (Van Wyk et al. 2015a, 2015b).

6.2.6 Polyphenols

Phenolics of several classes are ubiquitously found across the plant kingdom, having miscellaneous defensive action against plant diseases and predators. To date, ~10,000 structures of polyphenolics are identified from plants; structurally, they have 1 or more aromatic hydrocarbon ring with 1 or more hydroxyl groups attached (Elumalai and Eswariah 2012). Almost all the phenolic compounds with few exceptions are synthesized from a precursor that is formed by the phenylpropanoid biosynthetic pathway. A phenylpropane-derived compound represents a wide group of chemicals like cinnamic and caffeic acids, which has the highest oxidation state. Polyphenolics also include astringent tannins and hormone-mimicking phytoestrogens that have for hundreds of year been directed for gynecological disorders like fertility, menstrual, and menopausal (Elumalai and Eswariah 2012). Numerous polyphenolic extracts are retailed as nutritional supplements and cosmetics for beneficial health purposes without legal health claims, such as extract obtained from seeds of grapes, barks of olives or maritime pine and many more (European Food Safety Authority 2010). The astringent peel of the pomegranate has polyphenol termed punicalagins and is used as medicine in the Ayurvedic system (Jindal and Sharma 2004). In Ayurveda, curcumin, a polyphenol curcuminoid is known to accountable for the bright yellow color of turmeric (*Curcuma longa* L.), has been exploited for centuries for the treatment of alimentary system including inflammation (Ammon and Wahl 1991). Along with this, curcumin is can be used as a powerful suppression of bacteria, fungi, and viruses (Araujo and Leon 2001).

6.3 Development of Herbal Drug and Its Challenges

More than 50% of the drugs frequently used in world medicine are represented by herbal products and their derivatives. Among them, higher plants solitary contribute not less than 25% of total. During the last—three to four decades, at least a dozen potent drugs have been derived from flowering plants, including *Dioscorea* species, which produces diosgenin from which all ovulatory contraceptive agents have been derived. Two powerful anti-cancerous drugs vinblastine and vincristine which is obtained from the herbal plant *Catharanthus roseus* (Apocynaceae), laxative agents from *Cassia* species, a cardiotonic agent digitoxin to treat heart failure from *Digitalis* species and reserpine and other anti-hypertensive and tranquilizing alkaloids from *Rauwolfia* species. Tropical rainforests continuously provide numerous natural products along with invaluable compounds for the development of new drugs. Up to the present time, more than 50 drugs have come from tropical plants. The existence of undiscovered pharmaceuticals for modern medicine has often been cited as one of the most important reasons for protecting tropical forests (Abdin et al. 2003). Owing to the increasing tendency towards the self-medication and reduction in costs of the funded health care, the utilization of medicinal plants is expected to elevate globally. The elevation in utilization of medicinal plants denotes the rising pressure on wild plant

resources, hence the need for serious conservation efforts including the development of cultivation practices have never been more important. An average period of 10 years and more than $800 million are estimated for the process of drug discovery (Dickson and Gagnon 2004). Identification is the first step of drug development; it is predicted that only one in thousands of lead compounds will be approved for use through the scientific tribunal. The considerable tie has been taken in process of lead optimization and lead development including pharmacology, toxicology, pharmacokinetics, and also clinical trials (Jachak and Saklani 2007). In the broad spectrum, the natural products are typically isolated in small quantities that are inadequate for lead optimization and development as well as clinical trials. Consequently, to reconnoiter the possibilities of its overall synthesis there is a need to develop the collaborations with synthetic and medicinal chemists (Ley and Baxendale 2002; Federsel 2003; Lombardino and Lowe 2004). Modernization in instrumentation and biological assay provides the opportunity for the development of suitable quality control standards intended for the herbal drugs. For all drug discovery programs, the determination and implementation of appropriate, clinically relevant, high-quantity bioassays are difficult processes (Knowles and Gromo 2003; Kramer and Cohen 2004). The structural determination of fresh plant constituents can be performed with minimal delay by using a combination of sophisticated spectroscopic and X-ray crystallographic techniques (Kinghorn and Balandrin 1993). Bioassay challenges in screening the phytochemicals play an important part in the future of drug discovery from medicinal plants.

Despite the successfully running drug development programs from medicinal plants for 20 to 30 years, forthcoming endeavors face a lot of challenges. The issues regarding the standardization of raw materials and questions rose against the quality of natural products emerge as a major concern for herbal industries (Patwardhan et al. 2004; Yadav et al. 2014). Heavy metal contamination and adulteration can easily be done in herbal plants during growth, harvesting, drying, transportation and their processing. Therefore, scientists related to herbalism and pharmaceutical industries will need to constantly improve the quality and quantity of complexes going to enter into drug developing research programme (Clark 1996; Dash and Sharma 2001). Since the herbal remedies and plant extracts representing substantial sharing in the global market, the universally recognized guidelines for their quality-control matters are quite compulsory. WHO has documented the need to certify quality governing of the plant-derived products by using suitable standards and modern techniques. Pharmacopeias are designed on the basis of the chemical nature of medicine and pharmaceutical requirements with standards. Pharmacopeiasdo cover monograph and excellence control test for the herbal plants used in the particular countries, which include Indian Pharmacopoeia, Pharmacopoeia of Japanese, British, the Republic of China and United State Pharmacopoeia. Further, the problem remains constant by those formulations, which have the heterogeneous mixtures. For this, the use of cultivated plants in place of wild plants, which are often heterogeneous in respect, remove the causes of inconsistency (Gupta 2003). Few problems are there which are not applicable for mock-medication because hey influence the quality of herbal drugs. For instance

1. Mixtures of sundry constituents in herbal drugs.
2. Unknown active principles in most of the cases.
3. Unavailability of selective analytical methods or reference compounds.
4. Variations in chemical properties of plant materials.
5. The existence of chemo-varieties and chemo cultivars.
6. The variable raw material, their sources, and qualities.

6.4 Therapeutical Potential of Herbal Medicine

Herbs have been used as therapeutic drugs since ancient time and have progressed laterally with human civilization. Formerly, populaces have cycled native herbs and plants to extravagant multiple disorders and these have demonstrated flawless pharmacological activities. Knowledge of medicinal plant has been transferred from indigenous mythology available to families, tribes, and cultures, handed down from generation to generation. Remedial vegetation and its merchandise have been consumed by people since ancient time for diverse diseases and moved towards the discovery of many cherished drugs; for instance, anesthetics (morphine), antitussives (codeine), anti-hypertensives (reserpine), cardiotonic (digoxin), neoplastic resistant (vinblastine and taxol) and malaria resistant drugs (quinine and artemisinin). Drug discovery from therapeutic plants continuously delivers innovative and vital leads contrary to the numbers of pharmacological targets including cancer, malaria, cardiac illnesses and disorders related to the nervous system (Ramawat et al. 2009). Derivatives of natural products have been developed on a large scale to combat human diseases covering all therapeutic zones (Newman and Cragg 2007). Vigorous biochemical compounds obtained from remedial plants being used for managing various disorders in human beings are epitomized in Table 6.2.

Curative plants are rich in ancillary metabolites, and due to the presence of these compounds, these plants are termed as "medicinal" plants. A lot of research work has been done on these medicinal plants to analyze the actual biochemical properties of these drugs (called active principle) and, consequently, to obtain these amalgams through organic incorporation (Ramawat 2007). Presently, about 100 higher plants species have been used for the production of 125 clinically useful drugs. Along with this, approximately 5,000 plant species have been found as possible sources of new drugs (Tantry 2009). In the view of the above-mentioned data, less than 10% of the world's biodiversity

TABLE 6.2
Representing a Few Medicinal Plants Species with Their Vital Components Being Used for Managing Various Disorders in a Human Being

S. No.	Plant Name	Parts Studied	Active Components	Uses
1.	*Acorus calamus*	Rhizome	Alkaloids	Used medicinally for a wide variety of ailments, and its aroma makes calamus essential oil valued in the perfume industry.
2.	*Aegle marmelos*	Leaves	Alkaloids, Terpenoids, Saponins	Prescribed in a number of diseases such as gastrointestinal diseases, piles, edema, jaundice, vomiting, obesity, pediatric disorders, gynecological disorders.
3.	*Aloe vera*	Leaves	Vitamins A, C, E, Carotenoids, Saponins, Anthraquinone	Reduce pain through natural anti-inflammatory effects, effective as toothpaste in fighting cavities.
4.	*Andrographis paniculata*	Whole plant	Diterpenes, Lactoes	Used in the treatment of an array of diseases such as cancer, ulcer, diabetes, high blood pressure, leprosy, bronchitis, skin diseases, flatulence, colic, influenza, dysentery, dyspepsia, and malaria.
5.	*Azadirachta indica*	Leaves, fruits	Azadirachtin, Azadirone, Gedunin, Meliacarpin, Nimbin, Salannin, Vilasinin	Helps to cure hypertension, epilepsy. Also used as a sedative and analgesic. antifeedant, insect growth regulatory (IGR), fecundity and fitness reducing properties on insects.
6.	*Baccharis grisebachii*	Resinous exudate	Diterpenes, *p*-coumaric acid, Flavones	Showed activity towards dermatophytes and bacteria.
7.	*Boswellia serrata*	Gum resin	Pentacyclic triterpenes; Boswellic acid	Boswellic acid inhibits the leukotriene biosynthesis in neutrophilic granulocytes.
8.	*Carica papaya*	Leaves, fruits, and seeds	Terpenoids, Saponins, Tannins	Unripe fruit as a mild laxative or diuretic, the ripe fruit for rheumatism and alkalinizing the urine and the latex for psoriasis, ringworm, indigestion, also used as an antiseptic.
9.	*Cassia fistula*	Bark	Flavonoids	A mild laxative is suitable for children and pregnant women. It is also apurgative due to the wax aloin, used to treat many other intestinal disorders like healing ulcers.
10.	*Cassia podocarpa*	Leaf and flower	Glycosides, Anthraquinones, free aglycone	Optimum laxative activity and reduces the toxicity.

(Continued)

TABLE 6.2 (*Continued*)

Representing a Few Medicinal Plants Species with Their Vital Components Being Used for Managing Various Disorders in a Human Being

S. No.	Plant Name	Parts Studied	Active Components	Uses
11.	*Curculigo orchioides*	Rhizomes	Alkaloids, Flavonoids	Useful in general debility, deafness cough, asthma, piles, skin diseases, impotence, jaundice, urinary disorders, leucorrhoea, and menorrhagia.
12.	*Curcuma longa*	Rhizome	Flavonoids; Curcumin, Curcuminoids	Uses for bronchitis, asthma, sprains, skin diseases, and inflammation caused due to injuries.
13.	*Cinchona pubescens*	Bark	Quinine	Combats malaria.
14.	*Cyperus rotundus*	Rhizome	Saponin, Sesquiterpenoids	Highly beneficial to cure pulmonary problems like asthma, bronchitis, and cardiac problems.
15.	*Catharanthus roseus*	Root, leaves, latex	Theophylline	Opens bronchial passage, vincristine and vinblastine uses as anti-cancerous drugs.
16.	*Datura stramonium*	Seeds and leaves	Scopolamine	Eases motion sickness, useful in asthma and bone-setting.
17.	*Dalbergia sissoo*	Leaves, stem, and flower	Tannins	The remedy for gonorrhea and skin ailments. Leafy juice useful for eye ailments, the woody bark paste as anthelmintic, antipyretic and analgesic.
18.	*Dioscorea floribunda*	Root	Diosgenin	Contraceptive uses for curing various ailments such as a cough, cold, stomach ache, leprosy, burns, fungal diseases, skin diseases, contraceptive, dysentery, arthritis, rheumatism.
19.	*Digitalis purpurea*	Leaves	Digitoxin	Dropsy relieves heart congestion, also beneficial for asthma, epilepsy, tuberculosis, constipation, headache, and spasm.
20.	*Ephedra sinica*	Whole plant	Ephedrine	Prescribed for symptoms of cold and flu, including nasal congestion, cough, fever, and chills.
21.	*Emblica officinalis*	Seeds	Vitamine C, Tannins	Used as a laxative and for hyperacidity.
22.	*Filipendula ulmaria*	Seeds and rhizome	Aspirin, Salicine, Flavone-glycosides	Reduces pain and inflammations.
23.	*Ficus bengalensis*	Aerial root	Flavonoids, Tannins	Useful in the treatment of biliousness, ulcers, erysipelas, vomiting, vaginal complains, fever, inflammations, leprosy.
24.	*Hemidesmus indicus*	Stem	Alkaloids, Glycosides	Provides to cure chronic diseases such as Alzheimer's disease, malaria, and pain.
25.	*Hydratis canadensis*	Whole plant	Alkaloids; Barberine-etrahydroberberine and 8-oxoberberine	Exhibited vasodilator activity and also attributed to blockade of K^+channel and stimulation of Na^+-Ca^{++}exchanger.
26.	*Papaver somniferum*	Seeds	Codeine, Morphine	Eases pain, suppresses cough. The extract has been used as a sedative-analgesic and antitussive.
27.	*Psychotria ipecacuanha*	Root	Ipecac, Tannins, Isoquinoline and Glycosides	Controls vomiting, used as antinausea, expectorant, and diaphoretic, and was prescribed for bronchitis.
28.	*Pilocarpus jaborandi*	Leaves	Pilocarpine	Reduces pressure in the eyes, uses for convulsions, influenza, gonorrhea, fever, pneumonia, gastrointestinal inflammations, kidney disease, neurosis, and as an agent to promote sweating.
29.	*Magnifera indica*	Stembark, seeds	Reducing sugar, Flavonoids, Thiamine, Vitamin C, Riboflavin	The seeds are used in asthma and as an astringent.
30.	*Momordica charantia*	Fruit	Alkaloids, Saponins	Used to fight cancer, diabetes and many infectious diseases.
31.	*Plumbago zeylanica*	Root	Alkaloids, Glycosides	Used as a stimulant digestant, expectorant, laxative and in the treatment of muscular pain and rheumatic diseases.
32.	*Rauvolfia serpentina*	Plant extract	Reserpine	Lowers blood pressure, prescribed to cure anxiety, psychosis and epilepsy.
33.	*Solanum nigrum*	Fruit	Carotenoids, Ascorbic acid	Medicine considered to beantitumorigenic, antioxidant, anti-inflammatory, diuretic, hepatoprotective and antipyretic.
34.	*Syzygium cumini*	Bark and leaves	Triterpenoids, Ellagic acid	Useful in spleen enlargement and an efficient astringent in chronic diarrhea, juice of the ripe fruit is an agreeable stomachic and carminative and used as diuretic.
35.	*Zingiber officinale*	Rhizome	Polyphenolics; ginerol and 6-shogoal	Gingerols inhibits the growth of *Hylanthus pylori*.

has been examined for the possible biological activity, and a lot of more has to be explored for the use of the chief component. Scientists look forward to the new discoveries and have taken up the challenge of how to access this natural chemical diversity (Cragg and Newman 2005).

6.5 Worldwide Trend of Herbal Drugs

Herbal drugs are in great demand because of no side effects. The demand for herbal medications over past 20 years has been tremendously increased, yet is quite a substantial dearth of the investigation in the related field. A monograph on certain medicinal plants in three volumes has been printed by WHO since 1999. World's almost three-fourths of the population generally depends on herbal drugs for their well-being. Reports state that in industrialized countries like the United States and others, plant drugs constitute as much as 25% of the total drugs, while in fast-emerging countries such as India and China, the contribution is about 80%. Thus, herbal plants are more used in emerging countries. India has been well-known for a rich repository of medicinal plants. The forest cover of India has a huge number of therapeutic and aromatic plants, which are used as the raw materials for the preparation of drugs and perfumery products. The Ayurveda, Unani, Siddha, and Folk (tribal) prescriptions in India are the major systems of ethnic remedies. Among these systems, Ayurveda and Unani Medicine are most developed and widely practiced in India. Nearby 8,000 herbal remedies have been codified in AYUSH systems on 3/23/2018 and published in Introduction and Importance of Medicinal Plants and Herbs by National Health Portal of India.

6.6 Conclusion

Widespread variations of biologically active components that are obtained from herbal plants have been used extensively for the curing and treatment of various diseases. For a long time, plant-derived products have been documented as a source of therapeutic agents. Because of the presence of secondary metabolites, herbal plants have played a vital role in the finding of new chemical entities for drug development for which extensive effort has been done on medicinal plants. Being natural products herbs are relatively harmless, eco-friendly and easily available. Thus, there is a lot of demand for traditional remedies throughout the world to cure and treat various ailments.

Going forward, herbal medicine could become a new era of the medical system for controlling human diseases. Approximately, 80% of the world population relies on ethnomedicine for primary health care. Therefore, a lot of advance research has been needed for the advancement and characterization of a new medicine for the aid of better screening methods from plants and other natural sources. To solve the new-fangled encounters of the modern health care system, the use of herbal treatment becomes possible with the advancement in scientific research. These could be protracted for future investigation in the field of pharmacology, photochemistry, and ethnobotany for the discovery of drugs.

REFERENCES

Abdin, M.Z., Israr, M., Rehman, R.U. and Jain, S.K. 2003. Arteminisin, a novel antimalarial drug: Biochemical and molecular approaches for enhanced production. *Planta Medica*, 69, 289–299.

Ammon, H.P. and Wahl, M.A. 1991. Pharmacology of *Curcuma longa*. *Planta Med*, 57, 1–7.

Aniszewski, T. and Tadeusz, A. 2007. *Alkaloids: Secrets of Life*. Amsterdam, the Netherlands: Elsevier, p. 182.

Araujo, C. and Leon, L. 2001. Biological activities of *Curcuma longa* L. *Mem Inst Oswaldo Cruz*, 96, 723–728.

Chaurasia, S.C. and Vyas, K.K. 1997. In vitro effect of some volatile oil against *Phytophthora parasitica* var. piperina. *J Res Indian Med Yoga Homeopath*, 1, 24–26.

Clark, A.M. 1996. Natural products as a source for New Drugs. *Pharm Res*, 13(8), 1133–1141.

Cragg, G.M. and Newman, D.J. 2005. Biodiversity: A continuing source of novel drug leads. *Pure Appl Chem*, 77, 7–24.

Dash, B. and Sharma, B.K. 2001. *Charak Samhita*, 7th ed. Varanasi, India: Chaukhamba Sanskrit Series.

De Pasquale, A. 1984. Pharmacognosy: The oldest modern science. *J EthnoPharmacol*, 11, 1–16.

Debnath, P.K., Banerjee, S., Debnath, P., Mitra, A. and Mukherjee, P.K. 2015. Ayurveda e opportunity for developing safe and effective treatment choice for the future. In: Mukherjee, P.K., ed. *Evidence-based Validation of Herbal Medicine*. Amsterdam, the Netherlands: Elsevier, pp. 427–454.

Dewick, P.M. 2001. *Medicinal Natural Products*. Chichester, UK: Wiley.

Dickson, M. and Gagnon, J.P. 2004. Key factors in the rising cost of new drug discovery and development. *Nat Rev Drug Discov*, 3, 417–429.

Elumalai, A. and Eswariah, M.C. 2012. Herbalism: A review. *Int J Phytother*, 2(2), 96–105.

European Food Safety Authority. 2010. https://efsa.onlinelibrary. wiley.com/doi/abs/10. 2903/j.efsa.2010.1489. *EFSA Journal*, 8(2), 1489.

Fabricant, D.S. and Farnsworth, N.R. 2001. The value of plants used in traditional medicine for drug discovery. *Environ Health Perspect*, 109, 69–75.

Federsel, H.J. 2003. Logistics of process R&D: Transforming laboratory methods to manufacturing scale. *Nat Rev Drug Discov*, 2, 654–664.

Fessenden, R.J. and Fessenden, J.S. 1982. *Organic Chemistry*, 2nd ed. Boston, MA: Willard Grant Press, p. 139.

Fransworth, N.R., Akerele, O. and Bingel, A.S. 1985. Medicinal plants in therapy. *Bull World Health Org*, 63, 965–981.

Gupta, A.K. 2003. Quality standards of Indian medicinal plants, Vol. 1, published by ICMR, pp. 57–81.

Harborne J.R. 1993. *Introduction to Ecological Biochemistry*, 4th ed. London: Elsevier.

Harborne, J.B. and Baxter, H. 1993. *Phytochemical Dictionary: A Handbook of Bioactive Compounds from Plants*. London, UK: Taylor & Francis Group.

Jachak, S.M. and Saklani, A. 2007. Challenges and opportunities in drug discovery from plants. *Cur Sci*, 9, 92.

Jindal, K.K. and Sharma, R.C. 2004. Recent trends in horticulture in the Himalayas. New Delhi, India: Indus Publishing.

Kennedy, D.O. and Wightman, E.L. 2011. Herbal extracts and phytochemicals: Plant secondary metabolites and the enhancement of human brain function. *Adv Nutr*, 2, 32–50.

Kennedy, D.O. and Scholey, A.B. 2006. The psychopharmacology of European herbs with cognition-enhancing properties. *Curr Pharm Des*, 12, 4613–4623.

Kinghorn, A.D. and Balandrin, M.F. 1993. *Human Medicinal Agents from Plants*. Washington, DC: American Chemical Society.

Kleijnen, J. and Knipschild, P. 1992. *Gingko biloba. Lancet*, 340, 1136–1139.

Knowles, J. and Gromo, G. 2003. Target selection in drug discovery. *Nat Rev Drug Discov*, 2, 63–69.

Kramer, R. and Cohen, D. 2004. Functional genomics to new drug target. *Nat Rev Drug Discov*, 3, 965–972.

Kumar, N., Wani, Z.A. and Dhyani, S. 2015. Ethnobotanical study of the plants used by the local people of Gulmarg and its allied areas, Jammu & Kashmir, India. *Int J Curr Res Biosci Plant Biol*, 2(9), 16–23.

Ley, S.V. and Baxendale, I.R. 2002. New tools and concepts for modern organic synthesis. *Nat Rev Drug Discov*, 1, 573–586.

Lombardino, J.G. and Lowe, J.A. 2004. The role of the medicinal chemist in drug discovery-then and now. *Nat Rev Drug Discov*, 3, 853–862.

Newman, D.J. and Cragg, G.M. 2007. Natural products as sources of new drugs over the last 25 years. *J Nat Prod*, 70, 461–477.

Pan, S.Y., Chen, S.B., Dong, H.G., Yu, Z.L. and Dong, J.C. et al. 2011. New perspectives on Chinese herbal medicine (Zhong-Yao) research and development. *Evid Based Complement Alternat Med*. doi:10.1093/ecam/neq056.

Pan, S., Litscher, G. and Gao, S. 2014. Historical perspective of traditional indigenous medical practices: The current renaissance and conservation of herbal resources. *Evid Based Complement Altern Med*, 2014, 1e20. 21.

Patwardhan, B., Vaidya, A.D.B. and Chorghade, M. 2004. Ayurveda and natural products drugs discovery. *Curr Sci*, 86(6), 789–799.

Ramawat, K.G. 2007. Secondary plant products in nature. In: Ramawat, K.G. and Merillon, J.M., eds. *Biotechnology: Secondary Metabolites, Plants and Microbes*. Enfield. NH: Science Publisher, pp. 21–57.

Ramawat, K.G., Dass, S. and Mathur, M. 2009. The chemical diversity of bioactive molecules and therapeutic potential of medicinal plants. In: Ramawat, K.G., ed. *Herbal Drugs: Ethnomedicine to Modern Medicine*. Berlin, Germany: Springer-Verlag.

Rohmer, M. 1999. The discovery of a mevalonate-independent pathway for isoprenoid biosynthesis in bacteria, algae and higher plants. *Nat Prod Rep*, 16, 565–574.

Samy, R.P. and Gopalakrishnakone, P. 2010. Therapeutic potential of plants as anti-microbials for drug discovery. *Advance Access Publication*, 7(3), 283–294.

Sen, S. and Chakraborty, R. 2015. Toward the integration and advancement of herbal medicine: A focus on Traditional Indian medicine. *Bot Target Ther*, 5, 33–44.

Sen, S., Chakraborty, R., De, B., Ganesh, T., Raghavendra, H.G. and Debnath, S. 2010. Analgesic and inflammatory herbs: A potential source of modern medicine. *Int J Pharma Sci Res*, 1, 32–44.

Shakya, A.K. 2016. Medicinal plants: Future source of new drugs. *International Journal of Herbal Med*, 4(4), 59–64.

Shankar, D. and Majumdar, B. 1997. Beyond the biodiversity convention: The challenges facing the biocultural heritage of India's medicinal plants. In: Bodeker, G., Bhat, K.K.S., Burley, J., Vantomme, P., eds. *Medicinal Plants for Forest Conservation and Health Care*. Rome, Italy: Food and Agriculture Organization of the United Nations, pp. 87–99.

Tantry, M.A. 2009. Plant natural products and drugs: A comprehensive study. *Asian J Tradit Med*, 4(6), 241–249.

Taylor Leslie, N.D. 2000. Plant Based Drugs and Medicines, Rain Tree Nutrition. http://www.raintree.com/plantdrugs.htm#.VPGDGXyUeAA

Tchen, T.T. 1965. Reviewed work: The biosynthesis of steroids, Terpenes, and Acetogenins. American Scientist. Research Triangle Park, NC: Sigma Xi, *The Scientific Research Society*. 53(4), 499A–500A.

Tyler, V.E., Brady, L.R. and Robberts, J. E. 1988. *Pharmacognosy*, 9th ed. Philadelphia, PA: Lea and Febiger.

UNDP (United Nations Development Programme). 1994. Human Development Report 1994: New Dimensions of Human Security. New York: Oxford University Press.

Van Wyk, B.E. and Wink, M. 2015b. *Phytomedicines, Herbal Drugs and Poisons*. Cambridge, UK: Briza, Kew Publishing, Cambridge University Press.

Van Wyk, B.E., Wink, C. and Wink, M. 2015a. *Handbuch der Arzneipflanzen*, 3rd ed. Stuttgart, Germany: Wissenschaftliche Verlagsgesellschaft.

Wiart, C. 2014. Terpenes: Lead compounds from medicinal plants for the treatment of neurodegenerative diseases. pp. 189–284.

Wink, M. and Schimmer, O. 2010. Molecular modes of action of defensive secondary metabolites. In: Wink, M., ed. *Functions and Biotechnology of Plant Secondary Metabolites*. Annual Plant Reviews 39; London, UK: Wiley-Blackwell, pp. 21–161.

Wink, M., Schmeller, T. and Latz-Brüning, B. 1998. Modes of action of allelochemical alkaloids: Interaction with neuroreceptors, DNA and other molecular targets. *J Chem Ecol*, 24, 1881–1937.

Yadav, M., Chatterji, S., Gupta, S.K. and Watal, G. 2014. Preliminary phytochemical screening of six medicinal plants used in traditional medicine. *Int J Pharm Pharm Sci*, 6(5), 539–542.

7

Curcumin as a Potential Therapeutic for Alzheimer's Disease: A Multi-targeted Approach

Priyanka Velankanni, Satheeswaran Balasubramanian, Azhwar Raghunath, Frank Arfuso, Gautam Sethi, and Ekambaram Perumal

CONTENTS

7.1   Introduction

Alzheimer's disease (AD) is a customary, age-allied, dementing ailment clinicopathologically substantiated by progressive cognitive impairment synchronous with the aggregation of senile plaques (El-Desouki 2014). Alzheimer's, the common cause of dementia, makes up over half of dementia cases (Cummings and Benson 1992). The World Alzheimer report (2016) estimates that there were about 46.8 million people around the globe afflicted with dementia in 2015 and expects a rise to about 131.5 million by the year 2050 (Prince et al. 2016). A report by Brookmeyer et al. (2007) predicts that one person for every 85 individuals will suffer from AD by the year 2050.

In 1907, Alois Alzheimer narrated the symptoms of his patient named Auguste Deter, aged 51, with a five-year history of progressive cognitive impairment, hallucinations and delusion, and severely weakened social functioning. Alzheimer executed silver staining, a histological technique to examine her brain microscopically, and witnessed neuritic plaques, neurofibrillary tangles, and amyloid angiopathy, which have now become the hallmarks of the disease named after him by Kraepelin in 1910. On the basis of the age of onset, AD can be classified as early-onset AD affecting the population younger than 65, which only accounts of about 1%–6% of all cases; and late-onset AD, which affects the population older than 65 years and accounts for about 90% of the total cases (Brookmeyer et al. 2007). The ubiquity of AD is escalating swiftly in both developing and developed nations. The distribution of AD oscillates around the sphere based on socioeconomic and cultural differences.

The recent epoch of Alzheimer's research started only after the discernment that a great number of senile dementia cases had the plaque and tangle pathology similar to that of AD (Blessed et al. 1968; Tomlinson et al. 1968, 1970). The nosological difference between the AD and senile dementia of the Alzheimer's type was dropped, and AD research was given prime priority, followed by the formation of Alzheimer's association in the United States and later by other countries (Katzman 1976; Khachaturian 2006). Initially, the Alzheimer's research was mainly carried out based on three different aspects. The first was to understand the selective neurotransmitter loss in Alzheimer's, which emanated in the evolution of currently available therapies based on cholinergic cell loss in AD (Homykiewicz 2002). The second was largely based on the investigation of the pathogenic lesions, neuritic plaque, and the neurofibrillary tangle; and the third was to develop an understanding of disease pathogenesis (Hardy 2006).

This present chapter sheds a light on the therapeutic effects of curcumin in AD as evidenced from various *in vitro* and *in vivo* studies. The emphasis will be given to the molecular mechanisms of actions exhibited by curcumin and its potential efficacy against AD.

7.2   Prevalence

The Delphi consensus study revealed that the prevalence of dementia is higher in America and considerably lower in the less developed regions of the world such as Africa and Middle East (Ferri 2005). The prevalence of AD in the Eastern European nations seems to be relatively uniform and located between Japan and the USA. The variation in the statistical data obtained is, to an extent, contributed by the deprivation of monotonous methodologies. The prevalence of AD is significantly higher in the female than the male population. A report states that almost two-thirds of cases diagnosed with AD are women (Hebert et al. 2003). The global cost of dementia had a dramatic rise from the year 2010 to 2015 from US $604 billion to US $818 billion, which accounts for about 1.09% of the global GDP (Wimo et al. 2017).

7.3   Symptoms

AD symptoms differ with individuals; the initial symptoms include difficulty in remembering recent information, but as the disease progresses the patient experiences difficulties ranging from cognitive to functional disabilities. In advanced stages, the patient requires assistance, even for performing the basic activities of daily life. The rate of advancement of symptoms from mild to severe varies amongst sufferers (Prince et al. 2016). The most common symptoms of Alzheimer's are memory loss (dates, events, names, appointments, daily tasks, time, location, and visual images), vision impairment (cataract and glaucoma), and problems with speaking, writing, and vocabulary. Mild cognitive impairment is the major factor for Alzheimer's, where measurable changes in thinking abilities are noticeable by people around the patients but the patients do not realize it. They become suspicious and scared (Karlawish et al. 2017). Neuropsychiatric symptoms for AD include aggression, delusions, lack of sound sleep, hallucinations, apathy, depression, and sexually inappropriate mannerisms (Rosenberg et al. 2015).

7.4   Plaque Deposition

Amyloid plaques in the cerebral cortex are customarily associated with AD. Amyloid plaques are classified into two types: diffuse plaques and neuritic plaques. Diffuse plaques are characterized by the particulate amyloid depositions devoid of inclusion of thickened, dark-staining neuritis, and a dense central core. Neuritic plaques are those with the thickened, darkly stained, and contorted neuritis within or extending from the region of amyloid deposition and with or without the dense β-Amyloid core (Price et al. 1999).

The microvascular amyloid deposits (the primary component of the senile plaques) from the meninges of AD brains were initially purified, characterized, and a partial sequence of a ~4-kDa amyloid β (Aβ) protein was provided (Glenner and Wong 1984). The amyloid hypothesis states that the gradual accumulation of the micro hydrophobic peptides initiates a deadly cascade, eventually leading to synaptic alterations, microglial and astrocytic activation, alteration of the soluble tau protein into oligomers and then into insoluble paired helical filaments, accompanied with progressive neuronal loss associated with multiple neurotransmitter deficiencies and cognitive failure (Hardy and Higgins 1992).

The causative gene of AD encodes for the amyloid precursor protein (APP), a type-1 trans-membrane receptor-like glycoprotein that is ubiquitously expressed in neural and non-neural cells. The APP gene contains 18 exons and its size is more than 170 kb. The region coding for the Aβ sequence comprises of exons 16, 17, and 40 to 43 amino-acid residues extending from the ectodomain to the transmembrane protein domain (Yoshikai et al. 1990).

The Aβ peptide is physiologically produced by the proteolysis of a large, integral, membrane-bound precursor protein named the APP throughout life, and the APP was found to be in the circulation of extracellular fluids such as cerebrospinal fluid and plasma (Haass et al. 1992; Seubert et al. 1992). The APP is processed by α-, β-, and γ-secretase enzymes. The identification of these proteases has led to them being proposed as potential targets for treating AD.

APP is sequentially cleaved by the protease β-secretase (BACE1) and γ-secretase, resulting in the synthesis of isomers of Aβ peptides predominantly 40 or 42 amino acids in length ($A\beta_{40}$ and $A\beta_{42}$). The C-terminal variation is mainly associated with AD pathogenicity. $A\beta_{42}$ is the most toxic among the isomers. β-secretase produces the NH_2 terminal of Aβ and cleaves APP to generate a soluble form of APP (βAPPs) and a 99 residue carboxyl-terminal (CT99) that is bound to the cell membrane. On the other hand, α-secretase cleaves within the Aβ region to produce APPαs and an 83-residue COOH-terminal fragment (CT83). CT83 and CT99 act as substrates for γ-secretase. Proteolysis by γ-secretase is heterogeneous, resulting in the production of $A\beta_{40}$ and a small proportion of $A\beta_{42}$. Though $A\beta_{42}$ is a minor form of Aβ peptide, these longer and more hydrophobic species tend to form fibrils of insoluble β-sheet conformation and are eventually deposited in the diffuse senile plaques. $A\beta_{42}$ is the major species found in the cerebral plaques (Jarrett et al. 1993; Iwatsubo et al. 1994).

7.4.1 Aβ Aggregation

The aggregated Aβ peptides are the undeniable characteristic feature of AD, but the small Aβ oligomers are the most toxic species of Aβ. There are diverse mechanisms by which Aβ exhibits toxicity and involves the disruption of calcium homeostasis, altered membrane integrity, synaptic dysfunction, cholinergic dysfunction, inflammation, increased oxidative stress, and mitochondrial dysfunction.

7.4.2 Disruption of Calcium Homeostasis

Neuronal viability and function are widely intracellular Ca^{2+} signaling dependent. Increased cytosolic Ca^{2+} level initiate the neuronal cellular signaling pathways that are involved in neurotransmitter release, gene expression, cellular growth and differentiation, membrane excitability, cell death, and free radical species formation (Berridge et al. 2014). Aβ induced Ca^{2+} homeostasis impairment alters the characteristics of the neurotransmitter receptors, disrupts the membrane integrity, and induces the apoptotic signalling pathway, eventually leading to synaptic degeneration, cell death, and memory loss (Rovira et al. 2002). Aβ increases the Ca^{2+} influx through the activation of voltage-gated Ca^{2+} channels, thereby leading to enhanced postsynaptic responses (Snyder et al. 2005). The Ca^{2+} permeable receptors—the α-amino-3-hydroxy-5-methyl-4-isoxazoleproprionic acid (AMPA) and N-methyl-D-aspartate (NMDA) receptors are also encompassed in the AD pathology, predominantly by causing dynamin-1 degradation and promoting calpain, activity which is crucial for synaptic vesicle formation and functioning (Kelly and Ferreira 2006; Li et al. 2009). Alternatively, it is also reported that the Aβ oligomer downregulates the expression of the NMDA receptor and thereby causes reduced Ca^{2+} influx (Dewachter et al. 2009). Disregarding the mechanism by which Aβ causes disruption of Ca^{2+} homeostasis, the disruption of intracellular Ca^{2+} signaling will eventually result in synaptic breakdown, cell death, and memory loss.

7.4.3 Cholinergic Dysfunction

The deregulation of Ca^{2+} permeable n-acetylcholine receptor (nAChR) channels will lead to the malfunctioning of synaptic integrity. The neurons expressing high levels of nAChRs, in particular those containing the α7 subunit nicotinic receptor, are the most susceptible neurons in the AD brain (D'Andrea and Nagele 2006). Aβ binds with the α7 and α4β2 subunits of nAChRs present in the cortical and hippocampal synaptic membranes, thereby affecting the functioning of nAChRs (Buckingham et al. 2009). α7 and α4β2-nAChRs positively associate themselves with the Aβ accumulating neurons, and α7-nAChR colocalizes with the amyloid plaques. The exponential degeneration of cholinergic neurons and deficit of cholinergic neurotransmission leads to cognitive deterioration in AD patients (Bartus et al. 1982; Wevers et al. 2008).

7.4.4 Disruption of Membrane Integrity

Fluorescence studies have established the fact that Aβ binds strongly and quickly with all the cellular membranes and results in the elevation of the intercellular calcium levels, which in turn plays a significant role in various mechanisms of damage and cell death (Arispe et al. 1993, 2007). Aβ is said to have direct interactions with lipids such as gangliosides, phosphatidylcholine, phosphatidylglycerol, and phosphoinositides of the cellular membrane, forming pores or ionic channels that help in the Ca^{2+}, Na^+, and K^+ influx into the cell (Terzi et al. 1995; McLaurin and Chakrabartty 1996; Avdulov et al. 1997). Some studies have shown the insertion of $A\beta_{25-35}$ into the lipid membranes (Lau et al. 2007). The Aβ fragments are said to form faster and more effective non-specific channels than those of the full peptides (Kawahara and Kuroda 2000).

The composition of the cell membrane also affects Aβ toxicity by affecting its aggregation. The aggregation of $A\beta_{1-42}$ is affected by the varying composition of the cell membrane; however, $A\beta_{25-35}$ is not affected in its aggregation due to the alteration of the composition of the membrane lipids (Lau et al. 2007). The specific blockade of the ionic channels formed by Aβ reduces the calcium influx and neuronal damage significantly (Arispe et al. 2007). It has also been reported that the lack of presenilin 1/2 (PS1/2) also resulted in decreased membrane fluidity (Grimm et al. 2006).

7.4.5 Oxidative Stress

The Aβ peptide is furnished with some strong metal binding sites, especially Cu^{2+}, and also offers seemingly nearer affinity towards the best metallic chelants. The metal binding sites reside in its first 15 amino acid units, which is composed of histidines 6, 13, and 14, and a tyrosine in the 10th position (Kontush et al. 2001). The Cu^{2+} ions bind easily to the nitrogen atoms present in the histidines (Smith et al. 2007).

Aβ has the ability to reduce Cu^{2+} and Fe^{3+} into Cu^{+} and Fe^{2+} respectively. On reacting with molecular oxygen, the reduced metals produce superoxide anion, which then combine with the two hydrogen atoms to form hydrogen peroxide (H_2O_2). H_2O_2 thus formed later reacts with another reduced metal ion to form a hydroxyl radical by means of Fenton's reaction. The radical Aβ procures protons from the nearby lipids or proteins, generating lipid peroxides and carbonyls respectively (Smith et al. 2007). Copper-bound methionine sulfoxide was observed within amyloid plaques of AD, which in turn proved that the metals have a role in Aβ toxicity (Dong et al. 2003). Oxidative stress caused by Aβ can be increased by impairment of cellular iron homeostasis (Wan et al. 2011).

7.4.6 Mitochondrial Dysfunction

The presence of Aβ was spotted in membrane-bound intracellular structures such as the Golgi apparatus, endoplasmic reticulum, lysosomes, endosomes, and mitochondrial cristae and matrix (Wang et al. 2007). It was reported that Aβ is translocated into the mitochondria by using the mitochondrial outer membrane translocase enzyme (Petersen et al. 2008). The interaction of Aβ with the mitochondria causes a decrease in the respiratory states 3 and 4 and also causes a considerable downfall in the activity of cytochrome c oxidase and some other enzymes involved in the Krebs cycle, such as the pyruvate dehydrogenase complex and ketoglutarate dehydrogenase complex (Casley et al. 2002).

Aβ directly inhibits the generation of mitochondrial ATP and the proper functioning of the alpha subunit of ATP synthase (Schmidt et al. 2008). The chronic administration of Aβ in subtoxic doses is said to inhibit the transportation of nuclear proteins to the mitochondria, leading to the generation of reactive oxygen species (ROS) and alterations in the membrane potential (Sirk et al. 2007). Aβ induces the activation of the various enzymes such as nicotinamide adenine dinucleotide phosphate (NADPH) oxidase, xanthine oxidase, and the A2 phospholipases, which play a major role in the mitochondrial dysfunction and ROS production (Wang et al. 2007). The mitochondrial cascade hypothesis, which applies to sporadic AD, has mitochondria at the crest of the cascade.

7.4.7 Neuroinflammation

The activation of immunocompetent cells such as microglia and astrocytes leads to neuroinflammation, which is also induced by means of Aβ. Aβ stimulates the microglial production of

interleukin (IL)-1, IL-6, tumor necrosis factor alpha (TNF-α), macrophage inflammatory protein (MIP)-1α, monocyte chemotactic protein-1 (MCP-1) and superoxide free radicals (Akiyama et al. 2000). The excitotoxins produced by microglial cells are not sufficient enough to cause neuronal cell loss but however can cause cognitive impairment by means of dendritic pruning (Mattson and Barger 1993). Aβ stimulated microglia result in the activation of two parallel mitogen-activated protein (MAP) kinase cascades, the ERKs and p38 MAP kinase, which are directly responsible for the phosphorylation of transcription factors and the activation of pro-inflammatory gene expression (McDonald et al. 1998). Pattern recognition receptors and toll-like receptors are also involved in the Aβ activated microglial immune response (Giovannini et al. 2002; Walter et al. 2007; Salminen et al. 2009). The nuclear enzyme Poly (ADP-ribose) polymerase-1 (PARP) is reported to regulate the microglial activation, which is induced by Aβ. The activity of PARP is essential for the activation of Aβ induced NF-kB, nitric oxide (NO) release, TNF-α release, and neurotoxicity. Aβ also elicits an inflammatory response by the activation of astrocytes. The astrocytes in turn respond by the increased generation of the cytokine interleukin-1 (IL-1), NO, and TNF-α. In primary culture, astrocytes mediate Aβ-induced neurotoxicity and tau phosphorylation (White et al. 2005). Aβ is said to induce increased oxidative stress. The oxidative stress in turn is responsible for the release of cytokines and chemokines, which eventually results in the activation of microglia and astrocytes (Agostinho et al. 2010).

7.4.8 Synaptic Dysfunction

The degree of synaptic loss is directly proportional to the severity of cognitive impairment in AD patients (Selkoe 2008). The synaptic transmission disturbances occur long before the development of the other typical neuropathological lesions of AD (Oddo et al. 2003). Aβ is said to decrease the levels of postsynaptic density protein 95 (PSD-95) and negatively regulates the AMPA (GluR1) and NMDA glutamatergic receptors (Cerpa et al. 2008). In AD, there was a decrease of 25%–30% in the number of cortical synapses and a decrease of about 15%–35% in the number of neuron synapses (Reddy and Beal 2008).

7.5 Tangle Formation

Neurofibrillary lesions are a salient histopathological hallmark of AD. Neurofibrillary lesions are found in the cell bodies and in the apical dendrites as neurofibrillary tangles. In the distal dendrites, they are found as neuropil threads. The weighty proteinaceous component of the neurofibrillary lesions is composed of straight filaments or paired helical filaments. Neurofibrillary tangles are embodied by the bundles of paired helical filaments, which resemble two twisted ribbons with a cross-over spacing of approximately 80 nm and width ranging from 10 to 20 nm. In AD, the tangles formed are distinctive and unique from other neurological disorders. Filaments accumulate in the perikaryon region of the affected cell and have a characteristic paired helical structure. The paired helical fibers are also found in the dendritic neurites of neuritic plaques, yet another important hallmark of AD (Wischik et al. 1985).

Tau, a major microtubule-associated protein (MAP), is the core protein of these filaments. The major function of tau protein is the assembly of tubulin into microtubules and thereby stabilizing their structure (Cleveland et al. 1977). The tau protein gene is located over 100 kb on the long arm of chromosome 17 at band position 17q21 and has 16 exons. In the central nervous system, the alternative splicing of the exons 2, 3, and 10 results in 6 isoforms of tau proteins (τ3L, τ3S, τ3, τ4L, τ4S, and τ4), which are generally located in the neurons, particularly in the axons. The three isoforms lacking exon 10 have three microtubule binding domains while the other three isoforms have four binding domains with the microtubules, thereby making them more stable than the other forms (Buée et al. 2000).

The tau proteins are roughly divided into two regions, namely the projection domain and the microtubule-binding domain. Primary sequence analysis reveals that tau consists of multiple domains that include an acidic N-terminal domain and a central proline-rich domain, which contains basic repeat units responsible for binding to the microtubules, and a carboxyl terminal. The microtubule binding region comprises 3–4 repeats consisting of KXGS motifs based on the splicing of exon 10 (Kolarova et al. 2012). In an optimally active normal brain, tau has about 2–3 moles of phosphates per moles of protein, whereas tau in paired helical filaments and in the AD brain, which is highly hyperphosphorylated, has about 5–9 mole of phosphates per mole of protein (Kopke et al. 1993). There are multiple cellular processes that affect tau aggregation, which include various post-translational modifications (Fontaine et al. 2015).

7.5.1 Phosphorylation

The tau is phosphorylated and dephosphorylated as a regulatory mechanism by controlling the association and the dissociation of tau with the microtubules. Tau dephosphorylation promotes the microtubules' binding; meanwhile the tau dephosphorylation promotes the dissociation of tau from the microtubules by enhancing the production of tau phosphatases (Bramblett et al. 1993; Poppek et al. 2006). The hyperphosphorylation of tau leads to aggregation by means of neutralizing the innate antipolymerizing structural properties of tau protein and neutralizes the basic charges and anti-aggregation nature of tau within the acidic N-terminal inserts, thereby leading to the formation of an inert polymer tau and eventually leading to tau filament formation (Alonso et al. 2001). The abnormal hyperphosphorylation of tau results in neurodegenerative effects through the inhibition of microtubule function and by impairing neuronal axonal transport (Terwel et al. 2002).

The tau protein kinases, which modulate the phosphorylation of tau, belong to the enzyme group termed transferases as they pivot the transfer of a phosphate group from high energy donor molecules such as ATP/GTP to specific substrates. The three classes of tau protein kinases include: (i) Proline directed protein kinases (PDPK); (ii) Non-proline directed protein kinases (Non-PDPK); and (iii) Tyrosine protein kinases (TPK) (Martin et al. 2013). PDPK targets serine and threonine residues prior to the proline residues. This class involves the glycogen synthase kinase-3 (GSK-3), Cyclin-dependent protein kinase-5, and Mitogen-activated protein kinases (MAPKs) such as the Erk1/2, c-Jun N-terminal kinase (JNK), and p38.

The non-PDPK group consists of Tau-tubulin kinase 1 (TTBK1), Tau-tubulin kinase 2 (TTBK2), Casein kinase 1α/1δ/1ε/2, Dual specificity tyrosine phosphorylation regulated kinase 1A/2 (DYRK1A/2), Microtubule affinity-regulating kinases (MARK), Phosphorylase kinase (PhK), PKA, PKB/Akt, PKC, Protein kinase N, and Ca²⁺/calModulin-dependent protein Kinase II (CaMKII). TPK phosphorylates the tau protein at 5 tyrosine residues located at 18, 29, 197, 310, and 394 sites. TPK mainly consist of C-Abelson kinases and the Src family kinase members such as Src, Lymphocyte-specific protein tyrosine kinase (Lck), Spleen tyrosine kinase (syk), and Fyn, which are involved in tau phosphorylation at the tyrosine residue.

7.5.2 Acetylation

Apart from hyperphosphorylation, other pathogenic translational modifications such as acetylation are also responsible for tau aggregation. Tau acetylation is said to impede tau function by causing impaired tau-microtubule interaction and finally causing pathological tau aggregation. Acetylation was reported to be a preliminary stage involved in the accumulation of hyperphosphorylated tau, leading to fibrillary tangle formation. The mechanisms governing tau acetylation are complex in nature. Acetyltransferases p300, pCAF, and CBP are reported to acetylate tau, and deacetylases SIRT1 and HDAC6 are responsible for the deacetylation of tau. Some evidence also suggests that the tau itself is capable of autoacetylation (Min et al. 2010). There are about 23 lysine residues that are susceptible to acetylation, and mass spectrometry analysis has revealed that the primary acetylated residue of tau is K280. Tau that is acetylated at K280 tends to aggregate *in vitro* and is found in numerous tauopathies including AD (Cohen et al. 2011). Histone deacetylase is an enzyme that acts upon these residues, which on inhibition promotes tau clearance, which eventually leads to a conclusion that the enhanced deacetylation of tau within KXGS motifs will lead to the hyperphosphorylation of tau (Cook et al. 2013).

Apart from hyperphosphorylation and acetylation, there are many other post-translational modifications that are responsible for the process of tau aggregation. Some of the vital post-translational modifications include proteolytic cleavage, glycosylation, nitration, sumoylation, glycation, and poly-amination. The proteolytic cleavage of tau is mainly mediated by Caspase 3 and Caspase 6 at the D421 and D348 residues, while tau cleaved at the D421 residue is more prone to get aggregated than the full-length tau and is found in AD patients' brain. Apart from caspases, tau also can be cleaved by means of calpains, thrombin, and cathepsins (Rissman et al. 2004).

7.6 The Role of Acetylcholinesterase (AChE) in the AD

Acetylcholine-mediated neurotransmission is rudimentary for the impeccable function of the nervous system, and any aberrant or gradual impairment of this system would lead to the progressive deterioration of cognitive, autonomic, and neuromuscular functioning and can be also lethal. Initially, the pathogenesis of AD was linked with the deficiency of the acetylcholine neurotransmitter (Perry et al. 1978). The cholinergic hypothesis states that the loss of cholinergic function in the central nervous system contributes to the significant cognitive decline associated with AD and aging (Francis et al. 1999). The cholinergic replacement strategy has been used as a therapeutic approach to address the cognitive loss associated with AD. Extensively used therapeutics comprise muscarinic, nicotinic-cholinergic ligands, and AChE inhibitors (Perry et al. 1999). The acetylcholine dysfunction is not the primary pathological cause for AD but a mere consequence of the disease. An increase in AChE levels around the amyloid plaques and neurofibrillary tangles is a common feature of the neuropathology of AD.

Cholinesterase belongs to the class of serine hydrolases that catalyze the hydrolysis of the neurotransmitter acetylcholine into choline and acetic acid, a reaction which is essential to allow the cholinergic neuron to return to its resting state after activation. There are two forms of cholinesterases found in the vertebrates; they are the AChE and the butyrylcholinesterase. The major function of AChE is the hydrolysis of acetylcholine in the cholinergic synapses.

AChE is present in the nervous tissue, muscles, plasma, and blood cells (Brimijoin 1983). They possess a complex quaternary polymorphic structure. All the molecular forms of AChE are not equally affected in the AD brain (Massoulie et al. 1993). AChE molecular forms can be broadly classified into the asymmetric forms and the globular forms. The asymmetric forms are included in the extracellular matrix by means of a collagenic tail and are localised at the neuromuscular junction. The latter type, the globular forms, exist as monomers, dimers or as tetramers and are secreted as soluble forms and they are bound to the membrane by means of the hydrophobic domain; and are mainly expressed in the central nervous system (Talesa et al. 2001).

The AChE levels are influenced by the Aβ in the regions around the plaques, and the early increase in the AChE expression that takes place around the neurofibrillary tangles is a consequence of the disturbed tau phosphorylation (Perry et al. 1980; Silveyra et al. 2012a). The different localization of the various forms also determines and limits the potential interactions with other protein partners (Inestrosa et al. 1981). Some studies suggest that AChE directly interact with the Aβ in such a manner that increases the deposition of Aβ to form the plaques (Rees et al. 2003). It may play an important role in the pathogenesis of AD. It is said to modulate APP processing and Aβ production (Zhang et al. 2011b). AChE overexpression increases PS1 levels meanwhile AChE knockdown can lead to decreased PS1 in transfected cells (Silveyra et al. 2012b).

AChE inhibitors inhibit the cholinesterase enzyme from breaking down acetylcholine, thereby increasing the level and duration of the neurotransmitter action. Based on their mode of action, AChE inhibitors can be classified into reversible and irreversible inhibitors. The reversible inhibitors are associated with therapeutic applications while the irreversible inhibitors are generally associated with toxic effects. Donepezil, rivastigmine, and galantamine are reversible AChE inhibitors approved for the treatment of mild to moderate AD by the U.S. Food and Drug Administration and the European Medicines Agency. Tacrine was the first AChE inhibitor that was approved for the treatment of AD; however, its usage has been abandoned due to its side effects including hepatotoxicity (Birks et al. 2006; Colovic et al. 2013).

AD is currently an incurable neurodegenerative disorder that is highly prevalent among the elderly. The current existing treatments have been developed based on the various existing hypotheses of AD. There is an absolute necessity to develop promising therapeutics for the prevention as well as for impeding the progression of the disease. The treatments developed for AD can be broadly classified into two: the symptomatic treatments and the disease-modifying therapies. The symptomatic treatments are mainly developed for providing cognitive improvement through the neurotransmitter mechanisms. The disease-modifying therapies or treatments are meant for prevention and delaying the onset of disease, impeding disease progression, and also target the various underlying mechanisms that are involved in the ailment (Cummings et al. 2018).

7.7 Approved Symptomatic Therapies

Three cholinergic inhibitors are approved for the treatment of mild to moderate AD: they are donepezil (Pfizer, New York, USA) in 1996, rivastigmine (Novartis, Basel, Switzerland) in 1998, galantamine (Janssen, Beerse, Belgium) in 2001, tacrine in 1993, and an N-methyl-D-aspartate (NMDA) receptor AD antagonist—memantine—in 2003 (Cummings et al. 2014). There are multiple drugs developed for the treatment of AD, which are in various stages of clinical trials.

7.7.1 Inhibitors and Modulators of β-Secretase

BACE-1 inhibiting drug development is highly challenging, and lately several inhibitors of this kind have entered into clinical trials. Some of the very significant ones are mentioned below: LY2811376 [(S)-4-(2,4-difluoro-5-pyrimidine-5-yl-phenyl)-4-methyl-5,6-dihydro-4 H–[1,3] thiazin-2-ylamine] is the first non-peptidic inhibitor of its kind endorsed with oral bioavailability. However, the clinical development of this drug was discontinued in phase II clinical trials (May et al. 2011). MK8931 (Verubecestat): A novel BACE inhibitor that is strong and highly specific in nature. It was capable of preventing the formation of amyloid plaque deposits (Kennedy et al. 2016). The results of the clinical trial phase III indicated that the drug roughly reduced the amyloid plaques without any cognitive improvement even in the mildly symptomatic participants. Merck suspended its study thereafter in February 2018. LY2886721 exhibited higher potency than that of LY2811376 as it had improved protease selectivity (May et al. 2015). An abnormal liver enzyme elevation was observed in the clinical phase II study and then the drug was suspended from the clinical trials. E2609 interferes with the amyloid cascade even before Aβ generation. It is an orally active drug that exhibited reduction of cerebral Aβ levels in preclinical trials. A dose-dependent reduction of Aβ was observed in the cerebrospinal fluid (CSF) and plasma. No serious adverse effects were observed, and it has completed a clinical trial phase I (Kennedy et al. 2016). Lanabecestat (AZD3293) (LY3314814) is currently in phase III clinical trials. It is a brain-permeable oral inhibitor of BACE1. It reduces $A\beta_{40}$ and $A\beta_{42}$ concentration in the plasma, CSF, and in brain, and it also significantly decreased the soluble N terminal fragment of amyloid β precursor protein (Sims et al. 2017). HPP854 (TTP-854) is in a phase I clinical trial. It is an inhibitor for mild AD and cognitive impairment (Kuruva and Reddy 2017). RG7129 is an orally administered amino-oxazoline non-peptidic inhibitor and the clinical study of this molecule was terminated due to its hepatotoxicity (Jacobsen et al. 2014). CNP520 is an orally administered small molecule that is jointly developed by Novartis and Amgenb and it is currently in clinical trials (Yan 2016). AZ-4217 presented excellent efficiency in the cerebral amyloidosis mouse model (Meakin et al. 2018). AZD3839 (LY3314814) reduced the Aβ CSF level but the study was terminated in late phase I trial due to its interaction with the hERG ion channel (Yan 2016). JNJ-54861911 crossed the blood-brain barrier (BBB) and reduced all four forms of Aβ peptides but the study was discontinued in the later phase I trial due to eye-related damage (Yan 2016).

7.7.2 Inhibitors and Modulators of γ-Secretase

The γ-Secretase complex is very essential for the generation of $A\beta_{(1-40)}$ and $A\beta_{(1-42)}$. One of the important targets of γ-Secretase is the Notch protein, which is responsible for the regulation of cell fate. γ-Secretase is capable of cleaving APP, Notch, and other type I membrane proteins (Imbimbo et al. 2011). BMS-708163 (Avagacestat), an aryl sulfonamide, is a potent and highly selective, orally available γ-Secretase inhibitor developed by Bristol-Mayers Squibb. It is a Notch sparing γ-Secretase inhibitor that is capable of overcoming Notch-related toxicities (Barten and Albright 2008). It directly binds with the PS-I N-terminal fragment. The administration of this drug led to reduced Aβ40 levels for a sustained period in the brain, plasma, and CSF in rats and dogs during a clinical phase I trial (Gillman et al. 2010). The prodromal AD population that received avagacestat showed higher rates of clinical progression towards dementia and brain atrophy (Coric et al. 2015). Begacestat (GSI-953) effectively binds with the adenosine triphosphate terminal of the γ-Secretase complex. It is a selective γ-Secretase inhibitor that has a thiophene containing sulfonamide moiety that inhibits Aβ production and is synthesized by Bristol-Myers Squibb. It is highly selective for the inhibition of APP cleavage. A higher dosage of begacestat significantly reduces $A\beta_{40}$ levels in the brain, CSF, and in plasma, but at a lower dosage it significantly reduces the $A\beta_{40}$ levels in the brain and plasma but not in the CSF. It was also found to be capable of inhibiting the Notch intracellular domain and $A\beta_{42}$ production (Panza et al. 2010; McKee et al. 2013) and was capable of reversing contextual memory deficits. It has completed a phase I clinical trial (Silva et al. 2014). BMS-299897 is the first of its kind entering the clinical trial (Barten et al. 2005). Semagacestat (LY450139) dose-dependently decreased the $A\beta_{40}$ and $A\beta_{42}$ levels in the brain, plasma, and CSF. Later on, the study was discontinued due to transient bowel obstruction and other ill effects such as weight loss, skin cancers, and a higher risk of infections (Dockens et al. 2012; Tong et al. 2012a, 2012b). GSI MK-0752 is in phase II clinical trials, and it does not distinguish between APP and Notch (Rosen et al. 2006). MK-0752 was found to be safe and it considerably inhibited Aβ40 (Hoffman et al. 2015). PF-3084014 is a novel Notch sparing γ-Secretase inhibitor that reduces amyloid β production. It inhibited Notch-related T and B-cell maturation in an *in vitro* study. It also exhibited a dose-dependent reduction of Aβ in the brain, CSF, and plasma of Tg2576 mice and in guinea pigs (Imbimbo 2008; Booth and Kim 2014).

7.7.3 Selective γ-Secretase Modulators

These have been developed in such a way to avoid the adverse effects associated with general enzyme inhibition. The main aim of these treatments is to block the APP processing without interfering with other signaling pathways. CHF5074 is a non-steroidal anti-inflammatory drug devoid of cyclooxygenase (COX) inhibitory activity. It inhibited $A\beta_{42}$ production by blocking γ-Secretase complexes. It is also a microglial modulator as it reduces the amyloid burden and microglial activation. Trial II results showed improvements in several cognitive measures and a reduction in the inflammatory marker levels of CSF (Ross et al. 2013; Ronsisvalle et al. 2014). NIC5-15 (Pinitol) is a naturally occurring cyclic sugar alcohol that modulates γ-Secretase and reduces Aβ production without interfering with the substrate cleavage of Notch. It is said to improve cognitive function and memory in preclinical models of AD. It is a potential therapeutic agent against the treatment of AD as it does not intervene with Notch activity and it is also an insulin sensitizer (Pitt et al. 2013).

7.7.4 Inhibitors of Aβ Peptide Aggregation

3-APS, Alzhemed, interrupts the interaction of soluble Aβ with endogenous glycosaminoglycans; however, this drug was suspended due to the disappointing results of a phase III clinical trial (Aisen et al. 2007). Colostrinin is a proline-rich polypeptide complex that inhibits Aβ aggregation in *in vitro* assays and improves cognitive performance in murine models (Bilikiewicz and Gaus 2004). Scyllo-inositol (ELND005) is an orally administered amyloid anti-aggregation agent capable of reducing Aβ toxicity in the murine hippocampus (Salloway et al. 2011). The 8-HQ compound clioquinol and PBT2 are said to block the interaction between the base metals and brain Aβ peptide. It is also thought that these compounds could effectively prevent Aβ aggregation by restoring homeostasis in the cellular levels of copper and zinc ions. Unfortunately, these molecules failed in phase II and III clinical development (Matlack et al. 2014; Ryan et al. 2015).

7.7.5 Immunotherapy

AN1792 was the first vaccine of its kind tested on patients and it consisted of a 42 amino acid peptide. Later, it was found to cause adverse effects like aseptic meningoencephalitis and hence this drug was suspended from a phase II clinical trial (Gilman et al. 2005). CAD 106 is a second-generation vaccine designed by Novartis. It consisted of shorter peptide fragments of Aβ (1-6). It has completed a phase II clinical trial and has shown impressive results without causing adverse effects (Wiessner et al. 2011). Other vaccines such as ACI-24, MER5101, and AF205 are currently in various stages of preclinical trials (Folch et al. 2016).

Bapineuzumab is a humanized monoclonal antibody developed for the treatment of AD against the Aβ(1-6). A significant reduction in amyloid plaque and phosphorylated tau was reported but no significant cognitive improvements were observed in the patients. However, it has failed in two phase III clinical trials due to the lack of efficacy in patients with mild to the moderate AD (Salloway et al. 2014). Solanezumab—a humanized monoclonal antibody against Aβ (12-28)—improved cognition in mild AD, but statistical significance was not achieved. Currently, it is in phase III trials (Tayeb et al. 2013; Doody et al. 2014).

Gantenerumab is a fully humanized IgG1 monoclonal antibody designed to bind with high affinity to a conformational epitope on the β-amyloid fibers. Its use has been investigated in people who are at the risk of developing presenile AD caused due to genetic mutations (Novakovic et al. 2013; Jacobsen et al. 2014).

Crenezumab is another IgG4 humanized monoclonal antibody. It is currently in a clinical phase II trial aiming to evaluate its safety and efficacy in asymptomatic carriers of E280A, an autosomal-dominant mutation of PSEN1. Ponezumab (PF-04360365) targets the free carboxy terminal amino acids 33-40 of Aβ peptide. MABT5102A, GSK933776A, NI-101, SAR-228810, and BAN-2401 are some other important monoclonal antibodies that are in various phases of clinical development (Jindal et al. 2014).

7.7.6 Inhibitors of Tau Hyperphosphorylation

Tideglusib is an irreversible inhibitor of glycogen synthase kinase 3 beta (GSK3β), which has completed phase II trials, but due to the lack of efficacy, this drug was discontinued (Lovestone et al. 2015). Sodium selenite (VEL015) is a protein phosphatase 2 agonist that is in development. It reduces Tau phosphorylation both in cell culture and in murine models of the disease. On its administration to rodents, it resulted in significant cognitive improvements and a reduction of the neurodegenerative phenotype (Corcoran et al. 2010).

7.7.7 Inhibitors of Tau Aggregation

Rember is a methylene blue derivative that has been shown to stabilize AD progression in clinical trials, and has led to the generation of TRx 0237, which is another version of methylene blue. This compound inhibits tau protein aggregation and also dissolves the brain aggregates of Tau (Hochgrafe et al. 2015).

7.7.8 Microtubule Stabilizers

TPI 287 is a derivative of taxane and stabilizes microtubules by means of binding with tubulin (Fitzgerald et al. 2012).

7.7.9 Anti-Tau Immunotherapy

AADvac-1 consists of a synthetic peptide derived from the tau sequence coupled to keyhole limpet hemocyanin and uses aluminium hydroxide as an adjuvant. It has demonstrated a good preclinical safety profile (Kontesekova et al. 2014).

7.7.10 Fyn Kinase Inhibitors and Inhibitors of SCF/c-Kit Signaling

Saracatinib (AZD0530)—a Fyn kinase inhibitor—is currently in phase II clinical trials for mild to moderate AD. Masitinib (AB1010) is also a Fyn kinase inhibitor, which is currently in phase III clinical trials for mild to moderate AD. A nanomolar range of these compounds is capable of blocking Fyn. Oral masitinib administration along with any one of the AChE inhibitors

and/or memantine resulted in significant improvement in ADAS-Cog test. This compound may prevent neuroinflammation by blocking the activated Mast cells-microglia interactions (Piette et al. 2011; Folch et al. 2015).

7.7.11 5-HT6 Inhibitors

Lu-AE-58054 (SGS-518) and PF-05212365 (SAM-531) are being considered as possible treatments for mild-to-moderate AD. Other compounds that are in various stages of clinical research are SUVN-502, AVN-322, and PRX-07034 (Ramírez et al. 2011).

7.7.12 Antidiabetic Compounds in AD

Thiazolidinediones (TZDs, Rosiglitazone, Pioglitazone, Intranasal Insulin (Humulin R U- 100), Amylin, and pramlintide are some of the antidiabetic compounds that are in various stages of trial for the treatment of AD (Folch et al. 2016).

7.8 Natural Compounds for the Treatment of AD

Recently, the usage of natural products in medicine has gained attention. There are multiple plant-based compounds and their derivatives investigated for the treatment of various neurological disorders and disorders associated with aging. Some of the significant natural products used in the treatment of AD are discussed as below.

Codonopsis pilosula enhances learning and memory in rats (Weon et al. 2014). The extract of plants of this genus exhibit inhibitory effects on acetylcholinesterase (Lin et al. 2008). The extract contains hesperidin and atractylenolide. Hesperidin is an inhibitor of β-Secretase and prevents amyloid fibril formation, and also confers protection against aluminium chloride-induced cognitive dysfunction. It also suppresses inflammation by means of activating the Akt/Nrf2 signaling pathway and by inhibiting RAGE/NF-kB (Qi et al. 2011; Hong and An 2015). *Astragalus propinquus* extract enhanced learning and memory in a murine model of AD (Li et al. 2011). Astragaloside IV, a compound present in this plant, offers protection against amyloid β toxicity by protecting mitochondria and it also protects the BBB (Sun et al. 2014). *Dipsacus asper* contains saponins that safeguard neuronal cells from amyloid-β toxicity (Zhou et al. 2009) and also negotiates the loss of memory in rats. It alters the Akt and NF-kB signaling pathways and offers protection to the BBB (Yu et al. 2012; Yang et al. 2014). *Angelica* species are used in the treatment of AD. They contains several coumarins among which the umbelliferone 6-carboxylic acid and esculetin inhibit AChE and β secretase 1 (Ali et al. 2016). Ligustilide is an anti-inflammatory agent that decreases cortical and hippocampal nerve damage, decreases astrocyte activation, and also protects the BBB (Feng et al. 2012). Ferulic acid is an antioxidant and a free radical scavenger present in this plant species and has been shown to inhibit amyloid fibril formation (Sgarbossa et al. 2015). *Indigofera tinctoria* extract intercepts neuronal death in the hippocampus after intracerebroventricular injection of Aβ into mice (Balamurugan and Muralidharan 2010).

It has an antioxidant property (Kim et al. 2016). Gallic acid, quercitrin, and myricetin are also found in this plant (Bakasso et al. 2008). Gallic acid and catechins decreased amyloid fibril deposition and decreased brain inflammation in an AD murine model (Wang et al. 2009). Myricetin, a flavanoid found in this herb, inhibits β-secretase and is a neuroprotective agent (Shimmyo et al. 2008). *Glycyrrhiza glabra* shields mice from Aβ toxicity (Ahn et al. 2006). 2,2′,4′-Trihydroxychalcone is an active ingredient of this plant and inhibited β-secretase, improved memory, and decreased plaque formation in a murine model (Zhu et al. 2010). Liquiritin is a flavanone glucoside that is a neuroprotectant and modulates the ERK and Akt/GSK-3β signaling pathways (Teng et al. 2014). Glycyrrhizic acid is another neuroprotectant found in this plant, which inhibits oxidative stress and voltage-gated sodium channels (Asl and Hosseinzadeh 2008). *Heteromeles arbutifola* contains betulin, icariside E4, farrerol, and other active ingredients (Wang et al. 2016). Farrerol protects endothelial cells in the BBB (Li et al. 2013). Catechin stimulates the nonamyloidogenic cleavage of APP (Mandel and Youdim et al. 2004). Betulin prevents sterol regulatory element binding protein activation (Tang et al. 2011). *Crocus sativus*, also known as saffron, is compared to memantine and donepezil in the treatment of moderate to severe AD and mild to moderate AD respectively (Farokhnia et al. 2014; Moshiri et al. 2014). Crocin, an anti-inflammatory ingredient, inhibits amyloid fibril formation and crocetin inhibits acetylcholinesterase (Pitsika 2015). It decreases ceramide production and offers protection to the BBB. It also decreases microglial cell activation and inflammatory cytokine production by means of inhibiting the Notch signaling pathway (Wang et al. 2015). *Lycium barbarum* extracts offer protection to cultured neurons from Aβ toxicity (Yu et al. 2005). The polysaccharide compounds in the extracts decrease tau phosphorylation (Ho et al. 2010). The extracts of this plant are said to elicit hypoglycaemia in alloxan-induced diabetic or hyperlipidemic rabbits (Luo et al. 2004). *Panax notoginseng* is traditionally used in Chinese medicine and it contains ginsenoside Rg1, which is used for improving learning and memory function by targeting secretase activity in AD (Wang and Du 2009). Ginsenoside, another ingredient present in this plant, targets neprilysin (Yang et al. 2009). *Ginko biloba* plant extract is also used in traditional Chinese medicine for improving memory loss (Bäurle et al. 2009). *Dipsacus asper* contains Akebia saponin D and it inhibits Aβ toxicity (Zhou et al. 2009).

7.9 Curcumin

Turmeric, a culinary perennial herb, is rooted in Indian and Chinese medicine since the days of yore for a variety of medicinal uses. As an ancestral dietary spice, it is used as a colorant in food and as a home remedy for various ailments. Curcuminoids present in the rhizomes of *Curcuma spp* of the Zingiberaceae family is the color imparting polyphenol group subsuming 3%–6% of the total rhizome content. In the Curcuminoid group, Curcumin (77%), demethoxycurcumin (17%), and bisdemethoxycurcumin (3%) are the major ingredients. Curcumin is ratified to be the potent ingredient in an array of therapeutic uses. During the early 19th century, Vogel and Pelletier discovered curcumin from *Curcuma longa* (Turmeric), and its pure form was later prepared by Vogel in 1842 (Gupta et al. 2012). Milobedzka and Lampe's elucidation of its chemical structure in the year 1910 ensued its first synthesis in 1913. The first clinical trial based on curcumin was presented in the year 1937 and its biological characteristic as an antibacterial (Schraufstatter and Bernt 1949) provided the way for identifying its salubrious properties. In 75 years, almost 11,000 articles have been published related to curcumin. The most commonly identified traits include antioxidant, anti-inflammatory, anti-mutagenic, antimicrobial, and anti-cancer effects. The role of curcumin in AD was first exemplified when Lim et al. (2001) demonstrated the reduction of Aβ protein in an Alzheimer transgenic mouse. The various targets and mechanisms of action exhibited by curcumin in both in vitro and *in vivo* studies are summarized in Table 7.1.

TABLE 7.1

In vitro and *in vivo* Studies of Curcumin and its Analogues

Dose and Duration	Cell Lines/Animals	Targets/Mechanism of Action	References
SAHA (1 mM) + Curcumin (5 μm) added 1 h prior to adding the Aβ$_{25-35}$ fragments (20 mM) for 24 h	Rat pheochromocytoma PC12 cell	Aβ$_{25-35}$-induced neuronal damage (antioxidant pathway)	Meng et al. (2014)
Curcumin (0, 1.25, 5.0, 20.0 μmol/L) for 24 h and at 5.0 μmol/L for 0, 12, 24, and 48 h	SHSY5Y cells	Inhibiting the activity of GSK-3β, accompanied with the inhibition of γ-secretase	Zhang et al. (2011a)
Pretreated with curcumin I (5, 10, and 20 M) for 30 min and exposure to 6-OHDA at 25 uM for 24 h	SH-SY5Y cells	Inhibition of 6-OHDA-induced neurotoxicity	Jaisin et al. (2011)
ApoE3 mediated poly(butyl) cyanoacrylate nanoparticles containing curcumin (ApoE3-C-PBCA) (25, 50, and 100 nM) with Aβ (0.2 μM) for 24 h	SH-SY5Y cells	Aβ-induced cytotoxicity	Mulik et al. (2010)

(Continued)

TABLE 7.1 (*Continued*)

In vitro and *in vivo* Studies of Curcumin and its Analogues

Dose and Duration	Cell Lines/Animals	Targets/Mechanism of Action	References
Curcumin and curcumin I (10 mg/mL) for 24 h Aβ1-42 (10 mg/mL)	Human SK-N SH cells	Upregulation of human telomerase reverse transcriptase (hTERT) against Aβ induced toxicity	Xiao et al. (2014)
Curcumin (5 mM) pretreated for 24 h followed by paraquat (0.5 mM) for 24 h	SH-SY5Y1 cells	Reduction of amyloid precursor protein	Jaroonwitchawan et al. (2016)
Curcumin (0, 5, 10, and 15 μM) for 24 and 48 h	Cell line HT-22	Upregulation of TRPLM1	Zhang et al. (2017)
Curcumin at 0, 1.25, 5.0, 20.0 μmol/L for 24 h and with curcumin at 5.0 μmol/L for 0, and 12, 24, and 48 h	SH-SY5Y cells	Inhibiting GSK3β-mediated PS1 activation	Xiong et al. (2011)
Curcumin at 25 μM for 24 h	Primary cortical Neurons cell culture	Upregulation of co-chaperone BAG2 protein	Patil et al. (2013)
Curcumin 3 complex (0.1 μM) for 24 and 48 h	Differentiated macrophages	Enhanced Aβ uptake by macrophages	Zhang et al. (2006)
Curcumin analogues (0.1–10 nM) for 24 h	Peripheral blood mononuclear cells (PBMCs) human monocytic cell lines, U-937 and THP-1	Increased MGAT3, VDR, and TLR gene expression	Gagliardi et al. (2011)
Curcumin (1 mM) and 100 nM Aβ for 24 h	Hippocampal neuronal cells	Disassemble pre-formed Aβ aggregates	Varghese et al. (2010)
Curcumin and DMC (1–40 μM) for 24 h	Murine neuroblastoma (N2A) cells	CUR inhibiting the APP and Tau IRES activity	Villaflores et al. (2012)
Curcumin micelles (0.1 μM and 10 μM) for 1 h and stressed with sodium nitroprusside (0.5 mM) for 6 h	PC12 cells	Protecting PC12 cells form nitrosative stress	Hagl et al. (2015)
Curcumin-based diarylheptanoid Analogues 16a and 16b (0.3–30 μM) for 5 days	Human U87 MG GBM and (NB) SK-N-SH and SK-N-FI cells	Efficacious against neuroblastoma (SK-N-SH and SK-N-FI) and glioblastoma (U87MG) cell lines	Campos et al. (2013)
Curcumin (25 μM) for 2 h	Murine GLUTag L cell line	Stimulation of GLP-1 secretion in GLUTag cells and a significant increase in Ca MKII pathway	Takikawa et al. (2012)
Manganese complexes of curcumin (0.1–5 μg/mL) for 3 h	NG108-15 cells	Better neuroprotective in cell culture model	Vajragupta et al. (2003)
Pretreated with curcumin (1, 5, and 10 mM) for 4 h followed by 10 mM Aβ for 24 h	SH-SY5Y cells	Curcumin inhibits Aβ-induced Tau hyperphosphorylation involving PTEN/Akt/GSK3β pathway	Huang et al. (2014)
Curcumin-conjugated nanoliposomes (CnLs) (20 μg/mL) and 25 nM Aβ42 for 24 h	hAPPsw SH-SY5Y cells	Downregulation of Aβ peptide secretion in cells overexpressing hAPP	Lazar et al. (2013)
Curcumin-encapsulated biodegradable Poly(lactic-co glycolic acid) (PLGA) nanoparticles (Cur-PLGA-NPs) (0.1, 0.2, 0.5, 5, and 50 μM) for 24 h	Hippocampal neuronal stem cells	Induced adult neurogenesis through activation of the canonical Wnt/β-catenin pathway	Tiwari et al. (2013)
Curcuminoid submicron particle (CSP) (5 μM) co-incubated with oligomericAβ (oAβ) 5 μM for 48 h	SH-SY5Y cells	Protective effect against oligomeric Aβ-induced toxicity	Tai et al. (2018)
5-(4-Hydroxyphenyl)-3-oxo-pentanoic Acid [2-(5-Methoxy-1H-indol-3-yl)-ethyl]-amide 0.1 and 0.3 μM for 48 h	MC 65 cells and HT22 Cells	Inhibits the production of Aβ oligomers	Chojnacki et al. (2014)
Al(III)–curcumin complexes (5, 50, and 200 μL) for 60 h	PC12 cells	Remolded the preformed, mature, ordered Aβ 42 fibrils into the low toxic amorphous aggregates	Jiang et al. (2012)

(Continued)

TABLE 7.1 (*Continued*)

In vitro and *in vivo* Studies of Curcumin and its Analogues

Dose and Duration	Cell Lines/Animals	Targets/Mechanism of Action	References
Nps and poly (lactide-co-glycolide) (PLGA) based-nanoparticulate formulation (Nps-Cur) (5, 10, and 15 μL) for 24 h	SK-N-SH cells	Prevented the induction of the Nrf2	Doggui et al. (2012)
Curcumin derivative (30 μM) with Aβ for 24 h	SH-SY5Y cells	Inhibited of Aβ aggregation	Yanagisawa et al. (2015)
Palmitic Acid Curcumin Ester (0.2 μM) for 2 h pretreated and $A\beta_{40}$ 10 μM for 24 h	SH-SY5Y cells	Neuroprotective effects against Aβ insult by weakening the interaction between Aβ and cell membranes	Qi et al. (2017)
Tacrine-curcumin hybrids (0.1, 1, 5, and 10 μM) containing $A\beta_{42}$ (5 μM) or H_2O_2 (200 μM)	PC12 cells	Cholinesterase inhibitory activity and ion-chelating ability	Liu et al. (2017)
20 μM Cu(II) treated 1 h before followed by Curcumin (20 μM) for 24 h	Primary rat cortical neurons	Reversed Cu(II)-induced neuronal damage by scavenging intracellular ROS	Huang et al. (2011)
Curcumin (1–15 μM), with 50 μM H_2O_2 or 80 μM Fe-NTA for 24 h	Neuroblastoma N2a cells	Prevented tau self-assembly and microglial activation	Morales et al. (2017)
Curcumin-loaded nanoparticles (10 μM) with monomeric Aβ (1, 10, and 100 μM) for 24 h	Primary hippocampal cultures	Significant decrease of Aβ aggregates	Barbara et al. (2017)
Curcumin (Low dose-160 ppm; High dose-5000 ppm) orally for 6 months	Ten-month-old male and female APPSw Tg + and Tg – mice	Lowered the levels of oxidated proteins, IL-1B, soluble and insoluble Aβ, and plaque burden	Lim et al. (2001)
Curcumin 500 ppm, orally for 6 months	Tg2576 transgenic mice (17 month old)	Reduced plaque burden and soluble amyloid levels. Improved cognition deficits and memory impairments	Yang et al. (2005)
Curcumin (50,100, and 200 mg/kg), intra-hippocampal administration for 7 consecutive days	Male Sprague-Dawley rats	Improved cognitive function in a dose-dependent fashion. Upregulated BDNF-ERK signaling in the hippocampus	Zhang et al. (2015)
Curcumin hybrids 3,5 8c 50 mg/kg bw orally for 2 months	Adult female albino rats (16–18 weeks)	Increased brain Ach, GSH, paraoxenase, and BCL2 levels. Decreased brain AchE activity, urinary 8-OHG level, Serum caspase-3, and P53 level	Elmegeed et al. (2015)
Curcumin 80 mg/kg/day, oral administration for 3 months	Adult male Wistar rats	Reduced oxidative stress, ameliorated cognitive deficits	Samy et al. (2015)
Curcumin 0.1 mL/10 g bw/day (100, 200, and 300 mg/kg) orally for 3 months	APPswe/PS1dE9 double transgenic mice	Inhibited presenilin-2 and increased levels of degrading enzymes such as insulin degrading enzyme and neprilysin	Wang et al. (2014b)
Curcumin 0.1 mL/10 g bw/day (100, 200, and 400 mg/kg) orally for 3 months	APPswe/PS1dE9 double transgenic mice	Increased glucose metabolism and ameliorated impaired insulin signaling pathways in the brain. Upregulated (IGF)-1R, IRS-2, P13K, p-P13K, Akt, and p-Akt protein expression and downregulated IR and IRS-1	Wang et al. (2017)
Curcumin 50 mg/kg bw intraperitoneally injected for 2 and 5 days	5xFAD mice	Labeled Aβ plaques	Maiti et al. (2016)
150 mg/kg of Curcumin and 4 mg/kg of GW9662 intraperitoneally injected for 4 consecutive days	(APPswe/PS1Δ9) transgenic mice	Reduced the activation of microglia and astrocytes and cytokine production. Inhibited the NFkB signaling pathway. PPARγ is a potential target of curcumin to alleviate neuroinflammation	Liu et al. (2016)
Curcumin loaded polymersomes 15 mg/kg by caudal vein injection for 14 days	BALB/c mice, nude mice, and C57BL/6 mice	Offered neuroprotection and ameliorates cognitive dysfunction	Jia et al. (2016)
Curcumin −100, 200, and 400 mg/kg/day orally for 6 months	APPswe/PS1dE9	Regulated critical molecules in the brain insulin signaling pathways	Feng et al. (2016)
Curcumin −100, 200, and 400 mg/kg/day orally for 6 months	APPswe/PS1dE9	Regulated the synapse-related proteins PSD95 and Shank1	He et al. (2016)

7.10 Structure

1,6-heptadiene-3,5-dione-1,7-bis (4-hydroxy-3-methoxyphenyl)-(1E, 6E) or its synonym Diferuloylmethane ($C_{21}H_{20}O_6$) is a photosensitive, bright, yellow-colored, symmetrical compound with a molecular weight of 368.38 g/mol. In 1982, Tonneson et al. resolved its crystal structure and it was later determined by Parimita (2007). Curcumin, a hydrophobic molecule, has less solubility in hydrocarbon solvents such as cyclohexane and hexane, whereas it readily dissolves in polar solvents such as dimethylsulfoxide, methanol, and acetonitrile. The molar coefficient is 55,000 dm^3 mol^{-1} cm^{-1} at 425 nm (Lestari 2014). It has three reactive groups: two aromatic ring systems containing o-methoxy phenolic groups, connected by a seven-carbon linker consisting of a αβ unsaturated β diketo moiety. The two symmetrically arranged chromophores of the structural motif C=O–C=C and the conjugated double bond gives curcumin its characteristic color (Milobedzka 1910). Curcumin evinces keto-enol tautomerism depending on the acidity of the solution. It exists as the keto-form in acidic and neutral media, whereas in alkaline conditions it exists in the enol form. The enol form integrity is maintained by resonance-assisted hydrogen bonding (Parimita et al. 2007). The o-methoxyphenol group and methylenic hydrogen of curcumin are culpable for its various sanative attributes. The active hydrogen donor participation attributes to its varied biochemical reactions.

7.11 Stability

Curcumin undergoes rapid chemical degradation that escalates as the pH rises, leading to its low bioavailability (Priyadarshini 2013). Wang et al. (1997) showed that curcumin is unstable in physiological conditions *in vitro*, leading to rapid degradation of about 90% within 3 hours. The chemical breakdown of curcumin takes place through the hydrolysis of the αβ unsaturated β diketo moiety, forming Trans-6-(4'-hydroxy-Y-methoxyphenyl)-2, 4-dioxo-5-hexenal as a major product, and vanillin, ferulic acid, and feruloyl methane are the minor products formed. The degradation is notably downsized when it forms a conjugate with liposomes, lipids, albumins, and other macromolecular and microheterogeneous systems (Priyadarshini 2013). The metabolic products differ from the chemically degraded products. The metabolism takes place through enzymatic pathways such as O-conjugation and reduction involving phase I and phase II enzymes. Upon conjugation, curcumin forms curcumin glucuronide and curcumin sulfate (Phase I metabolism). Sulfonation takes place with the help of phenol sulfur-transferase enzymes. The conjugates get actively carried out of the erythrocytes through the multidrug resistance-related proteins MRP1 and MRP2, accounting for its low bioavailability. In reduction by hepatic reductases (Phase II metabolism), curcumin forms a large number of products, the major ones being tetrahydrocurcumin, hexahydrocurcumin, and octahydrocurcumin (Esatbeyoglu et al. 2012).

7.12 Biological Properties of Curcumin

Curcumin has been extensively hailed as the golden cure since it exhibits various therapeutic properties. Even before the scientific community understood its importance, curcumin has been used to treat various disorders in conventional medicine. Curcumin's multi-therapeutic potential was discovered by Western medicine in the 20th century (Schraufstatter and Bernt 1949) and since then knowledge of its effects has grown exponentially. The biological attributes of curcumin as include anti-inflammatory, antioxidant, anti-angiogenic, and its never-ending potency made way for the rise of curcumin based research (Somparn et al. 2007; Raghunath et al. 2018). Curcumin targets various signaling pathways to alleviate AD (Figure 7.1).

7.12.1 Anti-inflammatory

Inflammation is the physiological response stimulated by the immune system against deleterious situations such as tissue injury and infection. It involves a wide spectrum of physiological and pathological morbidities by altering various signaling pathways, and it increases inflammatory markers and levels of oxidants (Medzhitov 2008). Inflammation can be assorted as acute and chronic based on the duration of exposure. Acute inflammation is the initial stage that happens well within the site of infection or damage. It is caused by bacterial pathogens or damaged tissues. Granulocytes are the major immune cells involved. Resident macrophage signals initiate the inflammatory response and the signals are short-lived based on the exposure stimulus. If the infection or damage is not rectified by the acute response it will initiate the secondary stage- the chronic stage, which involves mononuclear cells and fibroblasts. It will lead to muscle degeneration, fibrosis, and necrosis, and may extend from many months to years. Chronic inflammation leads to oxidative stress and oxidative damage, which consequentially helps in the eradication of damaged and infected cells. The prolonged chronic inflammation leads to disorders such as obesity, diabetes, cardiovascular, neurodegenerative, and metabolic diseases. The causative factor is the low-grade inflammation provoked by the oxygen stress, which raises the level of inflammatory cytokines by activation of nuclear factor kappa-light-chain-enhancer of activated B cells (NF-κB) (Sikora 2010). The prominent pathways that participate in the inflammatory response are the NF-κB and TNF-α pathways. Curcumin has been widely utilized for wound healing as an anti-inflammatory agent since classic medicine. Curcumin suppresses inflammation through multiple pathways. Curcumin has been identified to repress inflammatory cytokines that include TNF-α, IL-1, -2, -6, -8, -12, MAPK, and JNK, as well as suppressing inducible nitric oxide synthase, COX-2, and lipoxygenase (LOX) in cancer cells (Shehzad et al. 2013).

Recently, it has been identified that curcumin inhibits NF-κB activation, matrix metalloproteinase (MMP-1, -9, and -13) secretion, COX-2 expression, and anti-apoptotic protein such as Bcl 2, as well as activating Bax and caspase-3. Curcumin also

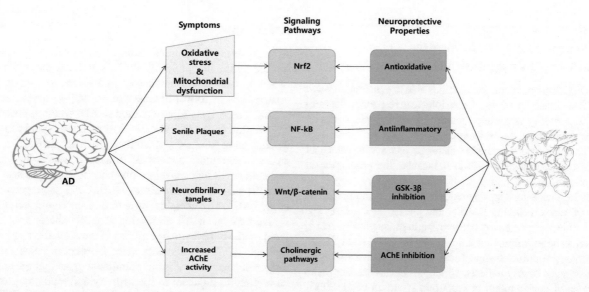

FIGURE 7.1 Targeting signaling pathways involved in AD by curcumin. Curcumin alleviates the symptoms of AD via antioxidative, antiinflammatory, GSK-3β and AChE inhibitory activities.

suppresses IL-1b-induced NF-κB activation and nuclear translocation as well as IL-1b-induced phosphatidylinositol 3-kinase (PI3K/Akt) activation through the reduced phosphorylation and deterioration of inhibitory κ B alpha (IκBα) (Buhrmann 2011). Curcumin nullifies the TNF-α-induced expression of intercellular adhesion molecule-1 and vascular cell adhesion molecule-1, disrupts TNF-a-induced secretion of IL-6 and -8, and MCP-1, and hinders NF-κB activity in endometriotic stromal cells (Kim 2012). Curcumin has also been proclaimed to reduce the expression of inflammatory markers such as NF-κB, COX-2, 5-LOX, MIP-1α, adhesion molecules, C-reactive protein, and chemokine receptor type 4 (Jurenka 2009). The anti-inflammatory mechanism of curcumin has been studied in obese mice in which inflammation in adipose liver steatosis was diminished through phosphorylation of the signal transducer and activator of transcription 3. The suppressor of cytokine signaling 3 and sterol regulatory element-binding protein-1c (SREPB-1c) were also downregulated. Curcumin inhibited the gene expression of mitochondrial DNA, nuclear respiratory factor 1 (NRF1), and mitochondrial transcription factor A, as well as reducing hepatic NFκ-B activities and the levels of thiobarbituric acid reactive substances (Kuo 2012).

7.12.2 Antioxidant

Oxidative stress can be defined as "an imbalance between oxidants and antioxidants in favor of the oxidants, leading to a disruption of redox signaling and control and/or molecular damage" (Sies 2007). Oxidative stress leads to the production of free radicals, which induces damage at both the cellular and molecular level. The oxidative stress leads to various degenerative disorders and has long been associated with aging. The major signaling pathway involved in the oxidative stress defence mechanism is the Nrf2/Keap pathway. The Nrf2/Keap

pathway induces the activation of various antioxidant enzymes, thereby maintaining a homeostatic level. Curcumin is a well-known antioxidant. Curcumin exerts antioxidant properties by acting upon the free radicals, chelating with the metal ions responsible for radical production or inducing the antioxidant enzymes. The biological activity of curcumin is based on its structural components. The presence of two Michael receptors and the phenolic and β-diketone, methoxy groups are known to contribute to its antioxidant scavenging activity. In the keto form, the methylene substituents on the β diketone account for its scavenging activity. Youssef et al. (2004) also showed that the radical-scavenging activity of curcumin increases when methylene and hydroxy groups are present in the para position. In free radical scavenging, curcumin has been shown to attach itself to free radicals, including the oxygen (Das and Das 2002) and superoxide anion radicals, hydroxyl radicals, and 1,1-diphenyl-2-picrylhydrazyl (DPPH), 2,2′-azino-bis(3-ethylbenzothiazoline-6-sulphonic acid) (ABTS), and dimethyl-4-phenylenediamine (DMPD) radicals (Ak and Gulcin 2008). Curcumin has been shown to indirectly reduce nitric oxide by scavenging nitrogen dioxide (Unnikrishnan and Rao 1995). Metal ions are responsible for the production of free radicals by Fenton's reaction. Curcumin chelates with Fe^{3+} (Borsari et al. 2002), Fe^{2+}, and Cu^{2+} (Baum and Ng 2004), and heavy metals such as lead and cadmium are also chelated (Daniel et al. 2004). Curcumin involves the antioxidant Nrf2 pathway due to its Michael's acceptors' reactivity with sulfhydryl groups of Keap, thereby inducing the activation of phase II enzymes. Some of the enzymes include hemeoxygenase (HO1) and paraoxanase1. Curcumin was also shown to depress the expression of xanthine oxidase, superoxide anion, lipid peroxides, and myeloperoxidase, while the status of superoxide dismutase, catalase, glutathione peroxidase, and glutathione-S-transferase activity was positively increased after curcumin administration (Kita et al. 2008).

7.13 Neuroprotective Mechanism of Curcumin against AD

Being a lipophilic molecule, curcumin is able to penetrate through the BBB to induce a series of physiological responses. Its neuroprotective ability is mainly factored by its key properties of being an antioxidant, anti-inflammatory, and metal chelator. AD, a well-known neurodegenerative disorder, has a wide range of mechanisms leading to its cause. It includes the production of senile plaques by amyloid β protein, neurofibrillary tangles by tau protein, metal ions causing oxidative stress in the brain, the defects of genes including BACE1, and presenilin 1 and 2. The primary cause of the neurodegeneration associated with AD is proposed to be inflammation followed by oxidative stress. In an epidemiological point of view, India has been shown to have a lower incidence of AD patients, which has been correlated with the increased consumption of curcumin (Ganguli et al. 2000). Curcumin, being a multifaceted drug, affects almost every factor involved in AD. The factors involved in AD and the protective effect of curcumin against them are described briefly in the following section.

7.13.1 Enzymes

AChE, a serine protease, hydrolyzes the neurotransmitter acetylcholine into acetic acid and choline, playing a major role in cholinergic neurotransmission. In AD, the cholinergic neurons deplete acetylcholine due to increased AChE activity, leading to cognitive impairments (Schliebs and Arendt 2006, 2011). The AChE has also been shown to promote Aβ oligomers to form plaques. Ahmed and Gilani (2009) investigated the effect of curcumin in both *in vitro* and *ex vivo* models and found that it inhibited AChE activity in a dose-dependent manner.

APP is an integral membrane protein prevalently expressed based on the cell's state. The APP is processed by two pathways: the amyloidogenic and non-amyloidogenic pathway. In the amyloidogenic pathway, APP is cleaved by β-site APP cleaving enzyme 1 (BACE1), which is further split into toxic oligomeric forms (Aβ 40/42) by γ secretases. The non-toxic fragments are produced through the non-amyloidogenic pathway in which APP is cleaved by α secretase followed by γ secretase. Curcumin has been shown to reduce the Aβ formation by repressing the genes involved in cleavage, including BACE1, Presenilin 1, and GSK3β, which are present in the gamma-secretase complex. Curcumin was found to reduce the production of Aβ with a simultaneous decrease in presenilin 1 gene expression, suggesting its role in γ-secretase modulation (Xiong et al. 2011).

Silent information regulator 1 (SIRT1), a member of the sirtuin family, is a nicotinamide adenine dinucleotide (NAD)-dependent histone deacetylase. It regulates cellular functions and activities including stress responses, cell apoptosis, and axonal degeneration, and the SIRT1 pathway plays an important role in the neuroprotective properties of curcumin against Aβ toxicity. Curcumin has been shown to activate SIRT1 that had been repressed by amyloid β toxicity in primary cortical neuron cultures (Sun et al. 2014).

7.13.2 Aβ

Aβ plaques are the characteristic feature of AD. Aβ plaques are formed by the insoluble Aβ fibrils, which are the major cause of neurotoxicity. APP is cleaved by the enzymes alpha β and γ secretases to form oligomers of 40 and 42 amino acids in length. The $A\beta_{42}$ is highly related to AD, which induces fibril aggregation due to its increased hydrophobicity and causes the generation of ROS. *In vitro* studies give a glimpse of the mechanism of curcumin against Aβ induced toxicity. In neuronal cell cultures, it induces microglial phagocytosis and heat shock proteins as a countermeasure against Aβ aggregates (Cole et al. 2007). The chemical structure of curcumin helps in the disaggregation of Aβ fibrillation. The Aβ sheets are destabilized by the polar hydroxyl groups in the aromatic rings, which interact with the polar pockets of Aβ peptide. Nuclear magnetic resonance spectroscopic studies suggest that the aromatic carbon atoms adjacent to the hydroxy and methoxy groups of curcumin interact with the carbon atoms of the Aβ peptide (Masuda et al. 2011). The neurotoxic effects are minimized by lowering the insertion rate of Aβ oligomers into the plasma membrane. The calcium influx induced by the Aβ oligomers, which is the major toxic effect, is prevented by curcumin by mediating the induced phosphorylation of the NMDA receptor. The cell death caused by calcium influx is prevented by reduction of Aβ-induced membrane disruption. Curcumin inhibits Aβ generation and induces autophagy by downregulating the PI3K/Akt/mTOR signaling pathway (Wang et al. 2014a). *In vivo* studies also suggest a protective effect of curcumin against Aβ induced neurotoxicity. The toxic effects produced by the intracerebral ventricular infusion of Aβ in rats are repressed by curcumin (Wang et al. 2009). The administration of curcumin resulted in a decrease in plaque formation followed by the reduction in oxidative stress (ROS and reactive nitrogen species) in mice (Begum et al. 2008).

7.13.3 Tau Protein

Tau proteins are responsible for microtubule arrangement, which is mediated by phosphorylation. The hyperphosphorylation of tau protein leads to the entanglement of tau proteins, leading to the formation of neurofibrillary tangles. The phosphorylation of tau is controlled by GSK3β, which is known to be repressed by curcumin, which inhibits the phosphatase and tensin homolog/protein kinase B (Akt)/GSK-3β pathway (Huang et al. 2014). The hyperphosphorylated tau proteins are degraded by the proteasome through BCL2 associated anthanogne 2 (BAG2). Curcumin stabilizes the internal assembly of microtubules by upregulation of the BAG2 gene and repairing the microtubule dissociation (Patil et al. 2013). Curcumin, even at the end stage in tangle-bearing hTau mice, reduces the development of soluble Tau oligomers and Fyn, perhaps through increasing heat shock protein (HSP)70, HSP90, and HSC70, which have the potential to clear misfolded tau (Ma et al. 2013).

7.13.4 Mitochondrial Dysfunction

Mitochondria are the primary site of ATP production and are involved in calcium homeostasis and intrinsic apoptosis.

Mitochondrial dysfunction has been identified in AD patients, leading to aberration of the electron transport system, increased ROS and reactive nitrogen species production, and indirectly leading to Aβ overproduction and deposition (Young et al. 2010). Curcumin has been shown to have protective effects against the mitochondrial dysfunction induced by Aβ toxicity. The harmful effects of energy metabolism are reduced upon administration of curcumin. The intrinsic apoptotic protein expression induced by the dysfunction, including BAX, Caspase-3, and cytochrome *C*, is suppressed by curcumin administration (Huang et al. 2012). Curcumin improved matrix metalloproteinases (MMP) and adenosine triphosphate (ATP) and restored mitochondrial fusion, probably by upregulating nuclear factor peroxisome proliferator-activated receptor gamma coactivator 1-α (PGC1α) protein expression in senescence-accelerated mice-prone 8 (Eckert et al. 2013).

7.13.5 Metal Chelation

Metal ions such as iron and especially copper have been linked with AD. Iron deregulation is associated with amyloid plaque formation, and increased serum copper levels are found in AD patients. Copper, like other metal ions, undergoes the Fenton reaction to produce ROS, thereby increasing oxidative stress which in turn increases Aβ production. Copper and zinc bind to the Aβ peptide forming β-sheet structures by interstrand histidine embrace. The effective binding of curcumin with these metal ions has been demonstrated by spectrophotometry (Baum et al. 2004). Curcumin chelates the metal ions directly and has been shown to chelate copper in the presence of the Aβ peptide (Picciano and Vaden 2013). The reduction of oxidative stress is based on curcumin's dosage. Curcumin in low dosage reduces copper-induced oxidative stress in primary rat cortical neurons, whereas an increased dosage leads to chromosomal aberration and cell damage (Huang et al. 2011).

7.14 Challenges and Future Directions

Curcumin has been shown to inhibit the progression of AD both *in vitro* and *in vivo*. The low bioavailability of curcumin is due to low absorption and rapid hepatic and internal metabolization, resulting in its rapid elimination in bile and urine. The insoluble nature of curcumin in water at acidic and neutral pH reduces its gastrointestinal absorption. The pharmacodynamic properties of curcumin suggest it acts more as a neuroprotective agent than as a treatment drug. The bioavailability of curcumin can be increased by encapsulation in polymeric micelles, liposomes, polymeric nanoparticles, lipid-based nanoparticles, and hydrogels. Various studies have shown an increase in the dispersion or solubilization of the synthesized particles. Curcumin-encapsulated poly(lactic-co-glycolic acid) nanoparticles induced neurogenesis in an Aβ-induced rat model and showed increased bioavailability compared to non-encapsulated curcumin. The prepared formulations have shown increased neuroprotective effects both *in vitro* and *in vivo* but the real challenge lies in human studies, for which until now only a handful of trials have been conducted that failed to demonstrate any significant benefit with curcumin. The obstacle lies in the clinical trials being solely focused on the treatment instead on prevention, along with the fact that the studies have been conducted over a short period of time rather than during the early stages of disease. Recent formulations have shown a promising future in the prevention of AD since the pharmacodynamic properties of curcumin suggest it will act more as a neuroprotective agent than as a treatment drug.

ACKNOWLEDGMENTS

Mr. Azhwar Raghunath acknowledges the UGC-BSR Senior Research Fellowship (UGC-BSR No. F7-25/2007) funded by UGC-BSR, New Delhi, India. The authors thank the Department of Science and Technology, Science and Engineering Research Board (EMR/2014/000600) and Empowerment and Equity Opportunities for Excellence in Science (SB/EMEQ-246/2014). The authors also thank the UGC-SAP DRS II: F-3-30/2013 and DST FIST: SR/FST/LSI-618/2014, New Delhi, India for their partial financial assistance.

REFERENCES

Agostinho, P., Cunha, R.A. and Oliveira, C., 2010. Neuroinflammation, oxidative stress and the pathogenesis of Alzheimer's disease. *Current Pharmaceutical Design*, 16(25), pp. 2766–2778.

Ahmed, T. and Gilani, A.H., 2009. Inhibitory effect of curcuminoids on acetylcholinesterase activity and attenuation of scopolamine-induced amnesia may explain medicinal use of turmeric in Alzheimer's disease. *Pharmacology Biochemistry and Behavior*, 91(4), pp. 554–559.

Ahn, J., Um, M., Choi, W., Kim, S. and Ha, T., 2006. Protective effects of Glycyrrhiza uralensis Fisch. on the cognitive deficits caused by β-amyloid peptide 25–35 in young mice. *Biogerontology*, 7(4), pp. 239–247.

Aisen, P.S., Gauthier, S., Vellas, B., Briand, R., Saumier, D., Laurin, J. and Garceau, D., 2007. Alzhemed: A potential treatment for Alzheimer's disease. *Current Alzheimer Research*, 4(4), pp. 473–478.

Ak, T. and Gülçin, İ., 2008. Antioxidant and radical scavenging properties of curcumin. *Chemico-Biological Interactions*, 174(1), pp. 27–37.

Akiyama, H., Arai, T., Kondo, H., Tanno, E., Haga, C. and Ikeda, K., 2000. Cell mediators of inflammation in the Alzheimer disease brain. *Alzheimer Disease and Associated Disorders*, 14(1), pp. S47–S53.

Ali, M.Y., Jannat, S., Jung, H.A., Choi, R.J., Roy, A. and Choi, J.S., 2016. Anti-Alzheimer's disease potential of coumarins from Angelica decursiva and Artemisia capillaris and structure-activity analysis. *Asian Pacific Journal of Tropical Medicine*, 9(2), pp. 103–111.

Alonso, A.D.C., Zaidi, T., Novak, M., Grundke-Iqbal, I. and Iqbal, K., 2001. Hyperphosphorylation induces self-assembly of τ into tangles of paired helical filaments/straight filaments. *Proceedings of the National Academy of Sciences*, 98(12), pp. 6923–6928.

Alzheimer's Association, 2011. 2011 Alzheimer's disease facts and figures. *Alzheimer's & Dementia: The Journal of the Alzheimer's Association*, 7(2), p. 208.

Arispe, N., Diaz, J.C. and Simakova, O., 2007. Aβ ion channels. Prospects for treating Alzheimer's disease with Aβ channel blockers. *Biochimica et Biophysica Acta (BBA)-Biomembranes*, *1768*(8), pp. 1952–1965.

Arispe, N., Rojas, E. and Pollard, H.B., 1993. Alzheimer disease amyloid beta protein forms calcium channels in bilayer membranes: Blockade by tromethamine and aluminum. *Proceedings of the National Academy of Sciences*, *90*(2), pp. 567–571.

Asl, M.N. and Hosseinzadeh, H., (2008) Review of pharmacological effects of Glycyrrhiza sp. and its bioactive compounds. *Phytotherapy Research*, *22*, pp. 709–724.

Avdulov, N.A., Chochina, S.V., Igbavboa, U., Warden, C.S., Vassiliev, A.V. and Wood, W.G., 1997. Lipid binding to amyloid β-peptide aggregates: Preferential binding of cholesterol as compared with phosphatidylcholine and fatty acids. *Journal of Neurochemistry*, *69*, pp. 1746–1752.

Bakasso, S., Lamien-Meda, A., Lamien, C.E., Kiendrebeogo, M., Millogo, J., Ouedraogo, A.G. and Nacoulma, O.G., 2008. Polyphenol contents and antioxidant activities of five Indigofera species (Fabaceae) from Burkina Faso. *Pakistan Journal of Biological Science*, *11*(11), p. 1429–1135.

Balamurugan, G. and Muralidharan, P., 2010. Effect of Indigofera tinctoria on β-amyloid (25-35) mediated Alzheimer's disease in mice: Relationship to antioxidant activity. *Bangladesh Journal of Pharmacology*, *5*, pp. 51–56.

Barbara, R., Belletti, D., Pederzoli, F., Masoni, M., Keller, J., Ballestrazzi, A., Vandelli, M.A., Tosi, G. and Grabrucker, A.M., 2017. Novel Curcumin loaded nanoparticles engineered for Blood-Brain Barrier crossing and able to disrupt Abeta aggregates. *International Journal of Pharmaceutics*, *526*(1–2), pp. 413–424.

Barten, D.M. and Albright, C.F., 2008. Therapeutic strategies for Alzheimer's disease. *Molecular Neurobiology*, 37, pp. 171–186.

Barten, D.M., Guss, V.L., Corsa, J.A., Loo, A., Hansel, S.B., Zheng, M., Munoz, B., Srinivasan, K., Wang, B., Robertson, B.J. and Polson, C.T., 2005. Dynamics of β-amyloid reductions in brain, cerebrospinal fluid, and plasma of β-amyloid precursor protein transgenic mice treated with a γ-secretase inhibitor. *Journl of Pharmacology and Experimental Therapeutics*, *312*(2), pp. 635–643.

Bartus, R.T., Dean, R.3., Beer, B. and Lippa, A.S., 1982. The cholinergic hypothesis of geriatric memory dysfunction. *Science*, *217*(4558), pp. 408–414.

Baum, L. and Ng, A., 2004. Curcumin interaction with copper and iron suggests one possible mechanism of action in Alzheimer's disease animal models. *Journal of Alzheimer's Disease*, *6*(4), pp. 367–377.

Bäurle, P., Suter, A. and Wormstall, H., 2009. Safety and effectiveness of a traditional ginkgo fresh plant extract–results from a clinical trial. *Complementary Medicine Research*, *16*(3), pp. 156–161.

Begum, A.N., Jones, M.R., Lim, G.P., Morihara, T., Kim, P., Heath, D.D., Rock, C.L., Pruitt, M.A., Yang, F., Hudspeth, B. and Hu, S., 2008. Curcumin structure-function, bioavailability, and efficacy in models of neuroinflammation and Alzheimer's disease. *Journal of Pharmacology and Experimental Therapeutics*, *326*(1), pp. 196–208.

Berridge, M.J., 1998. Neuronal calcium signaling. *Neuron*, *21*(1), pp. 13–26.

Bilikiewicz, A. and Gaus, W., 2004. Colostrinin 1 (a naturally occurring, proline-rich, polypeptide mixture) in the treatment of Alzheimer's disease. *Journal of Alzheimer's Disease*, *6*(1), pp. 17–26.

Birks, J., 2006. Cholinesterase inhibitors for Alzheimer's disease. *The Cochrane Database of Systematic Reviews*, *1*(1), p. CD005593.

Blessed, G., Tomlinson, B.E. and Roth, M., 1968. The association between quantitative measures of dementia and of senile change in the cerebral grey matter of elderly subjects. *The British Journal of Psychiatry*, *114*(512), pp. 797–811.

Booth, R. and Kim, H., 2014. Permeability analysis of neuroactive drugs through a dynamic microfluidic in vitro blood–brain barrier model. *Annals of Biomedical Engineering*, *42*(12), pp. 2379–2391.

Borsari, M., Ferrari, E., Grandi, R. and Saladini, M., 2002. Curcuminoids as potential new iron-chelating agents: Spectroscopic, polarographic and potentiometric study on their Fe (III) complexing ability. *Inorganica Chimica Acta*, *328*(1), pp. 61–68.

Bramblett, G.T., Goedert, M., Jakes, R., Merrick, S.E., Trojanowski, J.Q. and Lee, V.M., 1993. Abnormal tau phosphorylation at Ser396 in Alzheimer's disease recapitulates development and contributes to reduced microtubule binding. *Neuron*, *10*(6), pp. 1089–1099.

Brimijoin, S., 1983. Molecular forms of acetylcholinesterase in brain, nerve and muscle: Nature, localization and dynamics. *Progress in Neurobiology*, *21*(4), pp. 291–322.

Brookmeyer, R., Johnson, E., Ziegler-Graham, K. and Arrighi, H.M., 2007. Forecasting the global burden of Alzheimer's disease. *Alzheimer's & Dementia*, *3*(3), pp. 186–191.

Buckingham, S.D., Jones, A.K., Brown, L.A. and Sattelle, D.B., 2009. Nicotinic acetylcholine receptor signalling: Roles in Alzheimer's disease and amyloid neuroprotection. *Pharmacological Reviews*, *61*, pp. 39–61.

Buée, L., Bussiere, T., Buée-Scherrer, V., Delacourte, A. and Hof, P.R., 2000. Tau protein isoforms, phosphorylation and role in neurodegenerative disorders1. *Brain Research Reviews*, *33*(1), pp. 95–130.

Buhrmann, C., Mobasheri, A., Busch, F., Aldinger, C., Stahlmann, R., Montaseri, A. and Shakibaei, M., 2011. Curcumin modulates NF-κB-mediated inflammation in human tenocytes in vitro: Role of the phosphatidylinositol 3-kinase-Akt pathway. *Journal of Biological Chemistry*, pp. jbc-M111.

Campos, C.A., Gianino, J.B., Bailey, B.J., Baluyut, M.E., Wiek, C., Hanenberg, H., Shannon, H.E., Pollok, K.E. and Ashfeld, B.L., 2013. Design, synthesis, and evaluation of curcumin-derived arylheptanoids for glioblastoma and neuroblastoma cytotoxicity. *Bioorganic and Medicinal Chemistry Letters*, *23*(24), pp. 6874–6878.

Cashman, J.R., Gagliardi, S., Lanier, M., Ghirmai, S., Abel, K.J. and Fiala, M., 2012. Curcumins promote monocytic gene expression related to β-amyloid and superoxide dismutase clearance. *Neurodegenerative Diseases*, *10*(1–4), pp. 274–276.

Casley, C.S., Canevari, L., Land, J.M., Clark, J.B. and Sharpe, M.A., 2002. β-Amyloid inhibits integrated mitochondrial respiration and key enzyme activities. *Journal of Neurochemistry*, *80*(1), pp. 91–100.

Cerpa, W., Dinamarca, M.C. and Inestrosa, N.C., 2008. Structure-function implications in Alzheimer's disease: Effect of Aβ oligomers at central synapses. *Current Alzheimer Research*, *5*(3), pp. 233–243.

Chojnacki, J.E., Liu, K., Yan, X., Toldo, S., Selden, T., Estrada, M., Rodríguez-Franco, M.I., Halquist, M.S., Ye, D. and Zhang, S., 2014. Discovery of 5-(4-hydroxyphenyl)-3-oxo-pentanoic acid [2-(5-methoxy-1H-indol-3-yl)-ethyl]-amide as a neuroprotectant for Alzheimer's disease by hybridization of curcumin and melatonin. *ACS Chemical Neuroscience, 5*(8), pp. 690–699.

Cleveland, D.W., Hwo, S.Y. and Kirschner, M.W., 1977. Purification of tau, a microtubule-associated protein that induces assembly of microtubules from purified tubulin. *Journal of Molecular Biology, 116*(2), pp. 207–225.

Cohen, T.J., Guo, J.L., Hurtado, D.E., Kwong, L.K., Mills, I.P., Trojanowski, J.Q. and Lee, V.M., 2011. The acetylation of tau inhibits its function and promotes pathological tau aggregation. *Nature Communications, 2*, p. 252.

Cole, G.M., Teter, B. and Frautschy, S.A., 2007. Neuroprotective effects of curcumin. In *The Molecular Targets and Therapeutic Uses of Curcumin in Health and Disease* (pp. 197–212). Springer, Boston, MA.

Colovic, M.B., Krstic, D.Z., Lazarevic-Pasti, T.D., Bondzic, A.M. and Vasic, V.M., 2013. Acetylcholinesterase inhibitors: Pharmacology and toxicology. *Current Neuropharmacology, 11*(3), pp. 315–335.

Cook, C., Carlomagno, Y., Gendron, T.F., Dunmore, J., Scheffel, K., Stetler, C., Davis, M. et al., 2013. Acetylation of the KXGS motifs in tau is a critical determinant in modulation of tau aggregation and clearance. *Human Molecular Genetics, 23*(1), pp. 104–116.

Corcoran, N.M., Martin, D., Hutter-Paier, B., Windisch, M., Nguyen, T., Nheu, L., Sundstrom, L.E., Costello, A.J. and Hovens, C.M., 2010. Sodium selenate specifically activates PP2A phosphatase, dephosphorylates tau and reverses memory deficits in an Alzheimer's disease model. *Journal of Clinical Neuroscience, 17*(8), pp. 1025–1033.

Coric, V., Salloway, S., van Dyck, C.H., Dubois, B., Andreasen, N., Brody, M., Curtis, C. et al., 2015. Targeting prodromal Alzheimer disease with avagacestat: A randomized clinical trial. *JAMA Neurology, 72*(11), pp. 1324–1333.

Cummings, J., Lee, G., Ritter, A. and Zhong, K., 2018. Alzheimer's disease drug development pipeline: 2018. *Alzheimer's & Dementia: Translational Research & Clinical Interventions, 4*, pp. 195–214.

Cummings, J.L. and Benson, D.F., 1992. *Dementia: A Clinical Approach*. Boston, MA: Heinemann-Butterworths.

Cummings, J.L., Morstorf, T. and Zhong, K., 2014. Alzheimer's disease drug-development pipeline: Few candidates, frequent failures. *Alzheimer's Research and Therapy, 6*(4), p. 37.

D'Andrea, M.R. and Nagele, R.G., 2006. Targeting the alpha 7 nicotinic acetylcholine receptor to reduce amyloid accumulation in Alzheimer's disease pyramidal neurons. *Current Pharmaceutical Design, 12*(6), pp. 677–684.

Daniel, S., Limson, J.L., Dairam, A., Watkins, G.M. and Daya, S., 2004. Through metal binding, curcumin protects against lead-and cadmium-induced lipid peroxidation in rat brain homogenates and against lead-induced tissue damage in rat brain. *Journal of Inorganic Biochemistry, 98*(2), pp. 266–275.

Das, K.C. and Das, C.K., 2002. Curcumin (diferuloylmethane), a singlet oxygen (1O2) quencher. *Biochemical and Biophysical Research Communications, 295*(1), pp. 62–66.

Dewachter, I., Filipkowski, R.K., Priller, C., Ris, L., Neyton, J., Croes, S., Terwel, D. et al., 2009. Deregulation of NMDA-receptor function and down-stream signaling in APP [V717I] transgenic mice. *Neurobiology of Aging, 30*(2), pp. 241–256.

Dockens, R., Wang, J.S., Castaneda, L., Sverdlov, O., Huang, S.P., Slemmon, R., Gu, H. et al., 2012. A placebo-controlled, multiple ascending dose study to evaluate the safety, pharmacokinetics and pharmacodynamics of avagacestat (BMS-708163) in healthy young and elderly subjects. *Clinical Pharmacokinetics, 51*(10), pp. 681–693.

Doggui, S., Sahni, J.K., Arseneault, M., Dao, L. and Ramassamy, C., 2012. Neuronal uptake and neuroprotective effect of curcumin-loaded PLGA nanoparticles on the human SK-N-SH cell line. *Journal of Alzheimer's Disease, 30*(2), pp. 377–392.

Dong, J., Atwood, C.S., Anderson, V.E., Siedlak, S.L., Smith, M.A., Perry, G. and Carey, P.R., 2003. Metal binding and oxidation of amyloid-β within isolated senile plaque cores: Raman microscopic evidence. *Biochemistry, 42*(10), pp. 2768–2773.

Doody, R.S., Thomas, R.G., Farlow, M., Iwatsubo, T., Vellas, B., Joffe, S., Kieburtz, K. et al., 2014. Phase 3 trials of solanezumab for mild-to-moderate Alzheimer's disease. *New England Journal of Medicine, 370*(4), pp. 311–321.

Eckert, G.P., Schiborr, C., Hagl, S., Abdel-Kader, R., Müller, W.E., Rimbach, G. and Frank, J., 2013. Curcumin prevents mitochondrial dysfunction in the brain of the senescence-accelerated mouse-prone 8. *Neurochemistry International, 62*(5), pp. 595–602.

El-Desouki, R.A.K.M., 2014. New insights on Alzheimer's disease. *Journal of Microscopy and Ultrastructure, 2*(2), pp. 57–66.

Elmegeed, G.A., Ahmed, H.H., Hashash, M.A., Abd-Elhalim, M.M. and El-Kady, D.S., 2015. Synthesis of novel steroidal curcumin derivatives as anti-Alzheimer's disease candidates: Evidences-based on in vivo study. *Steroids, 101*, pp. 78–89.

Esatbeyoglu, T., Huebbe, P., Ernst, I.M., Chin, D., Wagner, A.E. and Rimbach, G., 2012. Curcumin—From molecule to biological function. *Angewandte Chemie International Edition, 51*(22), pp. 5308–5332.

Farokhnia, M., Shafiee Sabet, M., Iranpour, N., Gougol, A., Yekehtaz, H., Alimardani, R., Farsad, F., Kamalipour, M. and Akhondzadeh, S., 2014. Comparing the efficacy and safety of Crocus sativus L. with memantine in patients with moderate to severe Alzheimer's disease: A double-blind randomized clinical trial. *Human Psychopharmacology: Clinical and Experimental, 29*(4), pp. 351–359.

Feng, Z., Lu, Y., Wu, X., Zhao, P., Li, J., Peng, B., Qian, Z. and Zhu, L., 2012. Ligustilide alleviates brain damage and improves cognitive function in rats of chronic cerebral hypoperfusion. *Journal of Ethnopharmacology, 144*(2), pp. 313–321.

Feng, H.L., Dang, H.Z., Fan, H., Chen, X.P., Rao, Y.X., Ren, Y., Yang, J.D., Shi, J., Wang, P.W. and Tian, J.Z., 2016. Curcumin ameliorates insulin signalling pathway in brain of Alzheimer's disease transgenic mice. *International Journal of Immunopathology and Pharmacology, 29*(4), pp. 734–741.

Ferri, C.P., Prince, M., Brayne, C., Brodaty, H., Fratiglioni, L., Ganguli, M., Hall, K., Hasegawa, K., Hendrie, H., Huang, Y. and Jorm, A., 2005. Global prevalence of dementia: A Delphi consensus study. *The Lancet, 366*(9503), pp. 2112–2117.

Fitzgerald, D.P., Emerson, D.L., Qian, Y., Anwar, T., Liewehr, D.J., Steinberg, S.M., Silberman, S., Palmieri, D. and Steeg, P.S., 2012. TPI-287, a new taxane family member, reduces the brain metastatic colonization of breast cancer cells. *Molecular Cancer Therapeutics, 11*, pp. 959–967.

Folch, J., Petrov, D., Ettcheto, M., Pedros, I., Abad, S., Beas-Zarate, C., Lazarowski, A., Marin, M., Olloquequi, J., Auladell, C. and Camins, A., 2015. Masitinib for the treatment of mild to moderate Alzheimer's disease. *Expert Review of Neurotherapeutics, 15*(6), pp. 587–596.

Folch, J., Petrov, D., Ettcheto, M., Abad, S., Sánchez-López, E., García, M.L., Olloquequi, J., Beas-Zarate, C., Auladell, C. and Camins, A., 2016. Current research therapeutic strategies for Alzheimer's disease treatment. *Neural Plasticity, 2016.*

Fontaine, S.N., Sabbagh, J.J., Baker, J., Martinez-Licha, C.R., Darling, A. and Dickey, C.A., 2015. Cellular factors modulating the mechanism of tau protein aggregation. *Cellular and Molecular Life Sciences, 72*(10), pp. 1863–1879.

Francis, P.T., Palmer, A.M., Snape, M. and Wilcock, G.K., 1999. The cholinergic hypothesis of Alzheimer's disease: A review of progress. *Journal of Neurology, Neurosurgery & Psychiatry, 66*(2), pp. 137–147.

Gagliardi, S., Ghirmai, S., Abel, K.J., Lanier, M., Gardai, S.J., Lee, C. and Cashman, J.R., 2011. Evaluation in vitro of synthetic curcumins as agents promoting monocytic gene expression related to β-amyloid clearance. *Chemical Research in Toxicology, 25*(1), pp. 101–112.

Ganguli, M., Chandra, V., Kamboh, M.I., Johnston, J.M., Dodge, H.H., Thelma, B.K., Juyal, R.C., Pandav, R., Belle, S.H. and DeKosky, S.T., 2000. Apolipoprotein E polymorphism and Alzheimer disease: The Indo-US cross-national dementia study. *Archives of Neurology, 57*(6), pp. 824–830.

Gillman, K.W., Starrett Jr, J.E., Parker, M.F., Xie, K., Bronson, J.J., Marcin, L.R., McElhone, K.E. et al., 2010. Discovery and evaluation of BMS-708163, a potent, selective and orally bioavailable γ-secretase inhibitor. *ACS Medicinal Chemistry Letters, 1*(3), pp. 120–124.

Gilman, S., Koller, M., Black, R.S., Jenkins, L., Griffith, S.G., Fox, N.C., Eisner, L., Kirby, L., Rovira, M.B., Forette, F. and Orgogozo, J.M., 2005. Clinical effects of Aβ immunization (AN1792) in patients with AD in an interrupted trial. *Neurology, 64*(9), pp. 1553–1562.

Giovannini, M.G., Scali, C., Prosperi, C., Bellucci, A., Vannucchi, M.G., Rosi, S., Pepeu, G. and Casamenti, F., 2002. β-Amyloid-induced inflammation and cholinergic hypofunction in the rat brain in vivo: Involvement of the p38MAPK pathway. *Neurobiology of Disease, 11*(2), pp. 257–274.

Glenner, G.G. and Wong, C.W., 1984. Alzheimer's disease and Down's syndrome: Sharing of a unique cerebrovascular amyloid fibril protein. *Biochemical and Biophysical Research Communications, 122*(3), pp. 1131–1135.

Goozee, K.G., Shah, T.M., Sohrabi, H.R., Rainey-Smith, S.R., Brown, B., Verdile, G. and Martins, R.N., 2016. Examining the potential clinical value of curcumin in the prevention and diagnosis of Alzheimer's disease. *British Journal of Nutrition, 115*(3), pp. 449–465.

Grimm, M.O.W., Tschäpe, J.A., Grimm, H.S., Zinser, E.G. and Hartmann, T., 2006. Altered membrane fluidity and lipid raft composition in presenilin-deficient cells. *Acta Neurologica Scandinavica, 114*, pp. 27–32.

Gupta, S.C., Patchva, S., Koh, W. and Aggarwal, B.B., 2012. Discovery of curcumin, a component of golden spice, and its miraculous biological activities. *Clinical and Experimental Pharmacology and Physiology, 39*(3), pp. 283–299.

Haass, C., Schlossmacher, M.G., Hung, A.Y., Vigo-Pelfrey, C., Mellon, A., Ostaszewski, B.L., Lieberburg, I., Koo, E.H., Schenk, D., Teplow, D.B. and Selkoe, D.J., 1992. Amyloid β-peptide is produced by cultured cells during normal metabolism. *Nature, 359*(6393), p. 322.

Hagl, S., Kocher, A., Schiborr, C., Kolesova, N., Frank, J. and Eckert, G.P., 2015. Curcumin micelles improve mitochondrial function in neuronal PC12 cells and brains of NMRI mice–Impact on bioavailability. *Neurochemistry International, 89*, pp. 234–242.

Hardy, J., 2005. Expression of normal sequence pathogenic proteins for neurodegenerative disease contributes to disease risk: 'permissive templating'as a general mechanism underlying neurodegeneration. *Biochemical Society Transactions, 33*, pp. 578–581.

Hardy, J.A. and Higgins, G.A., 1992. Alzheimer's disease: The amyloid cascade hypothesis. *Science, 256*(5054), p. 184.

He, Y., Wang, P., Wei, P., Feng, H., Ren, Y., Yang, J., Rao, Y., Shi, J. and Tian, J., 2016. Effects of curcumin on synapses in APPswe/PS1dE9 mice. *International Journal of Immunopathology and Pharmacology, 29*(2), pp. 217–225.

Hebert, L.E., Scherr, P.A., Bienias, J.L., Bennett, D.A. and Evans, D.A., 2003. Alzheimer disease in the US population: Prevalence estimates using the 2000 census. *Archives of Neurology, 60*(8), pp. 1119–1122.

Ho, Y.S., Yu, M.S., Yang, X.F., So, K.F., Yuen, W.H. and Chang, R.C.C., 2010. Neuroprotective effects of polysaccharides from wolfberry, the fruits of Lycium barbarum, against homocysteine-induced toxicity in rat cortical neurons. *Journal of Alzheimer's Disease, 19*(3), pp. 813–827.

Hochgräfe, K., Sydow, A., Matenia, D., Cadinu, D., Könen, S., Petrova, O., Pickhardt, M., Goll, P., Morellini, F., Mandelkow, E. and Mandelkow, E.M., 2015. Preventive methylene blue treatment preserves cognition in mice expressing full-length pro-aggregant human Tau. *Acta Neuropathologica Communications, 3*(1), p. 1.

Hoffman, L.M., Fouladi, M., Olson, J., Daryani, V.M., Stewart, C.F., Wetmore, C., Kocak, M. et al., 2015. Phase I trial of weekly MK-0752 in children with refractory central nervous system malignancies: A pediatric brain tumor consortium study. *Child's Nervous System, 31*(8), pp. 1283–1289.

Hong, Y. and An, Z., 2015. Hesperidin attenuates learning and memory deficits in APP/PS1 mice through activation of Akt/Nrf2 signaling and inhibition of RAGE/NF-κB signaling. *Archives of Pharmacal Research*, pp. 1–9.

Hornykiewicz, O., 2002. Dopamine miracle: From brain homogenate to dopamine replacement. *Movement Disorders: Official Journal of the Movement Disorder Society, 17*(3), pp. 501–508.

Huang, H.C., Lin, C.J., Liu, W.J., Jiang, R.R. and Jiang, Z.F., 2011. Dual effects of curcumin on neuronal oxidative stress in the presence of Cu (II). *Food and Chemical Toxicology, 49*(7), pp. 1578–1583.

Huang, H.C., Tang, D., Xu, K. and Jiang, Z.F., 2014. Curcumin attenuates amyloid-β-induced tau hyperphosphorylation in human neuroblastoma SH-SY5Y cells involving PTEN/Akt/GSK-3β signaling pathway. *Journal of Receptors and Signal Transduction, 34*(1), pp. 26–37.

Huang, H.C., Xu, K. and Jiang, Z.F., 2012. Curcumin-mediated neuroprotection against amyloid-β-induced mitochondrial dysfunction involves the inhibition of GSK-3β. *Journal of Alzheimer's Disease, 32*(4), pp. 981–996.

Imbimbo, B.P., 2008. Alzheimer's disease: γ-secretase inhibitors. *Drug Discovery Today: Therapeutic Strategies, 5*(3), pp. 169–175.

Imbimbo, B.P. and Giardina, G.A., 2011. γ-secretase inhibitors and modulators for the treatment of Alzheimer's disease: Disappointments and hopes. *Current Topics in Medicinal Chemistry, 11*(12), pp. 1555–1570.

Inestrosa, N.C., Reiness, C.G., Reichardt, L.F. and Hall, Z.W., 1981. Cellular localization of the molecular forms of acetylcholinesterase in rat pheochromocytoma PC12 cells treated with nerve growth factor. *Journal of Neuroscience, 1*(11), pp. 1260–1267.

Ireson, C.R., Jones, D.J., Orr, S., Coughtrie, M.W., Boocock, D.J., Williams, M.L., Farmer, P.B., Steward, W.P. and Gescher, A.J., 2002. Metabolism of the cancer chemopreventive agent curcumin in human and rat intestine. *Cancer Epidemiology and Prevention Biomarkers, 11*(1), pp. 105–111.

Iwatsubo, T., Odaka, A., Suzuki, N., Mizusawa, H., Nukina, N. and Ihara, Y., 1994. Visualization of Aβ42 (43) and Aβ40 in senile plaques with end-specific Aβ monoclonals: Evidence that an initially deposited species is Aβ42 (43). *Neuron, 13*(1), pp. 45–53.

Jacobsen, H., Ozmen, L., Caruso, A., Narquizian, R., Hilpert, H., Jacobsen, B., Terwel, D., Tanghe, A. and Bohrmann, B., 2014. Combined treatment with a BACE inhibitor and anti-Aβ antibody gantenerumab enhances amyloid reduction in APPLondon mice. *Journal of Neuroscience, 34*(35), pp. 11621–11630.

Jaisin, Y., Thampithak, A., Meesarapee, B., Ratanachamnong, P., Suksamrarn, A., Phivthong-ngam, L., Phumala-Morales, N., Chongthammakun, S., Govitrapong, P. and Sanvarinda, Y., 2011. Curcumin I protects the dopaminergic cell line SH-SY5Y from 6-hydroxydopamine-induced neurotoxicity through attenuation of p53-mediated apoptosis. *Neuroscience Letters, 489*(3), pp. 192–196.

Jaroonwitchawan, T., Chaicharoenaudomrung, N., Namkaew, J. and Noisa, P., 2017. Curcumin attenuates paraquat-induced cell death in human neuroblastoma cells through modulating oxidative stress and autophagy. *Neuroscience Letters, 636*, pp. 40–47.

Jarrett, J.T. and Lansbury Jr, P.T., 1993. Seeding "one-dimensional crystallization" of amyloid: A pathogenic mechanism in Alzheimer's disease and scrapie? *Cell, 73*(6), pp. 1055–1058.

Jia, T., Sun, Z., Lu, Y., Gao, J., Zou, H., Xie, F., Zhang, G. et al., 2016. A dual brain-targeting curcumin-loaded polymersomes ameliorated cognitive dysfunction in intrahippocampal amyloid-β1–42-injected mice. *International Journal of Nanomedicine, 11*, p. 3765.

Jiang, T., Zhi, X.L., Zhang, Y.H., Pan, L.F. and Zhou, P., 2012. Inhibitory effect of curcumin on the Al (III)-induced Aβ42 aggregation and neurotoxicity in vitro. *Biochimica et Biophysica Acta (BBA)-Molecular Basis of Disease, 1822*(8), pp. 1207–1215.

Jindal, H., Bhatt, B., Sk, S. and Singh Malik, J., 2014. Alzheimer disease immunotherapeutics: Then and now. *Human Vaccines and Immunotherapeutics, 10*(9), pp. 2741–2743.

Jurenka, J.S., 2009. Anti-inflammatory properties of curcumin, a major constituent of Curcuma longa: A review of preclinical and clinical research. *Alternative Medicine Review, 14*(2).

Karlawish, J., Jack Jr, C.R., Rocca, W.A., Snyder, H.M. and Carrillo, M.C., 2017. Alzheimer's disease: The next frontier—Special Report 2017. *Alzheimer's & Dementia, 13*(4), pp. 374–380.

Katzman, R., 1976. The prevalence and malignancy of Alzheimer disease: A major killer. *Archives of Neurology, 33*(4), pp. 217–218.

Kawahara, M. and Kuroda, Y., 2000. Molecular mechanism of neurodegeneration induced by Alzheimer's β-amyloid protein: Channel formation and disruption of calcium homeostasis. *Brain Research Bulletin, 53*(4), pp. 389–397.

Kelly, B.L. and Ferreira, A., 2006. β-amyloid-induced dynamin 1 degradation is mediated by N-methyl-D-aspartate receptors in hippocampal neurons. *Journal of Biological Chemistry, 281*(38), pp. 28079–28089.

Kennedy, M.E., Stamford, A.W., Chen, X., Cox, K., Cumming, J.N., Dockendorf, M.F., Egan, M., Ereshefsky, L., Hodgson, R.A., Hyde, L.A. and Jhee, S., 2016. The BACE1 inhibitor verubecestat (MK-8931) reduces CNS β-amyloid in animal models and in Alzheimer's disease patients. *Science Translational Medicine, 8*(363), pp. 363ra150–363ra150.

Khachaturian, Z.S., 2006. Diagnosis of Alzheimer's disease: Two-decades of progress. *Journal of Alzheimer's Disease, 9*(s3), pp. 409–415.

Kim, J.B., Kopalli, S.R. and Koppula, S., 2016. Indigofera tinctoria Linn (Fabaceae) attenuates cognitive and behavioral deficits in scopolamine-induced amnesic mice. *Tropical Journal of Pharmaceutical Research, 15*(4), pp. 773–779.

Kim, K.H., Lee, E.N., Park, J.K., Lee, J.R., Kim, J.H., Choi, H.J., Kim, B.S., Lee, H.W., Lee, K.S. and Yoon, S., 2012. Curcumin attenuates TNF-α-induced expression of intercellular adhesion molecule-1, vascular cell adhesion molecule-1 and proinflammatory cytokines in human endometriotic stromal cells. *Phytotherapy Research, 26*(7), pp. 1037–1047.

Kita, T., Imai, S., Sawada, H., Kumagai, H. and Seto, H., 2008. The biosynthetic pathway of curcuminoid in turmeric (Curcuma longa) as revealed by 13C-labeled precursors. *Bioscience, Biotechnology, and Biochemistry, 72*(7), pp. 1789–1798.

Kolarova, M., García-Sierra, F., Bartos, A., Ricny, J. and Ripova, D., 2012. Structure and pathology of tau protein in Alzheimer disease. *International Journal of Alzheimer's Disease, 2012*.

Kontsekova, E., Zilka, N., Kovacech, B., Novak, P. and Novak, M., 2014. First-in-man tau vaccine targeting structural determinants essential for pathological tau–tau interaction reduces tau oligomerisation and neurofibrillary degeneration in an Alzheimer's disease model. *Alzheimer's Research and Therapy, 6*(4), p. 44.

Kontush, A., Berndt, C., Weber, W., Akopyan, V., Arlt, S., Schippling, S. and Beisiegel, U., 2001. Amyloid-β is an antioxidant for lipoproteins in cerebrospinal fluid and plasma. *Free Radical Biology and Medicine, 30*(1), pp. 119–128.

Köpke, E., Tung, Y.C., Shaikh, S., Alonso, A.D.C., Iqbal, K. and Grundke-Iqbal, I., 1993. Microtubule-associated protein tau. Abnormal phosphorylation of a non-paired helical filament pool in Alzheimer disease. *Journal of Biological Chemistry, 268*(32), pp. 24374–24384.

Kuo, J.J., Chang, H.H., Tsai, T.H. and Lee, T.Y., 2012. Positive effect of curcumin on inflammation and mitochondrial dysfunction in obese mice with liver steatosis. *International Journal of Molecular Medicine, 30*(3), pp. 673–679.

Kuruva, C.S. and Reddy, P.H., 2017. Amyloid beta modulators and neuroprotection in Alzheimer's disease: A critical appraisal. *Drug Discovery Today, 22*(2), pp. 223–233.

Lau, T.L., Gehman, J.D., Wade, J.D., Perez, K., Masters, C.L., Barnham, K.J. and Separovic, F., 2007. Membrane interactions and the effect of metal ions of the amyloidogenic fragment Aβ (25–35) in comparison to Aβ (1–42). *Biochimica et Biophysica Acta (BBA)-Biomembranes*, *1768*(10), pp. 2400–2408.

Lazar, A.N., Mourtas, S., Youssef, I., Parizot, C., Dauphin, A., Delatour, B., Antimisiaris, S.G. and Duyckaerts, C., 2013. Curcumin-conjugated nanoliposomes with high affinity for Aβ deposits: Possible applications to Alzheimer disease. *Nanomedicine: Nanotechnology, Biology and Medicine*, *9*(5), pp. 712–721.

Lestari, M.L. and Indrayanto, G., 2014. Curcumin. In *Profiles of Drug Substances, Excipients and Related Methodology* (Vol. 39, pp. 113–204). Academic Press.

Li, J.K., Ge, R., Tang, L. and Li, Q.S., 2013. Protective effects of farrerol against hydrogen-peroxide-induced apoptosis in human endothelium-derived EA. hy926 cells. *Canadian Journal of Physiology and Pharmacology*, *91*(9), pp. 733–740.

Li, S., Hong, S., Shepardson, N.E., Walsh, D.M., Shankar, G.M. and Selkoe, D., 2009. Soluble oligomers of amyloid β protein facilitate hippocampal long-term depression by disrupting neuronal glutamate uptake. *Neuron*, *62*(6), pp. 788–801.

Li, W.Z., Li, W.P., Zhang, W., Yin, Y.Y., Sun, X.X., Zhou, S.S., Xu, X.Q. and Tao, C.R., 2011. Protective effect of extract of astragalus on learning and memory impairments and neurons apoptosis induced by glucocorticoids in 12-month male mice. *The Anatomical Record: Advances in Integrative Anatomy and Evolutionary Biology*, *294*(6), pp. 1003–1014.

Lim, G.P., Chu, T., Yang, F., Beech, W., Frautschy, S.A. and Cole, G.M., 2001. The curry spice curcumin reduces oxidative damage and amyloid pathology in an Alzheimer transgenic mouse. *Journal of Neuroscience*, *21*(21), pp. 8370–8377.

Lin, H.Q., Ho, M.T., Lau, L.S., Wong, K.K., Shaw, P.C. and Wan, D.C., 2008. Anti-acetylcholinesterase activities of traditional Chinese medicine for treating Alzheimer's disease. *Chemico-Biological Interactions*, *175*(1–3), pp. 352–354.

Liu, Z., Fang, L., Zhang, H., Gou, S. and Chen, L., 2017. Design, synthesis and biological evaluation of multifunctional tacrine-curcumin hybrids as new cholinesterase inhibitors with metal ions-chelating and neuroprotective property. *Bioorganic and Medicinal Chemistry*, *25*(8), pp. 2387–2398.

Liu, Z.J., Li, Z.H., Liu, L., Tang, W.X., Wang, Y., Dong, M.R. and Xiao, C., 2016. Curcumin attenuates beta-amyloid-induced neuroinflammation via activation of peroxisome proliferator-activated receptor-gamma function in a rat model of Alzheimer's disease. *Frontiers in Pharmacology*, *7*, p. 261.

Lovestone, S., Boada, M., Dubois, B., Hüll, M., Rinne, J.O., Huppertz, H.J., Calero, M., Andrés, M.V., Gómez-Carrillo, B., León, T. and del Ser, T., 2015. A phase II trial of tideglusib in Alzheimer's disease. *Journal of Alzheimer's Disease*, *45*(1), pp. 75–88.

Luo, Q., Cai, Y., Yan, J., Sun, M. and Corke, H., 2004. Hypoglycemic and hypolipidemic effects and antioxidant activity of fruit extracts from Lycium barbarum. *Life Sciences*, *76*(2), pp. 137–149.

Lyketsos, C.G., Carrillo, M.C., Ryan, J.M., Khachaturian, A.S., Trzepacz, P., Amatniek, J., Cedarbaum, J., Brashear, R. and Miller, D.S., 2011. Neuropsychiatric symptoms in Alzheimer's disease. *Alzheimer's and Dementia*, *7*, pp. 532–539.

Ma, Q.L., Zuo, X., Yang, F., Ubeda, O.J., Gant, D.J., Alaverdyan, M., Teng, E. et al., 2013. Curcumin suppresses soluble tau dimers and corrects molecular chaperone, synaptic, and behavioral deficits in aged human tau transgenic mice. *Journal of Biological Chemistry*, *288*(6), pp. 4056–4065.

MacManus, A., Ramsden, M., Murray, M., Henderson, Z., Pearson, H.A. and Campbell, V.A., 2000. enhancement of 45ca2+ influx and voltage-dependent ca2+ channel activity by β-amyloid-(1–40) in rat cortical synaptosomes and cultured cortical neurons modulation by the proinflammatory cytokine interleukin-1β. *Journal of Biological Chemistry*, *275*(7), pp. 4713–4718.

Maiti, P., Hall, T.C., Paladugu, L., Kolli, N., Learman, C., Rossignol, J. and Dunbar, G.L., 2016. A comparative study of dietary curcumin, nanocurcumin, and other classical amyloid-binding dyes for labeling and imaging of amyloid plaques in brain tissue of 5x-familial Alzheimer's disease mice. *Histochemistry and Cell Biology*, *146*(5), pp. 609–625.

Mandel, S. and Youdim, M.B., 2004. Catechin polyphenols: Neurodegeneration and neuroprotection in neurodegenerative diseases. *Free Radical Biology and Medicine*, *37*(3), pp. 304–317.

Massoulié, J., Pezzementi, L., Bon, S., Krejci, E. and Vallette, F.M., 1993. Molecular and cellular biology of cholinesterases. *Progress in Neurobiology*, *41*(1), pp. 31–91.

Masuda, Y., Fukuchi, M., Yatagawa, T., Tada, M., Takeda, K., Irie, K., Akagi, K.I., Monobe, Y., Imazawa, T. and Takegoshi, K., 2011. Solid-state NMR analysis of interaction sites of curcumin and 42-residue amyloid β-protein fibrils. *Bioorganic and Medicinal Chemistry*, *19*(20), pp. 5967–5974.

Matlack, K.E., Tardiff, D.F., Narayan, P., Hamamichi, S., Caldwell, K.A., Caldwell, G.A. and Lindquist, S., 2014. Clioquinol promotes the degradation of metal-dependent amyloid-β (Aβ) oligomers to restore endocytosis and ameliorate Aβ toxicity. *Proceedings of the National Academy of Sciences*, p. 201402228.

Mattson, M.P. and Barger, S.W., 1993. Roles for calcium signaling in structural plasticity and pathology in the hippocampal system. *Hippocampus*, *3*(S1), pp. 73–87.

May, P.C., Dean, R.A., Lowe, S.L., Martenyi, F., Sheehan, S.M., Boggs, L.N., Monk, S.A. et al., 2011. Robust central reduction of amyloid-β in humans with an orally available, nonpeptidic β-secretase inhibitor. *Journal of Neuroscience*, *31*(46), pp. 16507–16516.

May, P.C., Willis, B.A., Lowe, S.L., Dean, R.A., Monk, S.A., Cocke, P.J., Audia, J.E. et al., 2015. The potent BACE1 inhibitor LY2886721 elicits robust central Aβ pharmacodynamic responses in mice, dogs, and humans. *Journal of Neuroscience*, *35*(3), pp. 1199–1210.

Mayadevi, M., Sherin, D.R., Keerthi, V.S., Rajasekharan, K.N. and Omkumar, R.V., 2012. Curcumin is an inhibitor of calcium/calmodulin dependent protein kinase II. *Bioorganic and Medicinal Chemistry*, *20*(20), pp. 6040–6047.

McDonald, D.R., Bamberger, M.E., Combs, C.K. and Landreth, G.E., 1998. β-Amyloid fibrils activate parallel mitogen-activated protein kinase pathways in microglia and THP1 monocytes. *Journal of Neuroscience*, *18*(12), pp. 4451–4460.

McKee, T.D., Loureiro, R.M., Dumin, J.A., Zarayskiy, V. and Tate, B., 2013. An improved cell-based method for determining the γ-secretase enzyme activity against both Notch and APP substrates. *Journal of Neuroscience Methods*, *213*(1), pp. 14–21.

McLaurin, J. and Chakrabartty, A., 1996. Membrane disruption by alzheimer β-amyloid peptides mediated through specific binding to either phospholipids or gangliosides implications for neurotoxicity. *Journal of Biological Chemistry, 271*(43), pp. 26482–26489.

Meakin, P.J., Jalicy, S.M., Montagut, G., Allsop, D.J., Cavellini, D.L., Irvine, S.W., McGinley, C. et al., 2018. Bace1-dependent amyloid processing regulates hypothalamic leptin sensitivity in obese mice. *Scientific Reports, 8*(1), p. 55.

Medzhitov, R., 2008. Origin and physiological roles of inflammation. *Nature, 454*(7203), p. 428.

Meng, J., Li, Y., Camarillo, C., Yao, Y., Zhang, Y., Xu, C. and Jiang, L., 2014. The anti-tumor histone deacetylase inhibitor SAHA and the natural flavonoid curcumin exhibit synergistic neuroprotection against amyloid-beta toxicity. *PLoS One, 9*(1), p. e85570.

Min, S.W., Cho, S.H., Zhou, Y., Schroeder, S., Haroutunian, V., Seeley, W.W., Huang, E.J. et al., 2010. Acetylation of tau inhibits its degradation and contributes to tauopathy. *Neuron, 67*(6), pp. 953–966.

Morales, I., Cerda-Troncoso, C., Andrade, V. and Maccioni, R.B., 2017. The natural product curcumin as a potential coadjuvant in Alzheimer's treatment. *Journal of Alzheimer's Disease, 60*(2), pp. 451–460.

Moshiri, M., Vahabzadeh, M. and Hosseinzadeh, H., 2015. Clinical applications of saffron (Crocus sativus) and its constituents: A review. *Drug Research, 65*(6), pp. 287–295.

Mulik, R.S., Mönkkönen, J., Juvonen, R.O., Mahadik, K.R. and Paradkar, A.R., 2010. Transferrin mediated solid lipid nanoparticles containing curcumin: Enhanced in vitro anticancer activity by induction of apoptosis. *International Journal of Pharmaceutics, 398*(1–2), pp. 190–203.

Novakovic, D., Feligioni, M., Scaccianoce, S., Caruso, A., Piccinin, S., Schepisi, C., Errico, F., Mercuri, N.B., Nicoletti, F. and Nisticò, R., 2013. Profile of gantenerumab and its potential in the treatment of Alzheimer's disease. *Drug Design, Development and Therapy, 7*, p. 1359.

Oddo, S., Caccamo, A., Shepherd, J.D., Murphy, M.P., Golde, T.E., Kayed, R., Metherate, R., Mattson, M.P., Akbari, Y. and LaFerla, F.M., 2003. Triple-transgenic model of Alzheimer's disease with plaques and tangles: Intracellular Aβ and synaptic dysfunction. *Neuron, 39*(3), pp. 409–421.

Oppenheimer, A., 1937. Turmeric (curcumin) in biliary diseases. *The Lancet, 229*(5924), pp. 619–621.

Panza, F., Frisardi, V., Imbimbo, B.P., Capurso, C., Logroscino, G., Sancarlo, D., Seripa, D., Vendemiale, G., Pilotto, A. and Solfrizzi, V., 2010. γ-Secretase inhibitors for the treatment of Alzheimer's disease: The current state. *CNS neuroscience and Therapeutics, 16*(5), pp. 272–284.

Parimita, S.P., Ramshankar, Y.V., Suresh, S. and Guru Row, T.N., 2007. Redetermination of curcumin:(1E, 4Z, 6E)-5-hydroxy-1, 7-bis (4-hydroxy-3-methoxyphenyl) hepta-1, 4, 6-trien-3-one. *Acta Crystallographica Section E: Structure Reports Online, 63*(2), pp. o860–o862.

Patil, S.P., Tran, N., Geekiyanage, H., Liu, L. and Chan, C., 2013. Curcumin-induced upregulation of the anti-tau cochaperone BAG2 in primary rat cortical neurons. *Neuroscience Letters, 554*, pp. 121–125.

Perry, E., Walker, M., Grace, J. and Perry, R., 1999. Acetylcholine in mind: A neurotransmitter correlate of consciousness? *Trends in Neurosciences, 22*(6), pp. 273–280.

Perry, E.K., Perry, R.H., Tomlinson, B.E., Blessed, G. and Gibson, P.H., 1980. Coenzyme A-acetylating enzymes in Alzheimer's disease: Possible cholinergic 'compartment'of pyruvate dehydrogenase. *Neuroscience Letters, 18*(1), pp. 105–110.

Perry, E.K., Tomlinson, B.E., Blessed, G., Bergmann, K., Gibson, P.H. and Perry, R.H., 1978. Correlation of cholinergic abnormalities with senile plaques and mental test scores in senile dementia. *British Medical Journal, 2*(6150), pp. 1457–1459.

Petersen, C.A.H., Alikhani, N., Behbahani, H., Wiehager, B., Pavlov, P.F., Alafuzoff, I., Leinonen, V. et al., 2008. The amyloid β-peptide is imported into mitochondria via the TOM import machinery and localized to mitochondrial cristae. *Proceedings of the National Academy of Sciences, 105*(35), pp. 13145–13150.

Picciano, A.L. and Vaden, T.D., 2013. Complexation between Cu (II) and curcumin in the presence of two different segments of amyloid β. *Biophysical Chemistry, 184*, pp. 62–67.

Piette, F., Belmin, J., Vincent, H., Schmidt, N., Pariel, S., Verny, M., Marquis, C. et al., 2011. Masitinib as an adjunct therapy for mild-to-moderate Alzheimer's disease: A randomised, placebo-controlled phase 2 trial. *Alzheimer's Research and Therapy, 3*(2), p. 16.

Pitsikas, N., 2015. The effect of Crocus sativus L. and its constituents on memory: Basic studies and clinical applications. *Evidence-Based Complementary and Alternative Medicine, 2015*.

Pitt, J., Thorner, M., Brautigan, D., Larner, J. and Klein, W.L., 2013. Protection against the synaptic targeting and toxicity of Alzheimer's-associated Aβ oligomers by insulin mimetic chiro-inositols. *The FASEB Journal, 27*(1), pp. 199–207.

Poppek, D., Keck, S., Ermak, G., Jung, T., Stolzing, A., Ullrich, O., Davies, K.J. and Grune, T., 2006. Phosphorylation inhibits turnover of the tau protein by the proteasome: Influence of RCAN1 and oxidative stress. *Biochemical Journal, 400*(3), pp. 511–520.

Price, J.L. and Morris, J.C., 1999. Tangles and plaques in nondemented aging and "preclinical" Alzheimer's disease. *Annals of Neurology: Official Journal of the American Neurological Association and the Child Neurology Society, 45*(3), pp. 358–368.

Prince, M., Comas-Herrera, A., Knapp, M., Guerchet, M. and Karagiannidou, M., 2016. World Alzheimer Report 2016: Improving healthcare for people living with dementia: Coverage, quality and costs now and in the future. Alzheimer's Disease International (ADI), London, UK.

Qi, H.Y., Wang, R., Liu, Y. and Shi, Y.P., 2011. Studies on the chemical constituents of Codonopsis pilosula. *Zhong yao cai= Zhongyaocai= Journal of Chinese Medicinal Materials, 34*(4), pp. 546–548.

Qi, Z., Wu, M., Fu, Y., Huang, T., Wang, T., Sun, Y., Feng, Z. and Li, C., 2017. Palmitic acid curcumin ester facilitates protection of neuroblastoma against oligomeric aβ40 insult. *Cellular Physiology and Biochemistry, 44*(2), pp. 618–633.

Raghunath, A., Sundarraj, K., Kanagaraj, V.V. and Perumal, E., 2018. Plant sources as potential therapeutics for Alzheimer's disease. *Medicinal Plants: Promising Future for Health and New Drugs*, p. 161.

Ramírez, M.J., 2013. 5-HT(6) receptors and Alzheimer's disease. *Alzheimer's Research and Therapy, 5*, p. 15.

Reddy, P.H. and Beal, M.F., 2008. Amyloid beta, mitochondrial dysfunction and synaptic damage: Implications for cognitive decline in aging and Alzheimer's disease. *Trends in Molecular Medicine, 14*(2), pp. 45–53.

Rees, T., Hammond, P.I., Soreq, H., Younkin, S. and Brimijoin, S., 2003. Acetylcholinesterase promotes beta-amyloid plaques in cerebral cortex. *Neurobiology of Aging, 24*(6), pp. 777–787.

Rissman, R.A., Poon, W.W., Blurton-Jones, M., Oddo, S., Torp, R., Vitek, M.P., LaFerla, F.M., Rohn, T.T. and Cotman, C.W., 2004. Caspase-cleavage of tau is an early event in Alzheimer disease tangle pathology. *The Journal of Clinical Investigation, 114*(1), pp. 121–130.

Ronsisvalle, N., Di Benedetto, G., Parenti, C., Amoroso, S., Bernardini, R. and Cantarella, G., 2014. CHF5074 protects SH-SY5Y human neuronal-like cells from amyloidbeta 25-35 and tumor necrosis factor related apoptosis inducing ligand toxicity in vitro. *Current Alzheimer Research, 11*(7), pp. 714–724.

Rosen, L.B., Stone, J.A., Plump, A., Yuan, J., Harrison, T., Flynn, M., Dallob, A. et al., 2006. O4-03-02: The gamma secretase inhibitor MK-0752 acutely and significantly reduces CSF Abeta40 concentrations in humans. *Alzheimer's & Dementia: The Journal of the Alzheimer's Association, 2*(3), p. S79.

Rosenberg, P.B., Nowrangi, M.A. and Lyketsos, C.G., 2015. Neuropsychiatric symptoms in Alzheimer's disease: What might be associated brain circuits? *Molecular Aspects of Medicine, 43*, pp. 25–37.

Ross, J., Sharma, S., Winston, J., Nunez, M., Bottini, G., Franceschi, M., Scarpini, E. et al., 2013. CHF5074 reduces biomarkers of neuroinflammation in patients with mild cognitive impairment: A 12-week, double-blind, placebo-controlled study. *Current Alzheimer Research, 10*(7), pp. 742–753.

Rossor, M.N., Emson, P.C., Mountjoy, C.Q., Roth, M. and Iversen, L.L., 1980. Reduced amounts of immunoreactive somatostatin in the temporal cortex in senile dementia of Alzheimer type. *Neuroscience Letters, 20*(3), pp. 373–377.

Rovira, C., Arbez, N. and Mariani, J., 2002. Aβ (25–35) and Aβ (1–40) act on different calcium channels in CA1 hippocampal neurons. *Biochemical and Biophysical Research Communications, 296*(5), pp. 1317–1321.

Ryan, T.M., Roberts, B.R., McColl, G., Hare, D.J., Doble, P.A., Li, Q.X., Lind, M. et al., 2015. Stabilization of nontoxic Aβ-oligomers: Insights into the mechanism of action of hydroxy-quinolines in Alzheimer's disease. *Journal of Neuroscience, 35*(7), pp. 2871–2884.

Salloway, S., Sperling, R., Fox, N.C., Blennow, K., Klunk, W., Raskind, M., Sabbagh, M. et al., 2014. Two phase 3 trials of bapineuzumab in mild-to-moderate Alzheimer's disease. *New England Journal of Medicine, 370*(4), pp. 322–333.

Salloway, S., Sperling, R., Keren, R., Porsteinsson, A.P., Van Dyck, C.H., Tariot, P.N., Gilman, S. et al., 2011. A phase 2 randomized trial of ELND005, scyllo-inositol, in mild to moderate Alzheimer disease. *Neurology*, pp. WNL-0b013e3182309fa5.

Salminen, A., Ojala, J., Kauppinen, A., Kaarniranta, K. and Suuronen, T., 2009. Inflammation in Alzheimer's disease: Amyloid-β oligomers trigger innate immunity defence via pattern recognition receptors. *Progress in Neurobiology, 87*(3), pp. 181–194.

Samy, D.M., Ismail, C.A., Nassra, R.A., Zeitoun, T.M. and Nomair, A.M., 2016. Downstream modulation of extrinsic apoptotic pathway in streptozotocin-induced Alzheimer's dementia in rats: Erythropoietin versus curcumin. *European Journal of Pharmacology, 770*, pp. 52–60.

Schliebs, R. and Arendt, T., 2006. The significance of the cholinergic system in the brain during aging and in Alzheimer's disease. *Journal of Neural Transmission, 113*(11), pp. 1625–1644.

Schliebs, R. and Arendt, T., 2011. The cholinergic system in aging and neuronal degeneration. *Behavioural Brain Research, 221*(2), pp. 555–563.

Schmidt, C., Lepsverdize, E., Chi, S.L., Das, A.M., Pizzo, S.V., Dityatev, A. and Schachner, M., 2008. Amyloid precursor protein and amyloid β-peptide bind to ATP synthase and regulate its activity at the surface of neural cells. *Molecular Psychiatry, 13*(10), p. 953.

Schraufstätter, E. and Bernt, H., 1949. Antibacterial action of curcumin and related compounds. *Nature, 164*(4167), p. 456.

Selkoe, D.J., 2008. Soluble oligomers of the amyloid β-protein: Impair synaptic plasticity and behavior. In *Synaptic Plasticity and the Mechanism of Alzheimer's Disease* (pp. 89–102). Springer, Berlin, Germany.

Seubert, P., Vigo-Pelfrey, C., Esch, F., Lee, M., Dovey, H., Davis, D., Sinha, S. et al., 1992. Isolation and quantification of soluble Alzheimer's β-peptide from biological fluids. *Nature, 359*(6393), p. 325.

Shehzad, A., Rehman, G. and Lee, Y.S., 2013. Curcumin in inflammatory diseases. *Biofactors, 39*, pp. 69–77.

Shimmyo, Y., Kihara, T., Akaike, A., Niidome, T. and Sugimoto, H., 2008. Multifunction of myricetin on Aβ: Neuroprotection via a conformational change of Aβ and reduction of Aβ via the interference of secretases. *Journal of Neuroscience Research, 86*(2), pp. 368–377.

Sies, H., 2007. Total antioxidant capacity: Appraisal of a concept. *Journal of Nutrition, 37*(6), pp. 1493–1495.

Sikora, E., Scapagnini, G. and Barbagallo, M., 2010. Curcumin, inflammation, ageing and age-related diseases. *Immunity and Ageing, 7*(1), p. 1.

Silva, T., Reis, J., Teixeira, J. and Borges, F., 2014. Alzheimer's disease, enzyme targets and drug discovery struggles: From natural products to drug prototypes. *Ageing Research Reviews, 15*, pp. 116–145.

Silveyra, M.X., García-Ayllón, M.S., de Barreda, E.G., Small, D.H., Martínez, S., Avila, J. and Sáez-Valero, J., 2012a. Altered expression of brain acetylcholinesterase in FTDP-17 human tau transgenic mice. *Neurobiology of Aging, 33*(3), pp. 624–e23.

Silveyra, M.X., García-Ayllón, M.S., Serra-Basante, C., Mazzoni, V., García-Gutierrez, M.S., Manzanares, J., Culvenor, J.G. and Sáez-Valero, J., 2012b. Changes in acetylcholinesterase expression are associated with altered presenilin-1 levels. *Neurobiology of Aging, 33*(3), pp. 627–e27.

Sims, J.R., Selzler, K.J., Downing, A.M., Willis, B.A., Aluise, C.D., Zimmer, J., Bragg, S. et al., 2017. Development review of the bace1 inhibitor lanabecestat (AZD3293/LY3314814). *The Journal of Prevention of Alzheimer's Disease, 4*(4), pp. 247–254.

Sirk, D., Zhu, Z., Wadia, J.S., Shulyakova, N., Phan, N., Fong, J. and Mills, L.R., 2007. Chronic exposure to sub-lethal beta-amyloid (Aβ) inhibits the import of nuclear-encoded proteins to mitochondria in differentiated PC12 cells. *Journal of Neurochemistry, 103*(5), pp. 1989–2003.

Smith, D.G., Cappai, R. and Barnham, K.J., 2007. The redox chemistry of the Alzheimer's disease amyloid β peptide. *Biochimica et Biophysica Acta (BBA)-Biomembranes, 1768*(8), pp. 1976–1990.

Snyder, E.M., Nong, Y., Almeida, C.G., Paul, S., Moran, T., Choi, E.Y., Nairn, A.C. et al., 2005. Regulation of NMDA receptor trafficking by amyloid-β. *Nature Neuroscience*, 8(8), p. 1051.

Somparn, P., Phisalaphong, C., Nakornchai, S., Unchern, S. and Morales, N.P., 2007. Comparative antioxidant activities of curcumin and its demethoxy and hydrogenated derivatives. *Biological and Pharmaceutical Bulletin*, 30(1), pp. 74–78.

Sun, Q., Jia, N., Wang, W., Jin, H., Xu, J. and Hu, H., 2014. Activation of SIRT1 by curcumin blocks the neurotoxicity of amyloid-β 25–35 in rat cortical neurons. *Biochemical and Biophysical Research Communications*, 448(1), pp. 89–94.

Tai, Y.H., Lin, Y.Y., Wang, K.C., Chang, C.L., Chen, R.Y., Wu, C.C. and Cheng, I.H., 2018. Curcuminoid submicron particle ameliorates cognitive deficits and decreases amyloid pathology in Alzheimer's disease mouse model. *Oncotarget*, 9(12), p. 10681.

Takikawa, M., Kurimoto, Y. and Tsuda, T., 2013. Curcumin stimulates glucagon-like peptide-1 secretion in GLUTag cells via Ca 2+/calmodulin-dependent kinase II activation. *Biochemical and Biophysical Research Communications*, 435(2), pp. 165–170.

Talesa, V., Romani, R., Antognelli, C., Giovannini, E. and Rosi, G., 2001. Soluble and membrane-bound acetylcholinesterases in Mytilus galloprovincialis (Pelecypoda: Filibranchia) from the northern Adriatic Sea. *Chemico-Biological Interactions*, 134(2), pp. 151–166.

Tang, J.J., Li, J.G., Qi, W., Qiu, W.W., Li, P.S., Li, B.L. and Song, B.L., 2011. Inhibition of SREBP by a small molecule, betulin, improves hyperlipidemia and insulin resistance and reduces atherosclerotic plaques. *Cell Metabolism*, 13(1), pp. 44–56.

Tayeb, H.O., Murray, E.D., Price, B.H. and Tarazi, F.I., 2013. Bapineuzumab and solanezumab for Alzheimer's disease: Is the 'amyloid cascade hypothesis' still alive? *Expert Opinion on Biological Therapy*, 13(7), pp. 1075–1084.

Teng, L., Meng, Q., Lu, J., Xie, J., Wang, Z., Liu, Y. and Wang, D., 2014. Liquiritin modulates ERK-and AKT/GSK-3β-dependent pathways to protect against glutamate-induced cell damage in differentiated PC12 cells. *Molecular Medicine Reports*, 10(2), pp. 818–824.

Terwel, D., Dewachter, I. and Van Leuven, F., 2002. Axonal transport, tau protein, and neurodegeneration in Alzheimer's disease. *Neuromolecular Medicine*, 2(2), pp. 151–165.

Terzi, E., Hölzemann, G. and Seelig, J., 1995. Self-association of β-amyloid peptide (1–40) in solution and binding to lipid membranes. *Journal of Molecular Biology*, 252(5), pp. 633–642.

Tiwari, S.K., Agarwal, S., Seth, B., Yadav, A., Nair, S., Bhatnagar, P., Karmakar, M. et al., 2013. Curcumin-loaded nanoparticles potently induce adult neurogenesis and reverse cognitive deficits in Alzheimer's disease model via canonical Wnt/β-catenin pathway. *ACS Nano*, 8(1), pp. 76–103.

Tomlinson, B.E., Blessed, G. and Roth, M., 1968. Observations on the brains of non-demented old people. *Journal of the Neurological Sciences*, 7(2), pp. 331–356.

Tomlinson, B.E., Blessed, G. and Roth, M., 1970. Observations on the brains of demented old people. *Journal of the Neurological Sciences*, 11(3), pp. 205–242.

Tong, G., Castaneda, L., Wang, J.S., Sverdlov, O., Huang, S.P., Slemmon, R., Gu, H. et al., 2012a. Effects of single doses of avagacestat (BMS-708163) on cerebrospinal fluid Aβ levels in healthy young men. *Clinical Drug Investigation*, 32(11), pp. 761–769.

Tong, G., Wang, J.S., Sverdlov, O., Huang, S.P., Slemmon, R., Croop, R., Castaneda, L. et al., 2012b. Multicenter, randomized, double-blind, placebo-controlled, single-ascending dose study of the oral γ-secretase inhibitor BMS-708163 (Avagacestat): Tolerability profile, pharmacokinetic parameters, and pharmacodynamic markers. *Clinical Therapeutics*, 34(3), pp. 654–667.

Tonnesen, H.H., Karlsen, J. and Mostad, A., 1982. Structural studies of curcuminoids: Part 1. The crystal structure of curcumin. *Chemischer Informationsdienst*, 13(50), pp. 475–479.

Unnikrishnan, M.K. and Rao, M.N.A., 1995. Curcumin inhibits nitrogen dioxide induced oxidation of hemoglobin. *Molecular and Cellular Biochemistry*, 146(1), pp. 35–37.

Vajragupta, O., Boonchoong, P., Watanabe, H., Tohda, M., Kummasud, N. and Sumanont, Y., 2003. Manganese complexes of curcumin and its derivatives: Evaluation for the radical scavenging ability and neuroprotective activity. *Free Radical Biology and Medicine*, 35(12), pp. 1632–1644.

Varghese, K., Molnar, P., Das, M., Bhargava, N., Lambert, S., Kindy, M.S. and Hickman, J.J., 2010. A new target for amyloid beta toxicity validated by standard and high-throughput electrophysiology. *PloS One*, 5(1), p. e8643.

Villaflores, O.B., Chen, Y.J., Chen, C.P., Yeh, J.M. and Wu, T.Y., 2012. Effects of curcumin and demethoxycurcumin on amyloid-β precursor and tau proteins through the internal ribosome entry sites: A potential therapeutic for Alzheimer's disease. *Taiwanese Journal of Obstetrics and Gynecology*, 51(4), pp. 554–564.

Wan, L., Nie, G., Zhang, J., Luo, Y., Zhang, P., Zhang, Z. and Zhao, B., 2011. β-Amyloid peptide increases levels of iron content and oxidative stress in human cell and Caenorhabditis elegans models of Alzheimer disease. *Free Radical Biology and Medicine*, 50(1), pp. 122–129.

Wang, C., Zhang, X., Teng, Z., Zhang, T. and Li, Y., 2014a. Downregulation of PI3K/Akt/mTOR signaling pathway in curcumin-induced autophagy in APP/PS1 double transgenic mice. *European Journal of Pharmacology*, 740, pp. 312–320.

Wang, P., Su, C., Li, R., Wang, H., Ren, Y., Sun, H., Yang, J. et al., 2014b. Mechanisms and effects of curcumin on spatial learning and memory improvement in APPswe/PS1dE9 mice. *Journal of Neuroscience Research*, 92(2), pp. 218–231.

Wang, K., Zhang, L., Rao, W., Su, N., Hui, H., Wang, L., Peng, C., Tu, Y., Zhang, S. and Fei, Z., 2015. Neuroprotective effects of crocin against traumatic brain injury in mice: Involvement of notch signaling pathway. *Neuroscience Letters*, 591, pp. 53–58.

Wang, X., Dubois, R., Young, C., Lien, E.J. and Adams, J.D., 2016. Heteromeles arbutifolia, a traditional treatment for Alzheimer's disease, Phytochemistry and Safety. *Medicines*, 3(3), p. 17.

Wang, X., Su, B., Perry, G., Smith, M.A. and Zhu, X., 2007. Insights into amyloid-β-induced mitochondrial dysfunction in Alzheimer disease. *Free Radical Biology and Medicine*, 43(12), pp. 1569–1573.

Wang, Y.H. and Du, G.H., 2009. Ginsenoside Rg1 inhibits β-secretase activity in vitro and protects against Aβ-induced cytotoxicity in PC12 cells. *Journal of Asian Natural Products Research*, 11(7), pp. 604–612.

Wang, Y.J., Pan, M.H., Cheng, A.L., Lin, L.I., Ho, Y.S., Hsieh, C.Y. and Lin, J.K., 1997. Stability of curcumin in buffer solutions and characterization of its degradation products. *Journal of Pharmaceutical and Biomedical Analysis*, 15(12), pp. 1867–1876.

Wang, Y.J., Thomas, P., Zhong, J.H., Bi, F.F., Kosaraju, S., Pollard, A., Fenech, M. and Zhou, X.F., 2009. Consumption of grape seed extract prevents amyloid-β deposition and attenuates inflammation in brain of an Alzheimer's disease mouse. *Neurotoxicity Research*, 15(1), pp. 3–14.

Wang, P., Su, C., Feng, H., Chen, X., Dong, Y., Rao, Y., Ren, Y., Yang, J., Shi, J., Tian, J. and Jiang, S., 2017. Curcumin regulates insulin pathways and glucose metabolism in the brains of APPswe/PS1dE9 mice. *International Journal of Immunopathology and Pharmacology*, 30(1), pp. 25–43.

Weon, J.B., Yun, B.R., Lee, J., Eom, M.R., Ko, H.J., Lee, H.Y., Park, D.S., Chung, H.C., Chung, J.Y. and Ma, C.J., 2014. Cognitive-enhancing effect of steamed and fermented Codonopsis lanceolata: A behavioral and biochemical study. *Evidence-Based Complementary and Alternative Medicine*, 2014, Article ID 319436.

Wevers, A., Monteggia, L., Nowacki, S., Bloch, W., Schütz, U., Lindstrom, J., Pereira, E.F.R. et al., 1999. Expression of nicotinic acetylcholine receptor subunits in the cerebral cortex in Alzheimer's disease: Histotopographical correlation with amyloid plaques and hyperphosphorylated-tau protein. *European Journal of Neuroscience*, 11(7), pp. 2551–2565.

White, J.A., Manelli, A.M., Holmberg, K.H., Van Eldik, L.J. and LaDu, M.J., 2005. Differential effects of oligomeric and fibrillar amyloid-β1–42 on astrocyte-mediated inflammation. *Neurobiology of Disease*, 18(3), pp. 459–465.

Wiessner, C., Wiederhold, K.H., Tissot, A.C., Frey, P., Danner, S., Jacobson, L.H., Jennings, G.T. et al., 2011. The second-generation active Aβ immunotherapy CAD106 reduces amyloid accumulation in APP transgenic mice while minimizing potential side effects. *Journal of Neuroscience*, 31(25), pp. 9323–9331.

Wimo, A., Guerchet, M., Ali, G.C., Wu, Y.T., Prina, A.M., Winblad, B., Jönsson, L., Liu, Z. and Prince, M., 2017. The worldwide costs of dementia 2015 and comparisons with 2010. *Alzheimer's & Dementia*, 13(1), pp. 1–7.

Wischik, C.M., Crowther, R.A., Stewart, M. and Roth, M., 1985. Subunit structure of paired helical filaments in Alzheimer's disease. *The Journal of Cell Biology*, 100(6), pp. 1905–1912.

Xiao, Z., Zhang, A., Lin, J., Zheng, Z., Shi, X., Di, W., Qi, W., Zhu, Y., Zhou, G. and Fang, Y., 2014. Telomerase: A target for therapeutic effects of curcumin and a curcumin derivative in Aβ1-42 insult in vitro. *PLoS One*, 9(7), p.e101251.

Xiong, Z., Hongmei, Z., Lu, S. and Yu, L., 2011. Curcumin mediates presenilin-1 activity to reduce β-amyloid production in a model of Alzheimer's disease. *Pharmacological Reports*, 63(5), pp. 1101–1108.

Yan, R., 2016. Stepping closer to treating Alzheimer's disease patients with BACE1 inhibitor drugs. *Translational Neurodegeneration*, 5(1), p. 13.

Yanagisawa, D., Ibrahim, N.F., Taguchi, H., Morikawa, S., Hirao, K., Shirai, N., Sogabe, T. and Tooyama, I., 2015. Curcumin derivative with the substitution at C-4 position, but not curcumin, is effective against amyloid pathology in APP/PS1 mice. *Neurobiology of Aging*, 36(1), pp. 201–210.

Yang, F., Lim, G.P., Begum, A.N., Ubeda, O.J., Simmons, M.R., Ambegaokar, S.S., Chen, P.P. et al., 2005. Curcumin inhibits formation of amyloid β oligomers and fibrils, binds plaques, and reduces amyloid in vivo. *Journal of Biological Chemistry*, 280(7), pp. 5892–5901.

Yang, L., Hao, J., Zhang, J., Xia, W., Dong, X., Hu, X., Kong, F. and Cui, X., 2009. Ginsenoside Rg3 promotes beta-amyloid peptide degradation by enhancing gene expression of neprilysin. *Journal of Pharmacy and Pharmacology*, 61(3), pp. 375–380.

Yang, X., Xiong, X., Wang, H. and Wang, J., 2014. Protective effects of panax notoginseng saponins on cardiovascular diseases: A comprehensive overview of experimental studies. *Evidence-Based Complementary and Alternative Medicine*, 2014.

Yoshikai, S.I., Sasaki, H., Doh-ura, K., Furuya, H. and Sakaki, Y., 1990. Genomic organization of the human amyloid beta-protein precursor gene. *Gene*, 87(2), pp. 257–263.

Young, K.J. and Bennett, J.P., 2010. The mitochondrial secret (ase) of Alzheimer's disease. *Journal of Alzheimer's Disease*, 20(s2), pp. S381–S400.

Youssef, K.M., El-Sherbeny, M.A., El-Shafie, F.S., Farag, H.A., Al-Deeb, O.A. and Awadalla, S.A.A., 2004. Synthesis of curcumin analogues as potential antioxidant, cancer chemopreventive agents. *Archiv der Pharmazie: An International Journal Pharmaceutical and Medicinal Chemistry*, 337(1), pp. 42–54.

Yu, M.S., Leung, S.K.Y., Lai, S.W., Che, C.M., Zee, S.Y., So, K.F., Yuen, W.H. and Chang, R.C.C., 2005. Neuroprotective effects of anti-aging oriental medicine Lycium barbarum against β-amyloid peptide neurotoxicity. *Experimental Gerontology*, 40(8–9), pp. 716–727.

Yu, X., Wang, L.N., Du, Q.M., Ma, L., Chen, L., You, R., Liu, L., Ling, J.J., Yang, Z.L. and Ji, H., 2012. Akebia Saponin D attenuates amyloid β-induced cognitive deficits and inflammatory response in rats: Involvement of Akt/NF-κB pathway. *Behavioural Brain Research*, 235(2), pp. 200–209.

Zhang, L., Fang, Y., Cheng, X., Lian, Y.-J., Xu, H., Zeng, Z.-S., & Zhu, H., 2017. Curcumin exerts effects on the pathophysiology of Alzheimer's disease by regulating pi(3,5)p2 and transient receptor potential mucolipin-1 expression. *Frontiers in Neurology*, 8, 531. doi:10.3389/fneur.2017.00531

Zhang, L., Fiala, M., Cashman, J., Sayre, J., Espinosa, A., Mahanian, M., Zaghi, J. et al., 2006. Curcuminoids enhance amyloid-β uptake by macrophages of Alzheimer's disease patients. *Journal of Alzheimer's Disease*, 10(1), pp. 1–7.

Zhang, X., Yin, W.K., Shi, X.D. and Li, Y., 2011a. Curcumin activates Wnt/β-catenin signaling pathway through inhibiting the activity of GSK-3β in APPswe transfected SY5Y cells. *European Journal of Pharmaceutical Sciences*, 42(5), pp. 540–546.

Zhang, Y.W., Thompson, R., Zhang, H. and Xu, H., 2011b. APP processing in Alzheimer's disease. *Molecular Brain*, 4(1), p. 3.

Zhang, L., Fang, Y., Xu, Y., Lian, Y., Xie, N., Wu, T., Zhang, H., Sun, L., Zhang, R. and Wang, Z., 2015. Curcumin improves amyloid β-peptide (1-42) induced spatial memory deficits through BDNF-ERK signaling pathway. *PLoS One*, 10(6), p. e0131525.

Zhou, Y.Q., Yang, Z.L., Xu, L., Li, P. and Hu, Y.Z., 2009. Akebia saponin D, a saponin component from Dipsacus asper wall, protects PC 12 cells against amyloid-β induced cytotoxicity. *Cell Biology International*, 33(10), pp. 1102–1110.

Zhu, Z., Li, C., Wang, X., Yang, Z., Chen, J., Hu, L., Jiang, H. and Shen, X., 2010. 2, 2', 4'-Trihydroxychalcone from Glycyrrhiza glabra as a new specific BACE1 inhibitor efficiently ameliorates memory impairment in mice. *Journal of Neurochemistry*, 114(2), pp. 374–385.

8

Black Cumin (Nigella sativa L.): Bioactive Compounds and Health Benefits

Tugba Ozdal and Sevcan Adiguzel

CONTENTS

8.1 Introduction

Black cumin (*Nigella sativa* L.) is an annual herbaceous plant that has been extensively cultivated in many countries, especially in the Eastern Mediterranean countries. The body of the plant is 20–50 cm long; its body is vertical, hairy, branched and sparse. The leaves of the black cumin plants are alternate and contain three parts. The flowers are long-stemmed and one-by-one and are found at the ends of the branches and blooms in June and July. The flowers are white or light blue and yellowish green ends, while fruit is in the form of a capsule that is very seed-bearing. Seeds are the most important part of the plant used widely, which are oval shaped, triangular, and 3 mm long in lengths. The *Ranunculaceae* plant family, one of the three species of *Nigella ginseng*, has been reported to have a rich historical and mystical background in the plants used for medicine and has been used for many years to protect food and enhance flavor. Black cumin is a traditional spice plant of ancient world civilizations. Black cumins are found in ancient Egypt in the grave of Tutankhamon in 1325 BC and also it was used as a spice in the time of Romans.

It has also been used by Hippocrates (460–370 BC), the founder of modern medicine, to strengthen the liver and eliminate digestive system complaints. It is mentioned that the seeds of the plant are used by Hippocrates for snake and scorpion insertions, old tumors, abscess treatment and skin rash, head and neck inflammation and colds. In Arabian/Greek medicine, seeds and oil of black cumin are used to improve health, lower high fever, relieve headaches and rheumatism, and reduce various microbial infections and intestinal parasites.

Black cumins have been used for thousands of years in many countries, especially in India, as spices. In addition to being used as a spice, it is also known that seeds of black cumin are used to decorate the bakery products in the food industry as well as to bring flavor to the products. Black cumin seeds are one of the known and widely used spice sources among the people and have been reported to be used in many bakery products and some cheese varieties as seasoning and aromatizer due to its flavor. It can be regarded as a promising natural antioxidant obtained from the black cumin plants. It is also emphasized that it has a potential to be used in different areas such as food, cosmetics and medicines.

8.2 Black Cumin Varieties and Chemical Compositions

In present days, one of the naturally occurring materials that have the characteristics that can be used as preservatives in foodstuffs is the substances found in the composition of the material. In the world, 16 different types of material are grown. The seeds of *Nigella sativa*, *Nigella damascena*, and *Nigella arvensi* are used more widely in folk medicine and as spices (Figure 8.1).

Chemical composition of black cumin varies according to the plant's harvest season, variety, the cultivation and the region of cultivation (Ramadan 2007). According to Shah and Kasturi (2003), black cumin contains 7% water, 23% protein, 39% oil, 15% starch, 5.4% raw fiber, 16% dietary fiber and 4.3% ash.

Black cumin seeds contain about 30%–40% raw oil, and it is reported that 50%–60% of this oil is composed of unsaturated fatty acids and contains 0.01%–0.1% alkaloid (nigellin), saponin (melantin) (Burits 2000). The seeds contain low levels of essential oils (0.5%–0.7%), vitamins A, B1, B2, B6, and C, minerals such as Mg, Zn, and Se, 18%–22% protein and 35%–40% carbohydrates (Burits 2000).

In recent years, black cumin oil is one of the most commonly used ingredients in health and food technology. But, the functional properties of this plant, its antioxidant active components, especially the essential oil composition, have not been adequately investigated. According to Ramadan (2007), 0.5%–1.6% of essential oil is found in the seed oil, compared with 0.4%–2.5% by Hosseizadeh and Parvardeh (2004). According to Muhammad et al. (2009) black cumin volatile oil contains 23.25% of thymoquinone, 3.84% of dihydrothymoquinone, 32.02% of p-cymene, 10.8% of carvacrol, 2.4% of α-thuene, 2.32% of thymol, 1.48% of α-pinene, 1.72% of β-pinene, 2.10% of t-anethol and 23.81% of minor components.

Benkaci-Ali et al. (2007) found that volatile oil was mainly composed of *p*-cymene, γ-terpinene, α-pinene, β-pinene, α-thuene, carvacrol, and thymoquinone. Earlier studies reported that black cumin contains nigellon, carvacrol, p-cymene, d-limonene, α and β-pinene as well as pharmacologically active basic components, thymoquinone, dithymyquinone, thymohydroquinone (Baytop 1984; Randhawa and Al-Ghamdi 2002). It has been determined that thymoquinone is the active compound with antioxidant activity which is very important in the black cumin (Hosseizadeh et al. 2004).

8.3 The Effects of *Nigella sativa* on Health

Many studies have been carried out on the biological activities and curative properties of black cumin. Black cumin is used in the treatment of many diseases worldwide. Biologically active compounds of black cumin include thymoquinone, thymohydroquinone, dithymoquinone (Bamosa 2002). It is now accepted that the nutraceutical substance in the black cumin is thymoquinone (Bourgou 2010). *Nigella sativa* L. seeds are known to contain nutritional components such as carbohydrates, fats, vitamins, minerals and proteins containing eight essential amino acid residues (Dehkordi 2008). Black cumin seeds are particularly rich in oleic acid, linolenic acid, arachidonic acid, palmitoleic acid, and stearic acid from fatty acids (Benkaci-Ali et al. 2007). The seeds also contain carotene, which is converted to vitamin A in the liver, and it is also quite rich in terms of potassium, phosphorus, calcium and iron (Muhammad 2009). Black cumin has been used for centuries in Asia and Middle East countries for the treatment of different diseases and is considered as an important drug (Randhawa et al. 2002) and has been recommended to use it regularly. Health benefits, of black cumin include its use in the treatment of cancer, cardiovascular disease and diabetes (Zaoui 2002). The health effects of black cumin are shown in Figure 8.2.

Nigella sativa L. contains wide variety of substances having different therapeutic activities such as anticarcinogenic, immunomodulation, antidiabetic, antihypertensive, antiallergic, anti-asthmatic, antidiarrheal, anti-inflammatory, therapeutic activity against stomach diseases, AIDS preventive, Kidney disease remedy, protect the heart and veins, cholesterol lowering, anti-rheumatic, anticoagulant, antimicrobial, antimycotic, antioxidant and neuroprotective activities (Ali et al. 2003).

8.3.1 Cardiovascular Effects of Black Cumin

Cardiovascular diseases are the leading cause of death in the world. It is known that phenolic compounds have a positive effect in protecting against coronary heart disease and cancer.

(a)

(b)

FIGURE 8.1 *Nigella sativa* plant (a) and seeds (b).

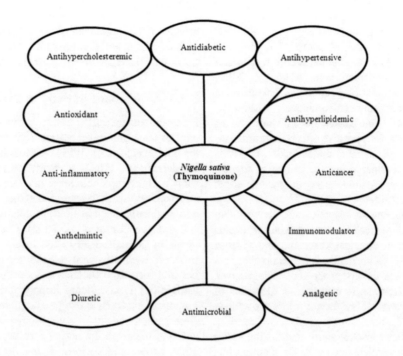

FIGURE 8.2 The health effects of *Nigella sativa* plant.

Nigella seed oil may be considered as a potential source of natural phenolic compounds because it has a higher phenolic content than other table oils, with the exception of olive oil, which is thought to be a rich source of phenolic compounds. Thus, there have been many recent studies on the effects of cardiovascular diseases (Shabana et al. 2013).

Atherosclerosis, primarily hypertension, LDL and high cholesterol are among the causes of cardiovascular disease and are increasing rapidly worldwide (Yusuf et al. 2001). The most important causes of cardiovascular diseases are known as lipid disorders (Sabzghabaee et al. 2012). From this point of view, there are studies that show that black cumin has positive effects on the lipid profile and significantly reduces serum triglycerides and LDL cholesterol (Gilani et al. 2004; Sabzghabaee et al. 2012). Gilani et al. (2004) investigated the effect of thyrokinin on serum lipid profile in rabbits, the experimental animals were fed with cholesterol-rich diet. As a result of this study it has been showed that black cumin significantly increases total cholesterol, LDL and triglyceride levels while increasing the HDL-cholesterol concentration (Gilani et al. 2004). Similar effects have been obtained in both powder and liquid forms of black cumin oil in other studies (Hawsawi et al. 2001).

Al-Naqeep et al. (2011) studied the effect of liquid form of black cumin on normal rat and they have observed that liquid form of black cumin reduce serum cholesterol level by 15.5% and serum triglyceride level by 22%. Furthermore, in a study conducted on humans, patients with high cholesterol levels were given 1 g/day (what was given) for two months. It is observed that the consumption of black cumin seed reduce LDL cholesterol and triglyceride levels and increase HDL cholesterol levels (Nader et al. 2010). In another similar study, it was observed that the use of black cumin oil improves the lipid profile in patients with high cholesterol levels and prevents heart disease (Zaoui

et al. 2002). It is thought that the maintenance of lipid profile due to the black cumin is mostly through thymoquinone, sterol and flavanoids (Bhatti et al. 2009). Furthermore, the lipid-lowering effect is thought to be due to inhibition of cholesterol synthesis and stimulation of bile acid secretion (Tasawar et al. 2011). However, there are also studies showing that the antioxidant effect of black cumin can also contribute to this effect (Bamosa et al. 1997).

Nasir et al. (2008) showed that black cumin significantly improved serum triglyceride levels in humans. In a study conducted in women with menopausal symptoms, it was observed that there were antidiabetic and hypolipidemic effects of black cumin (Nasir et al. 2008). It has been observed in animal studies, that diabetic rats significantly decreased fasting blood sugar and triglyceride levels and also significantly increased HDL levels (Alam et al. 2013; Ibrahim et al. 2014). In hypercholesterolemic rabbits, positive effects on the lipid profile due to black cumin were also shown (Alam et al. 2013). Earlier studies have shown that regeneration of the langerhans islets of the pancreas plays a role in the antidiabetic effect of black cumin in mice (Ali et al. 2012; Asgary et al. 2013).

8.3.2 Antidiabetic Effects of Black Cumin

Diabetes is a group of chronic metabolic diseases in which the blood sugar level is high. Untreated diabetes leads to many serious complications, especially cardiovascular diseases and renal failure (Sobhi et al. 2013). Many researchers report the antidiabetic and hypoglycemic activity of black cumin. It is thought that oxidative stress, which reduces the productivity of insulin production, is also one of the causes of diabetes and the productivity of cells in the pancreas, plays an important role. It has been shown that black cumin and its components are effective against diabetes by reducing oxidative stress and thus maintaining pancreatic beta cell integrity (Jamal et al. 2013). In many studies, it has been shown that the black cumin seed and oil have protective effects on pancreatic cells (Alam et al. 2013). It has been reported that the use of black cumin in diabetic mice is positively associated with degenerative changes in the pancreatic cells of the diabetics (Abu Khader 2012). Many studies have shown that black cumin has a positive effect on serum insulin, superoxide dismutase, glucose and malondialdehyde levels (Sowers et al. 1997). It is considered that the content of thymoquinone in black cumin is responsible for the antidiabetic effect of the plant (Shabana et al. 2013).

In a study of thymoquinone in diabetic rats, showed that thymoquinone at 80 mg/kg reduced the risk of type II diabetes and showed the activities of enzymes (exocrine and glucose 6-phosphate dehydrogenase) to normalize glucose uptake. In a study conducted in 60 patients, the first group of patients received statin and metformin for 6 weeks to demonstrate the effect of black cumin on insulin resistance syndrome, while the second group received 2 × 2.5 mL, black cumin seeds. As a result, a more significant improvement in cholesterol and fasting blood glucose levels in the group supplemented with black cumin this supplement study reveals the treatment potential of insulin resistance syndrome (Badary 2000). In a similar study, in addition to classical treatment, three different doses (1, 2, and 3 g) of black cumin were given and their effects on blood glucose, hemoglobin A1c and beta cell function were evaluated. These parameters were significantly improved in all three doses and the optimum dose was 2 g (Mansour 2002).

8.3.2.1 The Effects of Black Cumin on Blood Pressure and Endothelium

Black cumin has been reported with the positive effects of on both blood pressure and heart rate (Shabana et al. 2013). Studies have shown that black cumin has a relaxant effect in isolated animal vessels (Abdelmeguid et al. 2010). In another study, it was found that the use of thymoquinone in rabbits injected with anoxia in diabetes mellitus was effective in reducing diabetes-induced cardiac arrhythmias (Alenzi et al. 2010). Black cumin also has shown positive effects on high blood pressure in on patients with mild hypertension (Hawsawi et al. 2011). It has been suggested that antioxidant, calcium antagonist and diuretic effect may play a role in the blood pressure lowering effect. Endothel function plays a role in the pathogenesis of many cardiovascular diseases and may occur due to hypercholesterolemia, hypertension, obesity, septic shock, diabetes, and smoking (Zaoui et al. 1999). In some animal models, positive effects of endothel function were determined on various plants including *Nigella sativa*. It showed promising effect for patients with hypertension who have to use medications with lifetime side effects, such as antioxidants, calcium channel blockade, inhibition of endothel function and diuretic effects contribute to the treatment of hypertension. The highly active structure of the endothelium, regulates vascular homeostasis. Numerous herbicides and their isolated chemical components have been shown to have positive effects on the endothelium, possibly with antioxidant properties. In animal models, a group of botanicals, including thymoquinone, the active component of black cumin, has been shown to promote the development of age-related endothelial damage on isolated hearts, and the molecular mechanisms behind potential treatment effects are also studied (Meral et al. 2004).

8.3.3 Antioxidant Activity of Black Cumin

Free radicals are called atoms, groups of atoms or molecules with at least one unpaired electron in their outermost orbitals (Gutteridge et al. 1999). Antioxidants are responsible for the detrimental effects of oxidation in tissues. Free radicals show a highly unstable structure due to their non-associated electrons and want to acquire a stable structure by interacting with molecules such as lipid, protein or carbohydrate, causing cell damage (Kuhn et al. 2003). Certain protection mechanisms in the cells lead to the formation of free radicals for instance, liver uses free radicals for detoxification, and neutrophils form free radicals to exterminate pathogens (Lunec et al. 2002). Free radicals, largely of oxygen and nitrogen origin, occur naturally during metabolism however, free radicals from carbon and sulfur sources are also found in the cells. In addition to metabolic activity, external factors such as radiation, drugs, and various chemicals also cause free radical formation.

Antioxidants are divided into two different classes according to their mechanism of action:

1. Chain-breaking antioxidants, such as vitamin E and carotene, stabilize the free radicals by donating electrons and make them unreactive.
2. Enzymes that are protective antioxidants clean oxidants without initiating the oxidation chain (Kuhn et al. 2003).

Anti-oxidants are mechanisms that prevent radical formation, remove radicals, clean and repair damaged molecules. It has been reported that thymoquinone has antioxidant effects with different mechanisms. For example, in a study conducted by El-Dakhakhny et al. (2002), it has been shown that thymoquinone inhibits the formation of 5-lipoxygenase products such as 5-hydroxyieucosa-tetraenoic acid. These molecules are necessary for colon cancer cells to survive. Thymoquinone has been shown to have a radical scavenging effect on various oxygen species including superoxide radical anion and hydroxyl radicals (Kruk et al. 2000; Mansour et al. 2002; Badary et al. 2003). In addition, thymoquinone causes a significant decrease in hepatic antioxidant enzymes such as superoxide dismutase (SOD), catalase and glutathione peroxidase. Thymoquinone is able to effectively inhibit iron-dependent microsomal lipid peroxidation in doxorubicin-induced rats suffering from hyperlipidemic nephropathy (Badary et al. 2000). Thymoquinone has been shown to reduce cellular oxidative stress by stimulating glutathione in female Lewis rats suffering from Experimental Allergic Encephalomyelitis (Mohamed et al. 2003).

Many epidemiological studies have shown that the consumption of antioxidant rich foods can reduce the risk of cancer (Borek 2004). In experimental and clinical trials, oxidative stress has been shown to be effective in differentiation and progression of different cancers (Kim et al. 2004; Pathak et al. 2005). Badary studied the potential protective effect of thymoquinone on carcinogenesis in 1999 and 2007. Accordingly, thymoquinone inhibits the carcinogenesis process by modulating lipid peroxidation and cellular antioxidant environment.

In vitro studies have shown that *Nigella sativa* seed extract inhibits hemolytic activities of snake and scorpion poisons (Sallal et al. 1996). It protects erythrocytes against lipid peroxidation caused by hydrogen peroxide, protein degradation, loss of deformation and increased osmotic fragility (Suboh et al. 2004). It has also been shown to protect laryngeal cancer cells against apoptosis (programmed cell death) induced by cortisol or lipopolysaccharides (Corder et al. 2003). According to these results, *Nigella sativa* seed components are observed to have anti-toxic effects due to their antioxidant properties. Raw *Nigella sativa* oil and fractions (natural lipids, glycolipids, and phospholipids) show a radical scavenging effect due to total polyunsaturated fatty acids, non-saponifiables and phospholipids (Ramadan et al. 2003).

Hepotoxicity and nephrotoxicity are associated with changes in the levels and activities of mediators such as L-alanine aminotransferase (ALT), alkaline phosphatase (ALP), lipid peroxidase (LPD) and antioxidant cleansing enzyme system containing superoxide dismutase (GSH) and superoxide dismutase. The different liver and renal toxicities induced by *Nigella sativa*'s antioxidant effects, tert-butyl hydroperoxide, carbon tetrachloride, doxorubicin, gentamicin, methionine, potassium bromate, cisplatin or *Schistosoma mansoni* infection have been investigated *in vivo* on murine (rat) (Nagi et al. 1999; El-Dakhakhny et al. 2000; Meral et al. 2001; Türkdoğan et al. 2003; Kanter et al. 2003; Khan et al. 2003a, 2003b; Meral and Kanter 2003; Ali 2004). These *in vivo* studies reported the potential antioxidant activity *Nigella sativa* oil.

The treatment of the rat with prokinetically thymoquinone before carbon tetrachloride injections improves hepatotoxicity of carbon tetrachloride by lowering elevated serum enzyme levels and significantly increasing hepatic glutathione content (Burits and Bucar 2000; Enomoto et al. 2001). The treatment of rats with other essential oils has not changed the level of enzymes and glutathione. Thymoquinone has been shown to induce doxorubicin-induced nephrotoxicity, cardiotoxicity and oxidative stress in rats. Accordingly, the prevalence of thymoquinone on nephrotic hyperlipidemia and hyperproteinuria formation and the biomarkers of oxidative stress normalize (Badary et al. 2000).

In 2009, a cell-based test was developed by Girard-Lalancette et al. using DCFH-DA, a useful indicator of reactive oxygen species. The test is very sensitive in detecting antioxidant properties. Bourgou et al. (2010) investigated the antioxidant properties of *N. sativa* oil and the active ingredient thyroxine on the oil of cumin seeds in Tunisia, and found that *N. sativa* essential oil significantly inhibited ROS production. It has been shown in the same study that *N. sativa* oil has high antioxidant activity in *ex vivo* environment as opposed to other components of thymoquinone (IC50 value of 1.0 μM (0.2 μg/mL)), which is the main active compound of black cumin.

Nigella sativa L. seeds have been traditionally used depending on the antioxidant properties described above. *N. sativa* L. oil and its active ingredients seem to reduce the toxicity mediated by environmental or infection-related factors or by oxidative stress triggered by anti-cancer drugs. For example, chemotherapy, cyclophosphamide and other anti-cancer drugs

are used in preclinical and clinical trials in combination with anti-cancer therapy or cancer immunotherapy. Chemotherapy causes the immature granulocytes to expand considerably and produce large amounts of NO (nitric oxide). Thymoquinone has immunosuppressive effects against nitric oxide stimulated by chemotherapy.

8.3.4 Anti-inflammatory Effects of Black Cumin

The continuity and progression of the acute and chronic phases of inflammation is provided by cellular mediators, such as lytic enzymes secreted by inflammatory cells including eicosanoids, oxidants, cytokines, macrophages and neutrophils (Bohn et al., 1999). Reactive oxygen species, primarily NO, initiate toxic oxidative reactions in abundant amounts causing tissue damage and anti-inflammatory activities of *Nigella sativa* L. were determined by inhibition of cellular NO production capacity. In many tissues NO synthesized from L-arginine *via* nitric oxide synthase (NOS) is associated with many inflammatory diseases. Bourgou et al. (2010) studied *Nigella sativa* seed oil and its components from Tunisia. According to this study, *Nigella sativa* oil inhibits LPS-induced NO secretion by 90% and *N. sativa* is the main active compound, however, thymoquinone inhibited NO production by 95%. The inhibitory effect of thymoquinone on NO production is mediated through the reduction of inducible ROS mRNA and protein expression (El-Mahmoudy et al. 2002).

In addition to the inflammation induced by reactive oxygen species, there are two enzymes that cause inflammation; cyclooxygenase (COX) and lipoxygenase (LO) (Williams et al. 1999). COX is known to catalyze the formation of prostaglandins and thromboxane from arachidonic acid, while LO catalyzes the formation of leukotrienes. Prostaglandins and leukotrienes function as the main mediators of allergies and inflammations. Prostaglandins, thromboxanes and leukotrienes are all called eicosanoids (Gürdöl and Ademoğlu 2006). In many *in vitro* studies, the inhibitory effects of *Nigella sativa* oil and the active ingredients of the oil, in particular the mediators have been demonstrated. Thymoquinone and *N. sativa* L. have been shown to inhibit the COX and 5-LO pathways in arachidonic acid metabolism and the effect of the thymoquinone was found to be higher (Houghton et al. 1995; Mansour and Tomhamre 2004). Both substances inhibited non-enzymatic peroxidation in brain phospholipid liposomes, and again the effect of thymoquinone was found to be higher. In contrast, the inhibitory effect of *N. sativa* L. on the formation of eicosanoids and lipid peroxidation is greater than that of the thymoquinone, and components such as unsaturated fatty acids in the fat content are thought to play a role in the anti-eicosanoid and anti-oxidant effects of *N. sativa* L. oil. Furthermore, *N. sativa* L. oil with crude extract, nigellone or thymoquinone inhibits the formation of 5-LO products and 5-hydroxyiecosatetraenoic acid in a concentration dependent manner (El-Dakhakhny et al. 2002). Hence, the anti-inflammatory effect of *N. sativa* oil and its active components is observed to be mediated through the inhibition of COX and 5-LO pathways.

N. sativa oil components have been shown to have anti-inflammatory effects in some inflammatory diseases such as Experimental Autoimmune Encephalomyelitis (EAE) and colitis. EAE is a disease that destroys the myelin layer on the autoimmune nerve fibers that affect the central nervous system. The mediator of the disease is T cells, and oxidative stress plays a central role in the development and persistence of the disease (Chakrabarty et al. 2003). When humans with EAE were treated with thymoquinone, the glutathione level was elevated and perivascular inflammation disappeared with signs of disease compared to EAE untreated animals and animals treated with thymoquinone. In this case, it is understood that thymoquinone has a therapeutic potential on the EAE model. Accordingly, it is concluded that thymoquinone has a positive effect in the treatment of multiple sclerosis in humans (Burits et al. 2000). Ulcerative colitis is another anti-inflammatory disease characterized by hemorrhages in acute inflammatory cycles, ulceration and colonic mucosa. Although the pathogenesis of colitis is not fully understood, many mediators such as eicosanoids, leukotrienes, platelet activating factors, and oxygen producing free radicals are responsible for the pathogenesis of this disease (Nieto et al. 2000).

Studies by Campieri et al. (1991) and Gionchetti et al. (1991) have shown that antioxidant agents have a beneficial effect on the symptoms of the disease (Koch et al. 2000; Choudhary et al. 2001). Mahgoub (2003) investigated the effects of thymoquinone on acetic acid-induced colitis in rats. The investigator treated the animals with thymoquinone for 3 days prior to intracolonic injection of 3% acetic acid and showed that thymoquinone protected the acetic acid-induced colitis with a higher effect than sulfasalazine (an anti-colic agent). According to this study, anti-colitis effect of thymoquinone is provided by anti-oxidant and anti-histaminic activities.

The sharpness of the inflammatory immune response is controlled by the accumulation of inflammatory cells in the lesions. This process is achieved by the expression of certain inflammatory chemokines and adhesion molecules from inflammatory cells, ICAM-1, VCAM-1 and endothelial cells. The inflammatory chemokines were MCP-1 (CCL2), MIP-1a (CCL3), MIP-1h (CCL4) and RANTES (CCL5) (Bagglionini et al., 1994; Kallinich et al., 2005). Adhesion molecules are LFA-1, CD62L and CD44 (Cartier et al. 2005). The potential mechanism of the inhibitory effect of *Nigella sativa* on the immune response is thought to be the alteration of the functioning of inflammatory cells by modulating the expression of chemokines and adhesion molecules. Despite all this, there is no literature available on the effects of *Nigella sativa* on the inhibition of chemokines and adhesion molecules and IL-1 and TNF-inflammatory cytokines and also on the proliferation of IL-8 chemokines. Considering the potent anti-inflammatory activity of *Nigella sativa* seeds on different inflammatory diseases, the effects of *Nigella sativa* oil and its active components on the expression of chemokines and adhesion molecules of immunogens should be elucidated.

8.3.5 Anti-histaminic Effects of Black Cumin

Histidine is essential amino acid and converts histamine to decarboxylase through pyridoxal phosphate. Basophils are stored in mast cells, histamine vesicles synthesized in gastric mucosa, intestine and some brain parts and released by factors such as oxygen and temperature. Besides its ionic interaction with the heparin-protein complex, binds to the H1 receptor in the intestines and bronchi and causes smooth muscles to contract. It is secreted

in gastric mucosa cells by acetylcholine and gastrin effect and it binds to H2 receptors and stimulates hydrochloric acid secretion. Excessive secretion of histamine leads to allergic reactions in the body. Moreover, it is known that histamine edema, which leads to enlargement of capillaries by narrowing in small venules, is caused by an increase in the volume of vascular bed and consequently allergic shock due to the fall of blood pressure. In addition, histamine acts as a neurotransmitter in the brain (Gürdöl and Ademoğlu 2006). Histamine is secreted from the body tissues to form allergic reactions related to conditions such as bronchial asthma. The active ingredients of *Nigella sativa* seeds have been shown to have significant effect on the histamine-induced inflammatory diseases. Patients suffering from bronchial asthma have been shown to alleviate symptoms in the majority of patients when *N. sativa* seeds are given orally. This antihistaminic effect has been shown to be provided by dymythinone isolated from the volatile oil of *N. sativa* seeds and is called isolated dithymoquinone Nigellone (El-Dakhakhny 1965). Following this study, effective results were obtained in the treatment of Nigellone administered to children and adults suffering from bronchial asthma and no toxic effects were observed. In a clinical trial, people suffering from allergic diseases such as allergic internal nasal inflammation, bronchial asthma, and atopic eczema have been shown to reduce the amount of IgE, eosinophils, plasma and endocrine cortisol in the urine of *N. sativa* when treated with *N. sativa* L. (Kalus et al. 2003). Hence, *N. sativa* oil is helpful in the treatment of allergic diseases.

The antiallergic properties of the components of *N. sativa* seed may be associated with antihistaminic effects. It has been suggested that antihistaminic effects of thymocinone, in part, with the inhibition of lipoxygenase products of arachidonic acid metabolism, together with the selective blockade of histamine and serotonin receptors (Al-Majed et al. 2001). Preclinical and clinical studies have also shown anti-histaminic effects of *N. sativa* seeds. A gastric ulcer model induced by oral ethanol ingestion has shown a significant increase in the histamine content in mucosa. The amount of histamine in the gastric mucosa was measured in rats fed with *N. sativa* oil prior to ulcer induction and in ulcer induced rats and the histamine contents in the gastric mucosa were found to be lower in rats previously treated with *N. sativa* oil. Treatment with *N. sativa* oil showed a 53.56% protective effect against histamine release from the gastric mucosa (El-Dakhakhny et al. 2000). In another study, it has been shown that thymoquinone has a triggering effect on histamine (Tahir et al. 1993), as well as the histamine release-reducing effect demonstrated by El-Dakhakhny et al. (2000) taken together. Hence, it can be concluded that the different active ingredients in the *N. sativa* oil have different effects on histamine release. The Nigellone active ingredient (dithymyquinone) of the raw extract of *N. sativa* acts as a calcium channel blocker and, due to this effect, *N. sativa* is being used as a traditional therapeutic against diarrhea, asthma and hypertension.

8.3.6 Anti-microbial Effects of Black Cumin

Nigella sativa oil and its active ingredients have anti-microbial properties, including anti-bacterial, anti-fungal, anti-helminytic and anti-viral. Murine (rat) cytomegalovirus (MCMV) (Reynolds et al. 1993) and a human cytomegalovirus in humans with immunodeficiency (Moro et al. 1999), spread throughout the body in immunocompromised animals and cause a fatal disease. The anti-oxidant effect of *N. sativa* oil provides protection against these viruses through its anti-viral activity (Shah et al. 2003). The anti-viral effects of *N. sativa* oil against MCMV infection open new pathways for anti-viral therapy. With future studies, this effect should be confirmed on other viral models and the active ingredients of this anti-viral effect must be clarified.

Schistosomiasis is a tropical parasitic disease commonly seen in third world countries. Disease prevention is possible through cellular and humoral (due to body fluids) involvement. Despite many trials, chemotherapy is still the only option for the human host (Chitsulo et al. 2004). In a study conducted by Aboul-Ela (2002), the potential protective effect of *N. sativa* seed extract and thymoquinone against *S. mansoni* infection has been shown. The treatment of *S. freemenous* mice with *N. sativa* oil reduced the number of *S. masonic* worms in the liver and decreased the amount of eggs in the liver and intestines. The *N. sativa* oil has been shown to increase the efficacy of the named drug Praziquantel in the treatment of schistosomiasis (Mahmoud et al. 2002). In murine Schistosomiasis, various cytokines are involved as mediators of granulomatous inflammatory response. Accordingly, the modulation of cytokine levels regulates the sharpness of the inflammatory response. Similar to anti-schistosome effects, the volatile oil of *N. sativa* seeds exhibits anti-helminytic action against tapeworm, earthworm, nematode and intestinal worms even at 1:100 dilution (Agarwal et al. 1979; Akthar and Riffat 1991).

In addition to anti-viral and anti-helminitic effects, *N. sativa* oil, exhibit antibacterial activity against many bacterial strains such as *Escherichia coli*, *Bacillus subtilis*, *Streptococcus faecalis*, *Staphylococcus aureus* and *Pseudomonas aeruginosa*, *Candida albicans* and anti-fungal activity (El-Fatary 1975; Hanafy and Hatem 1991; Morsi 2000; Khan et al. 2003a, 2003b). A study by El-Fatatry (1975) showed that thymohydroquinone from the components of *N. sativa* showed anti-bacterial activity against gram-positive bacteria and the concentration-dependent inhibition of the diethyl ether extract against gram-positive bacteria *Staphylococcus aureus* and gram-negative bacteria *Pseudomonas aeruginosa* and *Escherichia coli* has also been observed. Furthermore, it has been shown that the ether extract increases the effect of many antibiotics (Hanafy and Hatem 1991). More importantly, the extract has been proven to be more effective against spontaneously resistant bacteria including *Vibrio cholerae*, *E. coli* and *Shigella dysenteriae* (Morsi 2000). *In vivo*, the *N. sativa* seed diethyl ether extract completely treats the infection when injected into the infection site in non-lethal subcutaneous staphylococcal infection in mice (Hanafy and Hatem 1991). Accordingly, the components of the *N. sativa* seed show bactericidal activity through different host factors *in vivo*. Bourgou et al. (2010) investigated antibacterial properties of oil and components of *N. sativa* seeds collected from Tunisia on *S. aureus* and *E. coli*. When chloramphenicol was used as a positive control, the antibacterial property of *N. sativa* oil was found to be mainly due to the thymoquinone content. They have found out IC50 value of 12.0 ± 4 g/mL and 62.0 ± 17 g/mL *Nigella sativa* oil against *S. aureus* and *E. coli*, respectively. Moreover, thymoquinone IC_{50} values were found to be 1.8 ± 0.6 M and

41.0 ± 19 M whereas chloramphenicol IC_{50} values were 7.0 ± 1.0 M and 0.8 ± 0.1 M, respectively.

Mouse grafting of *Candida albicans* results in *Candida albicans* colonies in the liver, spleen, kidneys and pharynx. Using this model, the anti-fungal effects of aqueous extracts of *N. sativa* seeds were investigated, 24 hours after inoculation of *C. albicans* (infected), the infected mice were treated with the aqueous extract of *N. sativa* seeds for 3 days (Khan et al. 2003a, 2003b). In this study, *N. sativa* L. oil and its components have shown the anti-viral, anti-helminytic, anti-bacterial and anti-fungal properties.

8.3.7 Anti-tumor Effects of Black Cumin

The fundamental change that leads to the development of cancer is the continuous and uncontrolled proliferation of cancer cells. These cells do not show a response to the cell's life cycle control signals and continue to multiply and divide uncontrollably, spreading throughout the body by invading normal tissues and organs. The uncontrolled proliferation of cells and the resulting cell populations are called tumors and are of two groups, such as malignant and benign. Benign tumors remain where they are formed without spreading to the surrounding tissue or other parts of the body. By contrast, malignant tumors spread throughout the body. The spread of cancer cells to surrounding tissues and then to the whole body is called metastasis.

Several *in vivo* and *in vitro* studies have demonstrated anti-tumor effects of *N. sativa* seeds and its active ingredients. The effects of volatile oil of *N. sativa* seeds were investigated on different cancers in humans and found to have cytotoxic effects against some of the fatty acids (Islam et al. 2004). As an example; It has been shown that MCF-7 breast cancer cells are treated with aqueous and alcoholic extracts of *N. sativa* alone or together with H_2O_2 (Swamy and Tan 2000) and found that *N. sativa* alone or in combination with oxidative stress is an effective anti-cancer agent. If tumor cells are thought to sustain metastases by bringing these factors into play, anti-tumor effects of *N. sativa* oil should be due to anti-angiogens effects through regional tumor invasion and *in vivo* inhibition of metastasis.

Besides the anti-tumor effects of the crude extract of *N. sativa*, thymoquinone, dithymyquinone and other active ingredients have been shown to exhibit cytotoxic effects. The *N. sativa* active compound, ahedrin, extracted from ethyl acetate column chromatographic fraction 5 (CC-5), showed anti-tumor effect against different cancer cell lines. Furthermore it has been observed that ahedrin is active against hepatocellular carcinoma, leukemic cell, Lewis lung carcinoma (Swamy and Tan 2000). Thymoquinone and dithymoquinone have equal cytotoxic effects against different cancer cell lines, such cancer as pancreatic adenocarcinoma, human uterine sarcoma and human leukemic cells (Salomi et al. 1992; Worthen et al. 1998). Thymoquinone and dithymoquinone stop the growth of these cells by triggering apoptosis while the cell cycle is in G1 phase. Stopping cell growth is achieved by increasing protein expression by expression of p53 gene and by inhibiting the anti-apoptotic Bcl-2 protein (Gali-Muhtasib et al. 2004). This means that the anti-neoplastic effect of thymoquinone is mediated by pro-apoptotic effects modulated by the expression of p53-dependent and Bcl-2 protein.

In vitro anti-tumor effects of *N. sativa* oil and its active components have been confirmed by *in vivo* studies on different tumor models. In an *in vitro* cytotoxic study by Salomi et al. (1992), crude methanol extract of *N. sativa* has shown a cytotoxic effect of approximately 50% on Ehrlich Assit Carcinoma (EAK), Dalton lymphoma (DLA-ascites) and Sarcoma-180 cells (S-180 cells). Also in Salomi's et al. (1991) study, topical application of *N. sativa* oil has been shown to inhibit the onset of croton oil-induced skin cancer in mice. *N. sativa* fat reduces tumor size, prevents liver metastasis, and P815 mastocytoma delays the death of tumor-bearing mice (Ait Mbarek et al. 2007). In a study conducted by El-Kadi et al. (1989), it was observed that 4 weeks of treatment with *N. sativa* oil in healthy volunteers increased the ratio of T helper cells to T suppressor cells and the function of natural killer cells (NK).

8.3.8 Hematological Effects of Black Cumin

Hematological effects of many natural products have been revealed. A study conducted by Enomoto et al. (2000) had shown the effects of *N. sativa* on platelet aggregation, blood coagulation and fibrinolysis activities. Platelets range in size from 2 to 4 μm in diameter. They are seedless, disc-shaped cell fragments that prevent blood clotting and help repair blood vessel cracks (Junqueira and Carneiro 2006). Hemostasis is the name given to the spontaneous arrest of a bleeding from a damaged and it occurs in two processes, primary and secondary. The first phase of primary hemostasis occurs in three stages; in the first stage hemorrhage is stopped in the vascular territory where the fibrin accumulates to form an insoluble bond. In the second stage, platelets adhere to the damaged vessel area and form plugs. In the last stage, vasoconstriction of the damaged vessel slows down the flow of blood outside the vein. Secondary hemostasis is the coagulation process involving about 20 plasma proteins.

Enomoto ct al. (2000) investigated the effect of *N. sativa* oil on platelet aggregation and inhibition on coagulation using fibrin layer and fibrinolysis using rabbit blood, firstly treated with n-hexane and methanol to solubilize *N. sativa* oil, arachidonic acid and adenosine diphosphate (ADP) and used as inducers of platelet aggregation in this study. This study has shown that *N. sativa* oil has a high inhibitory effect on blood clotting. It was later found that the methanol-soluble fraction of *N. sativa* oil was higher than the other parts of the inhibitory effect on blood clotting and platelet aggregation. The methanol-soluble portion of the *N. sativa* oil was separated into ten fractions using silica gel column chromatography. Each fraction was re-examined and fractions III and IV were found to exhibit high inhibitory activity in arachidonic acid-induced platelet aggregation. Fractions II and IX showed strong inhibition of blood clotting. In addition, all the fractions showed low inhibition of fibrinolysis and ADP-induced platelet aggregation (Enomoto et al. 2000).

In the proceedings of Enomoto et al. (2000), fractions III and IV were purified by column chromatography and high performance liquid chromatography (HPLC) and identified the compounds obtained. At the end of this study, two new inhibitors of arachidonic acid-induced platelet aggregation and two other previously known compounds were isolated. The new compound they have obtained has been identified as 2-(2-methoxypropyl)-5-methyl-1,4-benzenediol. And the other two compounds were thymol and carvacrol. The hydroxyl groups of the three compounds obtained thereafter were acetylated and the effect of the

five analog compounds, including thymoquinone, on arachidonic acid-induced platelet aggregation, with the corresponding acetylated compounds, was investigated. As a result, it has been found that the three afore mentioned compounds having aromatic hydroxyl groups and their acetylated forms have a 30-fold inhibitory effect on the aspirin, which is regarded as thrombosin resistance.

8.3.9 Immunomodulatory Effects of Black Cumin

Immunity can be divided into two types, natural immunity, and acquired immunity, (Medzhitov and Janeway 2000). Acquired immunity includes natural immunity, macrophages, natural killers, nonspecific cells such as granulocytes, while acquired immunity, include B-cell mediated secretory immunity, and CD4+ (helper) CD8+ (cytolytic) T cells mediated cellular immunity (Lucey et al. 1996). The CD4+ T helper cells are responsible for the regulation of the immune response; while CD8+ T cells are lytic cells that act against infections or cancer. These two types of T cells are cells that have critical prescription for eliminating infections and controlling cancer.

To support the *in vitro* studies, *in vivo* studies have also been conducted and shown that *N. sativa* oil has a triggering effect on T cell immunity. Abuharfeil et al. (2001) found that aqueous extracts of *N. sativa* seeds increased the number of spleen native killer cells by a factor of two, and that the cytotoxicity of these natural killer cells against YAC-1 tumor targets was greater than the cytotoxicity of normal control natural killer cells. *N. sativa* also has a healing effect on the age-related decline of T cell function. Consumption of *N. sativa* with food increases the immune response by altering the total amount and type of lipid taken in the diet in the elderly (Hummel 1993). *N. sativa* oil is rich in n-6 PUFA α-linoleic acid and n-3 PUFA α-linoleic acid and stearidonic acid (Laakso and Voutilainen 1996). *In vitro* studies have demonstrated the enhancing effects of T cell mediated immune response as well as the *N. sativa* components on a diminishing effect on B cell mediated immunity. The study by Islam et al. (2004) supports the hypothesis of betting *in vivo*. In particular, the effects of *N. sativa*'s volatile oil on antigen-specific responses in typhoid TH antigen-stimulated rats have been investigated. Compared with the control group, antibody production in rats exposed to *N. sativa* in response to typhoid awakening was approximately two-fold reduced. *In vivo* and *in vitro* studies show that *N. sativa* components trigger cellular uptake and increase humoral involvement. The prostaglandins (PGE), leukotriene B4 (LTB4) and oxidative stress mediators have an effect on lymphocyte proliferation (Meydani et al. 1990; Shapiro et al. 1993). *N. sativa* oil, however, reduces the production of these mediators, all of which suggests that the cell-mediated potentiating effect of *N. sativa* oil is partly related to anti-inflammatory effects (Salem 2005).

8.3.10 Gastroprotective, Radioprotective, and Hepatoprotective Effects

The protective effect of black cumin oil on stomach has been investigated in ethanol-induced ulcer in rats. The increase in glutathione levels, the decrease in histamine content in peptic ulcer formation and the increase in mucin content are indicative of the protective feature of black cumin oil (El-Ahbar et al. 2003). Further, black cumin oil has shown protective effect against DNA damage caused by irradiation (Rastogi et al. 2010). *N. sativa* oil has been shown to be effective against hepatotoxicity. *N. sativa* increased K^+ and Ca^{+2} levels in serum, reduced lipid peroxidation and liver enzymes, reduced erythrocyte, leukocyte, hemoglobin and antioxidant enzyme levels and corrected the serum lipid profile. It has significantly reduced liver damage (Salem 2005). Liver ischemia-reperfusion injury (Yıldız et al. 2010) and liver cirrhosis associated with bile duct involvement also have protective effects (Çoban et al. 2010).

8.3.11 Nephroprotective Effect

In a rat study, thymoquinone reduced urinary protein and albumin excretion, suggests that thymoquinone may be used as a protective agent for proteinuria and hyperlipidemia associated with nephrotic syndrome (Badary et al. 2000). The *N. sativa* extract, which is prophylactically applied in the study of renal oxidative damage caused by potassium bromate ($KBrO_3$) in rats, has been shown to significantly reduce lipid peroxidation and oxidative damage in the tissue and restore glutathione and antioxidant enzyme capacity (Khan et al. 2003a, 2003b). In the case of kidney ischemia-reperfusion injury, it also had a protective role by antioxidant effects (Yıldız et al. 2010).

8.3.12 Neuroprotective Effect

Nigella sativa oil and thymoquinone administration have been shown to be effective in the treatment of various global cerebral ischemia (Hosseinzadeh et al. 2007), subarachnoid hemorrhage (Erşahin et al. 2011), toluene-induced hippocampal neurodegeneration (Kanter 2008a), diabetic neuropathy (Kanter 2008b), protective effects against neuronal damage were demonstrated in experimental neurotrauma models. Subarachnoid hemorrhage-related brain damage reduced blood-brain barrier permeability, brain edema; showed protective effect against neuronal damage by antioxidant effect (Erşahin et al. 2011). Intra-cerebroventricular administration of thymoquinone showed anticonvulsant activity and reduced seizure formation (Hosseinzadeh et al. 2005).

8.3.13 Effect on Reproductive System

Nigella sativa seeds are also used as an enhancer of milk secretion among the population (Ali and Blunden 2003). In isolated organ bath studies it has been shown that they inhibit uterine oxytocin-induced contractions (Aqel and Shaheen 1996). In addition, estrogenic effects have also been reported (El-Halawany et al. 2011). Long-term (53 days) use of alcoholic extract has shown positive effects on male fertility in rats (Al Saaidi et al. 2009).

8.4 The Applications of Black Cumin in Food Products

Black cumin is one of the most important substances used because volatile oil is needed to prevent oxidative damage to food and to repair cell damage that can occur in the human body.

N. sativa seeds are used as aroma materials in the production of some foodstuffs, pastries, cheeses, pickles and bakery products. The components of the seeds are also functional and used in the preparation of cosmetic and dietary supplement products. Since the seeds of black cumin have a strong taste of pepper, they also find use in coffee, tea, bread and salads. According to Gordon et al. (2001), antioxidants maintain good quality of product when used in combination with good raw materials, a suitable production technique, packaging and storage methods, and eliminate oxidation problems in the grains. In addition, it has been found that as the amount of antioxidant increases within a certain limit, the protective property also increases. Black cumin is a very important component that can be used in the food industry due to its antibacterial and antifungal protective properties besides its nutritive and aroma richness characteristics.

8.5 Conclusion

Black cumin seeds have been used as spices and food preservatives for thousands of years. The seeds and oil of the plant show the potential drug properties and used in the traditional medicine field. The role of therapeutic treatment (healing) of the black cumin seed in the treatment of diseases affecting the body systems has been examined. The black cumin seed also has the antioxidant system action which cleans oxidant agents such as inflammatory (inflammatory) processes mediator (mediator) prostaglandins and leukotrienes by suppressing anti-inflammatory (anti-inflammatory) features. The immunomodulator properties of the plant are to contribute to the immune response (defense system) by increasing T cells and natural killer cells. It has anti-cancer and anti-microbial properties against various microorganisms and also against different cancers. It reduces DNA damage in colon tissue exposed to toxic agents and prevents carcinogenesis. Black cumin also has significant therapeutic effects on diabetes mellitus. In addition to enhancing glucose-induced insulin secretion, it also has an effect on reducing glucose absorption from the intestinal mucosa. *Nigella sativa* also regulates hepatic lipid peroxidation as well as being effective in protecting against harmful effects of hepatic tissue. It is thought that the effects of black cumins exhibit anti-inflammatory, antioxidant and antineoplastic (cancer regression, stop) effects combination. Black cumin is a folkloric medicine plant commonly used for a wide range of diseases. In addition to the positive and therapeutic effect it is also necessary to consider the undesirable effects due to improper and careless use.

REFERENCES

Abdelmeguid, N.E., Fakhoury, R., Kamal, S.M. and Al Wafai, R.J., 2010. Effects of *Nigella sativa* and thymoquinone on biochemical and subcellular changes in pancreatic-cells of streptozotocin induced diabetic rats. *J Diabetes, 2*, pp. 256–266.

Aboul-Ela, E.I., 2002. Cytogenetic studies on *Nigella sativa* seeds extract and thymoquinone on mouse cells infected with schistosomiasis using karyotyping. *Mutat Res Genet Toxicol Environ Mutagen, 516*, pp. 11–17.

Abu Khader, M.M., 2012. Thymoquinone: A promising antidiabetic agent. *Int J Diabetes Dev Ctries, 32*, pp. 65–68.

Abuharfeil, N.M., Salim, M. and Von Kleist, S., 2001. Augmentation of natural killer cell activity *in vivo* against tumour cells by some wild plants from Jordan. *Phytother Res, 15*, pp. 109–113.

Agarwal, R., Kharya, M.D. and Shrivastava, R., 1979. Antimicrobial & anthelmintic activities of the essential oil of *Nigella sativa* Linn. *Indian J Exp Biol, 17*, p. 1264.

Ahmad, A., Khan, R.M. and Alkharfy, K.M., 2013. Effects of selected bioactive natural products on the vascular endothelium. *J Cardiovasc Pharmacol., 62*, pp. 111–121.

Akhtar, M.S. and Riffat, S., 1991. Field trial of *Saussurea lappa* roots against nematodes and *Nigella sativa* seeds against cestodes in children. *J Pak Med Assoc, 41*, pp. 185–187.

Al-Majed, A.A., Daba, M.H., Asiri, Y.A., Al-Shabanah, O.A., Mostafa, A.A. and El-Kashef, H.A., 2001. Thymoquinone-induced relaxation of guinea-pig isolated trachea. *Res Commun Mol Pathol Pharmacol, 110*, pp. 333–345.

Alam, S., Reddy, S.K., Baig, A., Reddy, M.K., Mohiuddin, M., Reddy, M.V. et al., 2013. Evaluation of antidiabetic and anti-lipidimic potential of kalongi sugar powder water extract in stz induced diabetic rats. *Int J Pharm Sci, 5*, pp. 94–96.

Alenzi, F., El-Bolkiny, Y.S. and Salem M., 2010. Protective effects of *Nigella sativa* oil and thymoquinone against toxicity induced by the anti-cancer drug cyclophosphamide. *Br J Biomed Sci, 67*, pp. 20–28.

Ali, B.H. and Blunden, G., 2003. Pharmacological and toxicological properties of *Nigella sativa*. *Phytother Res, 17*, pp. 299–305.

Ali, S.A., Asghar, F., Nafees, M. and Tayyab, M., 2012. Effect of *Nigella sativa* (Kalonji) on serum lipid profile. *Annals, 18*, pp. 224–228.

Al-Naqeep, G., Al-Zubairi, A.S., Ismail, M., Amom, Z.H. and Esa, N.M., 2011. Antiatherogenic potential of *Nigella sativa* seeds and oil in diet-induced hypercholesterolemiain rabbits. *Evid Based Complement Alternat Med.*

Aqel, M. and Shaheen, R., 1996. Effects of the volatile oil of *Nigella sativa* seeds on the uterine smooth muscle of rat and guinea pig. *J Ethnopharmacol, 52*, pp. 23–26.

Asgary, S., Ghannadi, A., Dashti, G., Helalat, A., Sahebkar, A. and Najafi, S., 2013. *Nigella sativa* L. improves lipid profile and prevents atherosclerosis: Evidence from an experimental study on hypercholesterolemic rabbits. *J Functional Foods, 5*, pp. 228–234.

Badary, O.A., Abdel-Naim, A.B., Abdel-Wahab, M.H. and Hamada, F.M., 2000. The influence of TQ on doxorubicin-induced hyperlipidemic nephropathy in rats, *Toxicology, 143*, pp. 219–226.

Bamosa, A.O., Ali, B. and Sowayan, S., 1997. Effect of oral ingestion *Nigella sativa* seeds on some blood parameters. *Saudi Pharm J, 5*, pp. 126–129.

Bamosa, A.O., Ali, B.A. and al-Hawsawi, Z.A., 2002. The effect of thymoquinone on blood lipids in rats. *Indian J Physiol Pharmacol, 46*, pp. 195–201.

Baytop, T., 1984. Türkiye'de Bitkiler İle Tedavi. İ.Ü. Yayınları No: 3255.

Benkaci-Ali, F., Baaliouamer, A., Meklati, B.Y. and Chemat, F., 2007. Chemical composition of seed essential oils from Algerian *Nigella sativa* extracted by microwave and hydrodistillation. *Flavour Frag J, 22*, pp. 148–153.

Bhatti, I.U., Rehman, F.U., Khan, M. and Marwat, S., 2009. Effect of prophetic medicine kalonji (*Nigella sativa* L.) on lipid profile of human beings. An *in vivo* approach. *World Appl Sci J, 6*, pp. 1053–1057.

Bohn, L.M., Lefkowitz, R.J., Gainetdinov, R.R., Peppel, K., Caron, M.G. and Lin, F.T., 1999. Enhanced morphine analgesia in mice lacking β-arrestin 2. *Science, 286*, pp. 2495–2498.

Borek, C., 2004. Antioxidants and cancer. *Sci Med (Phila), 4*, pp. 51–62.

Bourgou, S., Pichette, A., Marzouk, B. and Legault, J., 2010. Bioactivities of black cumin essential oil and its main terpenes from Tunisia. *S Afr J Bot, 76*, pp. 210–216.

Burits, M. and Bucar, F., 2000. Antioxidant activity of Nigella sativa essential oil. *Phytother Res, 14*, pp. 323–328.

Campieri, M., Gionchetti, P., Belluzzi, A., Brignola, C., Tampieri, M., Iannone, P., Miglioli, M. and Barbara, L., 1991. Optimum dosage of 5-aminosalicylic acid as rectal enemas in patients with active ulcerative colitis. *Gut, 32*, pp. 929–931.

Cartier, L., Hartley, O., Dubois-Dauphin, M. and Krause, K.H., 2005. Chemokine receptors in the central nervous system: role in brain inflammation and neurodegenerative diseases. *Brain Res Rev, 48*, pp. 16–42.

Chakrabarty, A., Emerson, M.R. and LeVine, S.M., 2003. Hemeoxygenase-1 in SJL mice with experimental allergic encephalomyelitis. *Mult Scler J, 9*, pp. 372–381.

Choudhary, S., Keshavarzian, A., Yong, S., Wade, M., Bocckino, S., Day, B.J. and Banan, A., 2001. Novel antioxidants zolimid and AEOL11201 ameliorate colitis in rats. *Dig Dis Sci, 46*, pp. 2222–2230.

Corder, C., Benghuzzi, H., Tucci, M. and Cason, Z., 2003. Delayed apoptosis upon the treatment of Hep-2 cells with black seed. *Europe PMC, 39*, pp. 365–370.

Dehkordi, F.R. and Kamkhah, A.F., 2008. Antihypertensive effect of Nigella sativa seed extract in patients with mild hypertension. *Fundam Clini Pharmacol., 22*, pp. 447–452.

El-Dakhakhny, M., 1965. Studies on the Egyptian *Nigella sativa* L. part IV: Some pharmacological properties of the seed's active principle in comparison to its dihydro compound and its polymer. *Arzneim Forsch, 15*, pp. 1227–1229.

El-Dakhakhny, M., Barakat, M., El-Halim, M.A. and Aly, S.M., 2000. Effects of *Nigella sativa* oil on gastric secretion and ethanol induced ulcer in rats. *J Ethnopharmacol, 72*, pp. 299–304.

El-Dakhakhny, M., Madi, N.J., Lembert, N. and Ammon, H.P., 2002. *Nigella sativa* oil, nigellone and derived thymoquinone inhibit synthesis of 5-lipoxygenase products in polymorphonuclear leukocytes from rats. *J Ethnopharmacol, 81*, pp. 161–164.

El-Fatatry, H.M., 1975. Isolation and structure assignment of an antimicrobial principle from the volatile oil of *Nigella sativa* L. seeds. *Die Pharmazie, 30*, pp. 109–111.

El-Mahmoudy, A., Matsuyama, H., Borgan, M.A. and Takewaki, T., 2002. Thymoquinone suppresses expression of inducible nitric oxide synthase in rat. *Int J Immunopharmacol, 2*, pp. 1603–1611.

Enomoto, S., Asano, R., Iwahori, Y., Narui, T., Okada, Y., Singab, A.N. and Okuyama, T., 2001. Hematological studies on black cumin oil from the seeds of *Nigella sativa* L. *Biol Pharm Bull, 24*, pp. 307–310.

Erşahin, M., Toklu, H.Z., Akakin, D., Yuksel, M., Yeğen, B.Ç. and Sener, G., 2011. The effects of *Nigella sativa* against oxidative injury in a rat model of subarachnoid hemorrhage. *Acta Neurochir, 153*, pp. 333–341.

Gali-Muhtasib, H., Diab-Assaf, M., Boltze, C., Al-Hmaira, J., Hartig, R., Roessner, A. and Schneider-Stock, R., 2004. Thymoquinone extracted from black seed triggers apoptotic cell death in human colorectal cancer cells via a p53-dependent mechanism. *Int J Oncol, 25*, pp. 857–866.

Gilani, A., Jabeen, Q. and Khan, M., 2004. *Nigella sativa*'nın tıbbi kullanımları ve farmakolojik aktivitelerinin gözden geçirilmesi. *Pak J Biol Sci, 7,* pp. 441–451.

Gionchetti, P., Guarnieri C., Campieri, M., Belluzzi A., Brignola, C., Iannone, P., Miglioli, M. and Barbara, L., 1991. Scavenger effect of sulfasalazine, 5-aminosalicylic acid, and olsalazine on superoxide radical generation. *Dig Dis Sci, 36,* pp. 174–178.

Gutteridge, J.M.C., 1993. Free radicals in disease processes: A compilation of cause and consequence. *Free Radic Res Commun, 19,* pp. 141–158.

Gürdöl, F. and Ademoğlu, E., 2006. Biyokimya. Nobel Tıp Kitabevi, İstanbul. 1. Baskı.

Hanafy, M.S.M. and Hatem, M.E., 1991. Studies on the antimicrobial activity of *Nigella sativa* seed (black cumin). *J Ethnopharmacol, 34,* pp. 275–278.

Hawsawi, Z.A., Ali, B.A. and Bamosa, A.O., 2001. Effect of *Nigella sativa* (black seed) and thymoquinone on blood glucose in albino rats. *Ann Saudi Med, 21,* pp. 242–244.

Hosseinzadeh, H., Parvardeh, S., Asl, M.N., Sadeghnia, H.R. and Ziaee, T., 2007. Effect of thymoquinone and *Nigella sativa* seeds oil on lipid peroxidation level during global cerebral ischemia-reperfusion injury in rat hippocampus. *Phytomedicine, 14,* pp. 621–627.

Hosseinzadeh, H., Parvardeh, S., Nassiri-Asl, M. and Mansouri, M.T., 2005. Intracerebroventricular administration of thymoquinone, the major constituent of *Nigella sativa* seeds, suppresses epileptic seizures in rats. *Med Sci Monit, 11,* pp. 106–110.

Hosseizadeh, H. and Parvardeh, S., 2004. Anticonvulsant effect of thymoquinone, the majör constituent of *Nigella sativa* seeds, in mice. *Phytomedicine, 11,* pp. 56–64.

Houghton, P.J., Zarka, R., Heras, B. and Hoult, J.R.S., 1995. Fixed oil of *Nigella sativa* and derived thymoquinone İnhibit eicosanoid generation in leukocytes and membrane lipid peroxidation. *Planta Med, 61,* pp. 33–36.

Ibrahim, R.M., Hamdan, N.S., Mahmud, R., Imam, M.U., Saini, S.M., Rashid, S.N. et al., 2014. A randomised controlled trial on hypolipidemic effects of *Nigella sativa* seeds powder in menopausal women. *J Transl Med, 12,* p. 82.

Idris-Khodja, N. and Schini-Kerth, V., 2012. Thymoquinone improves aging-related endothelial dysfunction in the rat mesenteric artery. *Naunyn Schmiedebergs Arch Pharmacol, 385,* pp. 749–758.

Islam, S.K., Begum, P., Ahsan, T., Huque, S. and Ahsan, M., 2004. Immunosuppressive and cytotoxic properties of *Nigella sativa*. *Phytother Res, 18,* pp. 395–398.

Jamal, A., Hamza, A., Omar, E., Adnan, A. and Osman, M.T., 2013. *Nigella sativa* oil has significant repairing ability of damaged pancreatic tissue occurs in induced type 1 *diabetes mellitus*. *Global J Pharmacol, 7,* pp. 14–19.

Kallinich, T., Beier, K.C., Gelfand, E.W., Kroczek, R.A. and Hamelmann, E., 2005. Co-stimulatory molecules as potential targets for therapeutic intervention in allergic airway disease. *Clin Exp Allergy, 35,* pp. 1521–1534.

Kalus, U., Pruss, A., Bystron, J., Jurecka, M., Smekalova, A., Lichius, J.J. and Kiesewetter, H., 2003. Effect of *Nigella sativa* (black seed) on subjective feeling in patients with allergic diseases. *Phytother Res, 17,* pp. 1209–1214.

Kanter, M., 2008a. *Nigella sativa* and derived thymoquinone prevents hippocampal neurodegeneration after chronic toluene exposure in rats. *Neurochem Res, 33,* p. 579.

Kanter, M., 2008b. Effects of *Nigella sativa* and its major constituent, thymoquinone on sciatic nerves in experimental diabetic neuropathy. *Neurochem Res, 33,* pp. 87–96.

Kanter, M., Coskun, O. and Budancamanak, M., 2005. Hepatoprotective effects of *Nigella sativa* L. and *Urtica dioica* L. on lipid peroxidation, antioxidant enzyme systemsand liver enzymes in carbon tetrachloride treated rats. *World J Gastroenterol, 11,* pp. 6684–6688.

Khan, M.A.U., Ashfaq, M.K., Zuberi, H.S., Mahmood, M.S. and Gilani, A.H., 2003a. The *in vivo* antifungal activity of the aqueous extract from *Nigella sativa* seeds. *Phytother Res, 17,* pp. 183–186.

Khan, N., Sharma, S. and Sultana, S., 2003b. *Nigella sativa* (black cumin) ameliorates potassium bromate-induced early events of carcinogenesis: Diminution of oxidative stress, *Hum Exp Toxicol, 22,* pp. 193–203.

Koch, T.R., Yuan, L.X., Stryker, S.J., Ratliff, P., Telford, G.L. and Opara, E.C., 2000. Total antioxidant capacity of colon in patients with chronic ulcerative colitis. *Dig Dis Sci, 45,* pp. 1814–1819.

Laakso, P. and Voutilainen, P., 1996. Analysis of triacylglycerols by silver-ion high-performance liquid chromatography-atmospheric pressure chemical ionization mass spectrometry. *Lipids, 31,* pp. 1311–1322.

Mahgoub, A.A., 2003. Thymoquinone protects against experimental colitis in rats. *Toxicol Lett, 143,* pp. 133–143.

Mahmoud, M.R., El-Abhar, H.S. and Saleh, S., 2002. The effect of *Nigella sativa* oil against the liver damage induced by Schistosoma mansoni infection in mice. *J Ethnopharmacol, 79,* pp. 1–11.

Mansour, M.A., Nagi, M.N., El-Khatib, A.S. and Al-Bekairi, A.M., 2002. Effects of thymoquinone on antioxidant enzyme activities, lipid peroxidation and DT-diaphorase in different tissues of mice: A possible mechanism of action, *Cell Biochem Funct 20,* pp. 143–151.

Mansour, M. and Tornhamre, S., 2004. Inhibition of 5-lipoxygenase and leukotriene C4 synthase in human blood cells by thymoquinone. *J Enzyme Inhib Med Chem, 19,* pp. 431–436.

Medzhitov, R. and Janeway Jr, C., 2000. Innate immunity. *N Engl J Med, 343,* pp. 338–344.

Meral, I., Donmez, N., Baydas, B., Belge, F. and Kanter, M., 2004. Effect of *Nigella sativa* L. on heart rate and some haematological values of alloxan-induced diabetic rabbits. *Scand J Lab Anim Sci, 31,* pp. 49–53.

Meydani, S.N., Meydani, M. and Blumberg, J.B., 1990. Antioxidants and the aging immune response. *Adv Exp Med Biol, 262,* pp. 57–67.

Mohamed, A., Shoker, A., Bendjelloul, F., Mare, A., Alzrigh, M., Benghuzzi, H. and Desin, T., 2003. Improvement of experimental allergic encephalomyelitis (EAE) by thymoquinone; an oxidative stress inhibitor. *Europe PMC, 39,* pp. 440–445.

Moro, D., Lloyd, M.L., Smith, A.L., Shellam, G.R. and Lawson, M.A., 1999. Murine viruses in an island population of introduced house mice and endemic short-tailed mice in Western Australia. *J Wildl Dis, 35,* pp. 301–310.

Morsi, N.M., 2000. Antimicrobial effect of crude extracts of *Nigella sativa* on multiple antibiotics-resistant bacteria. *Acta Microbiol Pol, 49,* pp. 63–74.

Muhammad, T.S., Masood, S.B., Faqir, M.A., Amer, J., Saeed, A., Muhammad, N., 2009. Nutritional profile of indigenous cultivar of black cumin seeds andantioxidant potentional of its fixed and essential oil. *Pak J Bot, 41*(3), pp. 1321–1330.

Nader, M.A., El-Agamy, D.S. and Suddek, G.M., 2010. Protective effects of propolis and thymoquinone on development of atherosclerosis in cholesterol-fed rabbits. *Arch Pharmacol Res, 33*, pp. 637–643.

Nasir, A., Siddiqui, M.Y. and Mohsin, M., 2013. Efficacy of Saboose-Asapghol (*Plantago ovata*) and Kalonji (*Nigella sativa*) in the management of hypertriglyceridemia. *Int J Pharm India, 2*, pp. 560–568.

Nieto, N., Torres, M.I., Fernández, M.I., Girón, M.D., Ríos, A., Suárez, M.D. and Gil, A., 2000. Experimental ulcerative colitis impairs antioxidant defense system in rat intestine. *Dig Dis Sci, 45*, pp. 1820–1827.

Pathak, S.K., Sharm, R.A., Steward, W.P., Mellon, J.K., Griffiths, T.R.L. and Gescher, A.J., 2005. Oxidative stress and cyclooxygenase activity in prostate carcinogenesis: Targets for chemopreventive strategies. *Eur J Cancer, 41*, pp. 61–70.

Qidwai, W., Hamza, H.B., Qureshi, R. and Gilani, A., 2009. Effectiveness, safety, and tolerability of powdered *Nigella sativa* (kalonji) seed in capsules on serum lipid levels, blood sugar, blood pressure, and body weight in adults: Results of a randomized, doubleblind controlled trial. *J Altern Complement Med, 15*, pp. 639–644.

Ramadan, M.F., 2007. Nutritional value, functional properties and nutraceutical applications of black cumin (*Nigella sativa* L.) oilseeds: An overview. *Int J Food Sci Technol, 42*, pp. 1208–1218.

Randhawa, M.A. and Al-Ghamdi, M.S., 2002. A review of the pharmaco-therapeutic effects of *Nigella sativa. Pak J Med Res, 41*, pp. 77–83.

Rastogi, L., Feroz, S., Pandey, B.N., Jagtap, A. and Mishra, K.P., 2010. Protection against radiation-induced oxidative damage by an ethanolic extract of *Nigella sativa* L. *Int J Radiat Biol, 86*, pp. 719–731.

Reynolds, R.P., Rahija, R.J., Schenkman, D.I. and Richter, C.B., 1993. Experimental murine cytomegalovirus infection in severe combined immunodeficient mice. *Lab Anim Sci, 43*, pp. 291–295.

Sabzghabaee, A.M., Dianatkhah, M., Sarrafzadegan, N., Asgary, S. and Ghannadi, A., 2012. Clinical evaluation of *Nigella sativa* seeds for the treatment of hyperlipidemia: A randomized, placebo controlled clinical trial. *Med Arch, 66*, pp. 198–200.

Salem, M.L., 2005. Immunomodulatory and therapeutic properties of the *Nigella sativa* L. seed. *Int Immunopharmacol, 5*, pp. 1749–1770.

Sallal, A.K.J. and Alkofahi, A., 1996. Inhibition of the hemolytic activities of snake and scorpion venoms in vitro with plant extracts. *J Biomed Lett, 212*, pp. 211–215

Salomi, N.J., Nair, S.C., Jayawardhanan, K.K., Varghese, C.D. and Panikkar, K.R., 1992. Antitumour principles from *Nigella sativa* seeds. *Cancer Lett, 63*, pp. 41–46.

Salomi, M.J., Nair, S.C. and Panikkar, K.R., 1991. Inhibitory effects of *Nigella sativa* and saffron (*Crocus sativus*) on chemical carcinogenesis in mice. *Nutr Cancer, 16*, pp. 67–72.

Shabana, A., El-Menyar, A., Asim, M., Al-Azzeh, H. and Al Thani, H., 2013. Cardiovascular benefits of black cumin (*Nigella sativa*). *Cardiovascular Toxicol, 13*, pp. 9–21.

Shah, S. and Kasturi, S.R., 2003. Study on antioxidant and antimicrobial properties of black cumin (*Nigella sativa* Linn). *J Food Sci Technol, 40*, pp. 70–73.

Sobhi, W., Khettal, B., Belmouhoub, M., Atmani, D., Duez, P. and Benboubetra, M., 2013. Hepatotoxicity and Langerhans islets regenerative effects of polar and neutrallipids of *Nigella sativa* L. in nicotinamide/streptozotocin-induced diabetic rats. *Pteridines, 22*, pp. 97–104.

Sowers, J.R. and Epstein, M., 1995. *Diabetes mellitus* and associated hypertension, vascular disease, and nephropathy an update. *Hypertension, 26*, pp. 869–879.

Swamy, S.M.K. and Tan, B.K.H., 2000. Cytotoxic and immunopotentiating effects of ethanolic extract of *Nigella sativa* L. seeds. *J Ethnopharmacol, 70*, pp. 1–7.

Suboh, S.M., Bilto, Y.Y. and Aburjai, T.A., 2004. Protective effects of selected medicinal plants against protein degradation, lipid peroxidation and deformability loss of oxidatively stressed human erythrocytes. *Phytother Res, 18*, pp. 280–284.

Suddek, G.M., 2010. Thymoquinone-induced relaxation of isolated rat pulmonary artery. *J Ethnopharmacol, 127*, pp. 210–214.

Tahir, K.E., Ashour, M.M. and Al-Harbi, M.M., 1993. The respiratory effects of the volatile oil of the black seed (*Nigella sativa*) in guinea-pigs: Elucidation of the mechanism (s) of action. *Gen Pharmacol, 24*, pp. 1115–1122.

Tasawar, Z., Siraj, Z., Ahmad, N. and Lashari, M.H., 2011. The effects of *Nigella sativa* (Kalonji) on lipid profile in patients with stable coronary artery disease in multan, Pakistan. *Pak J Nutr, 10*, p. 162.

Türkdoğan, M.K., Ozbek, H., Yener, Z. and Tuncer, I., 2003. The role of *Urtica dioica* and *Nigella sativa* in the prevention of carbon tetrachloride-induced hepatotoxicity in rats. *Phytother Res, 17*, pp. 942–946.

Vanhoutte, P., Shimokawa, H., Tang, E. and Feletou, M., 2009. Endothelial dysfunction and vascular disease. *Acta Physiol, 196*, pp. 193–222.

Yildiz, F., Coban, S., Terzi, A., Savas, M., Bitiren, M., Celik, H. and Aksoy, N., 2010. Protective effects of *Nigella sativa* against ischemia-reperfusion injury of kidneys. *Ren Fail, 32*, pp. 126–131.

Yusuf, S., Reddy, S., Ôunpuu, S. and Anand, S., 2001. Global burdenof cardiovascular diseases part I: General considerations, the epidemiologic transition, risk factors, and impact of urbanization. *Circulation, 104*, pp. 2746–2753.

Zaoui, A., Cherrah, Y., Alaoui, K., Mahassine, N., Amarouch, H. and Hassar, M., 2002. Effects of *Nigella sativa* fixed oil on blood homeostasis in rat. *J Ethnopharmacol, 79*, pp. 23–26.

Zaoui, A., Cherrah, Y., Lacaille-Dubois, M., Settaf, A., Amarouch, H. and Hassar, M., 1999. Diuretic and hypotensive effects of *Nigella sativa* in the spontaneously hypertensive rat. *Therapie, 55*, pp. 379–382.

9

Indian Tropical Fruits and Role of Their Bioactive Compounds against Human Diseases

Pooja Rawat, Pawan Kumar Singh, and Vipin Kumar

CONTENTS

9.1 Introduction

Fruits are an integral food for every common man, both in India and across the globe. Every ingredient taken as food plays a vital role in maintaining and determining the human health. Numerous epidemiological, *in-vitro, in-vivo* and clinical trials data highlights the health benefits of fruits and the protective role of their phyto-constituents against multiple ailments. With the advancement of technological tools, increase in publications has been observed with overwhelming numbers of studies carried out by the researchers on the extraction and characterization of bioactive phytoconstituents from plants, including fruits. The health benefits from tropical fruits are evident from the outcomes of the detailed investigations done by several researchers across the globe. An abundant diversity of tropical fruit trees are found in all the tropical continents of globe. Africa is reported to hold 1200 tree species, followed by America (1,000), and Asia (500; of which 300 species are reported from Indian sub-continent). While, the local consumption of fresh fruits accounts for more than 90%, the demand for import of tropical fruits has seen a steady increasing trend (Sthapit and Scherr 2012). Most of the production occurs in developing countries (98%), while developed countries are the major importers (80%) of fruits. Many of the tropical fruits are very rich in nutrients as well as a good amount of anti-oxidants (Mahattanatawee et al. 2006; Lim et al. 2007). Many fruits contain high amount of anthocyanins (Sousa De Brito et al. 2007) as well total polyphenol and dietary fiber (Gorinstein et al. 1999), which are well known for their therapeutic values. Multiple studies have investigated the bioactive compounds in pulp and by-products of tropical fruits wherein it was observed that in some cases, fruit by-products contain higher bioactive content than pulp (Costa et al. 2013; Da Silva et al. 2014). The tropical fruits are rich in vitamin and minerals and therefore have health promoting effects (Hernández et al. 2006; Leterme et al. 2006; Mertz et al. 2009). Indian tropical fruits include mango (*Mangifera indica*), grape (*Vitis vinifera*),

aonla (*Phyllanthus emblica*), banana (*Musa paradisiaca*), sapota (*Manilkara zapota*), mangosteen (*Garcinia mangostana*), sweet orange (*Citrus sinensis*), mandarin (*Citrus reticulata*), carambola (*Averrhoa carambola*), jackfruit (*Artocarpus heterophyllus*), and custard apple (*Annona reticulata*) (Usha 2015). These tropical fruits have high nutritional value and have been thoroughly investigated for their phytochemical composition through multiple studies (Kawaii et al. 1999; Ma et al. 2003; Habib-ur-Rehman et al. 2007; Baliga et al. 2011; Pierson et al. 2014). These fruits offer great health benefits as evident from their traditional uses for treatment of multiple ailments (Jagtap and Bapat 2010; Debjit Bhowmik et al. 2012; Saghir et al. 2013; Rawat et al. 2017; Yadav et al. 2017). In this chapter, phytoconstituents isolated and characterized from selected Indian tropical fruits have been highlighted. These phytochemicals were isolated from different parts such as seed kernels, fruit peel, pericarp and pulp.

The main objective here is to identify the bioactive constituents of tropical fruits and review of their therapeutic roles in various diseases including chronic ones. Supportive research data highlights the importance of tropical fruit intake in daily diet and presents these as an important functional food.

9.2 Phytocompounds Derived from Tropical Fruits

An investigation of the published data on Indian tropical fruits revealed that these have been rigorously studied in terms of their phytochemical constituents. Compounds were fractionated from various parts such as fruit peel, pulp, seeds, juice, pericarp, and whole fruit also. However, few such as *Psidium guajava, Ananas comosus, Carica papaya* fruits have still not been explored much in terms of their phyto-constituents/compounds/molecules, though the studies on identification of chemicals from other parts of these plants can be found (Singh and Ali 2011).

Commercial usage of the fruits is mostly focused on the utilization of the pulp portion while other parts such as peels, kernels, and so on, are discarded during the process. Ongoing research studies have highlighted the importance of each part of fruit in terms of their important constituents. These studies highlighting presence of phytoconstituents in different portions of fruits may be referred to draw conclusions on their recovery from by-products. Several researchers have characterized and isolated large number of compounds from *Mangifera indica*. In one such study, phenolic compounds and anti-oxidant activities of four Brazilian mango varieties were investigated. Characterization of phyto-constituents was performed using liquid chromatography-electrospray ionization-tandem mass spectrometry (LC-ESI-MS) analysis. In the pulp, peels and seed kernels of *M. indica*, 12 flavonoids and xanthones were identified. The organically grown variety was found to have higher amount of these phytochemicals in comparison to conventionally produced variety (Ribeiro et al. 2008). Volatile constituents from fruit oils were analyzed from different mango cultivars Langra, Bombay, Desi (from India) using gas chromatography mass spectrometry (GC/MS). Predominant constituents reported were δ-3-carene (61%) in Langra, α-pinene (26%) and caryophyllene oxide (14%) in Bombay and *p*-cymend-ol (28%) in Desi cultivar (Ansari

et al. 1999). Study conducted by Berardini et al. (2004) showed that mango kernels are rich source of gallotannins, with very small amounts also present in peel and pulp. 18 gallotannins and five benzophenone derivatives were extracted from the peel and characterized using high pressure liquid chromatography electrospray ionization tandem mass spectroscopy (HPLC/ESI/MS) (Berardini et al. 2004). Mango peels have also been identified to be a rich source of pectin as well as polyphenolics, which provides rationale for commercial utilization of by-products. Also the characteristics profile of flavonol glycosides present in the peel may be utilized as an authenticity control of mango-based products produced from unpeeled fruits (Berardini et al. 2005).

Carotenoids present in *Artocarpus heterophyllus* kernels were extracted and identified as β-carotene, α-carotene, β-zeacarotene, α-zeacarotene and β-carotene-5,6-epoxide and crocetin. The study demonstrated the importance of *A. heterophyllus* as a good source of provitamin A carotenoids (Chandrika et al. 2005). Limited numbers of research studies on phytochemical investigations of *Musa paradisiaca*, *Manilkara zapota*, *Phyllanthus emblica*, *Annona reticulata*, *Vitis vinifera*, *Carica papaya*, and *Citrus sinensis* have also been carried out by different researchers. Important phytoconstituents characterized in the tropical fruits are summarized in Table 9.1.

TABLE 9.1

Phytocompounds Identified in Indian Tropical Fruits

Tropical Fruits	Botanical Name	Phytocompounds	References
Mango	*Mangifera indica*	Mangiferin (Pu, Pe)	Ribeiro et al. (2008)
		Isomangiferin (Pu, Pe)	
		Mangiferin gallate (Pu, Pe)	
		Isomangiferin gallate (Pu, Pe)	
		Quercetin 3-O-gal (Pu, Pe)	
		Quercetin 3-O-glc (Pu, Pe)	
		Quercetin 3-O-xy (Pu, Pe)	
		Quercetin 3-O-arap (Pu, Pe)	
		Quercetin 3-O-araf (Pu, Pe)	
		Quercetin 3-O-rha (Pu, Pe)	
		Kaempferol 3-O-glc (Pu, Pe)	
		Quercetin (Pu, Pe)	
		Rhamnetin 3-O-â-galactopyranoside (Pe)	Berardini et al. (2005)
		Rhamnetin 3-O-â-glucopyranoside (Pe),	
		Volatile oils-	Ansari et al. (1999)
		δ-3-carene (Pu+Pe)	
		α-pinene (Pu+Pe)	
		Caryophyllene oxide (Pu+Pe)	
		Humulene oxide (Pu+Pe)	
		p-cymen-8-ol (Pu+Pe)	
		Maclurin mono-O-galloyl-glucoside (Pe)	Berardini et al. (2004)
		Maclurin di-O-galloyl-glucoside (Pe)	
		Tetra-O-galloyl-glucose (Pe)	
		Iriflophenone di-O-galloyl-glucoside (Pe)	
		Maclurin tri-O-galloyl-glucoside (Pe)	
		Tetra-O-galloyl-glucose (Pe)	
		Penta-O-galloyl-glucose (Pe)	
		Hexa-O-galloyl-glucose (Pe)	
		Hepta-O-galloyl-glucose (Pe)	
		Octa-O-galloyl-glucose (Pe)	
		Nona-O-galloyl-glucose (Pe)	

(Continued)

TABLE 9.1 (*Continued*)

Phytocompounds Identified in Indian Tropical Fruits

Tropical Fruits	Botanical Name	Phytocompounds	References
Banana	*Musa paradisiaca*	Sitoindoside-I (Pu) Sitoindoside-II (Pu) Sitoindoside-III (Pu) Sitoindoside-IV (Pu) Sitosterol gentiobioside (Pu) Sitosterol myo-inosityl-β-D-glucoside (Pu)	Ghosal (1985)
Papaya	*Carica papaya*	Lycopene β-criptoxanthin β-carotene Ferulic acid p-coumaric acid Caffeic acid	Gayosso-García Sancho et al. (2011)
Pineapple	*Ananas comosus*	—	
Guava	*Psidium guajava*	—	
Sapota	*Manilkara zapota*	β- and α-amyrin esters Hexadecanoic acid Hexadecanoic acid ethyl esters Octadecanoic acid ethyl esters	Fernandes et al. (2013)
Aonla	*Phyllanthus emblica*	Gallic acid Ellagic acid 1-O-galloyl-beta-D-glucose 3,6-di-O-galloyl-D-glucose Chebulinic acid Quercetin Chebulagic acid Corilagin 3-ethylgallic acid (3-ethoxy-4,5-dihydroxy-benzoic acid), Isostrictiniin, 1,6-di-O-galloyl-beta-D-glucose	Zhang et al. (2003)
Custard apple	*Annona reticulata*	Annotemoyin-1 (S)	Usha (2015)
		Spathulenol (PeC) n-Hexadecanoic acid (PeC)	Chen et al. (2017)
Jack fruit	*Artocarpus heterophyllus*	Butyl acetate 3-Methylbutyl acetate Butanol Propyl isovalerate 3-Methylbutanol Butyl isovalerate Isoamyl isovalerate 3-Hydroxy 3-methyl butanoate Ethyl 3-hydroxy-3-methylbutanoate Hexyl isovalerate Benzaldehyde Isovaleric acid Methyl dodecanoate Octadecane 3-Phenylpropanol Isopropyl myristate Isovaleric acid, 3-phenylpropyl ester Hexadecanoic acid Stearic acid Oleic acid 4-Hexadecen-6-yne	Ong et al. (2006)
		β-carotene α-carotene β-zeacarotene α-zeacarotene β-carotene-5,6-epoxide Crocetin	Chandrika et al. (2005)

(*Continued*)

TABLE 9.1 (*Continued*)

Phytocompounds Identified in Indian Tropical Fruits

Tropical Fruits	Botanical Name	Phytocompounds	References
Grape	*Vitis vinifera*	Oleanolic acid Isoquercitrin Quercitrin 3-O-β-D-glucuronate sodium Quercitrin 3-O-β-Dglucuronopyranoside ethyl ester Gallic acid Caftaric acid	Liu et al. (2012)
Sweet orange	*Citrus sinensis*	Eriocitrin (J) Narirutin (J) Naringin (J) Hesperidin (J) Neohesperidin (J)	Ooghe et al. (1994)

Note: Pu—Pulp, Pe—Peel, S—Seed, PeC—Pericarp, J—Juice.

9.3 Biological Activities of Phytocompounds and Their Applications in Various Diseases

The nutritional and medicinal importance of fruits are evident from their compositional characteristics. Additionally, with substantial evidences available in the form of scientific data on the phytochemical profiles as well as their biological actions, the fruits and their processed products are being recognized as functional foods. Consumption of fruits is known for its various health benefits including physical and mental well-being (Van Duyn and Pivonka 2000; Johnson et al. 2017). Increasing numbers of exploratory research on various bioactive phytochemicals isolated from tropical fruits and outcomes of these studies corroborate these findings and reckons their importance in reducing disease risk. Health benefits associated with consumption of fruits and vegetables are thought to be due to the therapeutic actions of synergistic combinations of phytoconstituents (Liu 2003). Research studies highlight the protective efficacy of several fruit phytomolecules, in large numbers of chronic and acute diseases, such as cancer (Steinmetz and Potter 1991), diabetes and associated complications (Du et al. 2017), atherosclerosis (Steffen et al. 2003), cardiovascular diseases (Bazzano et al. 2002), allergy (Rosenlund et al. 2011), neuro-related disorders (McMartin et al. 2013) and many more. The compounds isolated from tropical fruits have also been investigated for their potential as anti-bacterial, anti-viral, and anti-parasitic. The bioactives from fruits are powerful anti-oxidants and anti-inflammatory agents. Considering the fact of inflammation and oxidation being the hallmark of majority of the chronic illnesses, anti-inflammatory properties of fruits may explain beneficial effects imparted in diverse range of diseases. Important phytochemicals, characterized and extensively investigated in Indian tropical fruits, include crocetin (*Artocarpus heterophyllus*), hesperidin, naringin (*Citrus sinensis*), mangiferin, quercetin, quercetin 3-O-gluc (*Mangifera indica*), corilagin, chebulagic acid, chebulnic acid, ellagic acid (*Phyllanthus emblica*), gallic acid, isoquercetrin, oleanolic acid (*Vitis vinifera*). These phytocompounds have been isolated from various parts of the fruits including peel, seed kernel, pericarp, fruit hull and pulp. The bioactive molecules demonstrate multiple biological activities, with

mechanisms of action elucidated in many cases. For example, role of mangiferin present in *Mangifera indica* have been implicated in numbers of diseases including diabetes, hyperlipidemia and atherogenesis (Muruganandan et al. 2005). The compound has been shown to have anti-oxidant potential (Sánchez et al. 2000), a chemopreventive agent (Yoshimi et al. 2001), antitumor, immunomodulatory and anti-human immunodeficiency virus (HIV) (Guha et al. 1996), anti-viral (Heng and Lu 1990), anthelminthic and antiallergic (García et al. 2003). Mangiferin has also been investigated for its gastroprotective effects (Carvalho et al. 2007), as anti-asthmatic (Rivera et al. 2011), for treatment of bone diseases (Ang et al. 2011a), anti-carcinogenic (Rajendran et al. 2008), in diabetic nephropathy (Li et al. 2010), as hepatoprotective (Das et al. 2012), in cerebral protection (Bhatia et al. 2008), hypolipidemic (Guo et al. 2011), cardiovascular diseases (Prabhu et al. 2006). Furthermore, the compound has also demonstrated its ability to promote improvement in recognition memory (Pardo Andreu et al. 2010), periodontal disease (Carvalho et al. 2009) and also in stress-elicited neuroinflammation and oxidative damage (Márquez et al. 2012). Apart from mangiferin, many other bioactive phytochemicals present in *Mangifera indica*, are reported to have multiple biological activities (Table 9.2).

Ellagic acid, a bioactive from *Phyllanthus emblica* has been investigated and exhibits efficacy in multiple diseases including asthma (Rogerio et al. 2008) and cancer (Malik et al. 2011), acts as an anti-depressant (Dhingra and Chhillar 2012), anti-fungal (Silva Junior et al. 2010), anti-hemorrhagic (Gopalakrishnan et al. 2014), anti-malarial (Soh et al. 2009) and anti-oxidant (Han et al. 2006). The compound shows anxiolytic (Girish et al. 2013), cardioprotective effects (Warpe et al. 2015) and is also found to have beneficial role in diabetic nephropathy (Ahad et al. 2014). Ellagic acid also acts as an estrogen-receptor modulator (Papoutsi et al. 2005), alleviates nephrotoxicity (Ateşşahín et al. 2007) and effective in reducing skin wrinkles and inflammation (Bae et al. 2010). Corilagin, chebulagic acid, and chebulinic acid are other important phytochemicals identified in *Phyllanthus emblica* and summarized in Table 9.3.

Similarly oleanolic acid, a bioactive from *Vitis vinifera*, has been well investigated and was found effective as an anti-bacterial and anti-parasitic (Szakiel et al. 2008; Kim et al. 2015), anti-cancer and anti-apoptotic (Zhu et al. 2015), anti-diabetic

TABLE 9.2

Biological Activity Reported for Various Phytoconstituents Identified in *Mangifera indica*

Phytochemical	Biological Activities (References)
Caryophyllene oxide (Pu+Pe)	Anti-cancer (Fidyt et al. 2016); Anti-fungal (Yang et al. 1999) Anti-inflammatory and analgesic (Chavan et al. 2010)
Isomangiferin (Pu, Pe)	Anti-viral (Heng and Lu 1990)
Mangiferin (Pu, Pe)	Anthelminthic and antiallergic (García et al. 2003); antidiabetic, antihyperlipidemic and antiatherogenic (Muruganandan et al. 2005); anti-oxidant (Sánchez et al. 2000); chemopreventive agent (Yoshimi et al. 2001) Antitumor, Immunomodulatory and Anti-HIV (Guha et al. 1996); anti-viral (Heng and Lu 1990); Gastroprotective effects (Carvalho et al. 2007) Asthma (Rivera et al. 2011); Bone diseases (Ang et al. 2011); carcinogenesis (Rajendran et al. 2008); diabetic nephropathy (Li et al. 2010); hepatoprotective (Das et al. 2012); cerebral protection (Bhatia et al. 2008); hypolipidemic (Guo et al. 2011); cardiovascular diseases (Prabhu et al. 2006); improvement of recognition memory (Pardo Andreu et al. 2010); periodontal disease (Carvalho et al. 2009); stress elicited neuroinflammation and oxidative damage (Márquez et al. 2012)
Quercetin (Pu, Pe)	Anti-hypertensive (Edwards et al. 2007); Anti-allergic (Mlcek et al. 2016); Anti-angiogenic (Igura et al. 2001); Anti-cancer (Murakami et al. 2008); Anti-oxidant and anti-inflammatory (Lesjak et al. 2018); Bronchodilator effect (Joskova et al. 2011); Metastatic (Caltagirone et al. 2000); Neuroprotective (Ossola et al. 2009; Dajas 2012); Obesity (Ahn et al. 2008)
Quercetin 3-O-gal/ glc/ xy/ rha (Pu, Pe)	Anti-oxidative (Liu et al. 2005); Anti-allergic (Makino et al. 2013); Anti-cancer (Lee et al. 2015); Anti-diabetic, anti-oxidative (Panda and Kar 2007); Wound healing (Süntar et al. 2010); Anti-inflammation (Carmen Recio et al. 1995); Immunostimulant (Lee et al. 2016)
δ-3-carene (Pu+Pe)	Anti-fungal (Cavaleiro et al. 2006)

TABLE 9.3

Biological Activity Reported for Various Phytoconstituents Identified in *Phyllanthus emblica*

Phytochemical	Biological Activities (References)
1,6-di-O-galloyl-beta-D-glucose	Anti-atherogenic (Duan et al. 2005)
1-O-galloyl-beta-D-glucose	Anti-inflammatory (Chang et al. 2013)
Chebulagic acid	Anti-angiogenic (Lu and Basu 2015); Anti-cancer (Reddy et al. 2009); Anti-diabetic (Shyni et al. 2014); Anti-inflammation (Reddy and Reddanna 2009); Anti-viral (Yang et al. 2013); Gastroprotective (Liu et al. 2017); Neuroprotective (Kim et al. 2014)
Chebulinic acid	Anti-angiogenic (Lu et al. 2012); Anti-hypertensive (Lin et al. 1993); Anti-ulcer (Mishra et al. 2013); Anti-viral (Kesharwani et al. 2017); Neuroprotective (Song et al. 2018); Retinopathy (Sivasankar et al. 2015)
Corilagin	Anti-atherogenic (Duan et al. 2005); Anti-cancer (Jia et al. 2013); Anti-cystic fibrosis (Gambari et al. 2012); Anti-hyperalgesic (Moreira et al. 2013); Anti-hypertensive (Cheng et al. 1995); Anti-microbial (Li et al. 2013a, 2013b); Anti-oxidant and hepatoprotective (Kinoshita et al. 2007); Anti-tumor (Jia et al. 2013); Anti-viral (Yeo et al. 2015); Cranial radiation induced inflammation (Tong et al. 2016); Neurodegenerative diseases (Chen and Chen 2011); Pulmonary fibrosis (Wang et al. 2014)
Ellagic acid	Anti-asthma (Rogerio et al. 2008); Anti-cancer (Malik et al. 2011); Anti-depressant (Dhingra and Chhillar 2012); Anti-fungal (Silva Junior et al. 2010); Anti-hemorrhagic (Gopalakrishnan et al. 2014); Anti-malarial (Soh et al. 2009); Anti-oxidant (Han et al. 2006); Anxiolytic effects (Girish et al. 2013); Cardio-protective (Warpe et al. 2015); Diabetic nephropathy (Ahad et al. 2014); Estrogen receptor modulator (Papoutsi et al. 2005); Nephrotoxicity (Ateşşahín et al. 2007); Skin wrinkles and inflammation (Bae et al. 2010)

(Teodoro et al. 2008; Castellano et al. 2013), anti-hepatitic (Kong et al. 2013), anti-hepatotoxic (Liu et al. 1998; Reisman et al. 2009) and anti-hypertensive (Bachhav et al. 2011). Oleanolic acid has also been found to have anti-inflammatory (Dharmappa et al. 2009), anti-obesity (Sung et al. 2010), anti-schizophrenic (Park et al. 2014), anxiolytic and antidepressant activities (Fajemiroye et al. 2014). Additionally, the role of this molecule has been recognized in case of cognitive dysfunction (Jeon et al. 2017), immunomodulation (Raphael and Kuttan 2003), insulin resistance (Li et al. 2014) and also in wound healing (Moura-Letts et al. 2006). Other compounds present in *Vitis vinifera* and their biological actions have been summarized in Table 9.4.

Naringin present in tropical fruit *Citrus sinensis* has been extensively studied for its biological activities against numbers of ailments. The compound exhibited anti-hypercholesteraemic effects (Jung et al. 2003), anti-metastatic (Tan et al. 2014), anti-oxidant, anti-genotoxic (Bacanli et al. 2015) and anti-inflammatory effects (Jain and Parmar 2011). Naringin has been found to be useful in case of periodontitis (Tsui et al. 2008), shows anxiolytic action (Fernandez et al. 2009), useful in asthma (Jiao et al. 2015), helps combat cognitive impairment (Kumar et al. 2010), acts as a neuroprotective agent (Gopinath et al. 2011), and exhibits beneficial effects in case of obesity, cardiovascular disorders (Alam et al. 2013) and osteoporosis (Ang et al. 2011b; Li et al. 2013, 2014) (Table 9.5).

TABLE 9.4

Biological Activity Reported for Various Phytoconstituents Identified in *Vitis vinifera*

Phytochemical	Biological Activities (References)
Caftaric acid	Hepatoprotective (Koriem and Soliman 2014)
Gallic acid	Anti-allergic (Kim et al. 2006); Anti-angiogenic (Lu et al. 2010); Anti-arthritic (Yoon et al. 2013); Anti-hyperglycaemic, anti-oxidant (Punithavathi et al. 2011); Anti-microbial (Sarjit et al. 2015); Anti-oxidant (Roidoung et al. 2016); Chemopreventive (Raina et al. 2008); Glucose uptake (Vishnu Prasad et al. 2010); Hepatoprotective (Rasool et al. 2010); Wound healing (Yang et al. 2016)
Isoquercitrin	Anti-asthmatic (Fernandez et al. 2005); Anti-cancer (Amado et al. 2014; Chen et al. 2015); Anti-colitis (Cibiček et al. 2016); Anti-inflammation (Rogerio et al. 2007); Anti-oxidant (Jung et al. 2010) Diuretic (Gasparotto Junior et al. 2011); Fungicidal (Yun et al. 2015) Hepatoprotective (Xie et al. 2016)
Oleanolic acid	Anti-bacterial and anti-parasitic (Szakiel et al. 2008; Kim et al. 2015); Anti-cancer and anti-apoptotic (Zhu et al. 2015); Anti-diabetic (Teodoro et al. 2008); Anti-hepatitis C (Kong et al. 2013); Anti-hepatotoxic (Reisman et al. 2009); Anti-hepatotoxic (Liu et al. 1998); Anti-hypertension (Bachhav et al. 2011); Anti-inflammatory (Dharmappa et al. 2009); Anti-obesity inflammation (Sung et al. 2010); Anti-schizophrenic (Park et al. 2014); Anxiolytic and antidepressant effect (Fajemiroye et al. 2014); Cognitive dysfunction (Jeon et al. 2017); Diabetes (Castellano et al. 2013); Immunomodulatory role (Raphael and Kuttan 2003); Insulin resistance (Li et al. 2014); Wound healing (Moura-Letts et al. 2006)

TABLE 9.5

Biological Activity Reported for Various Phytoconstituents Identified in *Citrus sinensis*

Phytochemical	Biological Activities (References)
Eriocitrin (J)	Anti-cancer (Wang et al. 2016); Anti-oxidative (Miyake et al. 1997); Hepatic steatosis (Hiramitsu et al. 2014)
Hesperidin (J)	Anti-cancer (Park et al. 2008); Anti-hypertensive (Ikemura et al. 2012); Anti-inflammation (Jain and Parmar 2011); Anti-osteoporotic (Chiba et al. 2003); Anti-oxidant (Etcheverry et al. 2008); Hepatoprotective (Tirkey et al. 2005); Hypoglycaemic (Jung et al. 2004)
Naringin (J)	Anti-cancer (Li et al. 2013); Anti-hypercholesterolemic effect (Jung et al. 2003); Anti-metastatic (Tan et al. 2014); Anti-oxidant and anti-genotoxic (Bacanli et al. 2015); Anti-oxidative and anti-inflammatory (Jain and Parmar 2011); Anti-periodontal pathogens (Tsui et al. 2008); Anxiolytic action (Fernandez et al. 2009) Asthma (Jiao et al. 2015); Cognitive impairment (Kumar et al. 2010); Neuroprotective (Gopinath et al. 2011); Obesity and cardiovascular effects (Alam et al. 2013); Osteoporosis (Li et al. 2013)
Narirutin (J)	Airway inflammation (Funaguchi et al. 2007); Anti-alzheimer (Chakraborty and Basu 2017); Hepatoprotective (Park et al. 2013)
Neohesperidin (J)	Anti-ulcer (Hamdan et al. 2014); Osteoporosis (Tan et al. 2017)

Except crocetin, none of the other compound such as β-carotene, α-carotene, β-zeacarotene, α-zeacarotene, β-carotene-5,6-epoxide, etc., present in *Artocarpus heterophyllus* has been explored much for its pharmacological effects (Table 9.6). Crocetin has been demonstrated to have anti-atherosclerosis effects (He et al. 2007). Multiple roles of this compound as possible anti-cancer agent (Dhar et al. 2009; Zhong et al. 2011), anti-hyperlipidemic (Lee et al. 2005), anti-hypertensive (Mancini et al. 2014), anti-inflammation (Nam et al. 2010), anti-oxidant (Tseng et al. 1995), as a neuroprotectant (Ahmad et al. 2005), for treatment of retinal degeneration (Yamauchi et al. 2011) have also been investigated through several scientific exploratory studies.

As indicated above, vast numbers of phytochemicals present in the above tropical fruits provide the essential protective benefits in fighting against chronic and acute diseases. Several epidemiological studies demonstrate that dietary anti-oxidants may lower the risk of chronic diseases. Scientific studies further validate the role of individual phytoconstituents present in tropical fruits, therefore provide an explanation for the observed health benefits associated with their regular consumption.

TABLE 9.6

Biological Activity Reported for Various Phytoconstituents Identified in *Artocarpus heterophyllus*

Phytochemical	Biological Activities (References)
Crocetin	Anti-atherosclerosis (He et al. 2007); Anti-cancer (Dhar et al. 2009; Zhong et al. 2011) Anti-hyperlipidemic (Lee et al. 2005); Anti-hypertensive (Mancini et al. 2014); Anti-inflammation (Nam et al. 2010) Anti-oxidant (Tseng et al. 1995); Neuroprotection (Ahmad et al. 2005); Retinal degeneration (Yamauchi et al. 2011)
Hexadecanoic acid	Anti-inflammation (Aparna et al. 2012)
Oleic acid	Anti-atherosclerosis (Parthasarathy et al. 1990)
α-Carotene	Anti-cancer (Murakoshi et al. 1989, 1992; Narisawa et al. 1996)
β-Carotene	Anti-oxidant (Terao 1989)

However, one of the crucial questions that still needs to be investigated in detail is whether individual phytoconstituents provide similar levels of protection against the diseases as that of the whole fruits. There is a growing belief in the scientific community that purified compound isolated from the fruit may not behave the way the whole fruit behaves and loss of bioactivity may be possible in isolated form. This belief is based on few scientific studies carried out on compounds purified from tropical fruits. β-carotene which is abundantly present in green leafy vegetables, has been clinically investigated for its effects on incidences of cancer and cardiovascular diseases. It was observed that supplementation with β-carotene for 12 years doesn't result into any improvement in the incidences of malignant neoplasms, cardiovascular disease, or death from all causes (Greenberg et al. 1990; Hennekens et al. 1996). Such outcomes don't correlate with the other studies showing beneficial effects of vegetables and fruits consumption on incidences of diseases (Veer et al. 2000). Though very limited numbers of such reports contradicting the protective efficacy of isolated phytoconstituents, are available, however this data opens up the opportunity for the researchers to investigate the feasibility of using purified phytochemicals so as to achieve healthy life. The fruit as a whole or their phyto-constituents alone or in various value added combinations may be developed into functional food forms to tackle the specific diseases of women, child or elderly.

9.4 Conclusion

Increasing awareness and interest of consumers towards use of herbal nutraceuticals for health wellness and proven reports on immense benefits associated with fruit consumption makes tropical fruits ideal candidates for development of functional food in modern forms. Whole fruit consumption has always been an excellent choice for maintaining healthy life. However, consumption of whole fruit becomes a limitation in case of certain diseases. For example, mangiferin, a bioactive present in mango, has been shown to be have anti-hyperglycaemic effects. The compound shows suppressive effects on blood lipids and also prevents diabetic induced nephropathy progression. However, this fruit is generally not recommended by medical practitioners to patients having diabetes and associated complications. In such cases, there is an enormous need for research studies focusing on the effects of intake of purified constituents from the fruits on target diseases. In addition, in order to increase its consumption, attempts should also be made to create multiple edible stable formulations of tropical fruits and their phytochemicals to overcome the limitations of seasonal and geographical availability. Exorbitant numbers of studies no doubt highlights the importance of tropical fruits as an obvious source of nutrients, however much needs to be done in the area.

ACKNOWLEDGMENT

Authors are grateful to Prof Anil Gupta, Executive Vice Chairperson, National Innovation Foundation, India, for his honorary guidance and encouragement for carrying out this research activity.

REFERENCES

Ahad, Amjid, Ajaz Ahmad Ganai, Mohd Mujeeb, and Waseem Ahmad Siddiqui. 2014. Ellagic Acid, an NF-κB Inhibitor, Ameliorates Renal Function in Experimental Diabetic Nephropathy. *Chemico-Biological Interactions* 219, pp. 64–75. doi:10.1016/j.cbi.2014.05.011.

Ahmad, Abdullah Shafique, Mubeen Ahmad Ansari, Muzamil Ahmad, Sofiyan Saleem, Seema Yousuf, Md Nasrul Hoda, and Fakhrul Islam. 2005. Neuroprotection by Crocetin in a Hemi-Parkinsonian Rat Model. *Pharmacology Biochemistry and Behavior* 81 (4), pp. 805–13. doi:10.1016/j.pbb.2005.06.007.

Ahn, Jiyun, Hyunjung Lee, Suna Kim, Jaeho Park, and Taeyoul Ha. 2008. The Anti-Obesity Effect of Quercetin Is Mediated by the AMPK and MAPK Signaling Pathways. *Biochemical and Biophysical Research Communications* 373 (4), pp. 545–49. doi:10.1016/j.bbrc.2008.06.077.

Alam, Md Ashraful, Kathleen Kauter, and Lindsay Brown. 2013. Naringin Improves Diet-Induced Cardiovascular Dysfunction and Obesity in High Carbohydrate, High Fat Diet-Fed Rats. *Nutrients* 5 (3), pp. 637–50. doi:10.3390/nu5030637.

Amado, Nathália G., Danilo Predes, Barbara F. Fonseca, Débora M. Cerqueira, Alice H. Reis, Ana C. Dudenhoeffer, Helena L. Borges, Fábio A. Mendes, and Jose G. Abreu. 2014. Isoquercitrin Suppresses Colon Cancer Cell Growth in Vitro by Targeting the Wnt/β-Catenin Signaling Pathway. *The Journal of Biological Chemistry* 289 (51), pp. 35456–67. doi:10.1074/jbc. M114.621599.

Ang, Estabelle, Qian Liu, Ming Qi, Hua G. Liu, Xiaohong Yang, Chen H. Honghui, Ming H. Zheng, and Jiake Xu. 2011a. Mangiferin Attenuates Osteoclastogenesis, Bone Resorption, and RANKL-Induced Activation of NF-kB and ERK. *Journal of Cellular Biochemistry* 112 (1), pp. 89–97. doi:10.1002/jcb.22800.

Ang, Estabelle S.M., Xiaohong Yang, Honghui Chen, Qian Liu, Ming H. Zheng, and Jiake Xu. 2011b. Naringin Abrogates Osteoclastogenesis and Bone Resorption via the Inhibition of RANKL-Induced NF-κB and ERK Activation. *FEBS Letters* 585 (17), pp. 2755–62. doi:10.1016/j.febslet.2011.07.046.

Ansari, S.H., Mohamed Ali, Arturo Velasco-Negueruela, and María José Pérez-Alonso. 1999. Volatile Constituents of the Fruits of Three Mango Cultivars, *Mangifera Indica* L. *Journal of Essential Oil Research* 11 (1), pp. 65–68. doi:10.1080/10412905.1999.9701073.

Aparna, Vasudevan, Kalarickal V. Dileep, Pradeep K. Mandal, Ponnuraj Karthe, Chittalakkottu Sadasivan, and Madathilkovilakathu Haridas. 2012. Anti-Inflammatory Property of N-Hexadecanoic Acid: Structural Evidence and Kinetic Assessment. *Chemical Biology and Drug Design* 80 (3), pp. 434–39. doi:10.1111/j.1747-0285.2012.01418.x.

Ateşşahín, Ahmet, Ali Osman Çeríbaşi, Abdurrauf Yuce, Özgür Bulmus, and Gürkan Çikim. 2007. Role of Ellagic Acid against Cisplatin-Induced Nephrotoxicity and Oxidative Stress in Rats. *Basic and Clinical Pharmacology and Toxicology* 100 (2), pp. 121–26. doi:10.1111/j.1742-7843.2006.00015.x.

Bacanli, Merve, A. Ahmet Başaran, and Nurşen Başaran. 2015. The Antioxidant and Antigenotoxic Properties of Citrus Phenolics Limonene and Naringin. *Food and Chemical Toxicology* 81, pp. 160–70. doi:10.1016/j.fct.2015.04.015.

Bachhav, Sagar S., Savita D. Patil, Mukesh S. Bhutada, and Sanjay J. Surana. 2011. Oleanolic Acid Prevents Glucocorticoid-Induced Hypertension in Rats. *Phytotherapy Research* 25 (10), pp. 1435–39. doi:10.1002/ptr.3431.

Bae, Ji Young, Jung Suk Choi, Sang Wook Kang, Yong Jin Lee, Jinseu Park, and Young Hee Kang. 2010. Dietary Compound Ellagic Acid Alleviates Skin Wrinkle and Inflammation Induced by UV-B Irradiation. *Experimental Dermatology* 19 (8). doi:10.1111/j.1600-0625.2009.01044.x.

Baliga, Manjeshwar Shrinath, Arnadi Ramachandrayya Shivashankara, Raghavendra Haniadka, Jerome Dsouza, and Harshith P. Bhat. 2011. Phytochemistry, Nutritional and Pharmacological Properties of *Artocarpus heterophyllus* Lam (Jackfruit): A Review. *Food Research International* 44 (7), pp. 1800–1811. doi:10.1016/j.foodres.2011.02.035.

Bazzano, Lydia A., Jiang He, Lorraine G. Ogden, Catherine M. Loria, Suma Vupputuri, Leann Myers, and Paul K. Whelton. 2002. Fruit and Vegetable Intake and Risk of Cardiovascular Disease in US Adults: The First National Health and Nutrition Examination Survey Epidemiologic Follow-up Study. *The American Journal of Clinical Nutrition* 76 (1), pp. 93–99. doi:10.1093/ajcn/76.1.93.

Berardini, Nicolai, Reinhold Carle, and Andreas Schieber. 2004. Characterization of Gallotannins and Benzophenone Derivatives from Mango (*Mangifera Indica* L. Cv. 'Tommy Atkins') Peels, Pulp and Kernels by High-Performance Liquid Chromatography/Electrospray Ionization Mass Spectrometry. *Rapid Communications in Mass Spectrometry* 18 (19), pp. 2208–16. doi:10.1002/rcm.1611.

Berardini, Nicolai, Ramona Fezer, Jürgen Conrad, Uwe Beifuss, Reinhold Carl, and Andreas Schieber. 2005. Screening of Mango (*Mangifera Indica* L.) Cultivars for Their Contents of Flavonol O- and Xanthone C-Glycosides, Anthocyanins, and Pectin. *Journal of Agricultural and Food Chemistry* 53 (5), pp. 1563–70. doi:10.1021/jf0484069.

Bhatia, Harsharan S., Eduardo Candelario-Jalil, Antonio C. Pinheiro de Oliveira, Olumayokun A. Olajide, Gregorio Martínez-Sánchez, and Bernd L. Fiebich. 2008. Mangiferin Inhibits Cyclooxygenase-2 Expression and Prostaglandin E2 Production in Activated Rat Microglial Cells. *Archives of Biochemistry and Biophysics* 477 (2), pp. 253–58. doi:10.1016/j.abb.2008.06.017.

Caltagirone, Sara, Cosmo Rossi, Andreina Poggi, Franco O. Ranelletti, Pier Giorgio Natali, Mauro Brunetti, Francesca B. Aiello, and Mauro Piantelli. 2000. Flavonoids Apigenin and Quercetin Inhibit Melanoma Growth and Metastatic Potential. *International Journal of Cancer* 87 (4), pp. 595–600. doi:10.1002/1097-0215(20000815)87:4<595::AID-IJC21>3.0.CO;2-5.

del Carmen Recio, Maria, Rosa Maria Giner, S. Manez, Amparo Talens, Laura Cubells, J. Gueho, H.R. Julien, K. Hostettmann, and J.L. Rios. 1995. Anti-Inflammatory Activity of Flavonol Glycosides from Erythrospermum Monticolum Depending on Single or Repeated Local TPA Administration. *Planta Medica* 61 (6), pp. 502–4. doi:10.1055/s-2006-959357.

Carvalho, Ana Carla S., Marjorie M. Guedes, Antonia L. De Souza, Maria T.S. Trevisan, Alana F. Lima, Flávia A. Santos, and Vietla S.N. Rao. 2007. Gastroprotective Effect of Mangiferin, a Xanthonoid from *Mangifera indica*, against Gastric Injury Induced by Ethanol and Indomethacin in Rodents. *Planta Medica* 73 (13), pp. 1372–76. doi:10.1055/s-2007-990231.

Carvalho, Roney Rick, Claudia Helena Pellizzon, Luis Justulin, Sergio Luis Felisbino, Wagner Vilegas, Fernanda Bruni, Mônica Lopes-Ferreira, and Clélia Akiko Hiruma-Lima. 2009. Effect of Mangiferin on the Development of Periodontal Disease: Involvement of Lipoxin A4, Anti-Chemotaxic Action in Leukocyte Rolling. *Chemico-Biological Interactions* 179 (2–3), pp. 344–50. doi:10.1016/j.cbi.2008.10.041.

Castellano, Jose M., Angeles Guinda, Teresa Delgado, Mirela Rada, and Jose A. Cayuela. 2013. Biochemical Basis of the Antidiabetic Activity of Oleanolic Acid and Related Pentacyclic Triterpenes. *Diabetes*. doi:10.2337/db12-1215.

Cavaleiro, C., E. Pinto, M.J. Gonçalves, and L. Salgueiro. 2006. Antifungal Activity of *Juniperus* Essential Oils against Dermatophyte, *Aspergillus* and *Candida* Strains. *Journal of Applied Microbiology* 100 (6), pp. 1333–38. doi:10.1111/j.1365-2672.2006.02862.x.

Chakraborty, Sandipan, and Soumalee Basu. 2017. Multi-Functional Activities of Citrus Flavonoid Narirutin in Alzheimer's Disease Therapeutics: An Integrated Screening Approach and in Vitro Validation. *International Journal of Biological Macromolecules* 103, pp. 733–43. doi:10.1016/j.ijbiomac.2017.05.110.

Chandrika, U.G., E.R. Jansz, and N.D. Warnasuriya. 2005. Analysis of Carotenoids in Ripe Jackfruit (*Artocarpus heterophyllus*) Kernel and Study of Their Bioconversion in Rats. *Journal of the Science of Food and Agriculture* 85 (2), pp. 186–90. doi:10.1002/jsfa.1918.

Chang, Kun Che, Brian Laffin, Jessica Ponder, Anna Énzsöly, János Németh, Daniel V. Labarbera, and J. Mark Petrash. 2013. Beta-Glucogallin Reduces the Expression of Lipopolysaccharide-Induced Inflammatory Markers by Inhibition of Aldose Reductase in Murine Macrophages and Ocular Tissues. *Chemico-Biological Interactions* 202, pp. 283–87. doi:10.1016/j.cbi.2012.12.001.

Chavan, M.J., P.S. Wakte, and D.B. Shinde. 2010. Analgesic and Anti-Inflammatory Activity of Caryophyllene Oxide from *Annona squamosa* L. Bark. *Phytomedicine* 17 (2), pp. 149–51. doi:10.1016/j.phymed.2009.05.016.

Chen, Quan, Ping Li, Yong Xu, Yang Li, and Bo Tang. 2015. Isoquercitrin Inhibits the Progression of Pancreatic Cancer in Vivo and in Vitro by Regulating Opioid Receptors and the Mitogen-Activated Protein Kinase Signalling Pathway. *Oncology Reports* 33 (2), pp. 840–48. doi:10.3892/or.2014.3626.

Chen, Ya-Yun, Chen-Xiao Peng, Yan Hu, Chen Bu, Shu-Chen Guo, Xiang Li, Yong Chen, and Jian-Wei Chen. 2017. Studies on Chemical Constituents and Anti-Hepatoma Effects of Essential Oil from *Annona squamosa* L. Pericarps. *Natural Product Research* 31 (11), pp. 1305–8. doi:10.1080/14786419.2016.1233411.

Chen, Yiyan, and Chonghong Chen. 2011. Corilagin Prevents Tert-Butyl Hydroperoxide-Induced Oxidative Stress Injury in Cultured N9 Murine Microglia Cells. *Neurochemistry International* 59 (2), pp. 290–96. doi:10.1016/j.neuint.2011.05.020.

Cheng, J.T, T.C Lin, and F.L Hsu. 1995. Antihypertensive Effect of Corilagin in the Rat. *Canadian Journal of Physiology and Pharmacology* 73 (10), pp. 1425–29. doi:10.1139/y95-198.

Chiba, Hiroshige, Mariko Uehara, Jian Wu, Xinxiang Wang, Ritsuko Masuyama, Kazuharu Suzuki, Kazuki Kanazawa, and Yoshiko Ishimi. 2003. Hesperidin, a Citrus Flavonoid,

Inhibits Bone Loss and Decreases Serum and Hepatic Lipids in Ovariectomized Mice. *The Journal of Nutrition* 133 (2002), pp. 1892–97. doi:10.1093/jn/133.6.1892.

Cibiček, Norbert, Lenka Roubalová, Jiří Vrba, Martina Zatloukalová, Jiří Ehrmann, Jana Zapletalová, Rostislav Večeřa, Vladimír Křen, and Jitka Ulrichová. 2016. Protective Effect of Isoquercitrin against Acute Dextran Sulfate Sodium-Induced Rat Colitis Depends on the Severity of Tissue Damage. *Pharmacological Reports* 68 (6), pp. 1197–1204. doi:10.1016/j.pharep.2016.07.007.

Costa, André Gustavo Vasconcelos, Diego F. Garcia-Diaz, Paula Jimenez, and Pollyanna Ibrahim Silva. 2013. Bioactive Compounds and Health Benefits of Exotic Tropical Red-Black Berries. *Journal of Functional Foods*. doi:10.1016/j.jff.2013.01.029.

Dajas, Federico. 2012. Life or Death: Neuroprotective and Anticancer Effects of Quercetin. *Journal of Ethnopharmacology*. doi:10.1016/j.jep.2012.07.005.

Das, Joydeep, Jyotirmoy Ghosh, Anandita Roy, and Parames C. Sil. 2012. Mangiferin Exerts Hepatoprotective Activity against D-Galactosamine Induced Acute Toxicity and Oxidative/nitrosative Stress via Nrf2-NFκB Pathways. *Toxicology and Applied Pharmacology* 260 (1), pp. 35–47. doi:10.1016/j.taap.2012.01.015.

Debjit Bhowmik, K.P. Sampath Kumar, M. Umadevi, and S. Duraivel. 2012. Traditional and Medicinal Uses of Banana. *Journal of Pharmacognosy and Phytochemistry* 1 (3), pp. 51–63.

Dhar, Animesh, Smita Mehta, Gopal Dhar, Kakali Dhar, Snigdha Banerjee, Peter Van Veldhuizen, Donald R. Campbell, and Sushanta K. Banerjee. 2009. Crocetin Inhibits Pancreatic Cancer Cell Proliferation and Tumor Progression in a Xenograft Mouse Model. *Molecular Cancer Therapeutics* 8 (2), pp. 315–23. doi:10.1158/1535-7163.MCT-08-0762.

Dharmappa, Kattepura K., Raju Venkatesh Kumar, Angaswamy Nataraju, Riyaz Mohamed, Holenarasipura V. Shivaprasad, and Bannikuppe S. Vishwanath. 2009. Anti-Inflammatory Activity of Oleanolic Acid by Inhibition of Secretory Phospholipase A2. *Planta Medica* 75 (3), pp. 211–15. doi:10.1055/s-0028-1088374.

Dhingra, Dinesh, and Ritu Chhillar. 2012. Antidepressant-like Activity of Ellagic Acid in Unstressed and Acute Immobilization-Induced Stressed Mice. *Pharmacological Reports* 64 (4), pp. 796–807. doi:10.1016/S1734-1140(12)70875-7.

Du, Huaidong, Liming Li, Derrick Bennett, Yu Guo, Iain Turnbull, Ling Yang, Fiona Bragg et al. 2017. Fresh Fruit Consumption in Relation to Incident Diabetes and Diabetic Vascular Complications: A 7-Y Prospective Study of 0.5 Million Chinese Adults. *PLOS Medicine* 14 (4), pp. e1002279. doi:10.1371/journal.pmed.1002279.

Duan, Weigang, Yun Yu, and Luyong Zhang. 2005. Antiatherogenic Effects of Phyllanthus Emblica Associated with Corilagin and Its Analogue. *Yakugaku Zasshi: Journal of the Pharmaceutical Society of Japan* 125 (7), pp. 587–91. doi:10.1248/yakushi.125.587.

Edwards, Randi L, Tiffany Lyon, Sheldon E Litwin, Alexander Rabovsky, J David Symons, and Thunder Jalili. 2007. Quercetin Reduces Blood Pressure in Hypertensive Subjects. *The Journal of Nutrition* 137 (11), pp. 2405–11.

Etcheverry, Susana Beatriz, Evelina Gloria Ferrer, Luciana Naso, Josefina Rivadeneira, Victoria Salinas, and Patricia Ana María Williams. 2008. Antioxidant Effects of the VO(IV)

Hesperidin Complex and Its Role in Cancer Chemoprevention. *JBIC Journal of Biological Inorganic Chemistry* 13 (3), pp. 435–47. doi:10.1007/s00775-007-0332-9.

Fajemiroye, James O., Pablinny M. Galdino, Iziara F. Florentino, Fabio F. Da Rocha, Paulo C. Ghedini, Prabhakar R. Polepally, Jordan K. Zjawiony, and Elson A. Costa. 2014. Plurality of Anxiety and Depression Alteration Mechanism by Oleanolic Acid. *Journal of Psychopharmacology* 28 (10), pp. 923–34. doi:10.1177/0269881114536789.

Fernandes, Caio P., Arthur L. Corrêa, Jonathas F.R. Lobo, Otávio P. Caramel, Fernanda B. De Almeida, Elaine S. Castro, Kauê F.C.S. Souza et al. 2013. Triterpene Esters and Biological Activities from Edible Fruits of *Manilkara subsericea* (Mart.) Dubard, Sapotaceae. *BioMed Research International*. doi:10.1155/2013/280810.

Fernandez, Jacquelina, Ricardo Reyes, Hector Ponce, Martha Oropeza, Marie Rose VanCalsteren, Christopher Jankowski, and Maria G. Campos. 2005. Isoquercitrin from Argemone Platyceras Inhibits Carbachol and Leukotriene D4-Induced Contraction in Guinea-Pig Airways. *European Journal of Pharmacology* 522 (1–3), pp. 108–15. doi:10.1016/j.ejphar.2005.08.046.

Fernandez, Sebastian P., Michael Nguyen, Tin Thing Yow, Cindy Chu, Graham A.R. Johnston, Jane R. Hanrahan, and Mary Chebib. 2009. The Flavonoid Glycosides, Myricitrin, Gossypin and Naringin Exert Anxiolytic Action in Mice. *Neurochemical Research* 34 (10), pp. 1867–75. doi:10.1007/s11064-009-9969-9.

Fidyt, Klaudyna, Anna Fiedorowicz, Leon Strządała, and Antoni Szumny. 2016. β-Caryophyllene and β-Caryophyllene Oxide—Natural Compounds of Anticancer and Analgesic Properties. *Cancer Medicine*. doi:10.1002/cam4.816.

Funaguchi, Norihiko, Yasushi Ohno, Bu Lin Bai La, Toshihiro Asai, Hideyuki Yuhgetsu, Masahiro Sawada, Genzou Takemura, Shinya Minatoguchi, Takako Fujiwara, and Hisayoshi Fujiwara. 2007. Narirutin Inhibits Airway Inflammation in an Allergic Mouse Model. *Clinical and Experimental Pharmacology and Physiology* 34 (8), pp. 766–70. doi:10.1111/j.1440-1681.2007.04636.x.

Gambari, Roberto, Monica Borgatti, Ilaria Lampronti, Enrica Fabbri, Eleonora Brognara, Nicoletta Bianchi, Laura Piccagli et al. 2012. Corilagin Is a Potent Inhibitor of NF-kappaB Activity and Downregulates TNF-Alpha Induced Expression of IL-8 Gene in Cystic Fibrosis IB3-1 Cells. *International Immunopharmacology* 13 (3), pp. 308–15. doi:10.1016/j.intimp.2012.04.010.

García, D., M. Escalante, R. Delgado, F.M. Ubeira, and J. Leiro. 2003. Anthelminthic and Antiallergic Activities of *Mangifera indica* L. Stem Bark Components Vimang and Mangiferin. *Phytotherapy Research* 17 (10), pp. 1203–8. doi:10.1002/ptr.1343.

Gasparotto Junior, Arquimedes, Francielly Mourão Gasparotto, Marcos Aurelio Boffo, Emerson Luiz Botelho Lourenço, Maria Élida Alves Stefanello, Marcos José Salvador, José Eduardo Da Silva-Santos, Maria Consuelo Andrade Marques, and Cândida Aparecida Leite Kassuya. 2011. Diuretic and Potassium-Sparing Effect of Isoquercitrin—An Active Flavonoid of *Tropaeolum majus* L. *Journal of Ethnopharmacology* 134 (2), pp. 210–15. doi:10.1016/j.jep.2010.12.009.

Gayosso-García Sancho, Laura E., Elhadi M. Yahia, and Gustavo Adolfo González-Aguilar. 2011. Identification and Quantification of Phenols, Carotenoids, and Vitamin C from

Papaya (Carica Papaya L., Cv. Maradol) Fruit Determined by HPLC-DAD-MS/MS-ESI. *Food Research International* 44 (5), pp. 1284–91. doi:10.1016/j.foodres.2010.12.001.

Ghosal, Shibnath. 1985. Steryl Glycosides and Acyl Steryl Glycosides from Musa Paradisiaca. *Phytochemistry* 24 (8), pp. 1807–10. doi:10.1016/S0031-9422(00)82556-X.

Girish, Chandrashekaran, Vishnu Raj, Jayasree Arya, and Sadasivam Balakrishnan. 2013. Involvement of the GABAergic System in the Anxiolytic-like Effect of the Flavonoid Ellagic Acid in Mice. *European Journal of Pharmacology* 710 (1–3), pp. 49–58. doi:10.1016/j.ejphar.2013.04.003.

Gopalakrishnan, Lalitha, Lakshmi Narashimhan Ramana, Swaminathan Sethuraman, and Uma Maheswari Krishnan. 2014. Ellagic Acid Encapsulated Chitosan Nanoparticles as Anti-Hemorrhagic Agent. *Carbohydrate Polymers* 111, pp. 215–21. doi:10.1016/j.carbpol.2014.03.093.

Gopinath, Kulasekaran, Dharmalingam Prakash, and Ganapasam Sudhandiran. 2011. Neuroprotective Effect of Naringin, a Dietary Flavonoid against 3-Nitropropionic Acid-Induced Neuronal Apoptosis. *Neurochemistry International* 59 (7), pp. 1066–73. doi:10.1016/j.neuint.2011.08.022.

Gorinstein, Shela, Marina Zemser, Ratiporn Haruenkit, Rachit Chuthakorn, Fernanda Grauer, Olga Martin-Belloso, and Simon Trakhtenberg. 1999. Comparative Content of Total Polyphenols and Dietary Fiber in Tropical Fruits and Persimmon. *Journal of Nutritional Biochemistry* 10 (6), pp. 367–71. doi:10.1016/S0955-2863(99)00017-0.

Greenberg, E. Robert, John A. Baron, Thérèse A. Stukel, Marguerite M. Stevens, Jack S. Mandel, Steven K. Spencer, Peter M. Elias et al. 1990. A Clinical Trial of Beta Carotene to Prevent Basal-Cell and Squamous-Cell Cancers of the Skin. *New England Journal of Medicine* 323 (12), pp. 789–95. doi:10.1056/NEJM199009203231204.

Guha, Surajit, Shibnath Ghosal, and Utpala Chattopadhyay. 1996. Antitumor, Immunomodulatory and Anti-HIV Effect of Mangiferin, a Naturally Occurring Glucosylxanthone. *Chemotherapy* 42 (6), pp. 443–51. doi:10.1159/000239478.

Guo, Fuchuan, Conghui Huang, Xilu Liao, Yemei Wang, Ying He, Rennan Feng, Ying Li, and Changhao Sun. 2011. Beneficial Effects of Mangiferin on Hyperlipidemia in High-Fat-Fed Hamsters. *Molecular Nutrition and Food Research* 55 (12), pp. 1809–18. doi:10.1002/mnfr.201100392.

Habib-ur-Rehman, Khawaja Ansar Yasin, Muhammad Aziz Choudhary, Naeem Khaliq, Atta-ur-Rahman, Muhammad Iqbal Choudhary, and Shahid Malik. 2007. Studies on the Chemical Constituents of *Phyllanthus emblica*. *Natural Product Research* 21 (9), pp. 775–81. doi:10.1080/14786410601124664.

Hamdan, Dalia I., Mona F. Mahmoud, Michael Wink, and Assem M. El-Shazly. 2014. Effect of Hesperidin and Neohesperidin from Bittersweet Orange (*Citrus aurantium* Var. *bigaradia*) Peel on Indomethacin-Induced Peptic Ulcers in Rats. *Environmental Toxicology and Pharmacology* 37 (3), pp. 907–15. doi:10.1016/j.etap.2014.03.006.

Han, Dong Hoon, Min Jeon Lee, and Jeong Hee Kim. 2006. Antioxidant and Apoptosis-Inducing Activities of Ellagic Acid. *Anticancer Research* 26 (5A), pp. 3601–6.

He, Shu Ying, Zhi Yu Qian, Na Wen, Fu Tian Tang, Guang Lin Xu, and Cheng Hua Zhou. 2007. Influence of Crocetin on Experimental Atherosclerosis in Hyperlipidamic-Diet Quails. *European Journal of Pharmacology* 554 (2–3), pp. 191–95. doi:10.1016/j.ejphar.2006.09.071.

Heng, M., and Z. Lu. 1990. Antiviral Effect of Mangiferin and Isomangiferin on Herpes Simplex Virus. *Chinese Medical Journal* 103 (2), pp. 160–65.

Hennekens, Charles H., Julie E. Buring, JoAnn E. Manson, Meir Stampfer, Bernard Rosner, Nancy R. Cook, Charlene Belanger et al. 1996. Lack of Effect of Long-Term Supplementation with Beta Carotene on the Incidence of Malignant Neoplasms and Cardiovascular Disease. *New England Journal of Medicine* 334 (18), pp. 1145–49. doi:10.1056/NEJM199605023341801.

Hernández, Yurena, M. Gloria Lobo, and Mónica González. 2006. Determination of Vitamin C in Tropical Fruits: A Comparative Evaluation of Methods. *Food Chemistry* 96 (4), pp. 654–64. doi:10.1016/j.foodchem.2005.04.012.

Hiramitsu, Masanori, Yasuhito Shimada, Junya Kuroyanagi, Takashi Inoue, Takao Katagiri, Liqing Zang, Yuhei Nishimura, Norihiro Nishimura, and Toshio Tanaka. 2014. Eriocitrin Ameliorates Diet-Induced Hepatic Steatosis with Activation of Mitochondrial Biogenesis. *Scientific Reports* 4. doi:10.1038/srep03708.

Igura, Koichi, Toshiro Ohta, Yukiaki Kuroda, and Kazuhiko Kaji. 2001. Resveratrol and Quercetin Inhibit Angiogenesis in Vitro. *Cancer Letters* 171 (1), pp. 11–16. doi:10.1016/S0304-3835(01)00443-8.

Ikemura, Miyako, Yasuto Sasaki, John C. Giddings, and Junichiro Yamamoto. 2012. Preventive Effects of Hesperidin, Glucosyl Hesperidin and Naringin on Hypertension and Cerebral Thrombosis in Stroke-Prone Spontaneously Hypertensive Rats. *Phytotherapy Research* 26 (9), pp. 1272–77. doi:10.1002/ptr.3724.

Jagtap, U.B., and V.A. Bapat. 2010. Artocarpus: A Review of Its Traditional Uses, Phytochemistry and Pharmacology. *Journal of Ethnopharmacology*. doi:10.1016/j.jep.2010.03.031.

Jain, Mandipika, and Hamendra Singh Parmar. 2011. Evaluation of Antioxidative and Anti-Inflammatory Potential of Hesperidin and Naringin on the Rat Air Pouch Model of Inflammation. *Inflammation Research* 60 (5), pp. 483–91. doi:10.1007/s00011-010-0295-0.

Jeon, Se Jin, Hong Ju Lee, Hyung Eun Lee, Se Jin Park, Yubeen Gwon, Haneul Kim, Jiabao Zhang, Chan Young Shin, Dong Hyun Kim, and Jong Hoon Ryu. 2017. Oleanolic Acid Ameliorates Cognitive Dysfunction Caused by Cholinergic Blockade via TrkB-Dependent BDNF Signaling. *Neuropharmacology* 113, pp. 100–109. doi:10.1016/j.neuropharm.2016.07.029.

Jia, L, H Jin, J Zhou, L Chen, Y Lu, Y Ming, and Y Yu. 2013. A Potential Anti-Tumor Herbal Medicine, Corilagin, Inhibits Ovarian Cancer Cell Growth through Blocking the TGF-Beta Signaling Pathways. *BMC Complementary and Alternative Medicine* 13 (1), pp. 33. doi:10.1186/1472-6882-13-33.

Jia, Luoqi, Hongyan Jin, Jiayi Zhou, Lianghua Chen, Yiling Lu, Yanlin Ming, and Yinhua Yu. 2013. A Potential Anti-Tumor Herbal Medicine, Corilagin, Inhibits Ovarian Cancer Cell Growth through Blocking the TGF-β Signaling Pathways. *BMC Complementary and Alternative Medicine* 13. doi:10.1186/1472-6882-13-33.

Jiao, Hao Yan, Wei Wei Su, Pei Bo Li, Yan Liao, Qian Zhou, Na Zhu, and Li Li He. 2015. Therapeutic Effects of Naringin in a Guinea Pig Model of Ovalbumin-Induced Cough-Variant Asthma. *Pulmonary Pharmacology and Therapeutics* 33, pp. 59–65. doi:10.1016/j.pupt.2015.07.002.

Johnson, R., W. Robertson, M. Towey, S. Stewart-Brown, and A. Clarke. 2017. Changes over Time in Mental Well-Being, Fruit and Vegetable Consumption and Physical Activity in a Community-Based Lifestyle Intervention: A before and after Study. *Public Health* 146, pp. 118–25. doi:10.1016/j.puhe.2017.01.012.

Joskova, M., S. Franova, and V. Sadlonova. 2011. Acute Bronchodilator Effect of Quercetin in Experimental Allergic Asthma. *Bratislava Medical Journal* 112 (1), pp. 9–12.

Jung, Sang Hoon, Beum Jin Kim, Eun Ha Lee, and Neville N. Osborne. 2010. Isoquercitrin Is the Most Effective Antioxidant in the Plant Thuja Orientalis and Able to Counteract Oxidative-Induced Damage to a Transformed Cell Line (RGC-5 Cells). *Neurochemistry International* 57 (7), pp. 713–21. doi:10.1016/j.neuint.2010.08.005.

Jung, U.J., H.J. Kim, J.S. Lee, M.K. Lee, H.O. Kim, E.J. Park, H.K. Kim, T.S. Jeong, and M.S. Choi. 2003. Naringin Supplementation Lowers Plasma Lipids and Enhances Erythrocyte Antioxidant Enzyme Activities in Hypercholesterolemic Subjects. *Clinical Nutrition* 22 (6), pp. 561–68. doi:10.1016/S0261-5614(03)00059-1.

Jung, Un Ju, Mi-Kyung Lee, Kyu-Shik Jeong, and Myung-Sook Choi. 2004. The Hypoglycemic Effects of Hesperidin and Naringin Are Partly Mediated by Hepatic Glucose-Regulating Enzymes in C57BL/KsJ-Db/db Mice. *The Journal of Nutrition* 134 (10), pp. 2499–503.

Kawaii, Satoru, Yasuhiko Tomono, Eriko Katase, Kazunori Ogawa, and Masamichi Yano. 1999. Quantitation of Flavonoid Constituents in *Citrus* Fruits. *Journal of Agricultural and Food Chemistry* 47 (9), pp. 3565–71. doi:10.1021/jf990153+.

Kesharwani, Ajay, Suja Kizhiyedath Polachira, Reshmi Nair, Aakanksha Agarwal, Nripendra Nath Mishra, and Satish Kumar Gupta. 2017. Anti-HSV-2 Activity of *Terminalia chebula* Retz Extract and Its Constituents, Chebulagic and Chebulinic Acids. *BMC Complementary and Alternative Medicine* 17 (1). doi:10.1186/s12906-017-1620-8.

Kim, Hee Ju, Joonki Kim, Ki Sung Kang, Keun Taik Lee, and Hyun Ok Yang. 2014. Neuroprotective Effect of Chebulagic Acid via Autophagy Induction in SH-SY5Y Cells. *Biomolecules and Therapeutics* 22 (4), pp. 275–81. doi:10.4062/biomolther.2014.068.

Kim, Sang-Hyun, Chang-Duk Jun, Kyongho Suk, Byung-Ju Choi, Hyunjeung Lim, Seunja Park, Seung Ho Lee, Hye-Young Shin, Dae-Keun Kim, and Tae-Yong Shin. 2006. Gallic Acid Inhibits Histamine Release and pro-Inflammatory Cytokine Production in Mast Cells. *Toxicological Sciences: An Official Journal of the Society of Toxicology* 91 (1), pp. 123–31. doi:10.1093/toxsci/kfj063.

Kim, Sejeong, Heeyoung Lee, Soomin Lee, Yohan Yoon, and Kyoung Hee Choi. 2015. Antimicrobial Action of Oleanolic Acid on Listeria Monocytogenes, *Enterococcus faecium*, and *Enterococcus faecalis*. *PLoS ONE* 10 (3). doi:10.1371/journal.pone.0118800.

Kinoshita, S., Y. Inoue, S. Nakama, T. Ichiba, and Y. Aniya. 2007. Antioxidant and Hepatoprotective Actions of Medicinal Herb, *Terminalia catappa* L. from Okinawa Island and Its Tannin Corilagin. *Phytomedicine* 14 (11), pp. 755–62. doi:10.1016/j.phymed.2006.12.012.

Kong, Lingbao, Shanshan Li, Qingjiao Liao, Yanni Zhang, Ruina Sun, Xiangdong Zhu, Qinghua Zhang et al. 2013. Oleanolic Acid and Ursolic Acid: Novel Hepatitis C Virus Antivirals That Inhibit NS5B Activity. *Antiviral Research* 98 (1), pp. 44–53. doi:10.1016/j.antiviral.2013.02.003.

Koriem, Khaled M.M., and Rowan E. Soliman. 2014. Chlorogenic and Caftaric Acids in Liver Toxicity and Oxidative Stress Induced by Methamphetamine. *Journal of Toxicology* 2014. doi:10.1155/2014/583494.

Kumar, Anil, Atish Prakash, and Samrita Dogra. 2010. Naringin Alleviates Cognitive Impairment, Mitochondrial Dysfunction and Oxidative Stress Induced by D-Galactose in Mice. *Food and Chemical Toxicology* 48 (2), pp. 626–32. doi:10.1016/j.fct.2009.11.043.

Lee, In-Ah, Jin Hee Lee, Nam-In Baek, and Dong-Hyun Kim. 2005. Antihyperlipidemic Effect of Crocin Isolated from the Fructus of *Gardenia jasminoides* and Its Metabolite Crocetin. *Biological & Pharmaceutical Bulletin* 28 (11), pp. 2106–10. doi:10.1248/bpb.28.2106.

Lee, Jisun, Ji Won Choi, Jae Kyung Sohng, Ramesh Prasad Pandey, and Yong Il Park. 2016. The Immunostimulating Activity of Quercetin 3-O-Xyloside in Murine Macrophages via Activation of the ASK1/MAPK/NF-κB Signaling Pathway. *International Immunopharmacology* 31, pp. 88–97. doi:10.1016/j.intimp.2015.12.008.

Lee, Jungwhoi, Song I. Han, Jeong Hun Yun, and Jae Hoon Kim. 2015. Quercetin 3-O-Glucoside Suppresses Epidermal Growth Factor–induced Migration by Inhibiting EGFR Signaling in Pancreatic Cancer Cells. *Tumor Biology* 36 (12), pp. 9385–93. doi:10.1007/s13277-015-3682-x.

Lesjak, Marija, Ivana Beara, Nataša Simin, Diandra Pintać, Tatjana Majkić, Kristina Bekvalac, Dejan Orčić, and Neda Mimica-Dukić. 2018. Antioxidant and Anti-Inflammatory Activities of Quercetin and Its Derivatives. *Journal of Functional Foods* 40, pp. 68–75. doi:10.1016/j.jff.2017.10.047.

Leterme, Pascal, André Buldgen, Fernando Estrada, and Angela M. Londoño. 2006. Mineral Content of Tropical Fruits and Unconventional Foods of the Andes and the Rain Forest of Colombia. *Food Chemistry* 95 (4), pp. 644–52. doi:10.1016/j.foodchem.2005.02.003.

Li, Fengbo, Xiaolei Sun, Jianxiong Ma, Xinlong Ma, Bin Zhao, Yang Zhang, Peng Tian, Yanjun Li, and Zhe Han. 2014. Naringin Prevents Ovariectomy-Induced Osteoporosis and Promotes Osteoclasts Apoptosis through the Mitochondria-Mediated Apoptosis Pathway. *Biochemical and Biophysical Research Communications* 452 (3), pp. 629–35. doi:10.1016/j.bbrc.2014.08.117.

Li, Hongzhong, Bing Yang, Jing Huang, Tingxiu Xiang, Xuedong Yin, Jingyuan Wan, Fuling Luo, Li Zhang, Hongyuan Li, and Guosheng Ren. 2013. Naringin Inhibits Growth Potential of Human Triple-Negative Breast Cancer Cells by Targeting β-Catenin Signaling Pathway. *Toxicology Letters* 220 (3), pp. 219–28. doi:10.1016/j.toxlet.2013.05.006.

Li, Na, Meng Luo, Yu Jie Fu, Yuan Gang Zu, Wei Wang, Lin Zhang, Li Ping Yao, Chun Jian Zhao, and Yu Sun. 2013. Effect of Corilagin on Membrane Permeability of *Escherichia coli*, *Staphylococcus aureus* and *Candida albicans*. *Phytotherapy Research* 27 (10), pp. 1517–23. doi:10.1002/ptr.4891.

Li, Nianhu, Yunpeng Jiang, Paul H. Wooley, Zhanwang Xu, and Shang You Yang. 2013. Naringin Promotes Osteoblast Differentiation and Effectively Reverses Ovariectomy-Associated Osteoporosis. *Journal of Orthopaedic Science* 18 (3), pp. 478–85. doi:10.1007/s00776-013-0362-9.

Li, Xuan, Xiaobing Cui, Xiaoyu Sun, Xiaodong Li, Quan Zhu, and Wei Li. 2010. Mangiferin Prevents Diabetic Nephropathy Progression in Streptozotocin-Induced Diabetic Rats. *Phytotherapy Research* 24 (6), pp. 893–99. doi:10.1002/ptr.3045.

Li, Ying, Jianwei Wang, Tieguang Gu, Johji Yamahara, and Yuhao Li. 2014. Oleanolic Acid Supplement Attenuates Liquid Fructose-Induced Adipose Tissue Insulin Resistance through the Insulin Receptor Substrate-1/phosphatidylinositol 3-kinase/Akt Signaling Pathway in Rats. *Toxicology and Applied Pharmacology* 277 (2), pp. 155–63. doi:10.1016/j.taap.2014.03.016.

Lim, Y.Y., T.T. Lim, and J.J. Tee. 2007. Antioxidant Properties of Several Tropical Fruits: A Comparative Study. *Food Chemistry* 103 (3), pp. 1003–8. doi:10.1016/j.foodchem.2006.08.038.

Lin, Ta Chen, Feng Lin Hsu, and Juei Tang Cheng. 1993. Antihypertensive Activity of Corilagin and Chebulinic Acid, Tannins from Lumnitzera, Racemosa. *Journal of Natural Products* 56 (4), pp. 629–32. doi:10.1021/np50094a030.

Liu, Rui Hai. 2003. Health Benefits of Fruit and Vegetables Are from Additive and Synergistic Combinations of Phytochemicals. *American Journal of Clinical Nutrition* 78. doi:10.1093/ajcn/78.3.517S.

Liu, Tao, Jun Zhao, Long Ma, Yusong Ding, and Deqi Su. 2012. Hepatoprotective Effects of Total Triterpenoids and Total Flavonoids from *Vitis vinifera* L against Immunological Liver Injury in Mice. *Evidence-Based Complementary and Alternative Medicine: eCAM* 2012, pp. 969386. doi:10.1155/2012/969386.

Liu, Wenxing, Peijin Shang, Tianlong Liu, Hang Xu, Danjun Ren, Wei Zhou, Aidong Wen, and Yi Ding. 2017. Gastroprotective Effects of Chebulagic Acid against Ethanol-Induced Gastric Injury in Rats. *Chemico-Biological Interactions* 278, pp. 1–8. doi:10.1016/j.cbi.2017.09.019.

Liu, Y., D.P. Hartley, and J. Liu. 1998. Protection against Carbon Tetrachloride Hepatotoxicity by Oleanolic Acid Is Not Mediated through Metallothionein. *Toxicology Letters* 95 (2), pp. 77–85.

Liu, Zhiyong, Xinyi Tao, Chongwei Zhang, Yanhua Lu, and Dongzhi Wei. 2005. Protective Effects of Hyperoside (Quercetin-3-O-Galactoside) to PC12 Cells against Cytotoxicity Induced by Hydrogen Peroxide and Tert-Butyl Hydroperoxide. *Biomedicine & Pharmacotherapy* 59 (9), pp. 481–90. doi:10.1016/j.biopha.2005.06.009.

Lu, Kai, and Sujit Basu. 2015. The Natural Compound Chebulagic Acid Inhibits Vascular Endothelial Growth Factor A Mediated Regulation of Endothelial Cell Functions. *Scientific Reports* 5. doi:10.1038/srep09642.

Lu, Kai, Debanjan Chakroborty, Chandrani Sarkar, Tingting Lu, Zhiliang Xie, Zhongfa Liu, and Sujit Basu. 2012. Triphala and Its Active Constituent Chebulinic Acid Are Natural Inhibitors of Vascular Endothelial Growth Factor-A Mediated Angiogenesis. *PLoS ONE* 7 (8). doi:10.1371/journal.pone.0043934.

Lu, Yong, Feng Jiang, Hao Jiang, Kalina Wu, Xuguang Zheng, Yizhong Cai, Mark Katakowski, Michael Chopp, and Shing Shun Tony To. 2010. Gallic Acid Suppresses Cell Viability, Proliferation, Invasion and Angiogenesis in Human Glioma Cells. *European Journal of Pharmacology* 641 (2–3), pp. 102–7. doi:10.1016/j.ejphar.2010.05.043.

Ma, Jun, Xiao Dong Luo, Petr Protiva, Hui Yang, Cuiying Ma, Margaret J. Basile, I. Bernard Weinstein, and Edward J. Kennelly. 2003. Bioactive Novel Polyphenols from the Fruit of *Manilkara zapota* (Sapodilla). *Journal of Natural Products* 66 (7), pp. 983–86. doi:10.1021/np020576x.

Mahattanatawee, Kanjana, John A. Manthey, Gary Luzio, Stephen T. Talcott, Kevin Goodner, and Elizabeth A. Baldwin. 2006. Total Antioxidant Activity and Fiber Content of Select Florida-Grown Tropical Fruits. *Journal of Agricultural and Food Chemistry* 54 (19), pp. 7355–63. doi:10.1021/jf060566s.

Makino, Toshiaki, Misaki Kanemaru, Shuji Okuyama, Ryosuke Shimizu, Hisashi Tanaka, and Hajime Mizukami. 2013. Anti-Allergic Effects of Enzymatically Modified Isoquercitrin (α-Oligoglucosyl Quercetin 3-O-Glucoside), Quercetin 3-O-Glucoside, α-Oligoglucosyl Rutin, and Quercetin, When Administered Orally to Mice. *Journal of Natural Medicines* 67 (4), pp. 881–86. doi:10.1007/s11418-013-0760-5.

Malik, Arshi, Sarah Afaq, Mohammad Shahid, Kafil Akhtar, and Abdullah Assiri. 2011. Influence of Ellagic Acid on Prostate Cancer Cell Proliferation: A Caspase-Dependent Pathway. *Asian Pacific Journal of Tropical Medicine* 4 (7), pp. 550–55. doi:10.1016/S1995-7645(11)60144-2.

Mancini, Andrea, Jessica Serrano-Díaz, Eduardo Nava, Anna Maria D'Alessandro, Gonzalo Luis Alonso, Manuel Carmona, and Sílvia Llorens. 2014. Crocetin, a Carotenoid Derived from Saffron (*Crocus sativus* L.), Improves Acetylcholine-Induced Vascular Relaxation in Hypertension. *Journal of Vascular Research* 51 (5), pp. 393–404. doi:10.1159/000368930.

Márquez, Lucía, Borja García-Bueno, José L.M. Madrigal, and Juan C. Leza. 2012. Mangiferin Decreases Inflammation and Oxidative Damage in Rat Brain after Stress. *European Journal of Nutrition* 51 (6), pp. 729–39. doi:10.1007/s00394-011-0252-x.

McMartin, Seanna E., Felice N. Jacka, and Ian Colman. 2013. The Association between Fruit and Vegetable Consumption and Mental Health Disorders: Evidence from Five Waves of a National Survey of Canadians. *Preventive Medicine* 56 (3–4), pp. 225–30. doi:10.1016/j.ypmed.2012.12.016.

Mertz, Christian, Anne Laure Gancel, Ziya Gunata, Pascaline Alter, Claudie Dhuique-Mayer, Fabrice Vaillant, Ana Mercedes Perez, Jenny Ruales, and Pierre Brat. 2009. Phenolic Compounds, Carotenoids and Antioxidant Activity of Three Tropical Fruits. *Journal of Food Composition and Analysis* 22 (5), pp. 381–87. doi:10.1016/j.jfca.2008.06.008.

Mishra, Vaibhav, Manali Agrawal, Samuel Adetunji Onasanwo, Gaurav Madhur, Preeti Rastogi, Haushila Prasad Pandey, Gautam Palit, and Tadigoppula Narender. 2013. Anti-Secretory and Cyto-Protective Effects of Chebulinic Acid Isolated from the Fruits of Terminalia Chebula on Gastric Ulcers. *Phytomedicine* 20 (6), pp. 506–11. doi:10.1016/j.phymed.2013.01.002.

Miyake, Yoshiaki, Kanefumi Yamamoto, and Toshihiko Osawa. 1997. Isolation of Eriocitrin (Eriodictyol 7-Rutinoside) from Lemon Fruit (*Citrus limon* BURM. F.) and Its Antioxidative Activity. *Food Science and Technology International, Tokyo* 3 (1), pp. 84–89. doi:10.3136/fsti9596t9798.3.84.

Mlcek, Jiri, Tunde Jurikova, Sona Skrovankova, and Jiri Sochor. 2016. Quercetin and Its Anti-Allergic Immune Response. *Molecules*. doi:10.3390/molecules21050623.

Moreira, Jeverson, Luiz Carlos Klein-Júnior, Valdir Cechinel Filho, and Fátima De Campos Buzzi. 2013. Anti-Hyperalgesic Activity of Corilagin, a Tannin Isolated from *Phyllanthus niruri* L. (Euphorbiaceae). *Journal of Ethnopharmacology* 146 (1), pp. 318–23. doi:10.1016/j.jep.2012.12.052.

Moura-Letts, Gustavo, Leon F. Villegas, Ana Marçalo, Abraham J. Vaisberg, and Gerald B. Hammond. 2006. In Vivo Wound-Healing Activity of Oleanolic Acid Derived from the Acid Hydrolysis of Anredera Diffusa. *Journal of Natural Products* 69 (6), pp. 978–79. doi:10.1021/np0601152.

Murakami, Akira, Hitoshi Ashida, and Junji Terao. 2008. Multitargeted Cancer Prevention by Quercetin. *Cancer Letters.* doi:10.1016/j.canlet.2008.03.046.

Murakoshi, M., J. Takayasu, O. Kimura, E. Kohmura, H. Nishino, A. Iwashima, J. Okuzumi et al. 1989. Inhibitory Effects of a-Carotene on Proliferation of the Human Neuroblastoma Cell Line GOTO. *JNCI Journal of the National Cancer Institute* 81 (21), pp. 1649–52. doi:10.1093/jnci/81.21.1649.

Murakoshi, Michiaki, Ryozo Iwasaki, Yoshiko Satomi, Junko Takayasu, Teiko Hasegawa, Harukuni Tokuda, and Akio Iwashima. 1992. Potent Preventive Action of a-Carotene against Carcinogenesis: Spontaneous Liver Carcinogenesis and Promoting Stage of Lung and Skin Carcinogenesis in Mice Are Suppressed More Effectively by a-Carotene Than by 0-Carotene. *Cancer Research* 52 (23), pp. 6583–87.

Muruganandan, S., K. Srinivasan, S. Gupta, P. K. Gupta, and J. Lal. 2005. Effect of Mangiferin on Hyperglycemia and Atherogenicity in Streptozotocin Diabetic Rats. *Journal of Ethnopharmacology* 97 (3), pp. 497–501. doi:10.1016/j.jep.2004.12.010.

Nam, Kyong Nyon, Young Min Park, Hoon Ji Jung, Jung Yeon Lee, Byung Duk Min, Seong Uk Park, Woo Sang Jung et al. 2010. Anti-Inflammatory Effects of Crocin and Crocetin in Rat Brain Microglial Cells. *European Journal of Pharmacology* 648 (1–3), pp. 110–16. doi:10.1016/j.ejphar.2010.09.003.

Narisawa, Tomio, Yoko Fukaura, Makiko Hasebe, Michiko Ito, Rika Aizawa, Michiaki Murakoshi, Shingo Uemura, Frederick Khachik, and Hoyoku Nishino. 1996. Inhibitory Effects of Natural Carotenoids, α-Carotene, β-Carotene, Lycopene and Lutein, on Colonic Aberrant Crypt Foci Formation in Rats. *Cancer Letters* 107 (1), pp. 137–42. doi:10.1016/0304-3835(96)04354-6.

Ong, B.T., S.A.H. Nazimah, A. Osman, S.Y. Quek, Y.Y. Voon, D. Mat Hashim, P.M. Chew, and Y.W. Kong. 2006. Chemical and Flavour Changes in Jackfruit (*Artocarpus heterophyllus* Lam.) Cultivar J3 during Ripening. *Postharvest Biology and Technology* 40 (3), pp. 279–86. doi:10.1016/j.postharvbio.2006.01.015.

Ooghe, Wilfried C., Sigrid J. Ooghe, Christ'l M. Detavernier, and André Huyghebaert. 1994. Characterization of Orange Juice (Citrus Sinensis) by Flavanone Glycosides. *Journal of Agricultural and Food Chemistry* 42 (10), pp. 2183–90. doi:10.1021/jf00046a020.

Ossola, Bernardino, Tiina M. Kääriäinen, and Pekka T. Männistö. 2009. The Multiple Faces of Quercetin in Neuroprotection. *Expert Opinion on Drug Safety* 8 (4), pp. 397–409. doi:10.1517/14740330903026944.

Panda, Sunanda, and Anand Kar. 2007. Antidiabetic and Antioxidative Effects of *Annona Squamosa* Leaves Are Possibly Mediated through Quercetin-3-O-Glucoside. *BioFactors* 31 (3–4), pp. 201–10. doi:10.1002/biof.5520310307.

Papoutsi, Zoi, Eva Kassi, Anna Tsiapara, Nikolas Fokialakis, George P Chrousos, and Paraskevi Moutsatsou. 2005. Evaluation of Estrogenic/antiestrogenic Activity of Ellagic Acid via the Estrogen Receptor Subtypes ERalpha and ERbeta. *Journal of Agricultural and Food Chemistry* 53 (20), pp. 7715–20. doi:10.1021/jf0510539.

Pardo Andreu, Gilberto L., Natasha Maurmann, Gustavo Kellermann Reolon, Caroline B. de Farias, Gilberto Schwartsmann, René Delgado, and Rafael Roesler. 2010. Mangiferin, a Naturally Occurring Glucoxilxanthone Improves Long-Term Object Recognition Memory in Rats. *European Journal of Pharmacology* 635 (1–3), pp. 124–28. doi:10.1016/j.ejphar.2010.03.011.

Park, H.J., M.J. Kim, E. Ha, and J.H. Chung. 2008. Apoptotic Effect of Hesperidin through caspase3 Activation in Human Colon Cancer Cells, SNU-C4. *Phytomedicine* 15 (1–2), pp. 147–51. doi:10.1016/j.phymed.2007.07.061.

Park, Ho Young, Sang Keun Ha, Hyojin Eom, and Inwook Choi. 2013. Narirutin Fraction from Citrus Peels Attenuates Alcoholic Liver Disease in Mice. *Food and Chemical Toxicology* 55, pp. 637–44. doi:10.1016/j.fct.2013.01.060.

Park, Se Jin, Younghwa Lee, Hee Kyong Oh, Hyung Eun Lee, Younghwan Lee, Sang Yoon Ko, Boseong Kim, Jae Hoon Cheong, Chan Young Shin, and Jong Hoon Ryu. 2014. Oleanolic Acid Attenuates MK-801-Induced Schizophrenia-like Behaviors in Mice. *Neuropharmacology* 86, pp. 49–56. doi:10.1016/j.neuropharm.2014.06.025.

Parthasarathy, S., J.C. Khoo, E. Miller, J. Barnett, J.L. Witztum, and D. Steinberg. 1990. Low Density Lipoprotein Rich in Oleic Acid Is Protected against Oxidative Modification: Implications for Dietary Prevention of Atherosclerosis. *Proceedings of the National Academy of Sciences* 87 (10), pp. 3894–98. doi:10.1073/pnas.87.10.3894.

Pierson, Jean T., Gregory R. Monteith, Sarah J. Roberts-Thomson, Ralf G. Dietzgen, Michael J. Gidley, and Paul N. Shaw. 2014. Phytochemical Extraction, Characterisation and Comparative Distribution across Four Mango (*Mangifera indica* L.) Fruit Varieties. *Food Chemistry* 149, pp. 253–63. doi:10.1016/j.foodchem.2013.10.108.

Prabhu, S., Mallika Jainu, K.E. Sabitha, and C.S. Shyamala Devi. 2006. Effect of Mangiferin on Mitochondrial Energy Production in Experimentally Induced Myocardial Infarcted Rats. *Vascular Pharmacology* 44 (6), pp. 519–25. doi:10.1016/j.vph.2006.03.012.

Punithavathi, Vilapakkam Ranganathan, Ponnian Stanely Mainzen Prince, Ramesh Kumar, and Jemmi Selvakumari. 2011. Antihyperglycaemic, Antilipid Peroxidative and Antioxidant Effects of Gallic Acid on Streptozotocin Induced Diabetic Wistar Rats. *European Journal of Pharmacology* 650 (1), pp. 465–71. doi:10.1016/j.ejphar.2010.08.059.

Raina, K., S. Rajamanickam, G. Deep, M. Singh, R. Agarwal, and C. Agarwal. 2008. Chemopreventive Effects of Oral Gallic Acid Feeding on Tumor Growth and Progression in TRAMP Mice. *Molecular Cancer Therapeutics* 7 (5), pp. 1258–67. doi:10.1158/1535-7163.MCT-07-2220.

Rajendran, P., G. Ekambaram, and D. Sakthisekaran. 2008. Effect of Mangiferin on Benzo(a)pyrene Induced Lung Carcinogenesis in Experimental Swiss Albino Mice. *Natural Product Research* 22 (8), pp. 672–80. doi:10.1080/14786410701824973.

Raphael, T.J., and G. Kuttan. 2003. Effect of Naturally Occurring Triterpenoids Glycyrrhizic Acid, Ursolic Acid, Oleanolic Acid and Nomilin on the Immune System. *Phytomedicine* 10 (6–7), pp. 483–89. doi:10.1078/094471103322331421.

Rasool, Mahaboob Khan, Evan Prince Sabina, Segu R. Ramya, Pranatharthiharan Preety, Smita Patel, Niharika Mandal, Punya P. Mishra, and Jaisy Samuel. 2010. Hepatoprotective and Antioxidant Effects of Gallic Acid in Paracetamol-Induced Liver Damage in Mice. *The Journal of Pharmacy and Pharmacology* 62 (5), pp. 638–43. doi:10.1211/jpp/62.05.0012.

Rawat, Pooja, Pawan Kumar Singh, and Vipin Kumar. 2017. Evidence Based Traditional Anti-Diarrheal Medicinal Plants and Their Phytocompounds. *Biomedicine and Pharmacotherapy* 96 (October), pp. 1453–64. doi:10.1016/j.biopha.2017.11.147.

Reddy, D. Bharat, and Pallu Reddanna. 2009. Chebulagic Acid (CA) Attenuates LPS-Induced Inflammation by Suppressing NF-κB and MAPK Activation in RAW 264.7 Macrophages. *Biochemical and Biophysical Research Communications* 381 (1), pp. 112–17. doi:10.1016/j.bbrc.2009.02.022.

Reddy, D. Bharat, T.C.M. Reddy, G. Jyotsna, Satish Sharan, Nalini Priya, V. Lakshmipathi, and Pallu Reddanna. 2009. Chebulagic Acid, a COX-LOX Dual Inhibitor Isolated from the Fruits of *Terminalia chebula* Retz., Induces Apoptosis in COLO-205 Cell Line. *Journal of Ethnopharmacology* 124 (3), pp. 506–12. doi:10.1016/j.jep.2009.05.022.

Reisman, Scott A., Lauren M. Aleksunes, and Curtis D. Klaassen. 2009. Oleanolic Acid Activates Nrf2 and Protects from Acetaminophen Hepatotoxicity via Nrf2-Dependent and Nrf2-Independent Processes. *Biochemical Pharmacology* 77 (7), pp. 1273–82. doi:10.1016/j.bcp.2008.12.028.

Ribeiro, S.M.R., L.C.A. Barbosa, J.H. Queiroz, M. Knödler, and A. Schieber. 2008. Phenolic Compounds and Antioxidant Capacity of Brazilian Mango (*Mangifera Indica* L.) Varieties. *Food Chemistry* 110 (3), pp. 620–26. doi:10.1016/j.foodchem.2008.02.067.

Rivera, Dagmar García, Ivones Hernández, Nelson Merino, Yilian Luque, Alina Álvarez, Yanet Martín, Aylin Amador, Lauro Nuevas, and René Delgado. 2011. *Mangifera indica* L. Extract (Vimang) and Mangiferin Reduce the Airway Inflammation and Th2 Cytokines in Murine Model of Allergic Asthma. *Journal of Pharmacy and Pharmacology* 63 (10), pp. 1336–45. doi:10.1111/j.2042-7158.2011.01328.x.

Rogerio, A.P., A. Kanashiro, C. Fontanari, E.V.G. Da Silva, Y.M. Lucisano-Valim, E.G. Soares, and L.H. Faccioli. 2007. Anti-Inflammatory Activity of Quercetin and Isoquercitrin in Experimental Murine Allergic Asthma. *Inflammation Research* 56 (10), pp. 402–8. doi:10.1007/s00011-007-7005-6.

Rogerio, Alexandre P., Caroline Fontanari, Érica Borducchi, Alexandre C. Keller, Momtchilo Russo, Edson G. Soares, Deijanira A. Albuquerque, and Lúcia H. Faccioli. 2008. Anti-Inflammatory Effects of Lafoensia Pacari and Ellagic Acid in a Murine Model of Asthma. *European Journal of Pharmacology* 580 (1–2), pp. 262–70. doi:10.1016/j.ejphar.2007.10.034.

Roidoung, Sunisa, Kirk D. Dolan, and Muhammad Siddiq. 2016. Gallic Acid as a Protective Antioxidant against Anthocyanin Degradation and Color Loss in Vitamin-C Fortified Cranberry Juice. *Food Chemistry* 210, pp. 422–27. doi:10.1016/j.foodchem.2016.04.133.

Rosenlund, Helen, Inger Kull, Göran Pershagen, Alicja Wolk, Magnus Wickman, and Anna Bergström. 2011. Fruit and Vegetable Consumption in Relation to Allergy: Disease-Related Modification of Consumption? *Journal of Allergy and Clinical Immunology* 127 (5), pp. 1219–25. doi:10.1016/j.jaci.2010.11.019.

Saghir, S.A.M., A. Sadikun, K.Y. Khaw, and V. Murugaiyah. 2013. Star Fruit (*Averrhoa Carambola* L.): From Traditional Uses to Pharmacological Activities. *Boletin Latinoamericano y del Caribe de Plantas Medicinales y Aromaticas* 12 (3), pp. 209–19.

Sánchez, G.M., L. Re, A. Giuliani, A.J. Núñez-Sellés, G.P. Davison, and O.S. León-Fernández. 2000. Protective Effects of *Mangifera indica* L. Extract, Mangiferin and Selected Antioxidants against TPA-Induced Biomolecules Oxidation and Peritoneal Macrophage Activation in Mice. *Pharmacological Research* 42 (6), pp. 565–73. doi:10.1006/phrs.2000.0727.

Sarjit, Amreeta, Yi Wang, and Gary A. Dykes. 2015. Antimicrobial Activity of Gallic Acid against Thermophilic Campylobacter Is Strain Specific and Associated with a Loss of Calcium Ions. *Food Microbiology* 46, pp. 227–33. doi:10.1016/j.fm.2014.08.002.

Shyni, Gangadharan Leela, Sasidharan Kavitha, Sasidharan Indu, Anil Das Arya, Sasidharan Suseela Anusree, Vadavanath Prabhakaran Vineetha, Sankar Vandana, Andikannu Sundaresan, and Kozhiparambil Gopalan Raghu. 2014. Chebulagic Acid from Terminalia Chebula Enhances Insulin Mediated Glucose Uptake in 3T3-L1 Adipocytes via PPARγ Signaling Pathway. *BioFactors* 40 (6), pp. 646–57. doi:10.1002/biof.1193.

Silva, Larissa Morais Ribeiro Da, Evania Altina Teixeira De Figueiredo, Nagila Maria Pontes Silva Ricardo, Icaro Gusmao Pinto Vieira, Raimundo Wilane De Figueiredo, Isabella Montenegro Brasil, and Carmen L. Gomes. 2014. Quantification of Bioactive Compounds in Pulps and by-Products of Tropical Fruits from Brazil. *Food Chemistry* 143, pp. 398–404. doi:10.1016/j.foodchem.2013.08.001.

Silva Junior, Iberê F., Marcela Raimondi, Susana Zacchino, Valdir Cechinel Filho, Vânia F. Noldin, Vietla S. Rao, Joaquim C.S. Lima, and Domingos T.O. Martins. 2010. Evaluation of the Antifungal Activity and Mode of Action of *Lafoensia pacari* A. St.-Hil., Lythraceae, Stem-Bark Extracts, Fractions and Ellagic Acid. *Revista Brasileira de Farmacognosia* 20 (3), pp. 422–28. doi:10.1590/S0102-695X2010000300021.

Singh, Onkar, and M. Ali. 2011. Phytochemical and Antifungal Profiles of the Seeds of *Carica papaya* L. *Indian Journal of Pharmaceutical Sciences* 73 (4), pp. 447–51. doi:10.4103/0250-474X.95648.

Sivasankar, Shanmuganathan, Ramu Lavanya, Pemaiah Brindha, and Narayanasamy Angayarkanni. 2015. Aqueous and Alcoholic Extracts of Triphala and Their Active Compounds Chebulagic Acid and Chebulinic Acid Prevented Epithelial to Mesenchymal Transition in Retinal Pigment Epithelial Cells, by Inhibiting SMAD-3 Phosphorylation. *PLoS ONE* 10 (3). doi:10.1371/journal.pone.0120512.

Soh, Patrice Njomnang, Benoît Witkowski, David Olagnier, Marie Laure Nicolau, Maria Concepcion Garcia-Alvarez, Antoine Berry, and Françoise Benoit-Vical. 2009. In Vitro and in Vivo

Properties of Ellagic Acid in Malaria Treatment. *Antimicrobial Agents and Chemotherapy* 53 (3), pp. 1100–1106. doi:10.1128/AAC.01175-08.

Song, Ji Hoon, Myoung Sook Shin, Gwi Seo Hwang, Seong Taek Oh, Jung Jin Hwang, and Ki Sung Kang. 2018. Chebulinic Acid Attenuates Glutamate-Induced HT22 Cell Death by Inhibiting Oxidative Stress, Calcium Influx and MAPKs Phosphorylation. *Bioorganic and Medicinal Chemistry Letters* 28 (3), pp. 249–53. doi:10.1016/j.bmcl.2017.12.062.

Sousa De Brito, Edy, Manuela Cristina Pessanha De Araújo, Ricardo Elesbão Alves, Colleen Carkeet, Beverly A. Clevidence, and Janet A. Novotny. 2007. Anthocyanins Present in Selected Tropical Fruits: Acerola, Jambolão, Jussara, and Guajiru. *Journal of Agricultural and Food Chemistry* 55 (23), pp. 9389–94. doi:10.1021/jf0715020.

Steffen, Lyn M., David R. Jacobs, June Stevens, Eyal Shahar, Teresa Carithers, and Aaron R. Folsom. 2003. Associations of Whole-Grain, Refined-Grain, and Fruit and Vegetable Consumption with Risks of All-Cause Mortality and Incident Coronary Artery Disease and Ischemic Stroke: The Atherosclerosis Risk in Communities (ARIC) Study. *American Journal of Clinical Nutrition* 78 (3), pp. 383–90.

Steinmetz, Kristi A., and John D. Potter. 1991. Vegetables, Fruit, and Cancer. I. Epidemiology. *Cancer Causes & Control* 2 (6), pp. 427–42. doi:10.1007/BF00051672.

Sthapit, S.R., and Sara J. Scherr. 2012. *Tropical Fruit Tree Species and Climate Change. Tropical Fruit Tree Species and Climate Change.* http://www.indiaenvironmentportal.org.in/files/file/Tropical_fruit_tree_species_and_climate_change.pdf.

Sung, Hye-Young, Sang-Wook Kang, Jung-Lye Kim, Jing Li, Eun-Sook Lee, Ju-Hyun Gong, Seoung Jun Han, and Young-Hee Kang. 2010. Oleanolic Acid Reduces Markers of Differentiation in 3T3-L1 Adipocytes. *Nutrition Research* 30 (12), pp. 831–39. doi:10.1016/j.nutres.2010.10.001.

Süntar, Ipek Peşin, Esra Küpeli Akkol, Funda Nuray Yalçin, Ufuk Koca, Hikmet Keleş, and Erdem Yesilada. 2010. Wound Healing Potential of *Sambucus ebulus* L. Leaves and Isolation of an Active Component, Quercetin 3-O-Glucoside. *Journal of Ethnopharmacology* 129 (1), pp. 106–14. doi:10.1016/j.jep.2010.01.051.

Szakiel, Anna, Dariusz Ruszkowski, Anna Grudniak, Anna Kurek, Krystyna I. Wolska, Maria Doligalska, and Wirginia Janiszowska. 2008. Antibacterial and Antiparasitic Activity of Oleanolic Acid and Its Glycosides Isolated from Marigold (*Calendula officinalis*). *Planta Medica* 74 (14), pp. 1709–15. doi:10.1055/s-0028-1088315.

Tan, Tzu Wei, Ying Erh Chou, Wei Hung Yang, Chin Jung Hsu, Yi Chin Fong, and Chih Hsin Tang. 2014. Naringin Suppress Chondrosarcoma Migration through Inhibition Vascular Adhesion Molecule-1 Expression by Modulating miR-126. *International Immunopharmacology* 22 (1), pp. 107–14. doi:10.1016/j.intimp.2014.06.029.

Tan, Zhen, Jianwen Cheng, Qian Liu, Lin Zhou, Jacob Kenny, Tao Wang, Xixi Lin, et al. 2017. Neohesperidin Suppresses Osteoclast Differentiation, Bone Resorption and Ovariectomised-Induced Osteoporosis in Mice. *Molecular and Cellular Endocrinology* 439, pp. 369–78. doi:10.1016/j.mce.2016.09.026.

Teodoro, Tracy, Liling Zhang, Todd Alexander, Jessica Yue, Mladen Vranic, and Allen Volchuk. 2008. Oleanolic Acid Enhances Insulin Secretion in Pancreatic Beta-Cells. *FEBS Letters* 582 (9), pp. 1375–80. doi:10.1016/j.febslet.2008.03.026.

Terao, J. 1989. Antioxidant Activity of β-Carotene-Related Carotenoids in Solution. *Lipids* 24 (7), pp. 659–61. doi:10.1007/BF02535085.

Tirkey, Naveen, Sangeeta Pilkhwal, Anurag Kuhad, and Kanwaljit Chopra. 2005. Hasperidin, a Citrus Bioflavonoid, Decreases the Oxidative Stress Produced by Carbon Tetrachloride in Rat Liver and Kidney. *BMC Pharmacology* 5. doi:10.1186/1471-2210-5-2.

Tong, Fan, Jian Zhang, Li Liu, Xican Gao, Qian Cai, Chunhua Wei, Jihua Dong, Yu Hu, Gang Wu, and Xiaorong Dong. 2016. Corilagin Attenuates Radiation-Induced Brain Injury in Mice. *Molecular Neurobiology* 53 (10), pp. 6982–96. doi:10.1007/s12035-015-9591-6.

Tseng, Tsui-Hwa, Chia-Yih Chu, Jin-Ming Huang, Song-Jui Shiow, and Chau-Jong Wang. 1995. Crocetin Protects against Oxidative Damage in Rat Primary Hepatocytes. *Cancer Letters* 97 (1), pp. 61–67. doi:10.1016/0304-3835(95)03964-X.

Tsui, V.W.K., R.W.K. Wong, and A. Bakr M Rabie. 2008. The Inhibitory Effects of Naringin on the Growth of Periodontal Pathogens in Vitro. *Phytotherapy Research* 22 (3), pp. 401–6. doi:10.1002/ptr.2338.

Usha, K. 2015. Orchard Layout and Establishment of Fruit Orchard—Principles and Practices. In *Fundamental of Fruit Production*, edited by K. Usha, Madhubala Thakre, Amit Kumar Goswami, and Nayan Deepak G., pp. 6–33. Indian Agricultural Research Institute New Delhi, India.

Van Duyn, M.A.S., and Pivonka, E. 2000. Overview of the Health Benefits of Fruit and Vegetable Consumption for the Dietetics Professional: Selected Literature. *Journal of the American Dietetic Association* 100 (12), pp. 1511–21.

Veer, P., M.C. Jansen, M. Klerk, and F.J. Kok. 2000. Fruits and Vegetables in the Prevention of Cancer and Cardiovascular Disease. *Public Health Nutrition* 3 (1), pp. 103–7.

Vishnu Prasad, C.N., T. Anjana, Asoke Banerji, and Anilkumar Gopalakrishnapillai. 2010. Gallic Acid Induces GLUT4 Translocation and Glucose Uptake Activity in 3T3-L1 Cells. *FEBS Letters* 584 (3), pp. 531–36. doi:10.1016/j.febslet.2009.11.092.

Wang, Zheng, Qiong Ya Guo, Xiao Ju Zhang, Xiao Li, Wen Ting Li, Xi Tao Ma, and Li Jun Ma. 2014. Corilagin Attenuates Aerosol Bleomycin-Induced Experimental Lung Injury. *International Journal of Molecular Sciences* 15 (6), pp. 9762–79. doi:10.3390/ijms15069762.

Wang, Ziyou, Hua Zhang, Jiahui Zhou, Xiangning Zhang, Liyong Chen, Kangxing Chen, and Zunnan Huang. 2016. Eriocitrin from Lemon Suppresses the Proliferation of Human Hepatocellular Carcinoma Cells through Inducing Apoptosis and Arresting Cell Cycle. *Cancer Chemotherapy and Pharmacology* 78 (6), pp. 1143–50. doi:10.1007/s00280-016-3171-y.

Warpe, Vikas S., Vishal R. Mali, S. Arulmozhi, Subhash L. Bodhankar, and Kakasaheb R. Mahadik. 2015. Cardioprotective Effect of Ellagic Acid on Doxorubicin Induced Cardiotoxicity in Wistar Rats. *Journal of Acute Medicine* 5 (1), pp. 1–8. doi:10.1016/j.jacme.2015.02.003.

Xie, Wenyan, Meng Wang, Chen Chen, Xiaoying Zhang, and Matthias F. Melzig. 2016. Hepatoprotective Effect of Isoquercitrin against Acetaminophen-Induced Liver Injury. *Life Sciences* 152, pp. 180–89. doi:10.1016/j.lfs.2016.04.002.

Yadav, Suraj Singh, Manish Kumar Singh, Pawan Kumar Singh, and Vipin Kumar. 2017. Traditional Knowledge to Clinical Trials: A Review on Therapeutic Actions of Emblica Officinalis. *Biomedicine & Pharmacotherapy* 93, pp. 1292–1302. doi:10.1016/j.biopha.2017.07.065.

Yamauchi, Mika, Kazuhiro Tsuruma, Shunsuke Imai, Tomohiro Nakanishi, Naofumi Umigai, Masamitsu Shimazawa, and Hideaki Hara. 2011. Crocetin Prevents Retinal Degeneration Induced by Oxidative and Endoplasmic Reticulum Stresses via Inhibition of Caspase Activity. *European Journal of Pharmacology* 650 (1), pp. 110–19. doi:10.1016/j.ejphar.2010.09.081.

Yang, Depo, Laura Michel, Jean Pierre Chaumont, and Joëlle Millet-Clerc. 1999. Use of Caryophyllene Oxide as an Antifungal Agent in an in Vitro Experimental Model of Onychomycosis. *Mycopathologia* 148 (2), pp. 79–82. doi:10.1023/A:1007178924408.

Yang, Dong Joo, Sang Hyun Moh, Dong Hwee Son, Seunghoon You, Ann W. Kinyua, Chang Mann Ko, Miyoung Song, Jinhee Yeo, Yun Hee Choi, and Ki Woo Kim. 2016. Gallic Acid Promotes Wound Healing in Normal and Hyperglucidic Conditions. *Molecules (Basel, Switzerland)* 21 (7). doi:10.3390/molecules21070899.

Yang, Yajun, Jinghui Xiu, Jiangning Liu, Li Zhang, Xiaoying Li, Yanfeng Xu, Chuan Qin, and Lianfeng Zhang. 2013. Chebulagic Acid, a Hydrolyzable Tannin, Exhibited Antiviral Activity in Vitro and in Vivo against Human Enterovirus 71. *International Journal of Molecular Sciences* 14 (5), pp. 9618–27. doi:10.3390/ijms14059618.

Yeo, Sang Gu, Jae Hyoung Song, Eun Hye Hong, Bo Ra Lee, Yong Soo Kwon, Sun Young Chang, Seung Hyun Kim, Sang Won Lee, Jae Hak Park, and Hyun Jeong Ko. 2015. Antiviral Effects of Phyllanthus Urinaria Containing Corilagin against Human Enterovirus 71 and Coxsackievirus A16 in Vitro. *Archives of Pharmacal Research* 38 (2), pp. 193–202. doi:10.1007/s12272-014-0390-9.

Yoon, Chong Hyeon, Soo Jin Chung, Sang Won Lee, Yong Beom Park, Soo Kon Lee, and Min Chan Park. 2013. Gallic Acid, a Natural Polyphenolic Acid, Induces Apoptosis and Inhibits Proinflammatory Gene Expressions in Rheumatoid Arthritis Fibroblast-like Synoviocytes. *Joint Bone Spine* 80 (3), pp. 274–79. doi:10.1016/j.jbspin.2012.08.010.

Yoshimi, Naoki, Kengo Matsunaga, Masaki Katayama, Yasuhiro Yamada, Toshiya Kuno, Zheng Qiao, Akira Hara, Johji Yamahara, and Hideki Mori. 2001. The Inhibitory Effects of Mangiferin, a Naturally Occurring Glucosylxanthone, in Bowel Carcinogenesis of Male F344 Rats. *Cancer Letters* 163 (2), pp. 163–70. doi:10.1016/S0304-3835(00)00678-9.

Yun, Jieun, Heejeong Lee, Hae Ju Ko, Eun Rhan Woo, and Dong Gun Lee. 2015. Fungicidal Effect of Isoquercitrin via Inducing Membrane Disturbance. *Biochimica et Biophysica Acta - Biomembranes* 1848 (2), pp. 695–701. doi:10.1016/j.bbamem.2014.11.019.

Zhang, Lan-Zhen, Wen-Hua Zhao, Ya-Jian Guo, Guang-Zhong Tu, Shu Lin, and Lin-Guang Xin. 2003. Studies on Chemical Constituents in Fruits of Tibetan Medicine *Phyllanthus emblica*. *Zhongguo Zhong Yao Za Zhi = Zhongguo Zhongyao Zazhi = China Journal of Chinese Materia Medica* 28 (10), pp. 940–43.

Zhong, Ying Jia, Fang Shi, Xue Lian Zheng, Qiong Wang, Lan Yang, Hong Sun, Fan He et al. 2011. Crocetin Induces Cytotoxicity and Enhances Vincristine-Induced Cancer Cell Death via p53-Dependent and -Independent Mechanisms. *Acta Pharmacologica Sinica* 32 (12), pp. 1529–36. doi:10.1038/aps.2011.109.

Zhu, Yue-Yong, Hong-Yan Huang, and Yin Wu. 2015. Anticancer and Apoptotic Activities of Oleanolic Acid Are Mediated through Cell Cycle Arrest and Disruption of Mitochondrial Membrane Potential in HepG2 Human Hepatocellular Carcinoma Cells. *Molecular Medicine Reports* 12 (4), pp. 5012–18. doi:10.3892/mmr.2015.4033.

10

Plant Alkaloids: Classification, Isolation, and Drug Development

Phurpa Wangchuk

CONTENTS

10.1 Introduction

Alkaloids are a group of naturally occurring phytochemicals that mostly contain nitrogen and are produced by plants as secondary metabolites for defense and protection against their enemies and predators (Ain et al. 2016). Consequently, plants produce higher concentration of alkaloids in places where herbivores usually attack, including the inflorescence, young shoots and peripheral cell layers of stems and roots. The content and the types of alkaloid also vary from species to species, different climatic conditions and regions or habitat. It is estimated that nearly 20% of plant species contain alkaloids (Cortinovis and Caloni 2015) and they are particularly observed in the plant families: Apocyanaceae, Papaveraceae, Papilionaceae, Ranunculaceae, Rutaceae, and Solanaceae. Since alkaloids are prepared by plants using intricate chemical reactions, which are quite often different from laboratory synthesis, their structural diversity is greater than that provided by most available laboratory or combinatorial approaches. Morphine was the first alkaloid isolated by Friedrich Serturner in the early 19th century, which laid foundation to the later discoveries of many alkaloids and modern drugs. Today, more than 20,000 alkaloids have been documented (Buckingham et al. 2010).

Humans have used alkaloid-containing plants for medicines, magic, murder, recreations, warfare, poisons for hunting food and depredating wild animals for thousands of years. For example, crude preparations from alkaloid-containing plants such as *Belladonna*, *Aconitum* and *Delphinium* species have long had a broad range of applications as arrowhead dips for hunting, an agent for euthanasia, and medical treatments for neuralgia, gout, hypertension and rheumatism in Asia, America and Europe. Mesopotamians first used the analgesic alkaloid-containing poppy juice for treating different ailments in 2600 BC (Cragg and Newman 2001). Indian Ayurvedic and Chinese traditional medicines use the alkaloid-containing plants such as

Datura, *Aconitum*, and *Sarcostemma* for treating various ailments including gastrointestinal diseases, arthritic pain, and as a febrifuge and bitter tonic (Maurya et al. 2015; He et al. 2017). Some alkaloid-containing plants are toxic to humans and animals, and thus require regulations for their use. For example, ingestion of pyrrolizidine alkaloids can result in acute or chronic liver toxicity and genotoxicity and consequently, this alkaloid-producing plant have been regulated in different European countries. However, many alkaloid-containing plants have been generally regarded as non-toxic when developed into drugs with right dosage. Generally, isolation and drug development from the alkaloid-containing plants require multidisciplinary approaches involving botany, natural products and medicinal chemistry, metabolomics, immunology and clinical trials. In-depth understanding of the plants and their families, and alkaloid classification and their chemical properties would facilitate efficient isolation and discovery of novel drug lead compounds. This chapter considers the naming, classification, isolation, and structure elucidation of alkaloids, and their application in pharmaceuticals with more focus on two categories of alkaloids—diterpenoids and isoquinolines.

10.2 Naming, Classification, and Biosynthesis of Plant Alkaloids

With alkaloid nomenclature, trivial naming is commonly observed; however, about four different types of naming are followed based on (Hesse 2002):

1. A systematic plant name, e.g., papaverine derived from *Papaver* species
2. The plant discoverer, e.g., spegazzinine from *Aspidosperma chakensi* SPEGAZZI

3. Geographical settings, e.g., tasmanine after Tasmania

4. A person's name, e.g., macrosalhine after the lady Salhiha

However, such trivial naming of alkaloids falls short when several new alkaloids are isolated from a single plant and the IUPAC naming system becomes complex for bigger molecules. Therefore, judicious naming can be followed based on the chemist's imagination, assigning sufixes like "-ine," "-idine," "- anine," "-aline," and "-inine" (Hesse 2002; Buckingham et al. 2010).

Similarly, the orderly classification of alkaloids is very difficult due to extreme heterogeneity in alkaloid structures and therefore their classification is not well established. Many alkaloids are classified based on their biogenesis and biological origins, structural relationships and spectroscopic properties (Fester 2010). Alkaloids can be mainly grouped into four classes:

1. Amino acid derivatives (e.g., ornithine, lysine, anthranilic acid, tyrosine, and tryptophan)
2. Purine derivatives (e.g., zanthine and caffeine)
3. Aminated terpenes (e.g., aconitine and solanine)
4. Polyketides (e.g., coniine and the coccinellines)

Alkaloids derived from ornithine are further sub-categorized into: simple alkaloids, tropane alkaloids, pyrrolizidine alkaloids, phenanthroindolizidine alkaloids, spermine, spermidine, and the miscellaneous alkaloids. Alkaloids resulting from lysine are sub-grouped into: simple piperidines, lobelia alkaloids, lycopodium alkaloids, lythraceae alkaloids and miscellaneous alkaloids. Alkaloids originated from anthranilic acid are sub-categorized into quinoline, furanoquinoline, acridine, and quinazoline groups. Alkaloids derived from tyrosine consist of simple β-phenylethylamine derivatives, Amaryllidaceae alkaloids, and the isoquinoline alkaloids. Alkaloids resulting from tryptophan constitute the largest alkaloid group and include tryptamine derivatives, carbazoles, ergot alkaloids, borreria alkaloids, monoterpenoid and indole alkaloids. Fester (2010) has beautifully described the biosynthetic origins and pathways for the selected alkaloids derived from the L-tryptophan and L-tyrosine amino acids. The β-Carboline alkaloids are biosynthetically derived from the amino acids and were later reported from other plant families: Leguminoseae, Malpighiaceae, Rubiaciae, Rutaceae, Graminae, Simaroubaceae, and Zygophyllaceae. Alkaloids derived from aminated terpenes include, monoterpenoid pyridine alkaloids, nuphar alkaloids, celastraceae alkaloids, erythrophleum alkaloids, diterpenoid alkaloids, steroidal alkaloids, and the daphniphyllum alkaloids.

All these alkaloids exhibit diverse biological activities. For example, cocaine is addictive, and aconitine is poisonous. Amongst different sub-classes of plant alkaloids, the diterpenoid and isoquinoline groups have been extensively studied for their chemical and pharmacological properties owing to their toxicity, intriguing structural diversity, and their potential medicinal values. These two alkaloid groups are further discussed in the following sections.

10.2.1 Diterpenoid Alkaloids

The diterpenoid alkaloids occur commonly either as amino alcohols or as esters of amino alcohols and these permutations and combinations thus give rise to many interesting bases. These alkaloids occur mainly in the families: Ranunculaceae (*Aconitum*, *Delphinium*, *Thalictrum*, and *Consolida* genera), Cornaceae (*Garrya* species), Rosaceae (*Spirea japonica*) and Compositae (*Inula royleana*) and possess widely recognised pharmacological and biological activities. More than 1100 diterpenoid alkaloids have been isolated to date from various species of *Aconitum*, *Consolida*, *Delphinium*, *Chamaecyparis*, *Clitocybe*, *Spiraea*, and *Tricalysia*. Over 197 of diterpenoid alkaloids were sequestered from 56 *Aconitum* species (Hamlin et al. 2014; Wangchuk et al. 2015). A number of reviews on diterpenoid alkaloids have been published in the past decades, and the most comprehensive ones are reported by Wang and co-workers (Wang et al. 2010) and Liang and his group (Liang et al. 2018). While a former review by Wang and his group covered the diterpenoid alkaloids isolated in between 1998–2008 and discussed about the classification, chemical reactions and biological activities; the later review by Liang and his co-workers described the most recent works on the diterpenoid alkaloids with cytotoxicity activities that were carried out in between 2008 and 2018. These two recent reviews classified the diterpenoid alkaloids into three main classes as C_{18}-, C_{19}-, and C_{20}-based on their chemical structures, which are briefly discussed here.

The C_{18}-diterpenoid alkaloids have mainly two structural types as lappaconitines and ranaconitines. While the lappaconitines contain methine unit at C-7 position, ranaconitines are characterized by the presence of oxygenated functional group at C-7. From 1998 to 2008, only 16 lappaconitine-type (e.g., piepunendine B) and 15 ranaconitines-type (e.g., tiantaishansine) alkaloids have been documented (Wang et al. 2010). Most of these C18-diterpenoid alkaloids were isolated from *Aconitum* and *Delphinium* species. Seven new C_{18}-diterpenoid alkaloids, anthriscifoltines A-G were isolated from *Delphinum anthriscifolium* var. *majus* (Shan et al. 2018). This same plant species has been reported to produce all C_{18}-, C_{19}-, and C_{20}- diterpenoid clases of alkaloids (Shan et al. 2017).

The C_{19}-diterpenoid alkaloids have three sub-types of basic skeletal ring systems: the aconitanes/aconitines, the lycoctonines, and the lactones/heteratisanes. The aconitane group, most of which possess an α-oriented hydroxyl group at C-6 position, accounts for majority of the C_{19}-diterpenoid class of alkaloids. Hemsleyatine (isolated from the plant *Aconitum hemsleyanum*) was the first aconitane-type C_{19} diterpenoid alkaloid bearing an amino group at the C-8 position (Zhou et al. 2003). Recently, more C_{19}-diterpenoid alkaloids including eight new arabinosides (aconicarmichosides E-L) and four known neo-line 14-O-arabinosides (only examples of glycosidic diterpenoid alkaloids so far) were isolated from *A. carmichaelii* (Guo et al. 2018). The un-substituted C-14 aconitane ring system with C-7 substitution occurs in the lycoctonine-type alkaloids. Its sub-type, the pyrodelphinine-type of the C_{19} diterpenoid alkaloids, originates as a result of C-8–C-15 double bond formation in the aconitane ring system. The oxidative fission of the C-13–C-14 bond of the aconitane framework forms lactone or heteratisane-type alkaloids as represented by heteratisine.

The C_{20} diterpenoid alkaloid is based on the atidane ring system or also known as atisine-type alkaloids. Atidine was first isolated from the Himalayan medicinal plant, *Aconitum heterophyllum*, along with the known alkaloids atisine, hetisine, heteratisine, and benzoylheteratisine. Hetisane, atisane and delnudine together represent the biggest sub-types of atidane-based C_{20} diterpenoid alkaloids, and they have the lowest toxicity compared to other classes of aconitum-diterpenoid alkaloids. The first representatives of hetisane-type diterpenoid alkaloids were paniculatine, hetisine and kobusine (Bessonova and Saidkhodzhaeva 2000). Orochrine, 2-*O*-acetylorochrine, and 2-*O*-acetyl-7α-hydroxyorochrine were isolated as new hetisane-type diterpenoid alkaloids from a Bhutanese medicinal plant, *A. orochryseum* (Wangchuk et al. 2007). The contraction of ring C in the hetisane forms complex derivatives belonging to the delnudine type of alkaloid as exemplified by delnudine. Four new C_{20}-diterpenoid alkaloids including rotundifosine F, which is a rare structure with quaternary ammonium salt, has been recently isolated from *A. rotundifolium* (Zhang et al. 2018). The other major C_{20} diterpenoid alkaloid is a veatchine skeletal type. *Delphinium* species (Ranunculaceae) generally contain veatchine or atisine-type diterpenoid alkaloids. A striking similarity can be observed in the alkaloid chemistry of atisine (atidane type) isolated from *Aconitum heterophyllum* (Ranunculaceae) and veatchine (veatchine type) isolated from *Garrya veatchii* (Garryaceae) although they belong to different families.

Varieties of bases that don't fit into any of the above structural types could also be formed by either modifications of the skeletal rings or by introduction of a new ring system (as a result of bond formation) into the main structural skeleton, as is the case with the icacine. They are grouped under miscellaneous diterpenoid alkaloids. Arcutinine, which has a C-5–C-20 bond instead of the normal C-11–C-17 and C-10–C-20 bridging bonds, was isolated from *Aconitum arcutum* and is still the only one of its kind (Saidkhodzhaeva et al. 2001a). Secokaraconitine, another new diterpenoid alkaloid that has no C-7–C-17 bond, was isolated from the Kyrgystan plant *Aconitum karacolicum* (Sultankhodzhaev et al. 2001b).

10.2.2 Isoquinoline Alkaloids

There are more than 1,200 isoquinoline alkaloids that have been isolated from 32 plant families (Buckingham et al. 2010). More than 190 isoquinoline alkaloids have been isolated from one family (papaveraceae) alone (Hesse 2002). Morphine was the first isoquinoline alkaloid to be isolated from *Papaver somniferum* (Papaveraceae) in 1805. The classification of isoquinoline alkaloids is not clearly defined. However, based on their structural types and biogenesis, they are commonly divided into 20 categories, each of which is further branched into different sub-classes. Each group presents striking and enormously varying structural types basically derived from tyrosine, which are briefly described as follows.

1. Simple isoquinolines, e.g. corydaldine
2. Emetines, e.g. emetine
3. Benzylisoquinolines, e.g. cularine
4. Proaporphines, e.g. orientalinone
5. Aporphines, e.g. liriodenine
6. Protoberberines, e.g. ophiocarpine
7. Benzophenanthridines, e.g. chelidonine
8. Dibenzopyrrocoline alkaloids, e.g. cryptaustoline
9. Protopines, e.g. protopine
10. Rhoeadine alkaloids, e.g. rhoeadine
11. Other minor categories include Spirobenzylisoquinolines, Bis-benzylisoquinolines, Phenethylisoquinolines, Phthalideisoquinolines, Morphines, Narceines, Pavines, Protostephanines, and Hasubanonine group

Although these alkaloids are quite often typical and restricted to certain genera of the family as in the case of rhoeadine alkaloids, which are encountered only in the *Papaver* genus, more than one class of alkaloids have been observed commonly even within a genus. The protoberberine alkaloids are distributed among the plant families Alangiaceae, Berberidaceae, Fumariaceae, Hydrastidaceae, Lauraceae, Leguminosae, Menispermaceae, Papaveraceae, Ranunculaceae, Rutaceae, and Annonaceae (Hesse 2002). However, these families also host large reservoirs of other classes of alkaloids. For examples, many protoberberine and aporphine alkaloids including corydine, annonaine, roemerine, norcorydine, norisocorydine, isocorydine, glaucine, dienone, aporphine and norlaureline are present in *Annona squamosa* of the Annonaceae family. Nine alkaloids belonging to aporphine, protoberberine and protopine alkaloids were isolated from *Dicranostigma leptopodum* (Papaveraceae) (Sun et al. 2014).

The protopine group of alkaloids are more common in the plant families of Fumariaceae, Papaveraceae, Hypecoaceae, Berberidaceae, Sapindaceae, Nandinaceae, and the Pteridophyllaceae. For example, protopine-type alkaloids, argemexicaines-A and B, were isolated from the family of Papaveraceae (*Argemone mexicana*) (Yuh-chwen et al. 2003). From 58 *Corydalis* species belonging to Fumariaceae, more than 119 different types of isoquinoline alkaloids were reported (Buckingham et al. 2010). Some examples of isoquonoline alkaloids isolated from the *Corydalis* species (*C. koidzumiana*, *C. calliantha*, *C, dubia*, and *C. crispa*) (Wangchuk et al. 2010) are protopine, α-allocryptopine, tetrahydropalmatine, corydaline, stylopine, scoulerine, capaurine, isocorypalmine, cheilanthifoline, corybulbine, sanguinarine, oxysanguinarine, dihydrosanguinarine, reticuline, sinoacutine, dubiamine, 13-oxoprotopine, coreximine, rheagenine, ochrobirine, sibiricine, bicuculline, capnoidine, corydecumbine and hydrastine and pallidine. From an unrelated Bhutanese medicinal plant, *Meconopsis simplicifolia*, a new protoberberine alkaloid, simplicifolianine, was isolated (Wangchuk et al. 2013). Recently, four new spirobenzylisoquinoline N-oxide alkaloids, hendersines C-F along with seven known isoquinoline alkaloids were isolated from *C. hendersonii* (Yin et al. 2018).

These alkaloids are related biosynthetically and many biosynthetic correlations have been studied for the isoquinolines. The biosynthetic pathway for protopine-type (e.g. protopine) and benzo[c]phenanthridine-type (e.g. chelidonine) of alkaloids were demonstrated in *Corydalis incisa* and *Chelidonium majus*, which suggested that chelidonine and protopine alkaloids were biosynthesised in plants from tyrosine. It has also been shown that the protopine-type alkaloids occupy a central position biosynthetically between the tetrahydroprotoberberine-*N*-metho salts and the benzophenanthridines and between the N-quaternary tetra-hydroberberines and the rhoedines. The benzophenanthridines type of alkaloids are derived from protoberberines via N-C$_6$ bond cleavage and the formation of C$_6$-C$_{13}$ bonds and it is one of the characteristics of the family Papaveraceae (Yu et al. 2014). The structurally related isoquinoline N-oxide alkaloids including tehauanine N-oxide, nigellimine N-oxide and jatamine N-oxides were isolated from *Pachycereus pringlei*, *Nigella sativa* and *Cocculus hirsutus*, respectively (Dembitsky et al. 2015).

10.3 Isolation, Structure Elucidation and Identification of Alkaloids

Generally, the isolation of alkaloids from a plant is achieved in five major steps, which includes preliminary physiochemical (color reactions) and Thin Layer Chromatography (TLC) plate loading tests (to detect alkaloids), extraction and fractionation with different solvents, purification using various chromatographic separation equipment and techniques, identification and structure elucidation of the isolated compounds applying Mass spectrometry, Nuclear Magnetic Resonance (NMR) and Infrared (IR) spectroscopies, and X-ray crystallography (Figure 10.1) (Wangchuk and Loukas 2018).

The first step is the physiochemical test, which can be carried out even in the field. When Dragendorf's and Mayer's reagents are added to the extracts of a plant, the solution would turn orange and white precipitate, respectively, in the presence of alkaloids. These two tests sometimes give false positive or ambiguous results and therefore, it is customary to confirm the result with TLC loading test method for alkaloid, which involves observing under ultraviolet lamp (254 and 354 wavelengths) and moderately spraying the plate with Dragendorf's reagent to obtain orange endpoint reaction spots. Standards (commercially

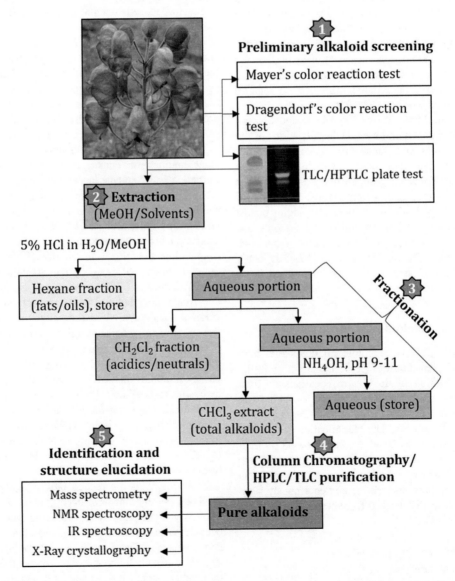

FIGURE 10.1 Schematic diagram of the isolation of alkaloids from a plant that can be achieved in five major steps. (1) Preliminary alkaloid screening. (2) Extraction of alkaloids using organic solvents. (3) Acid-base fractionation. (4) Purification of compounds using different chromatography techniques. (5) Identification and structure elucidation of compounds using spectroscopic and crystallographic techniques.

available) can be co-spotted in the TLC plate as reference point. An automated high-performance, thin-layer chromatography (HPTLC), which is an enhanced form of TLC, can be used to obtain reproducible and digital chromatogram.

The next step after preliminary physiochemical screening is the bulk extraction of crude extract from the plant. Although fresh wet plant materials can be used for extracting the alkaloids, it is much preferable to work with the shade/room temperature/freeze dried plant material. The dried plant material can be chopped into small pieces, made into coarse powder, soak in the solvent (MeOH), filter it and then the supernatant can be concentrated by Rotary Evaporator. The total alkaloids from the crude extract can be obtained using acid-base-solvent fractionation technique. While the alkaloids can be obtained without fractionation, acidification and basification processes, the yield of alkaloid is better using these step-wise fractionation processes. Common solvents used in the fractionation and isolation processes includes hexane, petroleum ether, dichloromethane, acetronitrile, methanol, butanol and water. The total crude alkaloid obtained through this fractionation processes is then separated using different chromatography and crystallization processes. Prior to mass separation of crude alkaloid extract, developing its TLC or analytical HPLC profiles are helpful for choosing the appropriate separation techniques and mobile phases. Separating the crude extract using the flash column chromatography (CC) or preparative thin layer chromatography (PTLC) helps in removing the unwanted materials including resins, chlorophyll and undissolved substances, which may affect the HPLC columns. The semi-purified fractions can be then separated by HPLC equipment. It is worthwhile to attempt growing crystals from each HPLC fractions since it is easier and reliable to elucidate the crystal structures using X-ray crystallography. Single crystals can be grown using single or two solvent systems with different volatility indices.

Once the alkaloids are obtained in pure form, their identities and structures must be established using MS, Infrared (IR) and NMR spectroscopies. Many known alkaloids can be identified by the GCMS library matching technique, which relies on the NIST compound library. The alkaloids identified using GCMS can be confirmed by comparing the physiochemical properties and 1D-NMR (^{1}H and ^{13}C-NMR) of the isolated compound with the reported literature. The physiochemical properties include colors, state of matter, melting point, optical rotation molecular weight and molecular formula. If the isolated alkaloids cannot be identified by GCMS library matching and 1D-NMR spectral peak (chemical shifts of the spectra) matching techniques, it is either new alkaloid or may have been wrongly identified before. In such cases, a step-wise and detailed MS, IR and NMR experiments are necessary. While the MS data is used for generating information on the mass, ionization pattern and the molecular formula (based on the abundance and the elemental composition) of a compound, the IR is used for determining the functional group present in its structure, and the NMR is used for elucidating the compound structure.

The NMR experiments comprise 1D-NMR and 2D-NMR (300–900 MHz NMR system of different makes—Bruker, Varian, Agilent Technologies, Oxford, Thermo Fisher Sientific) and these techniques have been recently described in techniques and technologies for natural products (Wangchuk and Loukas 2018). The most commonly used 1D-NMR experiments include

^{1}H-NMR, ^{13}C-NMR, APT, and DEPT. Similar to the MS data, 1D-NMR/proton experiments can provide information on the total number of hydrogens, carbons and the functional groups (CH, CH_2, and CH_3 groups) present in an alkaloid structure. The common 2D-NMR experiments that are used for defining the relative location of protons, carbons and functional groups include gCOSY, gNOESY, gTOCSY, gHSQC, and gHMBC. While gCOSY (^{1}H-^{1}H) and gTOCSY (^{1}H-^{1}H) can generate data on the coupled proton signals, gNOESY(^{1}H-^{1}H) shows the stereochemistry of a compound structure. On the other hand, gHSQC and gHMBC is used for obtaining information on chemical shifts correlation of ^{1}H-^{13}C. Depending upon the class and the structure of a new alkaloid, a new name is then given, which is itemized in a CAS registry.

10.4 Drug Development and Drug Leads Identification from Plant Alkaloids

While 25%–50% of the current pharmaceuticals, especially the antimalarial and anticancer drugs, were derived from plants, alkaloids represent only 16% of this chemotherapeutic pool (Wilkinson et al. 2002). They are poorly investigated for therapeutic purposes. Recently, 53 alkaloids have been listed in modern medicine for treating various diseases as anti-arrhythmic, anticholinergic, stimulant, analgesics, antigout, anthelmintic, uterotonic, antipyretic, antimalarial, acetylcholinesterase inhibitors, antihypertrnsive, muscle relaxant, antitumor, vasodilator, antimicrobial, anorectic, anesthetic, antidote, anticancer, antirheumatic, anti-emetic, anti-asmatic and aphrodisiac agents (Amirkia and Heinrich 2014). Some of the most popular drugs derived from alkaloids are: taxol, cocaine, quinine, codeine, morphine, atropine, ephedrine, reserpine, galanthamine, vincristine and vinblastine (Wangchuk 2018). Aconitine, which was isolated from *Aconitum napellus*, is used in many aconitine-containing liniments (Aconitysat™, Brinpax™, Etermol™, and Pectovox™), which are used against rheumatism, neuralgia and sciatica. Lappaconitine, is commonly used as a non-addictive postoperative analgesic for managing clinical cancer pain. An alkaloid drug called Provir/Virend/Crofelemer, which was approved by US FDA in 2012 for the treatment of diarrhea associated with anti-HIV drugs, was discovered from dragon's blood or the sap of the South American tree, *Croton lechleri* (Cottreau et al. 2012). An indole alkaloid, voacamine, which demonstrated significant antimalarial activity against *Plasmodium falciparum*, was isolated from Indonesian *Voacanga* species, and was approved for use as an antimalarial drug in several African countries (Federici et al. 2000).

Chemists and pharmaceutical companies are still using plants and plant alkaloids as one of the reliable sources of drugs and drug lead compounds. Among the isolated plant alkaloids, many biologically active isoquinoline and diterpenoid alkaloids with drug-like properties have been reported with some of them possessing significant pharmacological properties worthy for new drug development especially against cancer, malaria, Alzheimer's and microbial infections. While the reviews by Wang and group (Wang et al. 2010) and Liang and co-workers (Liang et al. 2018) provides comprehensive information on the biological activities of the diterpenoid alkaloids, a review by Iranshahy and co-workers (Iranshahy et al. 2014) sheds lights on the pharmacological activities of the isoquinoline alkaloids. Diterpenoid alkaloids are known for their anti-inflammatory, analgesic, antiparasitic, anticancer, anti-epileptiform, and cardiovascular action properties. Napelline and 1-*O*-benzoylnapelline exhibited strong antiarrhythmic activity, even exceeding that of Class I antiarrhythmic drugs novocainamide, quinidine and lidocaine (Shakhidoyatova et al. 2001). Atisine chloride, isoatisine and coryphine, which were isolated from *A. coreanum*, have been found to exhibit myorelaxant activity (Dzhakhangirov and Bessonova 2002). From 2008 to 2018, more than 250 diterpenoid alkaloids were tested against several cancer cell lines and some of them showed effective anticancer properties including cell growth inhibition, inducing apoptosis, and altering cell cycle and multidrug resistance (Liang et al. 2018). For examples, 8-*O*-azeloyl-14-benzoylaconine, lappaconitine hydro-bromide, neoline, 8-*O*-methylcolumbianine, 1,14-diacetylcardiopetaline, 18-*O*-demethylpubescenine, 14-deacetylpubescenine, pubescenine, 14-deacetylajadine, lycoctonine, browniine, delphatine, dehydrotakaosamine, and ajadelphinine showed efficient antitumor and selective cytotoxicity effects to cancerous cells. The assessment of various structural types of diterpenoid alkaloids and their derivatives for local anaesthetic activity identified 26 compounds with distinct activity for surface anesthesia in rabbit eye cornea, out of which 15 of them showed stronger activities than those of cocaine (Wang et al. 2010). 14-*O*-acetylneoline, which was isolated from *A. laciniatum* was identified as novel anti-inflammatory drug lead compounds against colitis (Wangchuk et al. 2015).

Iranshahy and co-workers (2014) reported that isoquinolne alkaloids possess various biological properties including cytotoxicity, antiproliferative, anticancer, acetylcholinesterase inhibition, antiplatelet aggregation, anti-inflammatory, antiviral, antihypertensive, cardio- and hepato-protective, antiplasmodial, anticonvulsant, sedative, antinociceptive, insecticidal antifungal and antifungal activities. For examples, while spallidamine and other 17 isoquinoline alkaloids showed strong cancer chemopreventative activity comparable with positive standard, dehydrocavidine exhibited hepatoprotective activity at a dose of 0.5 to 1 mg/kg. In this same review, they also predicted the *in silico* physiochemical/drug-like properties of some isoquinoline alkaloids using Marvin Sketch and "Lipinski rule of 5"(RO5), and they found that most of them satisfied the criteria for lead-like compounds. According to RO5, an orally bioavailable drug-like compounds should have a molecular weight of <500 Da, <10 hydrogen bond acceptors, <5 hydrogen bond donors, logP of <5 (octanol–water partition coefficient <5) and in addition to that <10 rotatable bonds (Lipinski 2004). Drug leads that conform to this RO5 are more likely to have lower attrition rates during clinical trials.

Amongst different classes of isoquinoline alkaloids isolated from the plants, protoberberine and protopine groups have shown a wide range of pharmacological activities including anti-acetylcholinesterase, anti-amnesic, inhibition of phospholipase and thromboxane synthetase, spasmolytic, anti-tumour, smooth muscle stimulant, bactericidal, antimalaial and sedative properties (Huang et al. 2018). While dehydrocorydaline, corydaline and tetrahydropalmatine exhibited pain killing and anti-ulcer properties; the benzo[c]phenanthridine-6-acetonyl-dihydro-chelerythrine

exhibited a significant anti-HIV activity in lymphocytes with EC_{50} value of 1.77 µg/mL (Yuh-chwen et al. 2003). A number of isoquinoline alkaloids, which were isolated from four Bhutanese, were identified as new drug leads against malaria, inflammation, and acetylcholinesterase. Simplicifolianine, protopine, cheilanthifoline, scoulerine and coreximine, which were isolated from the Bhutanese *Corydalis* species, were identified as novel antimalarial drug lead molecules against the chloroquine-antifolate sensitive strain, TM4/8.2 and multidrug resistant strain, K1CB1, of the *Plasmodium falciparum* (Wangchuk et al. 2012a, 2012b, 2013). Amongst four alkaloids (Scoulerine, 13-oxoprotopine, stylopine and ochrobirine) tested, scoulerine significantly inhibited acetylcholinesterase with a minimum inhibitory requirement (MIR) value of 0.0015 nmol, which markedly exceeded that of the anti-Alzheimer's drug, galanthamine (MIR value of 0.003 nmol) (Wangchuk et al. 2016). Therefore, it has potential to be a novel drug lead compound against Alzheimer's disease. Capnoidine (isolated from *C. crispa*) was identified as novel anti-inflammatory drug lead compounds against colitis (Shepherd et al. 2018).

10.5 Conclusions

Alkaloid-containing plants have played significant role in both traditional medicines and modern pharmaceuticals. The alkaloids are rich in structural diversity and displays arrays of pharmacological activities including toxicities and therapeutic properties. As a result of structural diversity, the nomenclature and classification of alkaloids have become complex. However, the "Dictionary of Alkaloids," which is considered an authoritative "alkaloid canon," has identified 20,000 alkaloids, which belong to four major groups/classes of alkaloids. The content analysis of this "alkaloid canon" revealed that only 53 alkaloids are presently used in pharmaceutical applications, which suggest that alkaloids are underrepresented in the global pharmaceutical armamentarium. The concerns over the potential toxicological effects of alkaloids may have hindered their applications in medicines. Other reasons could be the "alkaloid supply constraints" and the miss-match in resource distribution. For example, the poor developing countries possess rich biodiversity but lacks in financial resources, expertise and technologies to conduct demanding pharmaceutical product development. On the other hand, rich and developed countries have all the expertise, technologies and financial sources but are poor in biodiversity. This calls for productive collaborations between the biodiversity-rich countries and the technology-rich pharmaceutical companies. Unlike other classes of natural products, isolation of alkaloid is comparatively straightforward and the hit-rate of biological activity is much higher. The hit-rate of biological activity or the success rate of finding new drug leads becomes even greater when the ethno-directed bio-rational approach is employed in drug discovery programs. Or in other words, focusing on local traditional uses of plants, used by the indigenous people for hunting prey, culinary enhancement and healing, would provide better biological activity hit-rate than randomly screening plants from the forests for alkaloids. Chemo-rational approach based on the chemical information presented by plant families also provides better biological activity hit rates.

Therefore, future discovery of drug-like alkaloids must incorporate these two important strategies. As there are limited number of drugs especially against autoimmune diseases such as HIV, Alzheimer's disease, inflammatory bowel disease, neurodegenerative diseases, and also against neglected tropical diseases that affects poor developing countries, it is worthwhile to conduct lead optimization, structure-activity-relationship studies, analog development, pharmacodynamics and *in vivo* animal studies for these plant-derived drug-like alkaloids.

REFERENCES

Ain, Q.U., Khan, H., Mubarak, M.S. and Pervaiz, A., 2016. Plant alkaloids as antiplatelet agent: Drugs of the future in the light of recent developments. *Frontiers in Pharmacology*, 7, p. 292.

Amirkia, V. and Heinrich, M., 2014. Alkaloids as drug leads– A predictive structural and biodiversity-based analysis. *Phytochemistry Letters*, 10, pp. xlviii–liii.

Bessonova, I.A. and Saidkhodzhaeva, S.A., 2000. Hetisane-type diterpenoid alkaloids. *Chemistry of Natural Compounds*, 36(5), pp. 419–477.

Buckingham, J., Baggaley, K.H., Roberts, A.D. and Szabo, L.F., 2010. *Dictionary of Alkaloids, with CD-ROM*. CRC Press.

Chang, Y.C., Hsieh, P.W., Chang, F.R., Wu, R.R., Liaw, C.C., Lee, K.H. and Wu, Y.C., 2003. Two new protopines argemexicaines A and B and the anti-HIV alkaloid 6-acetonyldihydrochelerythrine from formosan Argemone mexicana. *Planta Medica*, 69(2), pp. 148–152.

Cortinovis, C. and Caloni, F., 2015. Alkaloid-containing plants poisonous to cattle and horses in Europe. *Toxins*, 7(12), pp. 5301–5307.

Cottreau, J., Tucker, A., Crutchley, R. and Garey, K.W., 2012. Crofelemer for the treatment of secretory diarrhea. *Expert Review of Gastroenterology and Hepatology*, 6(1), pp. 17–23.

Cragg, G. and Newman, D., 2001. Chemists' toolkit, Nature's bounty. *Chemistry in Britain*, 37(1), pp. 22–26.

Dembitsky, V.M., Gloriozova, T.A. and Poroikov, V.V., 2015. Naturally occurring plant isoquinoline N-oxide alkaloids: Their pharmacological and SAR activities. *Phytomedicine*, 22(1), pp. 183–120.

Dzhakhangirov, F.N. and Bessonova, I.A., 2002. Alkaloids of Aconitum coreanum. X. Curare-like activity-structure relationship. *Chemistry of Natural Compounds*, 38(1), pp. 74–77.

Federici, E., Palazzino, G., Nicoletti, M. and Galeffi, C., 2000. Antiplasmodial activity of the alkaloids of Peschiera fuchsiaefolia. *Planta Medica*, 66(1), pp. 93–95.

Fester, K., 2010. Plant alkaloids. In *Encyclopedia of Life Sciences (ELS)*. Chichester, UK: John Wiley & Sons, pp. 1–11. https://onlinelibrary.wiley.com/doi/pdf/10.1002/9780470015902.a0001914.pub2

Guo, Q., Xia, H., Meng, X., Shi, G., Xu, C., Zhu, C., Zhang, T. and Shi, J., 2018. C19-Diterpenoid alkaloid arabinosides from an aqueous extract of the lateral root of Aconitum carmichaelii and their analgesic activities. *Acta Pharmaceutica Sinica B*, 8(3), pp. 409–419.

Hamlin, A.M., Lapointe, D., Owens, K. and Sarpong, R., 2014. Studies on C20-diterpenoid alkaloids: Synthesis of the hetidine framework and its application to the synthesis of dihydronavirine and the atisine skeleton. *The Journal of Organic Chemistry*, 79(15), pp. 6783–6800.

He, F., Wang, C.J., Xie, Y., Cheng, C.S., Liu, Z.Q., Liu, L. and Zhou, H., 2017. Simultaneous quantification of nine aconitum alkaloids in Aconiti Lateralis Radix Praeparata and related products using UHPLC–QQQ–MS/MS. *Scientific Reports*, 7(1), p. 13023.

Hesse, M., 2002. *Alkaloids—Nature's Curse or Blessing?* Zurich, Switzerland: Verlag Helvetica Chimica Acta and Wiley-VCH.

Huang, Y.J., Cheng, P., Zhang, Z.Y., Tian, S.J., Sun, Z.L., Zeng, J.G. and Liu, Z.Y., 2018. Biotransformation and tissue distribution of protopine and allocryptopine and effects of Plume Poppy Total Alkaloid on liver drug-metabolizing enzymes. *Scientific Reports*, 8(1), p. 537.

Iranshahy, M., Quinn, R.J. and Iranshahi, M., 2014. Biologically active isoquinoline alkaloids with drug-like properties from the genus Corydalis. *RSC Advances*, 4(31), pp. 15900–15913.

Liang, X., Gao, Y. and Luan, S., 2018. Two decades of advances in diterpenoid alkaloids with cytotoxicity activities. *RSC Advances*, 8(42), pp. 23937–23946.

Lipinski, C.A., 2004. Lead-and drug-like compounds: The rule-of-five revolution. *Drug Discovery Today: Technologies*, 1(4), pp. 337–341.

Maurya, S.K., Seth, A., Laloo, D., Singh, N.K., Gautam, D.N.S. and Singh, A.K., 2015. Śodhana: An Ayurvedic process for detoxification and modification of therapeutic activities of poisonous medicinal plants. *Ancient Science of Life*, 34(4), p. 188.

Saidkhodzhaeva, S.A., Bessonova, I.A. and Abdullaev, N.D., 2001a. Arcutinine, a new alkaloid from *Aconitum arcuatum*. *Chemistry of Natural Compounds*, 37(5), pp. 466–469.

Shakhidoyatova, N.K., Dzhakhangirov, F.N. and Sultankhodzhaev, M.N., 2001b. Antiarrhythmic activity of diterpenoid alkaloids of the napelline type and their acylated derivatives. *Pharmaceutical Chemistry Journal*, 35(5), pp. 266–267.

Shan, L.H., Zhang, J.F., Gao, F., Huang, S. and Zhou, X.L., 2017. Diterpenoid alkaloids from *Delphinium anthriscifolium* var. majus. *Scientific Reports*, 7(1), p. 6063.

Shan, L.H., Zhang, J.F., Gao, F., Huang, S. and Zhou, X.L., 2018. C18-Diterpenoid alkaloids from *Delphinium anthriscifolium* var. majus. *Journal of Asian Natural Products Research*, 20(5), pp. 423–430.

Shepherd, C., Giacomin, P., Navarro, S., Miller, C., Loukas, A. and Wangchuk, P., 2018. A medicinal plant compound, capnoidine, prevents the onset of inflammation in a mouse model of colitis. *Journal of Ethnopharmacology*, 211, pp. 17–28.

Sultankhodzhaev, M.N., Tashkhodzhaev, B., Averkiev, B.B. and Antipin, M.Y., 2002. Secokaraconitine, a new diterpenoid alkaloid from *Aconitum karacolicum*. *Chemistry of Natural Compounds*, 38(1), pp. 78–82.

Sun, R., Jiang, H., Zhang, W., Yang, K., Wang, C., Fan, L., He, Q. et al., 2014. Cytotoxicity of aporphine, protoberberine, and protopine alkaloids from *Dicranostigma Leptopodum* (Maxim.) Fedde. *Evidence-Based Complementary and Alternative Medicine*, 2014.

Wang, F.P., Chen, Q.H. and Liu, X.Y., 2010. Diterpenoid alkaloids. *Natural Product Reports*, 27(4), pp. 529–570.

Wangchuk, P., 2018. Therapeutic applications of natural products in herbal medicines, biodiscovery programs, and biomedicine. *Journal of Biologically Active Products from Nature*, 8(1), pp. 1–20.

Wangchuk, P., Bremner, J.B. and Samosorn, S., 2007. Hetisine-type diterpenoid alkaloids from the Bhutanese medicinal plant *Aconitum orochryseum*. *Journal of Natural Products*, 70(11), pp. 1808–1811.

Wangchuk, P., Bremner, J.B., Rattanajak, R. and Kamchonwongpaisan, S., 2010. Antiplasmodial agents from the Bhutanese medicinal plant *Corydalis calliantha*. *Phytotherapy Research: An International Journal Devoted to Pharmacological and Toxicological Evaluation of Natural Product Derivatives*, *24*(4), pp. 481–485.

Wangchuk, P., Keller, P.A., Pyne, S.G., Lie, W., Willis, A.C., Rattanajak, R. and Kamchonwongpaisan, S., 2013. A new protoberberine alkaloid from Meconopsis simplicifolia (D. Don) Walpers with potent antimalarial activity against a multidrug resistant Plasmodium falciparum strain. *Journal of Ethnopharmacology*, *150*(3), pp. 953–959.

Wangchuk, P., Keller, P.A., Pyne, S.G., Sastraruji, T., Taweechotipatr, M., Rattanajak, R., Tonsomboon, A. and Kamchonwongpaisan, S., 2012a. Phytochemical and biological activity studies of the Bhutanese medicinal plant *Corydalis crispa*. *2012*, p. 575.

Wangchuk, P., Keller, P.A., Pyne, S.G., Willis, A.C. and Kamchonwongpaisan, S., 2012b. Antimalarial alkaloids from a Bhutanese traditional medicinal plant *Corydalis dubia*. *Journal of Ethnopharmacology*, *143*(1), pp. 310–313.

Wangchuk, P. and Loukas, A., 2018. Techniques and technologies for the biodiscovery of novel small molecule drug lead compounds from natural products. In *Natural Products and Drug Discovery* (pp. 435–465).

Wangchuk, P., Sastraruji, T., Taweechotipatr, M., Keller, P.A. and Pyne, S.G., 2016. Anti-inflammatory, anti-bacterial and anti-acetylcholinesterase activities of two isoquinoline alkaloids-scoulerine and cheilanthifoline. *Natural Product Communications*, *11*(12), pp. 1801–1804.

Wangchuk, P., Navarro, S., Shepherd, C., Keller, P.A., Pyne, S.G. and Loukas, A., 2015. Diterpenoid alkaloids of *Aconitum laciniatum* and mitigation of inflammation by 14-O-acetylneoline in a murine model of ulcerative colitis. *Scientific Reports*, *5*, p. 12845.

Wangchuk, P. and Olsen, A., 2010. Risk factors for the sustainability of medicinal plants in Bhutan. *Asian Medicine*, *6*(1), pp. 123–136.

Yin, X., Zhao, F., Feng, X., Li, J., Yang, X., Tu, P. and Chai, X., 2018. Four new spirobenzylisoquinoline N-oxide alkaloids from the whole plant of *Corydalis hendersonii*. *Fitoterapia*, *128*, pp. 31–35.

Yu, X., Gao, X., Zhu, Z., Cao, Y., Zhang, Q., Tu, P. and Chai, X., 2014. Alkaloids from the tribe Bocconieae (Papaveraceae): A chemical and biological review. *Molecules*, *19*(9), pp. 13042–13060.

Zhang, J.F., Li, Y., Gao, F., Shan, L.H. and Zhou, X.L., 2018. Four new C20-diterpenoid alkaloids from *Aconitum rotundifolium*. *Journal of Asian Natural Products Research*, *2018*, pp. 1–9.

Zhou, X.L., Chen, Q.H., Chen, D.L. and Wang, F.P., 2003. Hemsleyatine, a novel C19-diterpenoid alkaloid with 8-amino group from *Aconitum hemsleyanum*. *Chemical and Pharmaceutical Bulletin*, *51*(5), pp. 592–594.

11

The Role of Phytocompounds in Cosmeceutical Applications

Rabinarayan Parhi

CONTENTS

11.1 Introduction

Natural beauty is considered to be a blessing, and it is also a symbol of healthy life. However, there remains a persistent desire among all human beings to be beautiful and look youth (Charles et al. 2017). Therefore, it is important to look beautiful in the present world as it enhances, directly or indirectly, assertiveness, triumph, authority, satisfaction and moreover happiness in life. The Egyptians were among the first civilizations to use cosmetics, which is evident from a specific medical papyrus of herbal knowledge "Ebers Papyrus" written in c. 1550 BC (Ming et al. 2017). There was also the presence of documents of cosmetic substances in circa 2000 and 1550 BC owe to Indus valley civilization (Patkar 2008). In the process, a Greek physician, Galen, was first to invent a cold cream composed of beeswax, vegetable oil and water in the second century (Lin 2010).

There is a continuous growth in the use of cosmetics in parallel to drug that spreads across in almost each and every human society and civilization on the earth. Quality and safety are two important parameters of drugs and cosmetics that must be assured. Initially, these aspects were taken care of by the Food and Drug Administration (FDA, USA); it was further strengthened by the formation of the Food, Drug, and Cosmetic Act in 1938, and is revised as per the requirement to place effective supervision on different drugs and cosmetics (Millikan 2001). The FDA defines the terms like *drugs* and *cosmetics*. Drugs are the "products that cure, treat, mitigate or prevent disease, or that affect the structure or function of the human body" (Millikan 2001). Cosmetics, however, are defined differently by the FDA and European Directive. According to the FDA cosmetics are "articles intended to be rubbed, poured, sprinkled, or sprayed on, introduced into, or otherwise applied to human body for cleaning, beautifying, promoting attractiveness, or altering the appearance without affecting structure and function" (FDA 2001). European Directive 95/33/EEC defines it more elaborately as "any substance or preparation intended to be placed in contact with various external parts of human body (epidermis, hair system, nails, lips and external genital organs) or with teeth and the mucous membranes of the oral cavity with a view exclusively or mainly to cleaning them, perfuming them, changing their appearance and/or correcting body odours and/or protecting them or keeping them in good conditions" (http://eur lex.europa.eu/legalcontent/EN/NOT/?uri=CELEX:31993L0035&qid=1458440341871). All these definitions represent the term *cosmetics* as a product that enhances the appearance without any therapeutic benefits. Thus, the new term *cosmeceuticals* was first coined and used by Raymond Reed, founding member of the United States Society of Cosmetic Chemists, in 1961 (Joshi and Pawar 2015; Charles et al. 2017). According to Reed, the word *cosmeceutical* defined four complementary points: (i) a cosmeceuticals has desirable aesthetic properties; (ii) a cosmeceuticals is scientifically developed product aimed at external applications to the human body; (iii) a cosmeceuticals produces a useful, desired result; and finally (iv) a cosmeceuticals meets rigid chemical, physical and medical standards (Charles et al. 2017) The term was further used and popularized by Dr. Albert Kligman, who in 1984 stated that "Cosmeceuticals are topical agents which are distributed across wide spectrum of materials, present somewhere between pure drug such as antibiotics and corticosteroids and cosmetics such as lipstick and rouge (Gediya et al. 2011). It is unfortunate that as of this writing the FDA neither recognises nor defines the term cosmeceuticals. However, the term is universally understood and used as the products with additional health benefits along with cosmetic application. It may also be referred as a hybrid between cosmetics and pharmaceuticals. Thus, these products seem to fill the gap between pharmaceuticals that cure and heal and cosmetics that clean and beautify the human body.

The strong desire of consumers to constantly look flawless and youthful led to research and development in the cosmetic field (Ming et al. 2017). As a result, the cosmetic industry is one of the fastest-growing industrial sectors in the present scenario and significantly contributes to the world gross domestic products. The global market of cosmetic is categorized into various types, including skin, hair, colour, odour and fragrance (Carvalho et al. 2016). In addition, in the last decade gross spending on beauty and personal care products showed a 25.9% increase in sale share in Asia (Łopaciuk and Łoboda 2013). Based on a financial statistics report, the European cosmetic industry is estimated to be 70 billion euros (Secchi et al. 2016).

As a result of environmental limitations and constant consumer's demands, the cosmeceutical industry is in a persistent search for ingredients from natural sources with minimal toxic effect (Ramli 2015). This gives a novel opportunity to replace synthetic counterparts that can threaten to human health and the environment (Secchi et al. 2016). Botanicals obtained from plants are subsequently improved upon in laboratories to separate a large percentage of active ingredients called *phytocompounds*. These compounds are gaining popularity in the cosmeceuticals sector due to their major attributes, including mildness, low toxicity, biodegradability and efficacy (Deep and Saraf 2008). In this chapter, the uses of different phytocompounds in cosmeceuticals are discussed along with various methods used to deliver them across the upper layer of skin. In addition, formulation challenges and safety aspects are also presented.

11.2 Cosmeceuticals and Phytocompounds

Some cosmeceuticals are chemically synthesized whereas others are naturally derived; but all contain functional ingredients with therapeutic, disease-fighting or healing properties. Preferably, these cosmeceutical ingredients are novel, efficacious, safe, stable, biodegradable and affordable (Dooley 1997). In this context naturally derived ingredients or phytocompounds are considered as safe without any side effects, are efficacious, provide a wide selection, are compatible with all types of skin, and no requirement for animal study. Therefore, in roughly a decade, the natural cosmetic or cosmeceutical market has been a rapidly growing section in the personal care industry (Pande and Majeed 2015). In addition, more and more consumers are shifting towards phytocompounds in cosmeceuticals because of their ever-increasing awareness concerning healthy products with natural ingredients and better scientific knowledge of skin physiology (Charles et al. 2017). Different steps involved in cosmeceutical ingredient developmentare presented in Figure 11.1 (Draelos 2015; Lephart 2017).

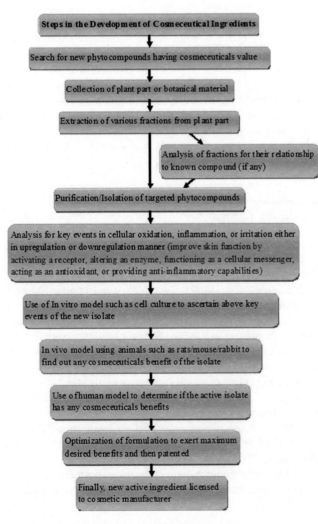

FIGURE 11.1   Steps in the development of cosmeceutical ingredients.

11.3 Types of Cosmeceuticals

Functional cosmetics can be divided in to four major categories: color cosmetics, cosmeceuticals, neurocosmeceuticals and nutraceuticals, as depicted in Figure 11.2. Further, the cosmeceuticals can be sub-classified into four types based on their function, the body part to which it is applied, products present in the market, and pharmaceutical ingredients (Giovanini 2016; Charles et al. 2017).

11.3.1 Skin Cosmeceuticals

The skin is the largest and most accessible organ of human body, and it covers the body surface area in excess of 2 cm² in an average adult. Skin varies in thickness across the body; the thicker or glabrous skin is present on the palms of the hands and soles of the feet, both lacking hair follicles and sebaceous gland, and hairy skin is found on remaining body surface with both hairs and sebaceous gland (Butler 2010). The pH of the skin is ranging from pH 4 to 6, which provides protection against chemical and microbial attack. The skin also plays major role in homeostasis and act as a barrier against ultraviolet ray (UV ray) from the sun and other sources (Martins et al. 2013). Human skin is developed into three physically well-defined layers, specifically

outermost epidermis, middle dermis and underlying subcutaneous or hypodermis (Jepps et al. 3013). Epidermis is considered as an avascular layer, consisting of an upper nonliving layer called as stratum corneum (SC) and lower living layers combined called as viable epidermis. In terms of cell thickness, epidermis is typically 50–150 μm (Holbrook and Odland 1974) and has around 35 cell layers out of which 15–20 layers constituting the stratum corneum. The dermis layer is subdivided into an outer papillary and inner reticular layer, which supports nerves, lymph and blood capillaries. The lowermost layer of skin is the subcutaneous layer, which acts as a reservoir/sink to the drug passed the upper layers of skin (Cevc and Vierl 2007). The extracellular matrix including collagen and elastin fibers have an important role in maintaining good skin health because the former provides structural framework to the skin whereas the later provide elasticity to the skin (Baxter 2008; Wend et al. 2012).

Any substance applied on the skin may act on skin (intradermal) or on systemic circulation (transdermal). In either case the substance has to permeate the stratum corneum, an impervious skin layer. An ingredient/molecule can permeate/cross the SC by either transappendageal route (i.e., across hair follicles, sebaceous and eccrine gland) or transepidermal route (i.e., across the lipid bilayer present in stratum corneum) (Parhi et al. 2015). Different factors including human races, skin age, place of application, and gender and skin condition along with

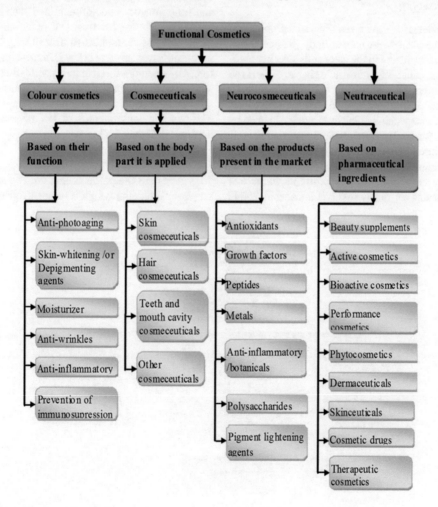

FIGURE 11.2 Classification of functional cosmetics.

physicochemical factors of the molecules are influencing the permeation of ingredients across the skin (Bary 2007).

The selection of skin cosmeceuticals depends mainly upon on skin types and gender. Generally, skin is categorized into four types based on the activity of sebum produced by the sebaceous gland present on the facial skin (Kumar et al. 2012). They are: (i) normal skin, which is free from visible pores or blemishes and has even tone and smooth; (ii) dry skin, which has a low level of sebum which is prone to develop chapping, cracking and erythema; (iii) oily skin, which looks shiny and thick due to excessive secretion of sebum and prone to pimples and blackhead; and (iv) a combination skin type, where the central part of the face such as forehead, nose and chin (denoted as T-zone) is oily and remaining part of face are dry. All types of skin need some kind of protection against aging, sunlight and microbial infections, but more care is essential in case of dry and oily skin. The facial skin of a male has many differences from the female counterpart: (i) males have more secretions of sebaceous, eccrine and apocrine glands; (ii) male skin is darker and thicker due to more melanin and collagen, respectively; (iii) transepidermal water loss is less in case of male skin; and (iv) male skin offers much more protection against photoaging due to facial hairs (Draelos 2010). All these factors push females to use more cosmetics or cosmeceuticals.

11.3.1.1 Anti-photoaging

Aging regardless of gender is a complex, multifactorial process that brings about progressive changes in human body (Sadick 2018). Particularly, the changes occurring in the cutaneous layer are associated with two distinct processes: intrinsic aging and extrinsic aging. Intrinsic or natural aging has a genetic origin and is remarkably affected by vascular as well as hormonal variations. Extrinsic aging is considered as the expedition of intrinsic aging triggered by various environmental factors. Out of all factors the principal one is the huge damage caused by chronic UV irradiation, a process called *photoaging* (Ichihashi et al. 2003). The subsequent manifestations of skin aging are loss of elasticity and texture, appearance of spots (pigmentation), drying out, and wrinkles (Lintner et al. 2009).

Abnormal pigmentation and wrinkles stand out as important factors in the event of photo-aging (Wulf et al. 2004).

The UV radiation, which reaches the surface of Earth, is described in different sub-regions as far-UV radiation (100–200 nm), UVC (200–280 nm), and UVA (320–400 nm). Among them, UVA and UVB radiations induce skin damage. UVA radiation penetrates up to the dermal layer of skin, resulting into both immediate and delayed skin pigmentation. The latter being considered as a tanning effect, and it also contributes about 15% of the sunburn reaction due to overexposure to sunlight. UVB is capable of penetrating up to the upper margin of dermis and induces photochemical reactions within living and actively dividing cells of the basal layer. This results in a number of pathological changes such as skin inflammation, edema, erythema and epidermal hyperplasia (Karthikeyan et al. 2016). Additionally, excessive subjection of skin to sunshine may lead to different hyperpigmentation ailments including melasma, age spots or solar lentigo (Vhatkar et al. 2018).

The underlying cause of tissue damage is oxidative stress mediated by reactive oxygen species (ROS) initiated by both UV exposure and cellular metabolism (Lintner et al. 2009). The ROS includes singlet oxygen ($^1O_2$), superoxide anion (O^{2-}), hydrogen peroxide (H_2O_2), nitric oxide (NO) and hydroxyl radical (.OH). These radicals damage DNA, collagen, elastin and fibroblast, resulting in photoaging (Aziz et al. 2017). Various potential anti-inflammatory agents, including glucocorticoids, non-steroidal anti-inflammatory and silfonamided such as silver sulfadiazine, are extensively used to treat UV radiation-induced skin injuries (Lopes and Mcmahon 2016). But in long-term use, these drugs may produce an adverse effect (Unzueta and Vargas 2013); therefore, phytocompounds can be a useful alternative to treat or protect skin from aging. Various phytocompounds in the form of sunscreens, antioxidants and growth factors are discussed below. The possible consequences of UV rays on the human skin are depicted in Figure 11.3.

11.3.1.1.1 Sunscreens

Sunscreens are the products that are applied on skin in order to screen or filter sunrays for a harmful wavelength (280–400 nm)

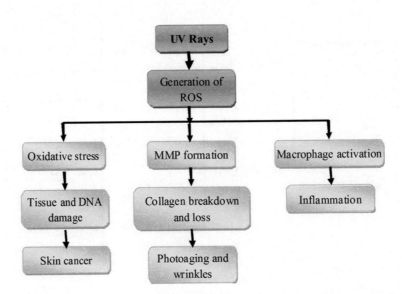

FIGURE 11.3 The possible concequences of UV rays on the human skin.

and allow higher wavelength to pass. Thus, sunscreens are products that protect the skin from UV rays and keep the skin healthy (Mishra et al. 2011). Sunscreens or UV filters have been broadly classified into physical blocker and chemical absorbers, depending on the mechanism of action. Physical blockers act by either reflecting or scattering UV radiation, whereas chemical sunscreens absorb high-intensity UV radiation with excitation to a higher energy state and lost energy with returning-to-ground state. All the phytocompounds are categorized under the chemical class.

Several phenolic groups such as phenolic acids, flavonoids (subclass—isoflavonoids), tannins and lignans, are used as sunscreens (Lephart et al. 2016). These compounds are obtained from natural sources, including plant, and phenolic acids among all are widely used. Again, phenolic acid has two groups, namely hydroxycinnamic acid group (caffeic acid, ferulic acid and sinapic acid) and hydroxybenzoic acid groups (Dias et al. 2016). Following are a few phytocompounds having photoprotective properties:

11.3.1.1.1.1 Unsaturated Fatty Acids (Essential Fatty Acids) The unsaturated fatty acids such as stearic and linoleic acids and phenolic compounds are isolated from the seeds of *Luffa cylindrical* (Linn) M. Roem. They act by scavenging free radicals formed as well as by inhibiting the inception of free radicals in the skin (Gupta and Sharma 2006; Prakash et al. 2009). The extract obtained from the seed of apple, blackberry, blackcurrant and strawberry contain unsaturated fatty acids viz., α-linolenic acid and linoleic acid. These ingredients screen UV radiation so that skin is protected from its side effects. The mechanisms of action are based on the balance among inflammatory, antioxidant and immune response exerted by these derivatives in the skin (Krasodomska and Jungnickel 2015).

11.3.1.1.1.2 Gallic Acid and Elaeocarpusin From ancient time amla (*Emblica officinalis* Gaertn.) has been ingested due to its rich dietary source, including vitamin C, minerals and amino acids (Berthakur and Arnold 1991). The water extract of dried amla powder is found to have 29.4% polyphenol such as gallic acid and elaeocarpusin and 2% of ascorbic acid (Anila and Vijayalakshmi 2002). All these ingredients exhibit potent antioxidant properties by providing protection to dermal fibroblast from oxidative stress. In addition, amla extract induces the production of procollagen and also enhances mitochondrial activity of human skin fibroblasts (Kim et al. 2001) and metalloprotinases (Yokozawa et al. 2007).

11.3.1.1.1.3 Tannin and Flavonoids Polyphenol, such as tannins, (e.g., catechin, epicatechin, procyanidin (Beltrame, 2005), catiguanine A and B (Tang et al. 2007), and flavonoids (e.g., stigmasterol, sitosterol, 3-O-d-glucopyranosyl sitosterol), are obtained from the leaves and bark extract of the catuaba plant (*Trichilia catigua* A. Juss Meliaceae). It was found that the crude extract from the plant increased the sun protection factor (Munhoz et al. 2012). This is due to the antioxidant ability of the polyphenol available in the extract (Longhini et al. 2017).

11.3.1.1.1.4 Carotene and Vitamin A The ingredient carotene is obtained from the root of carrot plant (*Daucus carota*, Family Apiaceae). The orange colour of carrots is mainly exerted by β-carotene and to a lesser extent by α- and γ-carotene. In the human body, α- and β-carotene partially metabolized into vitamin A (Strube and Dragsted 1999). Carrot seed oil was reported to have anti-aging, revitalizing and rejuvenating potential (Gediya et al. 2011).

11.3.1.1.1.5 β-glucans β-glucans are extracted from natural oats (*Arundinaria gigantea* Muhl Family: Poaceae), both wheat and baker's yeast. Chemically, these are polysachharides. β-glucans are potent moisturizer and are found to decrease formation of ROS and induce the Langerhans' cells in skin upon UV irradiation. Therefore, β-glucans of oats incorporated at 1% in topical creams was found to induce formation of collagen and ceramides, thereby enhancement of epidermal turnover rate and decrease in fine wrinkles (Mulder and Meinardi 2002; Southall et al. 2014).

11.3.1.1.1.6 Lycopene Acyclic hydrocarbon carotinoids were proved to be anti-inflammatory and anticarcinogenic. They are obtained from various fruits such as grapefruit, guava, watermelon, apricots, papaya and carrot (Draelos 2010; Oroian and Escriche 2015). They block UVB-induced apoptosis and also protect photodamage caused by UVB irradiation by inhibiting epidermal ornithine decarboxilase (Amer and Maged 2009).

11.3.1.1.1.7 Idebenone It is believed to inhibit post-UVB nuclear thymidine damage. It also reduces oxidation of cell membrane lipids (Pelle et al. 1999).

11.3.1.1.1.8 Boldine Boldine, an aporphine like alkaloids, is obtained from the leaves of *Peumus boldus* plant that belongs to the family, Monimiaceae (Peter et al. 2006). Previously, it was proved that boldine is a potent antioxidant when taken orally and led to various functions including cytoprotective, anti-inflammatory, anti-diabetic, and anti-tumour promoting (Griffiths et al. 1998). Then, it was found that boldine also has UV light-screening ability pertinent to photoprotective action. Even though, boldine is photo-unstable up to wavelength 300 nm, but exhibits photo-protective effects against UVB by preventing the rise in skin temperature of the tested rodents (Hidalgo et al. 1998). Later, boldine was found to protect skin on the back of human volunteers against erythema formation (Rancan et al. 2002).

11.3.1.1.1.9 Ferulic Acid Ferulic acid, 4-hydroxy-3-methoxy-cinnamic acid, is obtained from the *Parthenium hysterophorus L.* (Vashisth et al. 2015) plant, and from other sources such as vegetables like cabbage, potatoes, broccoli, spinach and grains like wheat and corn (Kumar and Pruthi 2014; Dey et al. 2016). It is found to exhibit higher radical-scavenging activity, which was proved by DPPH and 2, 2o-azino-bis(3-ethylbenzothiazolin-6-sulphonic acid (ABTS) assay methods (Lee et al. 2014). It also exhibits anti-collagenase activity due to the inhibition of expression of enzymes, matrix metalloproteinase 1 (MMP1) and MMP9 in mice when applied topically (Staniforth et al. 2012).

11.3.1.1.1.10 Catechin and Epicatechin These flavanoids are obtained from the flowers of the plant *Rosa damascene* L family Rosaceae. The hydroalcoholic extract of flowers are found to show UV-radiation absorption capability and thereby protect the skin from exposure to UV light when applied on the skin in the cream form (Tabrizi et al. 2003).

11.3.1.1.1.11 Root Extracts of Plant Coleus Forskohii L. Extract obtained from the plant *Coleus forskohii L.* showed UV protection and has the ability to induce lipolysis (Takshak and Agarwal 2015). On exposure to UV radiation, the ingredients in the extract convert into components such as flavonoids and phenolics, which have free-radical scavenger activity. Therefore, the extract may be used in cosmeceuticals for UV protection (Takshak and Agrawal 2016).

11.3.1.1.2 Antioxidants

Antioxidants are molecules or systems that can interact with free radicals and stop the reaction that can harm the important molecules in our system. In this process they oxidize themselves before other vital molecule (Oroian and Escriche 2015). Sunscreens protect the skin from UV-irradiation to a certain extent rather than offer complete protection. Further, they possess the risk of toxicity. Therefore, additional protection is needed. Note that skin has its own natural antioxidants, such as glutathione peroxidase, glutathione reductase, manganese superoxide dismutase, copper-zinc superoxide dismutase, catalase and extracellular superoxide dismutase to scavenge the free radicals produced by natural processes as well as UV irradiation (Shindo et al. 1994; Rabe et al. 2006). But, these herbal antioxidants are not sufficient when there is an excessive exposure of skin to UV irradiation leading to various photochemical reactions with chromophores such as DNA, protein and peptides and oxidize them to produce various oxygen free radicals. Therefore, topical antioxidants are necessary as adjuvant to overcome the damaging effect of free radicals generated because of excessive exposure of skin to UV irradiation (Amer and Maged 2009). Major classes of natural antioxidant are vitamins, polyphenol and caretonoids.

11.3.1.1.2.1 Vitamin C (L-ascorbic acid) Vitamin C is a major water-soluble antioxidant that humans unable to synthesize in their body; therefore, it has to be obtained from plant sources such as apple, citrus peel, bayberry, broccoli, peppermint, spearmint and citrus peel (Raska and Toropov 2006; Rozman et al. 2009; Oroian and Escriche, 2015). The antioxidant mechanism of vitamin C involves the hydroxylation of procollagen, proline and lysine (England and Seifter 1986). It was also found to stimulate collagen repair, thereby minimizing photoaging of skin. The main issue in the delivery of vitamin C topically is its instability leading to degradation (Gallarate et al. 1999).

11.3.1.1.2.2 Vitamin E Vitamin E is a lipid soluble antioxidant present in human plasma, tissues and membranes (Ekanayake-Mudiyanselage et al. 2005). The plant sources for vitamin E are oils obtained from olive, palm, and sunflower, and seeds of pumpkin and sunflower (Gaspar and Campos 2007; Rozman et al. 2009). Vitamin E constitutes eight types compounds, including four tocopherols and four tocotrienols. Topically applied, vitamin E penetrates into skin (SC and dermal layers) and protects the skin membrane from UV irradiation by terminating the lipid peroxidation process. This is due to the scavenging property of vitamin E against lipid peroxyl radicals (Kamal-Eldin and Appelqvist 1996). Vitamin E also has sunscreen property and protects skin from erythema, edema and sunburn when applied before UV irradiation (Kramer and Liebler 1997). Additionally, it has significant skin anti-wrinkling

properties (Jurkiewicz et al. 1995). In topical formulations the incorporated concentration of α-tocopherol is range from 0.1% to 1% (Ekanayake-Mudiyanselage et al. 2005).

11.3.1.1.2.3 Silibinin Silibinin (silybin or silibin) is a flavonolignan compound present in silymarin mixture is obtained from the extract of *Silybum marianum* (milk thistle plant) and *Cynara scolymus* (artichoke plat), Family Compositae. It has many isomers, including silybin A, silybin B, isosilibinin A and isosilibilin B (Davis-Searles et al. 2005). Silibinin is a string antioxidant and acts by scavenging free radicals produced by oxidation-induced lipid peroxidation (Carini et al. 1992). Additionally, it acts as an anti-inflammatory agent as it hinders the chemical induction of COX-2 expression without affecting the functioning of COX-1. It also protects skin from UVB-induced sunburn by decreasing the apoptic events in skin (Dhanalakshmi et al. 2004), photocarcinogenesis with the enhancement of apoptic cell death (Gu et al. 2005), epidermal hyperplasia by inhibiting cell proliferation and DNA damage by decreasing thymine dimmer formation (Dhanalakshmi et al. 2004).

11.3.1.1.2.4 Piperine The main source of piperine is the fruit (pepper) of *Piper longum*, Family Piperaceae. The fruits contain crystalline alkaloids (5%–95%), such as piperine and piperettine, along with resins (Koul and Kapil 1993). Piperine is found to have antioxidant potential and is therefore used in many herbal cosmetics (Koul et al. 1993; Balachandran and Govindarajan 2005).

11.3.1.1.2.5 Pycnogenol Pycnogenol, a polyphenolic bioflavonoid, is obtained from pine bark extract (*Pinus pinaster*, Family: Pinaceae). It is rich in proanthocyanidins (Mateos-Martín et al. 2012), and has antioxidant, anti-inflammatory, and anticarcinogenic potential. It is believed to act by stabilizing collagen and elastin due to its binding ability with elastin fibers. It also inhibits elastases and thereby decreases the rate of degeneration of elastin fibers. Anti-inflammatory effect of pycnogenol is due to blockage of release of inflammatory factors (Amer and Maged 2009).

11.3.1.1.2.6 Hydroxycinnamates Hydroxycinnamates is obtained from the golden rod plant (*Solidago virgaurea L* Family Asteraceae). Hydroxycinnamates is a potent antioxidant and also absorbs sun rays in the UV radiation range. Therefore, it is widely used in antiaging and sunscreen cosmeceuticals (Saluk-Juszczak et al. 2010; Jaiswal et al. 2011).

11.3.1.1.2.7 Methanolic Extract of Portulaca Oleracea The methanolic extract of purslane (*Portulaca oleracea*, Family Portulaceae) has shown antioxidant activity because of its ability to scavenge nitric oxide and super oxide (Sanja et al. 2009). Thus, it reduces skin wrinkling and the antiaging process and thereby provides photoprotection.

11.3.1.1.2.8 Essential Oil from Calendula Officinalis The essential oil obtained from calendula (*Calendula officinalis*) found to exhibit remarkable antioxidant and anti-inflammatory potential (Muley et al. 2009). The main component exists in the oil, which is responsible for the above activity, and includes α-thujene, α-pinene, 1,8-cineole, dihydrotagetone and T-muurolol (Okoh et al. 2008).

11.3.1.1.2.9 Genistein and Daidzen These are obtained from the fully matured dried seeds of the plants *Glyene-soja* and *Glycine max*, Family Laguminosae. Genistein is believed to protect against UV irradiation (Amer and Maged 2009).

11.3.1.1.2.10 Catechin from Grape Seed Extract Grape seed extract obtained from the grape tree has been found to induce vascular endothelium growth factor expression in keratinocytes. This induces the dermal wound healing process. It has a synergistic effect to sunscreens and additionally has free radical scavenging properties. All these beneficial properties are due to the presence of isoflavonoid catechin (Amer and Maged 2009).

11.3.1.1.2.11 Resveratrol Resveratrol is a potent polyphenolic antioxidant obtained from various plant sources, including grape berry skins and seeds, peanuts, red wine, and dried roots of *Polygonum cuspidatum* (Kuršvietienė et al. 2016). Its acts by stimulating the gene responsible in the production of antioxidant enzymes, including superoxide dismutase, thioredoxin reductase, and catalase, thereby protecting the skin by preventing the formation of ROS and suppressing oxidative stress (Lephart 2017).

11.3.1.1.2.12 Quercetin Different plant sources for quercetin are onions, black paper, broccoli, blueberry, tea, red wine, curly kale, leeks (Embuscado 2015; Park et al. 2015; Jeon et al. 2015), seed oil of Chia plant (*Salvia hispanica L.*) (Diwakar et al. 2014), and *Lens esculenta*, Family Fabaceae (Amarowicz et al. 2010). It has a wide application in skincare formulations because of its antioxidant attributes (Marineli et al. 2014).

11.3.1.1.2.13 Lutein It is generally obtained from the flowers of plant Mexican Marigold (*Tagetes erecta L.*, Family Asteraceae) (Siriamornpun et al. 2012). It is also obtained from spinach, leaf lettuce, peas, oranges, carrot and Zea mays (Oroian and Escriche, 2015; Mitri et al. 2011). It has good antioxidant properties as it suppresses NF kappa-B activation and also inhibits the manifestation of iNOS and COX-2 (Kijlstra et al. 2012). Additionally, lutein showed antiaging properties when the flower extract of *Tagetes erecta L.* is loaded in pharmaceutical cream. It is also used to tackle different skin ailments including sores, burns, wounds, ulcers and eczema (Charles et al. 2017).

11.3.1.1.2.14 Bixin Bixin is a caretonoid carboxillic acid, obtained from the extract of dried seeds of plant annatto (*Bixa orellana* L. Family: Bixaceae). Bixin is naturally present in *cis* form and gets converted into *trans* form during extraction process. Bixin is an antioxidant and therefore used topically to protect skin from harmful UV light (Kokate et al. 2008).

11.3.1.1.2.15 Pomegranete seed oil Oil obtained from the seed of pomegranate has potential antioxidant activities. Thus, it can be used in cosmeceutical preparations to prevent photodamage of skin induced by UV radiation (Baccarin et al. 2015a, 2015b).

11.3.1.1.3 Depigmentation and/or Whitening Compounds
The epidermis of human skin has two important varieties of cells: Langherhans' cells and melanocytes. The former acts as a first line of a non-specific defense system, whereas melenocytes

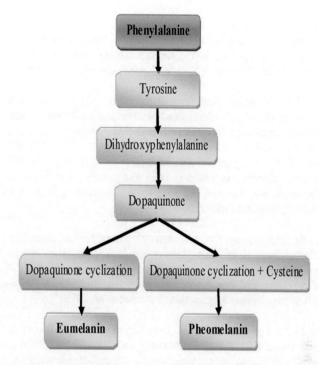

FIGURE 11.4 Schematic indicating melanin biosynthesis.

produce a pigment called as melanin as depicted in the Figure 11.4. Melanin is considered as a natural sunscreen because its main responsibility is to screen out harmful UV radiation and thereby protect the skin. An abnormal formation and accumulation of melanin pigment in the skin is the cause for different disorders related to pigmentation, including freckles, melasma and senile lentigo. This is due to enhanced melanogenesis by the activation of tyrosinase when skin is subjected to exposure of UV irradiation (Friedmann and Gilchrest 1987; Cichorek et al. 2013). As a preventive measure, sun avoidance or sunscreen containing SPF greater than 30 are recommended (Henry, 2018). After the formation of pigmentation disorder, treatment based on skin whitening or antimelanogenic compounds are the better choice for substantial ameliorative action in order to regain original colour of skin. The rate-limiting enzyme in the biosynthesis of melanin is tyrosinase (Figure 11.4) as it catalyzes two steps: hydroxylation of tyrosine to dihydroxyphenylalanine (L-DOPA) and subsequent oxidation of L-DOPA to dopaquinone (Hearing and Tsukamoto 1991). Therefore, by inhibiting tyrosinase enzyme, the formation of melanin can be prevented.

11.3.1.1.3.1 Polyphenol from Camellia Sinensis Polyphenols are obtained from the dried unfermented leaves of green tea tree (*Camellia sinensis*). Among them (-)-epicatechin-3-gallate (ECG, (-)-epigallocatechin-3-gallate (EGCG), and (-)-gallocatechin-3-gallate (GCG) are the major polyphenol that compose approximately 30% of the dry weight (Mukhtar and Ahmad 2000). The antimelanogenic effects of these polyphenols are due to the inhibition of tyrosinase manifestation or direct impediment of tyrosinase activity, both leading to the inhibition of melanin production (Roh et al. 2017). Antimelanogenic activity of the above polyphenols is related to the flavan-3-ol skeleton with a galloyl moiety at the 3-position (No et al. 1999). The extract of green tea is also found to exhibit anti-inflammatory activity induced by UVB irradiation, and the activity is believed to be due to EGCG (Katiyar et al. 1995; Katiyar and Elmets 2000).

11.3.1.1.3.2 Ellagic Acid from Mushroom Ellagic acid is a yellow-coloured powder obtained from mushroom extract. It has anti-tyrosinase activity apart from anti-collagenase, anti-elastase and anti-hyaluronnidase functions (Payeet al. 2007; Taofiq et al. 2016). Polyphenolic-rich fraction of *Rasa damascene* mill Family: Rosaceae is found to have ellagic acid. It is a potent tyrosinase inhibitor and can be used in cosmeceuticals to treat hyperpigmentation (Solimine et al. 2016).

11.3.1.1.3.3 Kaemferol This is a polyphenol present in the distillation fraction of *Rasoa damascene* Mill, along with ellagic acid and quercetin. The extract of the plant is an apotent inhibitor of tyrosinase and thus can be incorporated in cosmetics preparation to treat hyperpigmentation (Solimine et al. 2016).

11.3.1.1.3.4 p-Coumaric Acid from Oryza sativa L. It is naturally obtained from grains and ceramides of rice (*Oryza sativa* L. Family: Poaceae). *p*-coumaric acid has an anti-tyrosinase function (Jun et al. 2012).

11.3.1.1.3.5 Rosmarinic Acid and Methyl Rosmarinate Rosmarinic acid and its derivative methyl rosmarinate were obtained from plant *Salvia officinalis L Rosmarinus officinalis L menthe piperita L.* and *Rabdosia serra (Maxim.)*, respectively. In one study rosmarinic acid at 10 μM concentration was found to inhibit tyrosinase activity by 20% (Oliveira et al. 2013). In another study rosmarinic acid and methyl rosmarinate at 0.4 mM concentration inhibited the activity of tyrosinase by 19.80% and 37.10%, respectively (Lin et al. 2011).

11.3.1.1.3.6 Chlorogenic Acid This is present in the flower extract of *Chrysanthemum indicum L.* and *Helianthus annuus L* and root extract of *Cichorium intybus L.*, Family: Asteraceae (Charles et al. 2017). The chlorogenic acid was proved to inhibit activity on the tyrosinase enzyme in B16 melanoma cells, thus used as a hyperpigmentation-correcting ingredients in cosmeceuticals (Rivelli et al. 2010). The extract is also used as a free-radical scavenger, anti-aging moisturizer and skin-softening agent; therefore, the extract is incorporated into skin cream (Charles et al. 2017).

11.3.1.1.4 Moisturizer

This is the natural barrier to excessive moisture loss due evaporation and thereby preventing skin dehydration (Berardesca et al. 2001). Moreover, sebum emanated onto the skin surface acts as moisturizer. However, the variation of environmental condition along with skin diseases like rosacea, acne etc. lead to a higher loss of moisture from the skin surface. In those conditions, an additional moisturizer is essential to control the dehydration of skin or healing the barrier-damaged skin (e.g., rosacea) by minimizing transepidermal water loss (Draelos 1997).

11.3.1.1.4.1 Aloe Vera The obtained mucilage from the leaves of Aloe vera (*Aloe barbadensis*, Family: Liliaceae) is widely used in cosmeceuticals because of its moisturizing, revitalizing

(West and Zhu 2003) and anti-inflammatory action (Draelos 2017). Aloe vera gel is found to contain 99.5% of water, leading to moisturizing action and compounds such as aloin, aloe emodin, aletinic acid, chiline and cholin salicylate (Surjushe et al. 2008). The revitalizing action of the gel is due to its cellular repair along with digestion, assimilation of foods, vitamins and minerals in the skin (Ramachandra and Ramachandra 2008). Aloe vera gel has the ability to inhibit cyclooxygenase (COX) enzyme through cholin salicylate ingredients of the gel (Draelos 2017).

11.3.1.1.4.2 Hyaluronic Acid Hyaluronic acid or hyaluronan is chemically glycosaminoglycan and is an integral extracellular part of human skin. It performs a vital role in maintaining water contained in the skin and also facilitates collagen synthesis by upregulation of transforming growth factor β (TGF-β) (Elsner and Maibach 2000; Neudecker et al. 2005). Reduction of hyaluronic acid degradation by inhibiting hyaluronidase enzyme tends to increase the water-containing ability of the skin along with prevention of collagen-induced aging. Derivatives of rosmarinic acid obtained from the methanolic extract of *Meehania urticifolia* (Miq.) exhibited potential anti-hyaluronidase activity (Murata et al. 2011). Dimers of rosmarinic acid such as rashomonic acid and meehaniosides also exhibited anti-hyaluronidase potential. Another phytocompound clinopodic acid M is caffeic acid oligomers, extracted from *Clinopodium gracile* showed significant anti-hyaluronidase activity (Aoshima et al. 2012). The concentration of hyaluronic acid used in cosmetic preparation ranges from 0.025% to 0.05% (Aziz et al. 2017).

11.3.1.1.4.3 Glycerides of Lower Chain Fatty Acids This is a major component of the extract oil obtained by crushing dried kernels of the coconut tree (*Cocos nucifera*, Family: Aracaceae). It can be applied in both liquid and dried form on the skin as its melting point is 24°C–25°C. The oil is used as excellent skin moisturizer as it prevents excessive moisture evaporation from the skin surface along with a skin softening effect (Gediya et al. 2011).

11.3.1.1.4.4 Wax Esters from Simmondsia Chinensis The seed extract of jojoba shrub (*Simmondsia chinensis*, Family: Simmondsiaceae) is composed of long and linear chain liquid wax esters. The jojoba oil has a similar function as that of human sebum. It replenishes the sebum, which is generally removed by chemicals, pollutants, and moreover sun radiation, thereby maintaining a natural skin with natural skin pH balance. For better oxidative stability and without odour and colour, raw jojoba oil is refined and then used in cosmetics to provide a moisturization effect on the skin (Rabasco and Gonzalez 2000).

11.3.1.1.4.5 Allantoin This is naturally obtained from comfrey root and has potential moisturizer activity (Draelos 2001). It exists as crystalline powder and showed high solubility in hot water; therefore, it is easily incorporated into cream- and lotion-moisturizer cosmeceuticals. It is a skin protectant and used to reduce the redness of sensitive skin (Draelos 2017).

11.3.1.1.4.6 Hydroxy Acids Hydroxy acids such as α- and β-hydroxy acids, and α-keto acids obtained from sugarcane (*Panicum miliaceum L.*, Family: Fabaceae) and citrus fruit are found to have epidermal renewal properties, including reduction of corneocytes adhesion to the stratum cornem leading to shedding of coreneocytes rapidly, leaving smoother skin and increased skin hydration thereby acting as a moisturizer. It also reduces hyperpigmentation of skin. Lactic and glycolic acids are found to be most effective α-hydroxy acids and are used in 1%–5% in topical cosmetic products (Tagle et al. 2010; Feng et al. 2015).

11.3.1.1.4.7 Papain This is a mixture of proteolytic enzymes obtained from dried latex of green fruit of tropical melon tree (*Carica papaya L.*, Family: Caricaceae*). Because of its anti-inflammatory potential, papain is employed in wound healing (Starley et al. 1999). It is also used to exfoliate keratotic skin (Aziz et al. 2017).

11.3.1.1.4.8 Ceramide Ceramide is main component of skin and plays a vital role in maintaining the water level in the epidermis. The plant source of ceramide is the extract obtained from the kernel of the wild oats (*Avena sativa L*) and grains of rice plant (*Oryza sativa L*, Family: Poaceae). The moisturizing effect of the extract obtained from *Avena sativa* L was found to correlate with the ability of the plant to upregulate ceramide formation through the stimulation of peroxisome proliferator activated receptor (PPAR) (Southall et al. 2014). Ceramides obtained from *Oryza sativa L*. plant act as a moisturizer (Shimoda 2009). Ceramides obtained are widely used in the cosmeceuticals as a moisturizer (Smeetset al. 1997; Mulder and Meinardi 2002).

11.3.1.1.4.9 Oil Soluble Extract from Raspberry The seeds of raspberry (*Rubus idaeus* L., Family: Rosaceae) contain two major components—a phenolic component having anti-oxidant property, and an oil-soluble component showing moisturizer function. The moisturizer property was found to be its induction ability of genes responsible for skin hydration, and it stimulates the manifestation and function of enzyme glucocerebrosidase, too, which is involved in the synthesis of ceramides. Therefore, the extract is widely used in various lotions as a hydrating and moisturizing agent and face creams as antiaging ingredients (Tito et al. 2015).

11.3.1.1.4.10 Saffranol This is another active phytoconstituents, which shows skin moisturizing activities along with UV radiation protecting effects (Khameneh et al. 2015).

11.3.1.1.5 Anti-Wrinkles

Wrinkles are natural phenomenon but can be induced by UV irradiation. Skin aging is characteristically recognized as wrinkles on the face. Out of various extracellular matrix components in the skin, collagen constitutes majority of connective tissue (Lovell et al. 1987). Therefore, the breakdown of collagen induced by UV irradiation is the sole reason for the premature wrinkle formation. Collagen breakdown in the skin matrix is mediated by collgenolytic enzyme, metalloproteinases (MMP) and UV irradiation, and significantly enhances MMP formation. Consequently, this increases the splitting of collagen fibrils (Fisher et al. 2002; Fisher 2005). Furthermore, degradation of elastin fibers in the skin matrix adds to wrinkle formation in the skin (Hussain et al. 2013). Therefore, the main therapeutic approach for the prevention and treatment of UV irradiation-induced wrinkles is to inhibit the MMP enzyme (Roh et al. 2017).

11.3.1.1.5.1 EGCG Polyphenol Green tea polyphenols have antiwrinkle activity as they are potent inhibitors of collagenase. It was observed that the steric configuration present in 3-galloyl radical catechin is responsible for its enzyme inhibitory potential (Makimura et al. 1993). Additional galloyl unit and –OH groups of EGCG in comparison to catechin leading to improved H-bonding and hydrophobic interaction with collagenase. This resulted in a notable change in collagenase conformation and subsequently its inhibition (Madhan et al. 2007). Polyphenols obtained from green tea also inhibit not only collagenase expression but also its degradation (Bae et al. 2008), and generation and activity of MMPs (Demeule et al. 2000).

11.3.1.1.5.2 Polyphenols and Xyloglycans These are the main constituents of the tamarind seed (*Tamarindus indica* L.). Under the UV radiation, cultured human fibroblast treated with the tamarind extract exhibited a higher glutathione level and decreased the rate of cell death. The extract also decreased the secretion of MMP-1 following photodamage. Therefore, the seed extract of tamarind can be utilized as an antiaging agent cosmeceutic (Phetdee et al. 2014).

11.3.1.1.5.3 Extract of Malva sylvestris L. The extract of *Malva sylvestris* L., Family: malavaceae contains phyto constituents such as polyphenols, flavonoids, caretonoids and tocopherols. The total extract induces gene regulation at transcription level, and therefore it shows anti-wrinkle efficacy. It was also proved that the extract has the same gene modulation capacity as all trans-retinoic acid (Talbourdet et al. 2007).

11.3.1.1.6 Anti-inflammatory

The increase in the secretion of inflammatory mediators viz. interleukins (IL1β, IL-6, IL-8), COX-2, prostaglandins E2 (PGE2), tumor necrosis factor (TNF-α) and nitric oxide synthase (NOS) are increased by inflammatory cells, including macrophases and monocytes. The generation of ROS, due to the exposure of skin with UV irradiation, leads to the increase in the MMP enzymes expression that causes the degradation of collagen and elastin. This degradation pathway, in addition to other factors, has increased the manifestation of above inflammatory mediators. Thus, the phytocompounds having ROS scavenging property can act as an anti-inflammatory agent along with anti-oxidant effects (Pillai et al. 2005).

11.3.1.1.6.1 Polyphenol from Ginkgo biloba Leaves of ginkgo tree (*Ginkgo biloba*, Family: Ginkgoceae) contain polyphenols, which show anti-inflammatory effect. The polyphenol presents are terpenoids, including ginkgolides, bilobalides, flavinoids and flavonol glycosides (Svobodova et al. 2003). Antiradical and anti-lipoperoxidant effects are responsible for anti-inflammatory effect. Additionally, leaves of ginkgo reduce the capillary blood flow level and promote a vasomotor change in the arteriols of the subpapillary skin plexus resulting in the decline in redness of skin (Draelos 2017).

11.3.1.1.6.2 Parthenolide and Tanetin Leaves of the feverfew plant have the main ingredients parthenolide and tanetin, which are responsible for the anti-inflammatory activity. Reduction in the release of serotonin and prostaglandins is presumed to be the mechanism of action for its anti-inflammatory potential

(Martin et al. 2008). In addition, these ingredients also induce vasoconstriction. Thus, it is widely used in the reduction of redness in rosacea. But, parthenolide can induce allergic contact dermatitis, and therefore feverfew without parthenolide is marketed for the treatment of rosacea (Draelos 2017).

11.3.1.1.6.3 Licochalcone A   Licochalcone A is obtained from the root of the plant (*Glycerhhyza inflate*, Family: Licorice). It exhibits anti-inflammatory effect due to the inhibition of keratinocytes release of PGE2 in response to erythema induced by UVB. It also inhibits the lipopolysaccharides-induced release of PGE2 by fibroblast of skin (Weber et al. 2006). Therefore, licochalcone A is one of the widely used active agents in cosmeceuticals to reduce redness.

11.3.1.1.6.4 Curcuminoid pigments   Curcuminoid pigment is a deep yellow-orange powder obtained from the underground stem of *Curcuma longa*, Family: Zingiberaceae (Damalas 2011). This powder has many phytochemicals such as curcumin, demethoxucurcumin, bisdemethoxycurcumin, tetrahydrocurcumin, tryethylcurcumin, curcumol, zinngiberene, eugenol and turmeronols. Among all, curcumin is the main ingredients that exhibited anti-inflammatory, anti-oxidant, antispasmodic, and antiparasitic activities (Fujiyama 1992; Saikia et al. 2006; Filho et al. 2009; Hamzah 2011). Turmeric extract gel was found to reduce 30% of edema when applied on carrageenan-induced paw edema in albino rat (Hamzah 2011). Topically applied curumin has also proved to be a powerful inhibitor of induced skin tumor development. Thus, it is concluded that curcumin has the potential to treat solar radiation induced inflammation, wrinkles, and cancer in the skin (Gonçalves et al. 2014).

11.3.1.1.6.5 Boswellic acids   These are obtained from the resin of plant *Boswellia serrata* and proved to have anti-inflammatory potential. This is because of the inhibition of enzyme 5-lipoxygnease, responsible for inflammation and also inhibits enzyme-causing damage to the skin. Thus, it has been widely used in anti-inflammatory creams and lotions (Kokate et al. 2008).

11.3.1.1.6.6 Carnosic, Cosmarinic and Ursolic Acids   These acids are obtained from the extract of rosemary plant (*Rosemarimus officinalis*). They are mainly used as antioxidant to prevent tissue damage and to provide healthy status to the skin.

11.3.1.1.7 Anti-immunosuppression

The immune system of normal skin is capable of preventing skin infection and removing injured cells. However, persistent UV radiation can induce immunosuppression that could cause photoaging and skin cancer (Muller et al. 1995; Halliday 2005). This induction can happen through various mechanisms such as DNA damage, contact hypersensitivity suppression, infiltration of leukocytes and attenuation of antigen presenting capacity (Roh et al. 2017).

11.3.1.1.7.1 EGCG Polyphenol of Green Tea   EGCG obtained from green tea has shown skin protection potential from local immune suppression developed by CHC upon UV radiation exposure. It also inhibited the formation of UVB developed DNA damage through the formation of indicator cyclobutane pyrimidine dimmers (Katiyar et al. 1995; Katiyar 2003).

11.3.2 Hair Cosmeceuticals

11.3.2.1 Anti-dandruff Care

11.3.2.1.1 Lawsone
Lawsone is obtained from henna leaves (*Lawsonia inermis*, Family: Lythraceae). Lawsone was used as a strong hair dying agent. The leaves extract is also found to contain gallic acid, mannitol, fats, resins, mucilage and traces of alkaloids (Chaudhary et al. 2010). Lawsone was found to have significant antifungal activity, thus used in hair cosmetics (Dixit et al. 1980).

11.3.2.1.2 Neem
The extract of the neem plant (*Azadircta indica*, Family: Meliaceae) shows insect repellant, insecticide, nematicide, antifungal and antibacterial properties. The anti-dandruff property of neem extract is due to its antifungal and antibacterial properties (Anand et al. 2010). Herbal shampoo containing neem extract exhibited anti-dandruff activity against fungus *Candida albicans* and microbes such as *Staphylococcus aurous* and *Escherichia coli* (Halith et al. 2009).

11.3.2.1.3 Bala
Leaf extract of plant bala (*Sida cordifolia*, Family: Malavaceae) have excellent anti-dandruff property apart from antimicrobial property. The active ingredients present in leaf extract are ephedrine, pseudoephedrine, vasicinone, vasicine and vasicinol (Chandran et al. 2013). The aqueous extract showed a powerful antifungal potential against *Candida albicans* and *Cryptococcus neoformans* (Ternikar et al. 2010). Complete herbal shampoo with incorporated leaf extract of *Sida cordifolia* showed positive anti-fungal property on *Candida albicans* (Chandran et al. 2013).

11.3.2.1.4 Lemon
This is obtained from the cold expression of the peel of fruit lemon (*Citrus limon*, Family: Rutaceae). The extracted oil is composed of 70% limonene, terpinene, geraniol, linalol and many other ingredients (Evans 2006). The extract is used as an anti-dandruff agent, natural cleanser and pH modifier (Chandran et al. 2013). An herbal shampoo containing the juice of *Citrus limon* along with other herbal extract showed good anti-dandruff property (Golhani et al. 2015).

11.3.2.2 Haircare

11.3.2.2.1 Amla
Amla is obtained from the fresh and dried fruit of amla (*Emblica officinalis*, Family: Euphorbiace). Seeds and pulp of amla fruit contain plentiful vitamin C, tannin, precious oil and phyllemblin. The extract is used in the preparation of hair oil and shampoos to treat hair loss, dandruff and other scalp problems, and to strengthening hair (Chandran et al. 2013). The fixed oil present in the extract is believed to promote hair growth (hair tonic). Vitamin C is responsible for its oxygen free-radical scavenging property and therefore has been used as an antiaging ingredient in topical preparations (Pandey et al. 2010).

11.3.2.2.2 Shikakai

Shikakai is obtained from the fruit of the plant *Acacia concinna* Linn Family: Leguminosae. The powder extract of the fruit contains saponins, alkaloids, tannin, flavanoids and anthraquinone glycosides. It is used in hair oils and shampoos because of its hair growth improving, anti-dandruff and hair cleansing properties, respectively (Khanpara et al. 2012).

11.3.3 Teeth and Mouth Cavity Cosmeceuticals

There are two sets of teeth in any individual's lifetime: deciduous (baby teeth) and permanent. The primary role of teeth is to tear and grind solid foods. But, a set of clean and disease-free teeth and gums can also improve the cosmetic appearance. With the normal toothpaste available in the market, it is possible to only clean the teeth, but the disease generally occurs in the teeth and gums cannot be completely ruled out. Therefore, toothpaste with one more phytoconstituents having cosmeceutical potential should be added to protect our teeth and gums from various diseases such as dental caries and periodontal diseases. The following are a few phytoconstituents that could be useful in the cosmeceuticals related to teeth and the mouth cavity.

11.3.3.1 Essential Oil from Clove

Dried flower buds of *Eugenia caryophyllus* are known as clove, and the essential oil content is not less than 15% of total clove oil. The main constituent of essential oil is eugenol (70%–90%). It acts as dental analgesic and relieves a toothache when applied to a decayed tooth cavity (Alqareer et al. 2006).

11.3.3.2 Hydrolysable Tannins

Hydrolysable tannins such as chebulagic acid, chebulinic acid and corilagin are obtained from *Terminalia chebula*. It is useful in dental caries, bleeding gums and ulcerative oral cavity. It has also activity against fungus, bacteria and the HIV virus (Gupta et al. 2010).

11.3.3.3 Mangiferin

This is obtained from the leaves of *Mangiferaindica*. It is useful as a drug in preventing dental plaques. Additionally, it possesses anti-inflammatory and antibacterial potential against gram-positive bacteria (Chowdhury et al. 2013).

11.3.3.4 Perillyl Alcohol

This is a phytonutrient present in the leaf of *Mentha spicata*. It has been used for dental care, therefore can be incorporated into toothpaste. It also has anti-cancer properties and is found to prevent formation of skin, colon and lung cancer in animal studies (Chowdhury et al. 2013).

11.4 Cosmeceutical Delivery Systems

Drug delivery system (DDS) is essential to deliver active pharmaceutical ingredients through various routes as it delivers the drugs in a controlled manner and also sustains drug level in the plasma within therapeutic range (Coelho et al. 2010). Similarly, a suitable DDS for cosmeceuticals can serve many purposes, such as increases stability, prevents incompatibilities with other ingredients in the system/formulation, increases permeation of cosmeceuticals ingredients deeper in to the skin, increases duration of action and prevents undesirable local and systemic effects (Yapar 2016). In order to obtain maximum advantages, the preferred delivery systems include emulsion systems (multiple emulsion, microemulsion, nanoemulsion), vesicular systems (liposomes, niosomes, transferosomes, ethosomes, phytosomes, glycerosomes and hyalurosomes) and particulate systems (microsponges, polymeric nanoparticles, lipid nanoparticles such as solid lipid nanoparticles (SLN) and nanostructured lipid carriers (NLC), carbon nanoparticles viz. carbon nanotubes and fullerene) (Patravale and Mandawgade 2008; Saraf 2010; Bansal et al. 2012; Gökçe et al. 2016) are prepared (Figure 11.5). These systems are generally incorporated in different pharmaceutical preparations, such as emulsion, creams, and gels in order to prepare finished cosmeceuticals products.

11.4.1 Emulsion Delivery Systems

The conventional emulsion types that are used for cosmetic formulations include oil-in-water (O/W) emulsion, water-in-oil (W/O) emulsion, and double (O/W/O or W/O/W) emulsion. W/O emulsion type is considered as less elegant cosmetically compared to O/W type due to difficult in rubbing. Additionally, they can form skin stickiness. These outcomes are related to the waxes present in the formulations. Furthermore, water evaporation from the applied O/W emulsion leads to a cooling sensation. But, from a dermatological prospective, the W/O emulsion is suitable, since the formed lipid film on the skin favours the permeation of lipophilic drugs across the skin (Yapar 2017; Singh and Agarwal 2009; Spiess 1996; Costa and Santos, 2017). Emulsions are the most popular formulations among all types because they have the potential to cross SC and therefore improve their efficacy and durability. The primary components used in the preparation of emulsions are: (i) oil, (ii) water, (iii) emulsifier/surfactant, and (iv) a co-surfactant. Cosmeceuticals, if oil soluble, are added to the oil phase, whereas water soluble cosmeceuticals are dispersed or solubilized in the aqueous phase.

Microemulsions are stable formulations as they have the ability to encapsulate oil soluble cosmeceuticals such as antioxidants, antimicrobials and vitamins (Chen et al. 2006), and protect them from photo and thermal degradation. Microemulsions contain liquid crystals, which provide a long-lasting effect along with the stability of cosmeceuticals. In one study, naturally occurring whitening agents such as arbutin and kojic acid incorporated into microemulsions composed of lecithin and alkyl glucoside as mild and non-irritant surfactant, and their stability was compared with their aqueous solutions. The result showed that the stability of these agents were higher in microemulsion compared to solution (Gallarate et al. 2004). In another study, water proof effect of sunscreen agent octyimethoxycinnamate was obtained by combining it with soya-lecithin, surfactant blend and a polar phase (Carlotti et al. 2003). For ascorbic acid (an antioxidant molecule), different emulsions such as O/W and W/O emulsions, O/W microemulsions and multiple cosmetic emulsions were developed employing non-ionic, non-ethoxylated skin compatible emulsifier (Gallarate et al. 1999).

FIGURE 11.5 Different delivery systems for cosmeceuticals.

In the case of multiple emulsions, either O/W or W/O emulsion systems are dispersed inside outer water or oil medium. Between the two, the W/O/W variety is a wide application in cosmetics as skin moisturizer and also used to prolong the release of cosmeceuticals (Okochi and Nakano 2000). These have distinct benefits, including better protection of ingredients present inside and encapsulation of different natured ingredients in various compartments viz. lipophilic substance in oil globules and hydrophilic substance in water phase (Raynal et al. 1993). Thus, it can be used as successful DDS for various herbal cosmeceuticals; however, their stability issue restricted their widespread application (Agrawal et al. 2016).

Nanoemulsions are O/W dispersions having droplet sizes ranges between 20 to 300 nm. The appearance varies depending on the droplet size: nanoemulsion composed of droplets below 70 nm look transparent, droplet size between 70 and 100 nm appear opaque and greater than 100 nm look white (Chanchal and Saraf 2008). Due to its desired cosmetical requirement such as better penetration power and hydration properties, nanoemulsions are a better choice for herbal cosmeceuticals (Yapar 2017). Nanoemulsions based on rice bran oil were developed and found to have higher moisturizing potential to ameliorate skin diseases such as psoriasis and atopic dermatitis (Zhang et al. 2010). Protection of skin against photodamage along with antioxidant activities were observed for the nanoemulsion prepared with pomegranate seed oil (Baccarin et al. 2015a). The same group of researchers proved that pomegranate-seed-oil-based nanoemulsions have enhanced photo protection of lipid skin membrane

when studied on human erythrocytes (Baccarin et al. 2015b). Nanoemulsions with a droplet size between 92 and 233 nm based on *Opuntia ficus-indica* (L.) extract were found to enhance moisturizing effects (Ribeiro et al. 2015). The flavanons obtained from the leaves of *Eysenharditia platycarpa* developed into nanoemulsion of 70 nm droplet sizes and found to have enhanced antiaging activity (Domínguez-Villegas et al. 2014).

11.4.2 Vesicular Systems

Liposomes are microscopic spherical vesicular systems composed of bilayer of amphiphilic molecules. Liposomes have membrane-mimetic structures as they are made up of natural phospholipid lecithin, and size is in between 20 nm to few hundred mm, and able to penetrate the skin very efficiently. Further, the phospholipid (lecithin) obtained from plant sources have advantages due to the presence of poly-unsaturated fatty acid (linoleic acid). Liposomes have the ability to adhere to the corneocytes and thereby deliver the active cosmeceuticals either by diffusion or by fusion of liposomes with the lipids of stratum corneum. This adherence helps fix the active ingredients in contact with skin and thereby delay the removal by washing out (Cevc 1997). In addition it can encapsulate both hydrophilic and hydrophobic cosmetic actives without allowing them to reach systemic circulation, avoiding any undesirable effect (Bugaj 2015). Furthermore, liposomes and phospholipids enable the skin to maintain a moisture level and thereby restore the barrier functions of skin that maintain skin appearance (Niemiec et al. 1995).

In one study, liposomes incorporated with curcumin having size ranges from 213 to 320 nm were prepared and found to have enhanced penetration and promote growth of hairs by 70% (Jung et al. 2006). In another study, lauric acid-based nanosized liposomes (113 nm) exhibited higher antimicrobial activity against acne (Pornpattananangkul et al. 2013). Even though liposomes have a number of advantages, the lack of stability (both physical and chemical) that occurs owing to lipid oxidation and hydrolysis limits its use as delivery system (Wang et al. 2015).

In order to avoid the stability issue of liposomes, niosomes, a non-ionic surfactant-based vesicles system, was developed. They are biodegradable and biocompatible vesicles and able to encapsulate cosmetic actives and enhance their penetration through the skin (Marianecci et al. 2014). Furthermore, they were found to be less irritating than other colloidal vesicle carriers (Yapar 2017). The niosomal delivery system was developed by combining lauric acid and curcumin; the prepared niosomes exhibited improved antimicrobial activity against skin infections mediated by acne (Liu and Huang 2013). Larger niosomes with sizes between 471 and 565 nm incorporated with phytoconstituents including resveratrol, α-tocopherol and curcumin were prepared, and improved antioxidant activity to the skin was observed for cosmeceuticals applications (Tavano et al. 2014). In another study, nanoniosome of size range 91 nm of curcumin was developed and found to exhibit enhanced skin protective activity (Ascenso et al. 2015).

Transferosomes are deformable vesicular systems, prepared from phospholipids and an edge activator. Deformable property of transferosomes is particularly due to the edge activator (single-chain surfactant) having a high radius of curvature, which decreases bilayer stiffness of phospholipids and thereby enhances the penetration across the skin (Shuwaili et al. 2016). It shows better performance than liposomes as it can deliver a wide range of molecular cosmetic actives, compatible, efficacious and having high mechanical strength (Yapar 2017). Therefore, it has been considered for the delivery of small molecules as well as large molecules such as peptides and proteins (Benson 2006).

Basic components of ethosomes are phospholipids, high concentration of ethanol (20%–50%) and water (20%–50%) (Yapar 2017). Like transferosomes, ethosomes have high-deformability characteristics and can enhance the delivery of active ingredients including occlusive and non-occlusive conditions (Vijayakumar et al. 2010). Due to a high ethanol concentration, which is a penetration enhancer, it is able to penetrate and deliver cosmetic actives to the deeper layer of skin (Touitou et al. 2010). Polyphenolic extract obtained from the leaf and bark of *Fraxinus angustifolia* was encapsulated in ethosomal nanovesicles of size range 128 nm. This DDS showed enhanced skin protection and wound healing properties (Moulaoui et al. 2015).

Phytosomes or herbosomes are the combination of polar standardized or purified plant extract and phospholids. These are designed in order to improve the bioavailability of topically applied phytoconstituents resulting in better stability, increased efficacy and prevent undesirable effects (Suzuki 2010; Furmanowa et al. 1998). Phytosomes containing *Ginkgo biloba* terpenes exhibited enhanced soothing effects of persons suffering from contact reactions compared to other substances present in a topical delivery system (Loggia et al. 1996). Similarly, there was a six-fold increase in soothing activity of silymarin phytosomes compared to phytosomes without cosmetic actives

(Yanyu et al. 2006). Phytosomes of extracts obtained from *Citrus auranticum* and *Glycyrrhiza glabra* were developed and then incorporated into creams. This product was found to enhance the bioavailability of the above phytocompounds and are effective in skin aging (Damle et al. 2016). Likewise, phytoextracts rich in quercetin was developed into gold-nanoparticle-based phytosomes (100 nm) and were found to have enhanced quercetin activity (Demir et al. 2014). Similarly, rutin-based phytosomes showed higher bioavailability followed by enhanced antioxidant activity (Singh et al. 2012).

Glycerosomes and hyalurosomes are recently developed DDS containing glycerol and sodium hyalronate, respectively, in phospholipid. In one study, glycerosomes of quercetin were prepared, and particle size was found to be in the range of 80–110 nm. This is characterized by unilameller structure and found to enhance skin protective activity (Manca et al. 2014). Developed nanosized (100 nm) hyalurosomes of liquirice extract exhibited enhanced skin beauty (Castangia et al. 2015). The group of researchers prepared curcumin based nanohyalurosomes (112 nm) and found that the product has enhanced skin beauty activity (Manca et al. 2015).

11.4.3 Micro- and Nanoparticles

11.4.3.1 Microsponges

Microsponges are unique, highly cross-linked, microporous, polymeric microspheres beads with typically 10–25 μm in diameter. They release the active ingredients in a controlled manner, and release rate can also be enhanced with external stimuli including rubbing, temperature and pH. Presently, they are used in sunscreen cosmetic products. The advantages of microsponges DDS are minimal irritation, extended product stability and improved aesthetic property of topically applied phytoconstituents (Jadhav et al. 2013; www.lipochemicals.com).

11.4.3.2 Polymeric Nanoparticles

These are prepared from synthetic biodegradable and biocompatible polymers such as polylactic acid (PLA), poly (glycolic acid) (PGA), PLGA, and so on (Mishra et al. 2010). They are considered very stable and having high affinity to SC, therefore providing more bioavailability to the encapsulated phytoconstituents. The skin humidity was found to increase with the application of gel formulation composed of nanoparticles loaded with vitamin A and E derivatives. The above nanoparticles were capable of penetrating the upper layer of SC and fused with skin lipids resulting in the release of the cosmetic actives (Agrawal et al. 2016).

11.4.3.3 Lipid Based Nanoparticles

SLNs are prepared from lipids (solid at room temperature) and surfactants, and their mean diameter ranges from 50 to 1000 nm. They provide advantages including small-size, high-drug loading, large surface area, thus these are very attractive DDS not only for pharmaceuticals but also in cosmeceuticals drug delivery (Mishra et al. 2010; Mudshinge et al. 2011). Particle size of 182 nm was obtained for caffeine-loaded SLNs, and these particles exhibited an enhanced skin permeation with higher skin protection (Puglia et al. 2016). A prolonged skin health was maintained with SLN loaded with phyto constituents lutein due to protection of skin from UV light and antioxidant activity (Mitri et al. 2011). SLN loaded with resveratrol with a particle size of 180 nm exhibited better skin protection from sunlight-induced skin damage (Teskac and Kristl 2010). Enhanced skin moisture retention and skin occlusive effects in porcine skin were observed with vitamin-A-loaded SLN of particle size 210 nm (Jenning et al. 2000). Curcumin-loaded SLN was developed having a particle size of 210 nm, and this drug loaded SLN exhibited improved activity in hydrogels for induced inflammatory activity in porcine skin (Zamarioli et al. 2015).

NLCs are another lipid-based drug carrier system composed of both solid and liquid lipid. They are prepared in different forms such as imperfect, formless and multiple types containing bioactive constituents without drug expulsion during preparation and storage. Therefore, these have wide use in delivering phyto constituents for skin care and protection (Mudshinge et al. 2011). In one study, E-resveratrol loaded NLC showed better skin protection against UV irradiation (Detoni et al. 2012). In another study, NLC of resveratrol of size 110 nm were developed and found to exhibit enhanced skin protection from oxidation (Gokce et al. 2012). NLC loaded with squalene with particle size between 208 to 265 nm were used to treat hair follicles of individuals infected with alopecia areata (Lin et al. 2013). A quercetin containing NLC was prepared with particle size 282 nm and exhibited enhanced antioxidative property, thereby improved skin beauty (Bose et al. 2013). Topical delivery of lycopene was successfully carried out from the developed NLC loaded with lycopene (Okonogi et al. 2015). Similarly, lutein-based NLC with particle size 166–350 nm was developed and showed higher UV protection tested on pig ear skin (Mitri et al. 2011).

11.4.3.4 Carbon Based Nanoparticles

Fullerene is made up entirely of carbon and has various shapes such as hallow sphere, ellipsoid and tube. The spherical shape is called as fullerene whereas cylindrical one is called as nanotubes. Fullerene has identical structure to graphite and consists of 60 carbon (Holister et al. 2003). Fullerene nanocapsules with incorporated ascorbic acid and vitamin E were developed (Ito et al. 2010). Both of the formulations were found to have protective action against premature skin aging, and the antioxidant activities are responsible for the above results. A photodynamic therapy based on curcumin loaded fullerene for the treatment of skin-based diseases was developed (Yin and Hamblin 2015; Zhang et al. 2015). Carbon nanotube containing curcumin was also developed for its enhanced delivery (Li et al. 2014). Recently, hyaluronic acid was successfully delivered after loading into carbon nanotube (Tripodo et al. 2015).

11.5 Formulation Challenges

The major challenges faced by phyto-based cosmeceuticals in its preparations are contamination, instability and lack of enough amounts of different ingredients with hectic collection and purification process (Singh et al. 2014; Ganesan and Choi, 2016).

The herbal ingredients obtained from the wild are generally contaminated with different toxic residues released by insects and fungi. These toxic residues not only make the cosmeceuticals unstable but also damage skin and hairs like hair falls (Singh et al. 2014). Due to urbanization based on deforestation, there is unavailability of the desired amount of different ingredients for the preparation of various phyto-based cosmeceuticals. Furthermore, people are fond of easy-to-use products and not at all interested to go deep into the hectic process of preparation of cosmeceuticals (Singh et al. 2014).

Likewise, in usage of phyto-based cosmeceuticals the concerns include poor skin penetration and lower final quality like less whitening effects (Singh et al. 2014). Lower penetration of phyto-constituents into the skin is because of the presence of impervious skin layer called SC. To be therapeutically beneficial, this impervious must be overcome (Koizumi et al. 2004). Poor quality of the finished product is due to size and solubility of the phyto-ingredients (Kidd 2009; Krausz et al. 2014).

11.6 Safety Aspects of Cosmeceuticals

Lack of information regarding efficacy as well as safety about the phyto-constituents creates a great deal of confusion both in consumer and dermatologist. As such, there is a huge share of phyto-comeceuticals in the beauty market, but many of them are not supported by scientific evidence for their safety and efficacy. According to one recent report, clinical trials have been performed with only few, such as soy, black and green tea, pomegranate and date as antiaging agents (Thornfeldt 2005; Singh and Agarwal 2009). It is sure that phyto-cosmeceuticals may be used as adjunct therapy, but for that a minimum criteria such as efficacy, safety and scientific validation is essential.

11.7 Conclusions and Future Prospective

There are many phyto-constituents that are used to develop cosmeceuticals, and the list of both phyto-constituents and cosmeceuticals are extending further with time. Despite the widespread use of phyto-constituents-based cosmeceuticals, challenges such as formulation and its use followed by the lack of scientific evidence are the major drawbacks. Formulation challenges such as contamination can be avoided by properly identifying and purifying the constituents, and instability can be prevented by judicious selection of other excipients. Poor permeation across the skin can be improved with the help of penetration enhancers from plant sources or by the use of physical methods including sonophoresis, iontophoresis and electroporation. In addition, novel DDS, such as nanoemulsion and nanoparticles will improve the penetration capability of phyto-constituents. Similarly, preclinical and clinical studies on each phytoconstituents and on their cosmeceuticals products must be conducted in order to acertain their safety and efficacy. Therefore, an integrative approach must be contemplated including formulation process, consumer requirements and perceptions in addition to well-planned marketing strategies, in order to develop a successful product.

REFERENCES

Agrawal, A., Kulkarni, S., and Sharma, S., 2016. Recent advancements and applications of multiple emulsions. *International Journal of Advances in Pharmaceutics*, 4, pp. 94–103.

Alqareer, A., Alyahya, A., and Andersson, L., 2006. The effect of clove and benzocaine versus placebo as topical anesthetics. *Journal of Dentistry*, 34, pp. 747–750.

Amarowicz, R., Estrella, I., Hernández, T., Robredo, S., Troszyńska, A., Kosińska, A. and Pegg, R.B., 2010. Free radical-scavenging capacity, antioxidant activity, and phenolic composition of green lentil (*Lens culinaris L.*). *Food Chemistry*, 121, pp. 705–711.

Amer, M., and Maged, M., 2009. Cosmeceuticals versus pharmaceuticals. *Clinics in Dermatology*, 27, pp. 428–430.

Anand, N., Aquicio, J.M., and Anand, A., 2010. Antifungal properties of Neem (*Azadirachta indica*) leaves extract to treat Hair Dandruff. *E-International Scientific Research Journal*, 2, pp. 244–252.

Anila, L., and Vijayalakshmi N.R., 2002. Flavonoids from Emblica officinalis and *Mangifera indica*-effectiveness for dyslipidemia. *Journal of Ethnopharmacology*, 79, pp. 81–87.

Aoshima, H., Miyase, T., and Warashina, T., 2012. Caffeic acid oligomers with hyaluronidase inhibitory activity from *Clinopodium gracile*. *Chemical and Pharmaceutical Bulletin*, 60, pp. 499–507.

Ascenso, A., Raposo, S., Batista, C., Cardoso, P., Mendes, T., Praça, F.G., Bentley, M.V.L.B. and Simões, S., 2015. Development, characterization, and skin delivery studies of related ultra-deformable vesicles: Transfersomes, ethosomes, and trans-ethosomes. *International Journal of Nanomedicine*, 10, pp. 5837–5851.

Aziz, A.A., Taher, Z.M., Muda, R., and Aziz, R., 2017. Cosmeceuticals and Natural Cosmetics. In: *Rosnani Hashams Recent Trends in Research into Malaysian Medicinal Plants*, 3rd Ed., Penerbit UTM Press, Johor, Malaysia, pp. 126–175.

Baccarin, T., Mitjans, M., Lemos-Senna, E., and Vinardell, M.P., 2015a. Protection against oxidative damage in human erythrocytes and preliminary photosafety assessment of *Punica granatum* seed oil nanoemulsions entrapping polyphenol-rich ethyl acetate fraction. *Toxicology in Vitro*, 30(1 Pt B), pp. 421–428.

Baccarin, T., Mitjans, M., Ramos, D., Lemos-Senna, E., and Vinardell, M.P., 2015b. Photoprotection by Punica granatum seed oil nanoemulsion entrapping polyphenol-rich ethyl acetate fraction against UVB-induced DNA damage in human keratinocyte (HaCaT) cell line. *Journal of Photochemistry and Photobiology B*, 153, pp. 127–136.

Bae, J.Y., Choi, J.S., Choi, Y.J., Shin, S.Y., Kang, S.W., Han, S.J. and Kang, Y.H., 2008. (-) Epigallocatechin gallate hampers collagen destruction and collagenase activation in ultraviolet-B-irradiated human dermal fibroblasts: Involvement of mitogen-activated protein kinase. *Foog Chemistry Toxicology*, 46, pp. 1298–1307.

Balachandran, P., and Govindarajan, R., 2005. Cancer-an ayurvedic perspective. *Pharmacological Research*, 51, pp. 19–30.

Bansal, S., Kashyap, C.P., Aggarwal, G. and Harikumar, S.L., 2012. A comparative review on vesicular drug delivery system and stability issues. *International Journal of Research in Pharmacy and Chemistry*, 2, pp. 704–713.

Bary, B.W., 2007. Transdermal drug delivery. In *Aultons Pharmaceutics: The design and Manufacturing of Medicines*. New Work, Churchill Livingstone, pp. 565–597.

Baxter, R.A., 2008. Anti-aging properties of resveratrol: Review and report of a potent new antioxidant skin care formulation. *Journal of Cosmetic Dermatology*, 7, pp. 2–7.

Beltrame, F.L., 2005. Caracterizac̦ão de duas espécies vegetais (*Trichilia catiguae* Anemopaegma arvense) usadas como Catuaba por métodos cromatográficos hifenados a métodos espectroscópicos e análise multivariada, Química. In Universidade Federal de São Carlos, São Carlos, p. 153.

Benson, H.A., 2006. Transfersomes for transdermal drug delivery. *Expert Opinion on Drug Delivery*, 6, pp. 727–737.

Berardesca, E., Barbareschi, M., Veraldi, S., and Pimpinelli, N., 2001. Evaluation of efficacy of a skin lipid mixture in patients with irritant contact dermatitis, allergic contact dermatitis or atopic dermatitis: A multicenter study. *Contact Dermatitis*, 45, pp. 280–285.

Berthakur, N, N., and Arnold, N.N., 1991. Chemical analysis of the emblic (*Phyllanthus emblica L.*) and its potential as food source. *Scientia Horticulturae*, 47, pp. 99–105.

Bose, S., and Michniak-Kohn, B., 2013. Preparation and characterization of lipid-based nanosystems for topical delivery of quercetin. *European Journal of Pharmaceutical Sciences*, 48, pp. 442–452.

Bugaj, A.M., 2015. Intradermal delivery of active cosmeceutical ingredients. In Donnelly, R.F Singh, T.R.R. (Eds.), *Novel Delivery Systems for Transdermal and Intradermal Drug Delivery*. London, UK, John Wiley & Sons, pp. 209–230.

Butler, H., 2010. *Poucher's Perfumes, Cosmetics and Soaps*. New Delhi, India, Springer India private limited, pp. 394–395.

Carini, R., Comoglio, A., Albano, E., and Poli, G., 1992. Lipid peroxidation and irreversible damage in the rat hepatocyte model: Protection by thesilybin-phospholipid complex IdB 1016. *Biochemical Pharmacology*, 43, pp. 2111–2115.

Carlotti, M.E., Gallarate, M., and Rossatto, V., 2003. O/W microemulsion as a vehicle for sunscreens. *Journal of Cosmetic Science*, 54, pp. 451–462.

Carvalho, I.T., Estevinho, B.N., and Santos, L., 2016. Application of microencapsulated essential oils in cosmetic and personal healthcare products-A review. *International Journal of Cosmetic Science*, 38, pp. 109–119.

Castangia, I., Caddeo, C., Manca, M.L., Casu, L., Latorre, A.C., Díez-Sales, O., Ruiz-Saurí, A., Bacchetta, G., Fadda, A.M. and Manconi, M., 2015. Delivery of liquorice extract by liposomes and hyalurosomes to protect the skin against oxidative stress injuries. *Carbohydrate Polymers*, 134, pp. 657–663.

Cevc, G., 1997. Drug delivery across the skin. *Expert Opinion on Investigational Drugs*, 12, pp. 1887–1937.

Cevc, G., and Vierl, U., 2007. Spatial distribution of cutaneous microvasculature and local drug clearance after drug application on the skin. *Journal of Controlled Release*, 118, pp. 18–26.

Chanchal, D., and Swarnlata, S. Novel approaches in herbal cosmetics. *Journal of Cosmetic Dermatology*, 7, pp. 89–95.

Chandran, S., Vipin, K.V., Augusthy, A.R., Lindumol, K.V. and Shirwaikar, A., 2013. Development and evaluation of antidandruff shampoo based on natural sources. *Journal of Pharmacy and Phytotherapeutics*, 1, p. 4.

Charles, A.I.D., Amalraj, A., Gopi, S., Varma, K., and Anjana, S.N., 2017. Novel cosmeceuticals from plants—An industry guided review. *Journal of Applied Research on Medicinal and Aromatic Plants*, 7, pp. 1–26.

Chaudhary, G., Goyal, S., and Poonia, P., 2010. *Lawsonia inermis* Linnaeus: A phytopharmacological review. *International Journal of Pharmaceutical Sciences and Drug Research*, 2, pp. 91–98.

Chen, H., Weiss, J., and Shahid, F., 2006. Nanotechnology in nutraceuticals and functional foods. *Food Technology*, 60, pp. 30–36.

Chowdhury, B.R., Garai, A., Deb, M., and Bhattacharya, S., 2013. Herbal toothpaste-A possible remedy for oral cancer. *Journal of Natural Products*, 6, pp. 44–55.

Cichorek, M., Wachulska, M., Stasiewicz, A., and Tyminska, A., 2013. Skin melanocytes: Biology and development. *Advances in Dermatology and Allergology*, 30, pp. 30–41.

Coelho, J.F., Ferreira, P.C., Alves, P., Cordeiro, R., Fonseca, A.C., Góis, J.R. and Gil, M.H., 2010. Drug delivery systems: Advanced technologies potentially applicable in personalized treatments. *EPMA Journal*, 1, pp. 164–209.

Costa, R., and Santos, L., 2017. Delivery systems for cosmetics–From manufacturing to the skin of natural antioxidants. *Powder Technology*, 322, pp. 402–416.

Damalas, C.A., 2011. Potential uses of turmeric (*curcuma longa*) products as alternative means of pest management in crop production. *Plant Omics*, 4, pp. 136–141.

Damle, M., and Mallya, S., 2016. Development and evaluation of a novel delivery system containing phytophospholipid complex for skin aging. *AAPS PharmSciTech*, 17, pp. 607–617.

Davis-Searles, P.R., Nakanishi, Y., Kim, N.C., Graf, T.N., Oberlies, N.H., Wani, M.C., Wall, M.E., Agarwal, R. and Kroll, D.J., 2005. Milk thistle and prostate cancer: Differential effects of pure flavonolignans from *Silybum marianum* on antiproliferative end points in human prostate carcinoma cells. *Cancer Research*, 65, pp. 4448–4457.

Deep, C., and Saraf, S., 2008. Novel approaches in herbal cosmetics. *Journal of Cosmetic Dermatology*, 7, pp. 89–95.

Demeule, M., Brossard, M., Pagé, M., Gingras, D. and Béliveau, R., 2000. Matrix metalloproteinase inhibition by green tea catechins. *Biochimica et Biophysica Acta*, 1478, pp. 51–60.

Demir, B., Barlas, F.B., Guler, E., Gumus, P.Z., Can, M., Yavuz, M., Coskunol, H. and Timur, S., 2014. Gold nanoparticle loaded phytosomal systems: Synthesis, characterization and in vitro investigations. *RSC Advances*, 4, pp. 34687–34695.

Detoni, C.B., Souto, G.D., da Silva, A.L.M., Pohlmann, A.R. and Guterres, S.S., 2012. Photostability and skin penetration of different E-resveratrol-loaded supramolecular structures. *Photochemistry and Photobiology*, 88, pp. 913–921.

Dey, T.B., Chakraborty, S., Jain, K.K., Sharma, A. and Kuhad, R.C., 2016. Antioxidant phenolics and their microbial production by submerged and solid state fermentation process: A review. *Trends in Food Science & Technology*, 53, pp. 60–74.

Dhanalakshmi, S., Mallikarjuna, G.U., Singh, R.P. and Agarwal, R., 2004. Silibinin prevents ultraviolet radiation-caused skin damages in SKH-1 hairless mice via a decrease in thymine dimer positive cells and an up-regulation of p53-p21/Cip1 in epidermis. *Carcinogenesis*, 25, pp. 1459–1465.

Dias, M.I., Sousa, M.J., Alves, R.C. and Ferreira, I.C., 2016. Exploring plant tissue culture to improve the production of phenolic compounds: A review. *Industrial Crops and Products*, 82, pp. 9–22.

Diwakar, G., Rana, J., Saito, L., Vredeveld, D., Zemaitis, D. and Scholten, J., 2014. Inhibitory effect of a novel combination of *Salvia hispanica* L. (Chia) seed and *Punica granatum* L. (pomegranate) fruit extracts on melanin. *Fitoterapia*, 97, pp. 164–171.

Dixit, S.N., Srivastava, H.S. and Tripathi, R.D., 1980. Lawsone, The antifungal antibiotic from leaves of *lawsonia inermis* and some aspects of its mode of action. *Indian Phytopathology*, 31, pp. 131–133.

Domínguez-Villegas, V., Clares-Naveros, B., García-López, M.L., Calpena-Campmany, A.C., Bustos-Zagal, P. and Garduño-Ramírez, M.L., 2014. Development and characterization of two nano-structured systems for topical application of flavanones isolated from *Eysenhardtia platycarpa*. *Colloids and Surfaces B: Biointerfaces*, 116, pp. 183–192.

Dooley, T.P., 1997. Is there room for a moderate regulatory oversight? In Hori, W. (Ed.), *Drug Discovery Approaches for Developing Cosmeceuticals*: *Advanced Skincare and Cosmetics Products*. Southborough, MA, IBC Library Series.

Draelos, Z.D., 1997. Sensitive skin, pp. Perceptions, evaluation, and treatment. *American Journal of Contact Dermatitis*, 8, pp. 67–78.

Draelos, Z.D., 2001. Botanicals as topical agents. *Clinics in Dermatology*, 19, pp. 474–477.

Draelos, Z.D., 2010. Nutrition and enhancing youthful-appearing skin. *Clinics in Dermatology*, 28, pp. 400–408.

Draelos, Z.D. 2015. *Cosmeceuticals*. Elsevier, pp. 267–268.

Draelos, Z.D., 2017. Cosmeceuticals for rosacea. *Clinics in Dermatology*, 35, pp. 213–217.

Ekanayake-Mudiyanselage, S., Tavakkol, A., Polefka, T.G., Nabi, Z., Elsner, P. and Thiele, J.J., 2005. Vitamin E delivery to human skin by a rinse-off product, pp. penetration of α-tocopherol versus wash-out effects of skin surface lipids. *Skin Pharmacology and Physiology*, 18, pp. 20–26.

Elsner, P., and Maibach, H.I., 2000. *Cosmeceuticals: Drugs vs Cosmetics*. New York, Marcel Dekker Publications.

Embuscado, M.E., 2015. Spices and herbs: Natural sources of antioxidants–A mini review. *Journal of Functional Foods*, 18, pp. 811–819.

Englard, S., and Seifter, S., 1986. The biochemical function of ascorbic acid. *Annual Review of Nutrition*, 6, pp. 365–406.

Evans, W.C., 2006. *Trease and Evance Pharmacognosy*. NewDelhi, India, Rajkamal Electric Press, p. 466.

Feng, S., Luo, Z., Tao, B., and Chen, C., 2015. Ultrasonic-assisted extraction and purification of phenolic compounds from sugarcanes (*Saccharum officinarum L.*) rinds. *LWT-Food Science and Technology*, 60, pp. 970–976.

Filho, C.R.M.S., Souza, A.G., Conceição, M.M., Silva, T.G., Silva, T.M.S., and Ribeiro, A.P.L., 2009. Avaliação da bioatividade dos extratos de cúrcuma (*Curcuma longa L Zingiberaceae*) em *Artemia salina* e *Biomphalaria glabrata*. *Revista Brasileira de Farmacognosia*, 19, pp. 919–923.

Fisher, G.J., 2005. The pathophysiology of photoaging of the skin. *Cutis*, 75, pp. 5–8.

Fisher, G.J., Kang, S., Varani, J., Bata-Csorgo, Z., Wan, Y., Datta, S. and Voorhees, J.J., 2002. Mechanisms of photoaging and chronological skin aging. *Archives of Dermatology*, 138, pp. 1462–1470.

Friedmann, P.S., and Gilchrest, B.A., 1987. Ultraviolet radiation directly induces pigment production by cultured human melanocytes. *Journal of Cellular Physiology*, 133, pp. 88–94.

Fujiyama, Y.F., 1992. Effects of sesamin and curcumin on delta 5-desaturation and chain elongation of polyunsaturated fatty acid metabolism in primary cultured fatty acid metabolism in primary cultured rat hepatocytes. *Journal of Nutritional Science and Vitaminology*, 38, pp. 353–363.

Furmanowa, M., Skopinska, R.E., Rogala, E., and Malgorzata, H., 1998. *Rhodiola rosea* in vitro culture, pp. phytochemical analysis and antioxidant action. *Acta Societis Botanicorum Poloniae*, 67, pp. 69–73.

Gallarate, M., Carlotti, M.E., Trotta, M., and Bovo, S., 1999. On the stability of ascorbic acid in emulsified systems fortopical and cosmetic use. *International Journal of Pharmaceutics*, 188, pp. 233–241.

Gallarate, M., Carlotti, M.E., Trotta, M., Grande, A.E. and Talarico, C., 2004. Photostability of naturally occurring whitening agents in cosmetic microemulsions. *Journal of Cosmetic Science*, 55, pp. 139–148.

Ganesan, P., and Choi, D.-K., 2016. Current application of phytocompound-based nanocosmeceuticals for beauty and skin therapy. *International Journal of Nanomedicine*, 11, pp. 1987–2007.

Gaspar, L.R., and Campos, P.M.B.G.M., 2007. Photostability and efficacy studies of topical formulations containing UV-filters combination and vitamins A, C and E, *International Journal of Pharmaceutics*, 343, pp. 181–189.

Gediya, S.K., Mistry, R.B., Patel, U.K., Blessy, M., and Jain, H.N., 2011. Herbal plants, pp. used as cosmetics. *Journal of Natural Product and Plant Resources*, 1, pp. 24–32.

Giovanini, F., 2006. Cosmeceuticals come of age. *Household and Personal Care Today*, pp. 54–56.

Gokce, E.H., Korkmaz, E., Dellera, E., Sandri, G., Bonferoni, M.C., and Ozer, O., 2012. Resveratrol-loaded solid lipid nanoparticles versus nanostructured lipid carriers: Evaluation of antioxidant potential for dermal applications. *International Journal of Nanomedicine*, 7, pp. 1841–1850.

Gökçe, E.H., Yapar, E.A., Tuncay, T.S., and Özer, Ö., 2016. Nanocarriers in cosmetology. In, Grumezescu, A. (Ed.), *Nanobiomaterials in Galenic Formulations and Cosmetics: Applications of NanoBioMaterials*. London, UK, Elsevier, pp. 363–393.

Golhani, D., Pandey, V., Shukla, A., and Shukla, R., 2015. Formulation and comparative evaluation of herbal shampoo with marketed products. *Mintage Journal of Pharmaceutical and Medical Sciences*, 4, pp. 3–6.

Gonçalves, G.M.S., da Silva, G.H., Barros, P.P., Srebernich, S.M., Shiraishi, C.T.C., de Camargos, V.R., and Lasca, T.B., 2014. Use of *Curcuma longa* in cosmetics: Extraction of curcuminoid pigments, development of formulations, and in vitro skin permeation studies. *Brazilian Journal of Pharmaceutical Sciences*, 50, pp. 885–893.

Griffiths, H.R., Mistry, P., Herbert, K.E., and Lunec, J., 1998. Molecular and cellular effects of ultraviolet light-induced genotoxicity. *Critical Reviews in Clinical Laboratory Sciences*, 35, pp. 189–237.

Gu, M., Dhanalakshmi, S., Singh, R.P., and Agarwal, R., 2005. Dietary feeding of silibinin prevents early biomarkers of UVB radiation-induced carcinogenesis in SKH-1 hairless mouse epidermis. *Cancer Epidemiology, Biomarkers & Prevention*, 14, pp. 1344–1349.

Gupta, A., Mishra, A.K., Bansal, P., Singh, R., Kumar, S., and Gupta, V., 2010. Phytochemistry and pharmacological activities of Haritaki-A review. *Journal of Pharmacy Research*, 3, pp. 417–424.

Gupta, V.K., and Sharma, S.K., 2006. Plants as natural antioxidants-A review. *Natural Product Radiance*, 5, pp. 326–334.

Halith, S.M., Abirami, A., Jayaprakash, S., Karthikeyini, C., Pillai, K.K., and Mohamed, P.U., 2009. Firthouse Effect of *Ocimum sanctum* and *Azadiracta indica* on the formulation of antidandruff herbal shampoo powder. *Der Pharmacia Lettre*, 1, pp. 68–76.

Halliday, G.M., 2005. Inflammation, gene mutation and photoimmunosuppression in response to UVR-induced oxidative damage contributes to photocarcinogenesis. *Mutation Research*, 571, pp. 107–120.

Hamzah, M.M., 2011. Evaluation of topical preparations containing curcuma, acacia and lupinus extracts as an anti-inflammatory drugs. *International Journal of Applied Research in Natural Products*, 4, pp. 19–23.

Hearing, V.J., and Tsukamoto, K., 1991. Enzymatic control of pigmentation in mammals. *THE FASEB Journal*, 5, pp. 2902–2909.

Henry, M., 2018. Cosmetic concerns among ethnic men. *Dermatologic Clinics*, 36, pp. 11–16.

Hidalgo, M.E., Gonzalez, I., Toro, F., Fernandez, E., Speisky, H., and Jimenez, I., 1998. Boldine as a sunscreen, its photoprotector capacity against UVB radiation. *Cosmetics & Toiletries*, 113, pp. 59–66.

Holbrook, K.A., and Odland, G.F., 1974. Regional differences in the thickness (cell layers) of the human stratum corneum: An ultrastructural analysis. *Journal of Investigative Dermatology*, 62, pp. 415–422.

Holister, P., Cristina, R.V., and Fullerenes, H.T., 2003. Nanoparticles. *Technology White Papers. Cientifica*, pp. 1–12.

Hussain, S.H., Limthongkul, B., and Humphreys, T.R., 2013. The biomechanical properties of the skin. *Dermatologic Surgery*, 39, pp. 193–203.

Ichihashi, M., Ueda, M., Budiyanto, A., Bito, T., Oka, M., Fukunaga, M., Tsuru, K., and Horikawa, T., 2003. UV-induced skin damage. *Toxicology*, 189, pp. 21–39.

Ito, S., Itoga, K., Yamato, M., Akamatsu, H., and Okano, T., 2010. The co-application effects of fullerene and ascorbic acid on UV-B irradiated mouse skin. *Toxicology*, 267, pp. 27–38.

Jadhav, N., Patel, V., Mungekar, S., Bhamare, G., Karpe, M., and Kadams, V., 2013. Microsponge delivery system: An updated review, current status and future prospects. *Journal of Scientific and Innovative Research*, 2, pp. 1097–1110.

Jaiswal, R.R., Kiprotich, J., and Kuhnert, N., 2011. Determination of the hydroxycinnamate profile of 12 members of the Asteraceae Family. *Phytochemistry*, 72, pp. 781–790.

Jenning, V., Schäfer-Korting, M., and Gohla, S., 2000. Vitamin A-loaded solid lipid nanoparticles for topical use: Drug release properties. *Journal of Controlled Release*, 66, pp. 115–126.

Jeon, S., Yoo, C.Y., and Park, S.N., 2015. Improved stability and skin permeability of sodium hyaluronate-chitosan multilayered liposomes by layer-by-layer electrostatic deposition for quercetin delivery. *Colloids Surface B: Biointerfaces*, 129, pp. 7–14.

Jepps, O.G., Dancik, Y., Anissimov, Y.G., and Roberts, M.S., 2013. Modeling the human skin barrier-Towards a better understanding of dermal absorption. *Advanced Drug Delivery Reviews*, 65, pp. 152–168.

Joshi, L.S., and Pawar, H.A., 2015. Herbal cosmetics and cosmeceuticals: An overview. *Natural Products Chemistry and Research*, 3, p. 170.

Jun, H.J., Lee, J.H., Cho, B.R., Seo, W.D., Kim, D.W., Cho, K.J. and Lee, S.J., 2012. p-Coumaric acid inhibition of CREB phosphorylation reduces cellular melanogenesis. *European Food Research and Technology*, 235, pp. 1207–1211.

Jung, S., Otberg, N., Thiede, G., Richter, H., Sterry, W., Panzner, S. and Lademann, J., 2006. Innovative liposomes as a transfollicular drug delivery system: Penetration into porcine hair follicles. *Journal of Investigative Dermatology*, 126, pp. 1728–1732.

Jurkiewicz, B.A., Bissett, D.L., and Buettner, G.R., 1995. Effect of topically applied tocopherolbon on ultraviolet radiationmediated free radical damage in skin. *Journal of Investigative Dermatology*, 104, pp. 484–488.

Kamal-Eldin, A., and Appelqvist, L.A., 1996. The chemistry and antioxidant properties of tocopheols and tocotrienols. *Lipids*, 31, pp. 671–701.

Karthikeyan, R., Kanimozhi, G., Prasad, N.R., Agilan, B., Ganesan, M., Mohana, S. and Srithar, G., 2016. 7-Hydroxycoumarin prevents UVB-induced activation of NF-κB and subsequent overexpression of matrix metalloproteinases and inflammatory markers in human dermal fibroblast cells. *Journal of Photochemistry and Photobiology B: Biology*, 161, pp. 170–176.

Katiyar, S.K., 2003. Skin photoprotection by green tea: Antioxidant and immunomodulatory effects. *Current Drug Targets-Immune, Endocrine and Metabolic Disorders*, 3, pp. 234–242.

Katiyar, S.K., and Elmets, C.A., 2000. Green tea and skin. *Archives of Dermatology*, 136, pp. 989–994.

Katiyar, S.K., Elmets, C.A., Agarwal, R., and Mukhtar, H., 1995. Protection against ultraviolet-B radiation-induced local and systemic suppression of contact hypersensitivity and edema responses in C3 H/HeN mice by green tea polyphenols. *Photochemistry and Photobiology*, 62, pp. 855–861.

Khameneh, B., Halimi, V., Jaafari, M.R., and Golmohammadzadeh, S., 2015. Safranal-loaded solid lipid nanoparticles: Evaluation of sunscreen and moisturizing potential for topical applications. *Iranian Journal of Basic Medical Sciences*, 18, pp. 58–63.

Khanpara, K., Renuka, V., Shukla, J., and Harsha, C.R., 2012. A Detailed investigation of shikakai (*Acacia concinna* Linn) fruit. *Journal of Current Pharma Research*, 9, pp. 6–10.

Kidd, P.M., 2009. Bioavailability and activity of phytosome complexes from botanical polyphenols: The silymarin, curcumin, green tea, and grape seed extracts. *Alternative Medicine Review*, 14, pp. 226–246.

Kijlstra, A., Tian, Y., Kelly, R.E., and Berendschot, T.J.M.T., 2012. Lutein: More than just a filter for blue light. *Progress in Retinal and Eye Research*, 31, pp. 303–315.

Kim, J., Hwang, J.S., Cho, Y.K., Han, Y., Jeon, Y.H., and Yang, K.H., 2001. Protective effects of (-)-epigallocatechin-3-gallate on UVA- and UVB-induced skin damage. *Skin Pharmacology and Applied Skin Physiology*, 14, pp. 11–19.

Koizumi, A., Fujii, M., Kondoh, M., and Watanabe, Y., 2004. Effect of Nmethyl-2-pyrrolidone on skin permeation of estradiol. *European Journal of Pharmaceutics and Biopharmaceutics*, 57, pp. 473–478.

Kokate, C.K., Purohi, A.P., and Gokhale, S.B., 2008a. *Pharmacognosy.* Pune, India: Nirali Prakashan, 11.97.

Kokate, C.K., Purohi, A.P., and Gokhale, S.B., 2008b. *Pharmacognosy.* Pune, India: Nirali Prakashan, 16.14.

Koul, I.B., and Kapil, A., 1993. Evaluation of the liver protective potential of piperine. *Planta Medica*, 59, pp. 413–417.

Koul, I.B., Kapil, A., Barthakur, M.N.N., and Arnold, N.P., 1993. Evaluation of the liver protective potential of piperine, an active principle of black and long peppers. *Planta Medica*, 59, pp. 413–417.

Kramer, K.A., and Liebler, D.C., 1997. UVB induced photooxidation of vitamin E. *Chemical Research in Toxicology*, 10, pp. 219–224.

Krasodomska, O., and Jungnickel, C., 2015. Viability of fruit seed oil O/W emulsions in personal care products. *Colloids and Surfaces A: Physicochemical and Engineering Aspects*, 481, pp. 468–475.

Krausz, A., Gunn, H., and Friedman, A., 2014. The basic science of natural ingredients. *Journal of Drugs in Dermatology*, 13, pp. 937–943.

Kumar, N., and Pruthi, V., 2014. Potential applications of ferulic acid from natural sources. *Biotechnology Reports*, 4, pp. 86–93.

Kumar, S., Swarankar, V., Sharma, S., and Baldi, A., 2012. Herbal cosmetics: Used for skin and hair. *Inventi Rapid*, 4, pp. 1–7.

Kuršvietienė, L., I. Stanevičienė, A. Mongirdienė and J. Bernatonienė. 2016. Multiplicity of effects and health benefits of resveratrol. *Medicina*, 52, pp. 148–55.

Lee, D.S., Woo, J.Y., Ahn, C.B., and Je, J.Y., 2014. Chitosan-hydroxycinnamic acid conjugates: Preparation, antioxidant and antimicrobial activity. *Foog Chemistry*, 148, pp. 97–104.

Lephart, E.D., 2016. Skin aging and oxidative stress: Equol's anti-aging effects via biochemical and molecular mechanisms. *Ageing Research Reviews*, 31, pp. 36–54.

Lephart, E.D., 2017. Resveratrol, 40 Acetoxy Resveratrol, R-equol, Racemic Equol or S-equol as cosmeceuticals to improve dermal health. *International Journal of Molecular Sciences*, 18, 1193, pp. 1–21.

Li, H., Zhang, N., Hao, Y., Wang, Y., Jia, S., Zhang, H., Zhang, Y. and Zhang, Z., 2014. Formulation of curcumin delivery with functionalized single-walled carbon nanotubes: Characteristics and anticancer effects in vitro. *Drug Delivery*, 21, pp. 379–387.

Lin, L., Dong, Y., Zhao, H., Wen, L., Yang, B. and Zhao, M., 2011. Comparative evaluation of rosmarinic acid, methyl rosmarinate and pedalitin isolated from *Rabdosia serra (MAXIM.)* HARA as inhibitors of tyrosinase and α-glucosidase. *Food Chemistry*, 129, pp. 884–889.

Lin, T.J., 2010. Evolution of cosmetics: Increased need for experimental clinical medicine. *Journal of Experimental & Clinical Medicine*, 2, pp. 49–52.

Lin, Y.K., Al-Suwayeh, S.A., Leu, Y.L., Shen, F.M. and Fang, J.Y., 2013. Squalene containing nanostructured lipid carriers promote percutaneous absorption and hair follicle targeting of diphencyprone for treating alopecia areata. *Pharmaceutical Research*, 30, pp. 435–446.

Lintner, K., Mas-Chamberlin, C., Mondon, P., Peschard, O. and Lamy, L., 2009. Cosmeceuticals and active ingredients. *Clinics in Dermatology*, 27, pp. 461–468.

Liu, C.H., and Huang, H.Y., 2013. In vitro anti-propionibacterium activity by curcumin containing vesicle system. *Chemical and Pharmaceutical Bulletin*, 61, pp. 419–425.

Loggia, R.D., Sosa, S., Tubaro, A., Morazzoni, P., Bombardelli, E. and Griffini, A., 1996. Anti-inflammatory activity of some *Gingko biloba* constituents and of their phospholipids-complexes. *Fitoterapia*, 3, pp. 257–273.

Longhini, R., Lonni, A.A., Sereia, A.L., Krzyzaniak, L.M., Lopes, G.C. and Mello, J.C.P.D., 2017. *Trichilia catigua*: Therapeutic and cosmetic values. *Revista Brasileira de Farmacognosia*, 27, pp. 254–271.

Łopaciuk, A., and Łoboda, M., 2013. Global beauty industry trends in the 21st century. In *Management, Knowledge and Learning International Conference*, pp. 19–21.

Lopes, D.M., and Mcmahon, S.B., 2016. Ultraviolet radiation on the skin: A painful experience? *CNS Neuroscience & Therapeutics*, 22, pp. 118–126.

Lovell, C.R., Smolenski, K.A., Duance, V.C., Light, N.D., Young, S. and Dyson, M., 1987. Type I and III collagen content and fibre distribution in normal human skin during aging. *British Journal of Dermatology*, 117, pp. 419–428.

Madhan, B., Krishnamoorthy, G., Rao, J.R. and Nair, B.U., 2007. Role of green tea polyphenols in the inhibition of collagenolytic activity by collagenase. *International Journal of Biological Macromolecules*, 41, pp. 16–22.

Makimura, M., Hirasawa, M., Kobayashi, K., Indo, J., Sakanaka, S., Taguchi, T. and Otake, S., 1993. Inhibitory effect of tea catechins on collagenase activity. *Journal of Periodontology*, 64, pp. 630–636.

Manca, M.L., Castangia, I., Zaru, M., Nácher, A., Valenti, D., Fernàndez-Busquets, X., Fadda, A.M. and Manconi, M., 2015. Development of curcumin loaded sodium hyaluronate immobilized vesicles (hyalurosomes) and their potential on skin inflammation and wound restoring. *Biomaterial*, 71, pp. 100–109.

Manca, M.L., Castangia, I., Caddeo, C., Pando, D., Escribano, E., Valenti, D., Lampis, S., Zaru, M., Fadda, A.M. and Manconi, M., 2014. Improvement of quercetin protective effect against oxidative stress skin damages by incorporation in nanovesicles. *Colloids Surfaces B Biointerface*, 123, pp. 566–574.

Marianecci, C., Di Marzio, L., Rinaldi, F., Celia, C., Paolino, D., Alhaique, F., Esposito, S. and Carafa, M., 2014. Niosomes from 80s to present: The state of the art. *Advances in Colloid and Interface Science*, 205, pp. 187–206,

Marineli, R.D.S., Moraes, É.A., Lenquiste, S.A., Godoy, A.T., Eberlin, M.N. and Maróstica Jr, M.R., 2014. Chemical characterization and antioxidant potential of Chilean chia seeds and oil (*Salvia hispanica L.*). *LWT- Food Science and Technology*, 59, pp. 1304–1310.

Martin, K., Sur, R., Liebel, F., Tierney, N., Lyte, P., Garay, M., Oddos, T., Anthonavage, M., Shapiro, S. and Southall, M., 2008. Parthenolide-depleted feverfew (*Tanacetum parthenium*) protects skin from UV irradiation and external aggression. *Archives of Dermatology Research*, 300, pp. 69–80.

Martins, M., Azoia, N.G., Ribeiro, A., Shimanovich, U., Silva, C. and Cavaco-Paulo, A., 2013. In vitro and computational studies of transdermal perfusion of nanoformulations containing a large molecular weight protein. *Colloids Surfaces B: Biointerfaces*, 108, pp. 271–278.

Mateos-Martín, M.L., Fuguet, E., Quero, C., Pérez-Jiménez, J. and Torres, J.L., 2012. New identification of proanthocyanidins in cinnamon (*Cinnamomum zeylanicum L.*) using MALDI-TOF/TOF mass spectrometry". *Analytical and Bioanalytical Chemistry*, 402, pp. 1327–1336.

Millikan, LE., 2001. *Cosmetology, Cosmetics, Cosmeceuticals: Definitions and Regulations*. New York, Elsevier Science, pp. 371–374.

Ming, L.C., Ang, W.C., Yang, Q., Thitilertdecha, P., Wong, T.W. and Khan, T.M., 2017. Cosmeceuticals: Safety, efficacy and potential benefits. In Keservani, R.K., Sharma, A.K., Kesharwani, R.K. (Eds.), *The Recent Advances in Drug Delivery Technology*. Hershey PA, IGI Global, pp. 287–288.

Mishra, A.K., Mishraand, A., and Chattopadhyay, P., 2011. Herbal cosmeceuticals for the photoprotection from Ultraviolet B radiation: A review. *Tropical Journal of Pharmaceutical Research*, 10, pp. 351–360.

Mishra, B., Patel, B.B., and Tiwari, S., 2010. Colloidal nanocarriers: A review on formulation technology, types and applications toward targeted drug delivery. *Nanomedicine: Nanotechnology, Biology and Medicine*, 6, pp. 9–24.

Mitri, K., Shegokar, R., Gohla, S., Anselmi, C. and Müller, R.H., 2011. Lipid nanocarriers for dermal delivery of lutein: Preparation, characterization, stability and performance. *International Journal of Pharmaceutics*, 414, pp. 267–275.

Moulaoui, K., Caddeo, C., Manca, M.L., Castangia, I., Valenti, D., Escribano, E., Atmani, D., Fadda, A.M. and Manconi, M., 2015. Identification and nanoentrapment of polyphenolic phytocomplex from *Fraxinus angustifolia*: In vitro and in vivo wound healing potential. *European Journal of Medicinal Chemistry*, 89, pp. 179–188.

Mudshinge, S.R., Deore, A.B., Patil, S. and Bhalgat, C.M., 2011. Nanoparticles: Emerging carriers for drug delivery. *Saudi Pharmaceutical Journal*, 19, pp. 129–141.

Mukhtar, H., and Ahmad, N., 2000. Tea polyphenols: Prevention of cancer and optimizing health. *The American Journal of Clinical Nutrition*, 71, pp. 1698S–1702S; discussion 1703S–1694S.

Mulder, W.M.C., and Meinardi, M.M.H.M., 2002. Dermatological drugs and topical agents. *Side Effects of Drugs Annual*, 25, pp. 175–182.

Muley, B.P., Khadabadi, S.S., and Banaase, N.B., 2009. Phytochemical constituents and pharmacological activities of *calendula officinalis Linn (Asteraceae)*: A review. *Tropical Journal of Pharmaceutical Research*, 8, pp. 455–465.

Mueller, G., Saloga, J., Germann, T., Schuler, G., Knop, J. and Enk, A.H., 1995. IL-12 as mediator and adjuvant for the induction of contact sensitivity in vivo. *The Journal of Immunology*, 155, pp. 4661–4668.

Munhoz, V.M., Lonni, A.A.S.G., Mello, J.C.P.D. and Lopes, G.C., 2012. Avaliacão do fator de protecão solar em fotoprotetores acrescidos com extratos da flora brasileira ricos em substâncias fenólicas. *Revista de Ciências Farmacêuticas Básica e Aplicada*, 33, pp. 225–232.

Murata, T., Miyase, T., and Yoshizaki, F., 2011. Hyaluronidase inhibitory rosmarinic acid derivatives from *Meehania urticifolia*. *Chemical and Pharmaceutical Bulletin*, 59, pp. 88–95.

Neudecker, B.A., Maibach, H.I., and Stern, R., 2005. Hyaluronan: The Natural Skin Moisturizer. In Elsner, P., Maibach, H.I. (Eds.), *Cosmeceuticals and Active Cosmetics: Drugs Versus Cosmetics, Cosmetics Science and Technology Series*. Boca Raton, FL, Taylor & Francis Group.

Niemiec, S.M., Ramachandran, C., and Weiner, N., 1995. Influence of nonionic liposomal composition on topical delivery of peptide drugs into pilosebaceous units: An in vivo study using the hamster ear model. *Pharmaceutical Research*, 8, pp. 1184–1188.

No, J.K., Kim, Y.J., Shim, K.H., Jun, Y.S., Rhee, S.H., Yokozawa, T. and Chung, H.Y., 1999. Inhibition of tyrosinase by green tea components. *Life Sciences*, 65, pp. PL241–246.

Okochi, H., and Nakano, M., 2000. Preparation and evaluation of w/o/w type emulsions containing vancomycin. *Advanced Drug Delivery Reviews*, 45, pp. 5–26.

Okoh, O.O., Sadimenko, A.P., Asekeen, O.T., and Afolayan, A.J., 2008. The effects ofDrying on the chemical components of Essential oils of *Caledula officinalis L. African Journal of Biotechnology*, 7, pp. 1500–1502.

Okonogi, S., and Riangjanapatee, P., 2015. Physicochemical characterization of lycopene-loaded nanostructured lipid carrier formulations for topical administration. *International Journal of Pharmaceutics*, 478, pp. 726–735.

Oliveira, K.B., Palú, É., Weffort-Santos, A.M. and Oliveira, B.H., 2013. Influence of rosmarinic acid and *Salvia officinalis* extracts on melanogenesis of B16f10 cells. *Brazilian Journal of Pharmacognsosy*, 23, pp. 249–258.

Oroian, M., and Escriche, I., 2015. Antioxidants: Characterization, natural sources, extraction and analysis. *Food Research International*, 74, pp. 10–36.

Pande, A., and Majeed, M., 2015. "Multi-Functional Botanicals For Topical Applications." in. Rosen, M.R. (ed.). *Harry's Cosmeticology*, pp. 654–690.

Pandey, S., Meshya, N., and Viral, D., 2010. Herbs play an important role in the field of cosmetics. *International Journal of PharmTech Research*, 2, pp. 632–639.

Parhi, R., Suresh, P., and Patnaik S., 2015. Physical means of stratum corneum barrier manipulation to enhance transdermal drug delivery. *Current Drug Delivery*, 12, pp. 122–138.

Park, E.J., Kim, J.-Y., Jeong, M.S., Park, K.Y., Park, K.H., Lee, M.W., Joo, S.S., and Seo, S.J., 2015. Effect of topical application of quercetin-3-O-(2″-gallate)-α-l-rhamnopyranoside on atopic dermatitis in NC/Nga mice. *Journal of Dermatological Science*, 77, pp. 166–172.

Patkar, K.B., 2008. Herbal cosmetics in ancient India. *Indian Journal of Plastic Surgery: Official Publication of the Association of Plastic Surgeons of India*, 41 (Suppl.), pp. S134–S137.

Patravale, V.B., and Mandawgade, S.D., 2008. Novel cosmetic delivery systems: An application update. *International Journal of Cosmetic Science*, 30, pp. 19–33.

Paye, M., Barel, A.O., and Maibach, H.I., 2007. *Handbook of Cosmetic Science and Technology*. New York, Informa Healthcare, p. 313.

Pelle, E., Muizzuddin, N., Mammone, T., Marenus, K., and Maes, D., 1999. Protection against endogenous and UVB-induced oxidative damage in stratum corneum lipids by an antioxidant-containing cosmetic formulation. *Photodermatology, Photoimmunology & Photomedicine*, 15, p. 115.

Peter, O.B., Catalina, C.P., and Herman, S., 2006. Boldine and its antioxidant or health promoting properties. *Chemico-Biological Interactions*, 159, pp. 1–17.

Phetdee, K., Rakchai, R., Rattnamanee, K., Teaktong, T., and Viyoch, J., 2014. Preventive effects of tamarind seed coat extract on UVA induced alterations in human skin fibroblasts. *Journal of Cosmetic Science*, 65, pp. 11–24.

Pillai, S., Oresajo, C., and Hayward, J., 2005. Ultraviolet radiation and skin aging: Roles of reactive oxygen species, inflammation and protease activation, and strategies for prevention of inflammation induced matrix degradation- A review. *International Journal of Cosmetic Science*, 27, pp. 17–34.

Pornpattananangkul, D., Fu, V., Thamphiwatana, S., Zhang, L., Chen, M., Vecchio, J., Gao, W., Huang, C.M. and Zhang, L., 2013. In vivo treatment of Propionibacterium acnes infection with liposomal lauric acids. *Advanced Healthcare Materials*, 2, pp. 1322–1328.

Prakash, Y.G., Ilango, K., Kumar, S. and Elumalai, A., 2009. In vitro antioxidant activity of *Luffa cylindrical* seed oil. *Journal of Global Pharma Technology*, 2, pp. 93–97.

Puglia, C., Offerta, A., Tirendi, G.G., Tarico, M.S., Curreri, S., Bonina, F. and Perrotta, R.E., 2016. Design of solid lipid nanoparticles for caffeine topical administration. *Drug Delivery*, 23, pp. 36–40.

Rabasco A.A.M., and Gonzalez, R.M.L., 2000. Lipids in pharmaceutical and cosmetic preparations. *Grasas y Aceites*, 51, pp. 74–96.

Rabe, J.H., Mamelak, A.J., McElgunn, P.J., Morison, W.L. and Sauder, D.N., 2006. Photoaging: Mechanism and repair. *Journal of the American Academy of Dermatology*, 55, pp. 1–19.

Ramachandra, C.T., and Ramachandra, P., 2008. Processing of Aloe vera leaf Gel: A review. *American Journal of Agricultural and Biological Sciences*, 3, pp. 502–510.

Ramli, N.S., 2015. Immigrant entrepreneurs on the world's successful global brands in the cosmetic industry. *Procedia - Social and Behavioral Sciences*, 195, pp. 113–122.

Rancan, F., Rosan, S., Boehm, K., Fernández, E., Hidalgo, M.E., Quihot, W., Rubio, C., Boehm, F., Piazena, H. and Oltmanns, U., 2002. Protection against UVB irradiation by natural filters extracted from lichens. *Journal of Photochemistry and Photobiology*, 68, pp. 133–139.

Raska, I., and Toropov, A., 2006. Comparison of QSPR models of octanol/water partition coefficient for vitamins and non-vitamins. *European Journal of Medicinal Chemistry*, 41, pp. 1271–1278.

Raynal, S., Grossiord, J.L., Seiller, M. and Clausse, D., 1993. A topical w/o/ w multiple emulsion containing several active substances: Formulation, characterization and study of release. *Journal of Controlled Release*, 26, pp. 129–140.

Ribeiro, R.C.D.A., Barreto, S.M.A.G., Ostrosky, E.A., Rocha-Filho, P.A.D., Veríssimo, L.M. and Ferrari, M., 2015. Production and characterization of cosmetic nanoemulsions containing *Opuntia ficus-indica* (L.) mill extract as moisturizing agent. *Molecules*, 20, pp. 2492–2509.

Rivelli, D.P., Filho, C.A., Almeida, R.L., Ropke, C.D., Sawada, T.C. and Barros, S.B., 2010. Chlorogenic acid UVA–UVB photostability. *Photochemistry and Photobiology*, 86(5), pp. 1005–1007.

Roh, E., Kim, J.E., Kwon, J.Y., Park, J.S., Bode, A.M., Dong, Z. and Lee, K.W., 2017. Molecular mechanisms of green tea polyphenols with protective effects against skin photoaging, *Critical Reviews in Food Science and Nutrition*, 57, pp. 1631–1637.

Rozman, B., Gasperlin, M., Tinois-Tessoneaud, E., Pirot, F. and Falson, F., 2009. Simultaneous absorption of vitamins C and E from topical microemulsions using reconstructed human epidermis as a skin model. *European Journal of Pharmaceutics and Biopharmaceutics*, 72(1), pp. 69–75.

Sadick, N.S., 2018. The pathophysiology of the male aging face and body. *Dermatologic Clinics*, 36, pp. 1–4.

Saikia, A.P., Ryakala, V.K., Sharma, P., Goswami, P. and Bora, U., 2006. Ethnobotany of medicinal plants used by Assamese people for various skin ailments and cosmetics. *Journal of Ethnopharmacology*, 106(2), pp. 149–157.

Saluk-Juszczak, J., Pawlaczyk, I., Olas, B., Kołodziejczyk, J., Ponczek, M., Nowak, P., Tsirigotis-Wołoszczak, M., Wachowicz, B. and Gancarz, R., 2010. The effect of polyphenolic-polysaccharide conjugates from selected medicinal plants of Asteraceae Family: On the peroxynitrite-induced changes in blood platelet proteins. *Journal of Controlled Release*, 47(5), pp. 700–705.

Sanja, S.D., N.R. Sheth, N.K. Patel, D. Patel and B. Patel. 2009. Characterization and evaluation of antioxidant activity of Portulaca oleracea. *International Journal of Biological Macromolecules*, 1, pp. 5–10.

Saraf, S. 2010. Applications of novel drug delivery system for herbal formulations. *Fitoterapia*, 81, pp. 680–689.

Secchi, M., Castellani, V., Collina, E., Mirabella, N. and Sala, S., 2016. Assessing eco-innovations in green chemistry: Life Cycle Assessment (LCA) of a cosmetic product with a bio-based ingredient. *Journal of Cleaner Production*, 129, pp. 269–281.

Shindo, Y., Witt, E., Han, D., Epstein, W. and Packer, L., 1994. Enzymic and non-enzymic antioxidants in epidermis and dermis of human skin. *Journal of Investigative Dermatology*, 102(1), pp. 122–124.

Shuwaili, A.H.A., Rasool, B.K.A. and Abdulrasool, A.A., 2016. Optimization of elastic transfersomes formulations for trans-dermal delivery of pentoxifylline. *European Journal of Pharmaceutics and Biopharmaceutics*, 102, pp. 101–114.

Singh, D., SM Rawat, M., Semalty, A. and Semalty, M., 2012. Rutin-phospholipid complex: An innovative technique in novel drug delivery system-NDDS. *Current Drug Delivery*, 9(3), pp. 305–314.

Singh, R.P. and Agarwal, R., 2009. Cosmeceuticals and silibinin. *Clinics in Dermatology*, 27(5), pp. 479–484.

Singh, S.R., Phurailatpam, A.K. and Senjam, P., 2014. Identification of plant use as natural herbal shampoo in Manipur. *African Journal of Traditional, Complementary and Alternative Medicines*, 11(1), pp. 135–139.

Siriamornpun, S., Kaisoon, O. and Meeso, N., 2012. Changes in colour, antioxidant activities and carotenoids (lycopene, β-carotene, lutein) of marigold flower (Tagetes erecta L.) result-ing from different drying processes. *Journal of Functional Foods*, 4(4), pp. 757–766.

Smeets, J.W., De Pater, R.M. and Lambers, J.W., Gist-Brocades NV, 1997. Enzymatic synthesis of ceramides and hybrid cerami-des. U.S. Patent 5,610,040.

Solimine, J., Garo, E., Wedler, J., Rusanov, K., Fertig, O., Hamburger, M., Atanassov, I. and Butterweck, V., 2016. Tyrosinase inhibi-tory constituents from a polephenol enriched fraction of rose oil distillation wastewater. *Fitoterapia*, 108, pp. 13–19.

Southall, M., Pappas, A., Suhyoun, C. and Earland, R., 2014. Oat (Avena sativa) oil activates the PPAR? and PPAR β/δ path-ways, resulting in keratinocyte differentiation and upregula-tion of ceramide synthesis. *Journal of the American Academy of Dermatology*, 70, pp. AB64 (Sul.1).

Spiess, E., 1996. Raw materials. In Williams, D.F Schmitt, W.H. (Eds.), *Chemistry and Technology of the Cosmetics and Toiletries Industry*. London, UK, Blackie Academic & Professional, pp. 1–34.

Staniforth, V., Huang, W.C., Aravindaram, K. and Yang, N.S., 2012. Ferulic acid, a phenolic phytochemical, inhibits UVB-induced matrix metalloproteinases in mouse skin via posttranslational mechanisms. *The Journal of Nutritional Biochemistry*, 23, pp. 443–451.

Starley, I.F., Mohammed, P., Schneider, G. and Bickler, S.W., 1999. The treatment of paediatric burns using topical papaya. *Burns*, 25, pp. 636–639.

Strube, M. and Dragsted, O.L., 1999. Naturally occuring antitu-mourigens. Iv. Carotenoids except β-carotene. Nordic Council of Ministers.

Surjushe, A., Vasani, R. and Saple, D.G., 2008. Aloe vera: A short review. *Indian Journal of Dermatology*, 53, pp. 163–166.

Suzuki, D., 2010. The "Dirty Dozen" Ingredients Investigated in the David Suzuki Foundation Survey of Chemicals in Cosmetics. *Backgrounder*, pp. 1–19.

Svobodová, A., Psotova, J. and Walterová, D., 2003. Natural pheno-lics in the prevention of UV-induced skin damage: A review. *Biomedical Papers of the Medical Faculty of the University Palacky, Olomouc, Czechoslovakia*, 147, pp. 137–145.

Tabrizi, H., Mortazavi, S.A. and Kamalinejad, M., 2003. An in vitro evaluation of various *Rosa damascena* flower extracts as a natural antisolar agent. *International Journal of Cosmetic Science*, 25, pp. 259–265.

Tagle, J.M., Macchetto, P.C. and Páramo, R.M.D., 2010. Clinical performance of a dermal filler containing natural glycolic acid and a polylactic acid polymer: Results of a clinical trial in human immunodeficiency virus subjects with facial lipoat-rophy. *The Journal of Clinical and Aesthetic Dermatology*, 3, pp. 42–47.

Takshak, S. and Agarwal, S.B., 2015. Defence strategies adopted by the medicinal plant *Coleus forskohlii* against supplemental ultraviolet-B irradiation: Augmentation of secondary metabo-lites and antioxidants. *Plant Physiology and Biochemistry*, 97, pp. 124–138.

Takshak, S. and Agrawal, S.B., 2016. The role of supplemental ultra-violet-B radiation in altering the Metabolite profile, essential oil content and composition, and free radical scavenging activities of *Coleus forskohlii*, an indigenous medicinal plant. *Environmental Science and Pollution Research International*, 23, pp. 7324–7337.

Talbourdet, S., Sadick, N.S., Lazou, K., Bonnet-Duquennoy, M., Kurfurst, R., Neveu, M., Heusèle, C., Andre, P., Schnebert, S., Draelos, Z.D. and Perrier, E., 2007. Modulation of gene expression as a new skin anti-aging strategy. *Journal of Drugs in Dermatology*, 6 (Suppl. 6), pp. S25–S33.

Tang, W., Hioki, H., Harada, K., Kubo, M. and Fukuyama, Y., 2007. Antioxidant phenylpropanoid-substituted epicatechins from *Trichilia catigua*. *Journal of Natural Products*, 70, pp. 2010–2013.

Taofiq, O., González-Paramás, A.M., Martins, A., Barreiro, M.F. and Ferreira, I.C., 2016. Mushrooms extracts and compounds in cosmetics, cosmeceuticals and nutricosmetics-A review. *Industrial Crops and Products*, 90, pp. 38–48.

Tavano, L., Muzzalupo, R., Picci, N. and de Cindio, B., 2014. Co-encapsulation of lipophilic antioxidants into niosomal carri-ers: Percutaneous permeation studies for cosmeceutical applica-tions. *Colloids and Surfaces B: Biointerfaces*, 114, pp. 144–149.

Ternikar, S.G., Alagawadi, K.R., Pasha, I., Dwivedi, S., Rafi, M.A. and Sharma, T., 2010. Evaluation of antimicrobial and acute antiinflammatory activity of *Sida cordifolia* Linn Seed Oil. *Cell and Tissue Research*, 10, pp. 2385–2388.

Teskac, K. and Kristl, J., 2010. The evidence for solid lipid nanoparticles mediated cell uptake of resveratrol. *International Journal of Pharmaceutics*, 390, pp. 61–69.

The Council of The European Communities Official Journal of the European Commission–L series 151. "Directive 93/35/EEC" (online). Retrieved from http://eur lex.europa.eu/legalcontent/EN/NOT/?uri=CELEX:31993L0035&qid=1458440341871 (Accessed March 10, 2018).

Thornfeldt, C., 2005. Cosmeceuticals containing herbs: Fact, fiction, and future. *Dermatologic Surgery*, 31, pp. 873–880.

Tito, A., Bimonte, M., Carola, A., De Lucia, A., Barbulova, A., Tortora, A., Colucci, G. and Apone, F., 2015. An oil-soluble extract of Rubus idaeus cells enhances hydration and water homeostasis in skin cells. *International Journal of Cosmetic Science*, 37, pp. 588–594.

Touitou, E., Dayan, N., Bergelson, L., Godin, B. and Eliaz, M., 2000. Ethosomes- Novel vesicular carriers for enhanced delivery: Characterization and skin penetration properties. *Journal of Controlled Release*, 65, pp. 403–418.

Tripodo, G., Trapani, A., Torre, M.L., Giammona, G., Trapani, G. and Mandracchia, D., 2015. Hyaluronic acid and its derivatives in drug delivery and imaging: Recent advances and challenges. *European Journal of Pharmaceutics and Biopharmaceutics*, 97(Pt B), pp. 400–416.

U.S. Food and Drug Administration Centre for Food Safety and Applied Nutrition. 2002. "CFSAN/Office of Cosmetics and Colors Fact Sheet: Is it a Cosmetic, a drug or both? (or it is Soap?)" (online). Retrieved from http://www.cfsan.fda.gov/dms/cos-218.html (Accessed March 10, 2018).

Unzueta, A. and Vargas, H.E., 2013. Nonsteroidal anti-inflammatory drug-induced hepatoxicity. *Clinical Liver Disease*, 17, pp. 643–656.

Vashisth, P., Kumar, N., Sharma, M. and Pruthi, V., 2015. Biomedical applications of ferulic acid encapsulated electrospun nanofibers. *Biotechnology Reports*, 8, pp. 36–44.

Vhatkar, N., Raut, S., Pore, M., Dhope, S., Foscolo, I. and Mali, A.S., 2018. Design and development of cosmeceutical cream for hyperpigmentation and anti-aging. *International. Journal of Advanced Community Medicine*, 1, pp. 27–31.

Vijayakumar, M.R., Sathali, A.H. and Arun, K., 2010. Formulation and evaluation of diclofenac potassium ethosomes. *International Journal of Pharmaceutics Pharmaceutical Sciences*, 2, pp. 2–6.

Wang, L., Hu, X., Shen, B., Xie, Y., Shen, C., Lu, Y., Qi, J., Yuan, H. and Wu, W., 2015. Enhanced stability of liposomes against solidification stress during freeze-drying and spray-drying by coating with calcium alginate. *Journal of Drug Delivery Science and Technology*, 30, pp. 163–170.

Weber, T.M., Ceilley, R.I., Buerger, A., Kolbe, L., Trookman, N.S., Rizer, R.L. and Schoelermann, A., 2006. Skin tolerance, efficacy, and quality of life of patients with red facial skin using a skin care regimen containing licochalcone A. *Journal of Cosmetic Dermatology*, 5, pp. 227–232.

Wend, K., Wend, P. and Krum, S.A., 2012. Tissue-specific effects of loss of estrogen during menopause and aging. *Frontiers in Endocrinology*, 3, pp. 1–9.

West, D.P. and Zhu, Y.F., 2003. Evaluation of Aloe vera gel gloves in the treatment of dry skin associated with occupational exposure. *American Journal of Infection Control*, 31, pp. 40–42.

Wulf, H.C., Sandby-Møller, J., Kobayasi, T. and Gniadecki, R., 2004. Skin aging and natural photoprotection. *Micron*, 35, pp. 185–191.

Yanyu, X., Yunmei, S., Zhipeng, C. and Qineng, P., 2006. The preparation of silybin-phospholipid complex and the study on its pharmacokinetics in rats. *International Journal of Pharmaceutics*, 307, pp. 77–82.

Yapar, E.A., 2016. Cosmetovigilance and global approaches. *Clinical and Experimental Health Sciences*, 6, pp. 93–97.

Yapar, E.A., 2017. Herbal Cosmetics and Novel Drug Delivery Systems. *Indian Journal of Pharmaceutical Education Research*, 51.

Yin, R. and Hamblin, M.R., 2015. Antimicrobial photosensitizers: drug discovery under the spotlight. *Current Medicinal Chemistry*, 22, pp. 2159–2185.

Yokozawa, T., Kim, H.Y., Kim, H.J., Tanaka, T., Sugino, H., Okubo, T., Chu, D.C. and Juneja, L.R., 2007. Amla (*Emblica officnalis* Gaertn.) attenuates age-related renal dysfunction by oxidative stress. *Journal of Agricultural and Food Chemistry*, 55, pp. 7744–7752.

Zamarioli, C.M., Martins, R.M., Carvalho, E.C. and Freitas, L.A., 2015. Nanoparticles containing curcuminoids (*Curcuma longa*): Development of topical delivery formulation. *Revista Brasileira de Farmacognosia*, 25, pp. 53–60.

Zhang, T., Jin, J.H., Yang, S.L., Li, G. and Jiang, J.M., 2010. Preparation and characterization of poly(p-phenylene benzobisoxazole) (PBO) fiber with anti-ultraviolet aging. *Acta Chimica Sinica*, 68, pp. 199–204.

Zhang, Z., Chen, X., Rao, W., Long, F., Yan, L. and Yin, Y., 2015. Preparation of novel curcumin-imprinted polymers based on magnetic multi-walled carbon nanotubes for the rapid extraction of curcumin from ginger powder and kiwi fruit root. *Journal of Separation Science*, 38, pp. 108–114.

12

Rosmarinic Acid: Sources, Properties, Applications and Biotechnological Production

Sandra Gonçalves and Anabela Romano

CONTENTS

12.1 Introduction

Plant kingdom is a repository of high-value metabolites from which phenolics are probably the most relevant group due to their health promotion effects. Rosmarinic acid (RA) (Figure 12.1) is an abundant phenolic compound widely distributed in the plant kingdom particularly found in plants of the Lamiaceae and Boraginaceae families (Petersen 2013). It is an ester of caffeic acid and 3,4-dihydroxyphenyl lactic acid that is biosynthesized from the amino acids L-phenylalanine and L-tyrosine, with the participation of eight enzymes (Petersen and Simmonds 2003). RA gained interest in recent years for therapeutic and food applications due to its diverse range of biological functions, namely antioxidant, anti-inflammatory, neuroprotective, chemopreventive, cardioprotective, anti-diabetic among other. Several *in vitro* and *in vivo* models have been used to elucidate its health effects and therapeutic potential. Moreover, the usefulness of RA to protect food from oxidative degradation and in the development of nutraceutical products with nutritional and health benefit value has been also recently studied.

Bioavailability and tissue distribution are important factors to be considered in the development of new functional food, nutraceutical, and pharmaceutical products based on phenolic compounds (Nunes et al. 2017). Many studies demonstrated that the beneficial effects of phenolics can be limited by their low bioaccessibility, bioavailability, and its biochemical interaction with biological components of food matrices. Therefore, a delivery system could be essential to protect these compounds (including RA) from these events, improving bioavailability, stability and retaining their beneficial biological properties.

The increasing demand of RA boosted the application of biotechnological approaches to improve its production by plant cell and hairy root cultures combining empirical strategies with metabolic engineering tools. Indeed, plant biotechnology can provide a valid alternative for the production of important metabolites when the supply of a phytochemical becomes difficult due to a scant distribution in nature (Khojasteh et al. 2014).

The investigation on RA increased largely in the last years and, therefore, the aim of this chapter is to summarize the information about: the key biological functions of RA; its therapeutic and nutraceutical applications; bioavailability studies and drug delivery strategies tested; and its production using biotechnological tools.

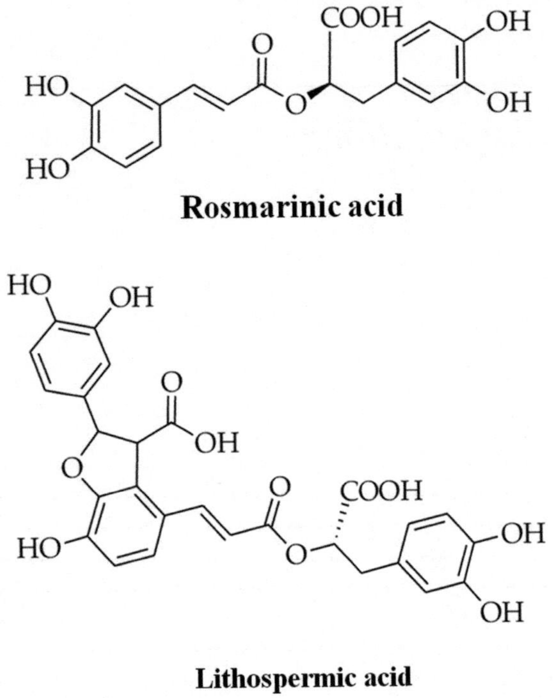

Rosmarinic acid

Lithospermic acid

FIGURE 12.1 Chemical structure of rosmarinic and lithospermic acids.

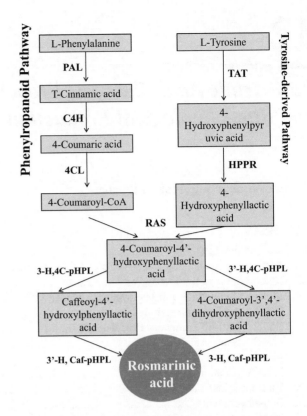

FIGURE 12.2 Metabolic pathway of rosmarinic acid biosynthesis in *Coleus blumei*. Enzymes abbreviations: PAL, phenylalanine ammonialyase; C4H, cinnamic acid 4-hydroxylase; 4CL; 4-coumarate-CoA ligase; TAT, tyrosine aminotransferase; HPPR, 4-hydroxyphenylpyruvic acid; RAS, hydroxycinnamoyl-CoA, hydroxyphenyllactate hydroxycinnamoyl transferase; 3-H and 3′-H, 4C-pHPL, 4-coumaroyl-4′-hydroxyphenyllactate 3/3′-hydroxylases; 3-H and 3′-H, Caf-pHPL, caffeoyl-4′-hydroxyphenyllactate 3/3′-hydroxylase. (Adapted from Petersen, M. et al., The biosynthesis of rosmarinic acid in suspension cultures of *Coleus blumei*, in *Primary and Secondary Metabolism of Plants and Cell Cultures III*, pp. 171–179, Springer, Dordrecht, the Netherlands, 1994.)

12.2 Rosmarinic Acid: Chemical Aspects and Biosynthesis

RA was isolated for the first time by Scarpati and Oriente (1958) from *Rosmarinus officinalis* L. (Lamiaceae). RA and its derivatives are synthesized from phenylalanine through the esterification of caffeic acid and from tyrosine through 3,4-dihydroxyphenyllactic acid (danshansu) (Ellis and Towers 1970; Petersen and Simmonds 2003). Other related caffeic acid esters are chlorogenic acid, an ester of caffeic acid and quinic acid, and caffeoylshikimic acid (Petersen 2013). The biosynthesis of RA involves two synthetic routes to yield the two monomeric units, caffeic acid and danshansu. The biosynthesis steps start from the primary metabolites phenylalanine and tyrosine and eight enzymes have been identified in *Coleus blumei* Benth. as being involved in this process: phenylalanine ammonia lyase (PAL); cinnamic acid 4-hydroxylase (CAH); 4-coumaric acid CoA-ligase (4CL); tyrosine aminotransferase (TAT); hydroxyphenylpyruvate reductase (HPPR); rosmarinic acid synthase (RAS), 3-H and 3′-H, 4-coumaroyl-4′-hydroxyphenyllactate 3/3′-hydroxylases (3-H and 3′-H, 4C-pHPL); 3-H and 3′-H, caffeoyl-4′-hydroxyphenyllactate 3/3′-hydroxylase (3-H and 3′-H, Caf-pHPL) (Petersen et al. 1994). The process is summarized in Figure 12.2 and more details can be found in the reviews (Petersen et al. 2009; Petersen 2013).

12.3 Natural Sources and Derivatives

RA is found in 39 plant families, ranging from one of the most primitive groups of land plants (hornworts) to highly evolved species of mono- and dicotyledonous plants (Petersen 2013). RA has not been reported in gymnosperms and has been particularly found in species of the Lamiaceae and Boraginaceae families (Petersen et al. 2009; Petersen 2013). All species of the Boraginaceae family, which are already investigated are shown to contain RA. In the Lamiaceae family, mainly the members of the subfamily, Nepetoideae contain this compound (Petersen et al. 2009), but it is rarely found in species of other sub-families. RA is also found in ferns (Blechnaceae family), lower plants such as the hornworts, and in monocotyledonous plants like the sea grass (Zosteraceae), the related Potamogetonaceae as well as the Cannaceae (Park et al. 2008).

Besides RA, several related derivatives can be found including among other lithospermic, yunnaneic, salvianolic, and melitric acids. Lithospermic acid A, a conjugate of RA and caffeic acid, is probably the best-known (Kim et al. 2015) (Figure 12.1). Methyl ethers of RA or its derivatives as methylmelitric acid or clerodendranoic acid, are also frequently found. These more complex compounds are formed from RA or RA-derivatives and other phenylpropanoids (Petersen 2013).

12.4 Biological Functions and Therapeutic Applications

A broad range of biological activities and/or health promoting effects have been reported for RA, namely antioxidant, anti-inflammatory, neuroprotective, as well as chemopreventive, cardioprotective, anti-diabetic among other. In the following subsections some recent studies reporting the biological and therapeutic applications of RA are reviewed (Tables 12.1 and 12.2). There are interesting reviews on this subject that can be consulted for more details (e.g., Khojasteh et al. 2014; Kim et al. 2015; Nunes et al. 2017; Habtemariam 2018).

12.4.1 Antioxidant Activity

Antioxidant activity is one of the most important biological properties of RA and is mainly associated with its free radical scavenging capacity inducing membrane stabilization and oxidative damage protection (Perez-Fons et al. 2010). Numerous reports have been published in the last years demonstrating the antioxidant effects of RA. Adomako-Bonsu et al. (2017) investigated the antioxidant activity of RA in non-cellular and cellular assays, by evaluating the DPPH radical scavenging capacity and the protection from an oxidant stress (by t-butyl hydroperoxide), respectively. It was observed that radical-scavenging activity of RA was comparable to that of quercetin although in the cellular assay RA was less potent.

Accumulating evidences from several investigations demonstrated that the free radical scavenging properties of RA against oxidative stress mediated several disorders/pathologies. Zhou et al. (2017) observed that RA alleviates the endothelial dysfunction induced by hydrogen peroxide (H_2O_2) in rat aortic rings and that this effect is related with the activation of AMPK-eNOS signaling pathway. Results from Lu et al. (2017) suggested that this compound is a promising candidate for the prevention and treatment of liver fibrosis; it counteracts activation of hepatic stellate cells mainly due to its antioxidant capability.

12.4.2 Anti-inflammatory Properties

The anti-inflammatory properties of RA have been also extensively described, particularly that concerning the inhibition of lipoxygenases, cyclooxygenases and the interference with the complement cascade, the inhibitory action on cellular pathways

TABLE 12.1

Examples of Recent Studies Reporting Some of the Many Biological Functions of Rosmarinic Acid

Biological Function	Model(s)	Outcome/Key Notes	Reference
Antioxidant	Chemical and cellular assays	The activity of RA was equivalent to that of quercetin in chemical assay but lower in cellular assay	Adomako-Bonsu et al. (2017)
Antifibrogenic potential in the prevention and treatment of liver fibrosis	Hepatic stellate cells	RA stimulate ARE promoter activity by inducing translocation of Nrf2 into the nucleus and subsequent GCLc up-regulation resulting elevation of GSH content, consequently ROS depletion inhibits NF-κB-dependent MMP-2 expression leading to inhibition of cells activation	Lu et al. (2017)
Inhibitory effect on pterygium (a common tumor-like ocular disease)	Pterygium epithelial cells	RA enhance the antioxidant system and scavenge ROS production by promoting nuclear Nrf2 protein expression and elevating HO-1, NQO1, SOD, and CAT expression levels; Inhibit cell viability through regulation of extrinsic and intrinsic apoptosis pathways	Chen et al. (2017)
Anti-inflammatory	Lipopolysaccharide-induced mastitis in mice	RA effectively attenuates lipopolysaccharide-induced mastitis by inhibiting the TLR4/MyD88/NF-κB signaling pathway	Jiang et al. (2018)
Neuroprotective effect	Doxorubicin-induced neurotoxicity in rats	RA significantly mitigate the neural changes induced by doxorubicin, it ameliorate pro-inflammatory cytokines and attenuate oxidative stress biomarkers and brain monoamines	Rizk et al. (2017)
Anticancer	Colorectal cancer cells	RA induce cell death in metastatic colorectal cancer cells and inhibited their metastatic properties by activating AMPK	Han et al. (2018)
Cardioprotective effect	Isoproterenol-induced acute myocardial infarction in rats	The cardioprotective effects of RA against acute myocardial infarction and arrhythmia are related with its ability to enhance expression of plasma antioxidant enzymes and genes involved in Ca^{2+} homeostasis $SERCA_2$ and RyR_2	Javidanpour et al. (2017)

TABLE 12.2

Selected Examples of Approaches Studied for Rosmarinic Acid Delivery

Approach	Material	Method	Tests/Administration Form	Outcome/Potential Application	Reference
Solid lipid nanoparticles	Witepsol wax	Hot-melt ultrasonication	Characterization of the particles	Food, nutraceuticals	Campos et al. (2014)
Solid lipid nanoparticles	Glycerol monostearate	Hot homogenization	*In vivo* (nasal administration to rats)	Management of Huntington's disease; Increased behavioral alterations, oxidative damage	Bhatt et al. (2015)
Phospholipid complex	Phospholipid	Solvent evaporation	*In vitro* (Caco-2 cell permeability) and *in vivo* (oral administration to rats)	Enhance oral bioavailability and bioefficacy (prophylactic effects on oxidative stress and hepatic damage in rats)	Yang et al. (2015)
Solid lipid nanoparticles	Witepsol and Carnauba waxes	Hot-melt ultrasonication	*In vitro* and *in vivo* (oral administration to rats)	Particles are safe when loaded with a moderate concentration of RA	Madureira et al. (2016a)
Solid lipid nanoparticles	Witepsol and carnauba waxes	Hot-melt ultrasonication	*In vitro* gastrointestinal simulation	Protect antioxidant activity, controlled release and significant absorption at intestinal cells	Madureira et al. (2016b)
Nanoparticles	Chitosan and sodium tripolyphosphate	Ionic gelation	*In vitro* ocular cell-based models	Particles are safe and showed mucoadhesive properties	Silva et al. (2016)
Microparticles	Chitosan and modified chitosan	Spray drying	*In vitro* (skin delivery models)	Microparticles are efficient vehicles to encapsulate RA to be used in cosmetic formulations	Casanova et al. (2016)
Nanoparticles	Polyacrylamide-chitosan-poly(lactide-co-glycolide) nanoparticles grafted with cross-reacting material 197 and apolipoprotein E	Microemulsion, solvent diffusion, grafting, and surface modification	*In vitro* (human astrocyte and brain-microvascular endothelial cells and *in vivo* (intravenous injection to rats)	Promising formulation to deliver RA to Aβ-insulted neurons in the pharmacotherapy of Alzheimer's disease	Kuo and Rajesh (2017)

and on the activity of proinflammatory cytokines. The capacity of RA to prevent or delay the progression of inflammation has been demonstrated in several cell and animal models. Rocha et al. (2015) proved the capacity of RA against local (carrageen-induced paw edema model) and systemic inflammation (liver ischaemia–reperfusion and thermal injury models). The results of these authors suggest that anti-inflammatory activity of RA is related with its antioxidant properties, inhibition of neutrophil activity and of MMP-9 activity, and modulation of the NF-kB pathway.

Mast cells played an important role in inflammatory allergic reactions since they release histamine, prostaglandins, leukotrienes and inflammatory cytokines. In addition, thymic stromal lymphopoietin (TSLP) accelerates the mast cell-mediated allergic inflammatory reactions. TSLP promotes mast cell proliferation via the regulation of apoptotic and anti-apoptotic factors by murine double minute 2 (MDM2). Results from Yoou et al. (2016) showed that RA has a significant anti-inflammatory effect on TSLP-induced inflammatory reaction in human mast cell line (HMC-1), by through down-regulating MDM2 expression and significantly reduced TSLP levels in the short ragweed pollen-induced allergic conjunctivitis mouse model. These results evidenced the anti-allergic inflammatory effects of RA. Additionally, using a murine model of asthma, Liang et al. (2016) also demonstrated that RA may effectively delay the progression of airway inflammation at a lower dose. Sepsis is an imbalance between pro and anti-inflammatory responses, Bacanlı et al. (2016) demonstrated that RA attenuates sepsis-induced oxidative damage in rats not only by decreasing the DNA damage in lymphocytes and liver and kidney tissues but also by increasing the antioxidant status and DNA repair capacity.

The anti-inflammatory effects and underlying molecular mechanism of RA in mice with dextran sulphate sodium (DSS)-induced colitis was investigated by Jin et al. (2017). It was observed that RA significantly reduced the inflammatory-related cytokines and protein levels of COX-2 and iNOS in mice with DSS-induced colitis and inhibited nuclear factor-kappa B and signal transducer and activator of transcription 3 activation. Consequently, RA reduced the severity of ulcerative colitis that is an important chronic inflammatory disorder of the colon (Jin et al. 2017).

Recently, it was demonstrated that RA effectively attenuates lipopolysaccharide-induced mastitis by inhibiting the TLR4/MyD88/NF-κB signaling pathway (Jiang et al. 2018). In addition, RA and several analogs were also investigated for their anti-inflammatory potential against lipopolysaccharide-induced alveolar macrophages (Thammason et al. 2018). Among tested analogues, ethyl rosmarinate exerted potent inhibitory activity against lung inflammation by inhibiting key inflammatory mediators. Also, RA attenuates the development and existing chronic constriction injury-induced painful peripheral neuropathy due to its anti-inflammatory, glia activation inhibition and anti-apoptotic properties (Rahbardar et al. 2018).

12.4.3 Neuroprotective Effects

The neuroprotective properties of rosmarinic acid have been investigated in several disease models *in vitro* and *in vivo*. Overall, it has been demonstrated that RA is effective in the treatment of neurodegenerative diseases and improves cognitive performance. Hwang et al. (2016) evaluated the influence of RA on cellular functions in neuronal cultures. The results suggest the possibility that RA can enhance neural plasticity by modulating glutamatergic signaling pathways, as well as providing neuro-protection with reduced cholinergic activity. RA isolated from *Blechnum brasiliense* showed a multifunctional profile related with neurodegeneration, including antioxidant effects (against hydroxyl and nitric oxide radicals and inhibition of lipid peroxidation), and inhibition of the enzyme's monoamine oxidases and catechol-*O*-methyl transferase (Andrade et al. 2016). In the same study RA (up to 5 mM) showed no cytotoxicity on polymorphonuclear rat cells.

RA prevents memory deficits induced by cerebral ischemia in mice and this was related with its anti-inflammatory and synaptogenic properties as well as its antioxidant and antiapoptotic effects (Fonteles et al. 2016). Recent results showed that this compound can protect against doxorubicin induced neurotoxicity in rats being the mechanisms underlying the neuroprotective effect again associated with its antioxidant, anti-inflammatory and antiapoptotic properties (Rizk et al. 2017). This is an important result since doxorubicin is a chemotherapeutic agent widely used in human malignancies and its long-term use cause neurobiological side effects.

Results from Lee et al. (2016) using an amyloid-β_{25-35}-injected mouse model suggest that RA have beneficial effects on cognitive improvement and may help preventing and delaying the progression of memory impairment observed in Alzheimer's disease patients. Bayomy et al. (2017) demonstrated the nephroprotective properties of RA since it showed beneficial effects versus gentamicin caused renal damage in adult rats. This was accomplished by safeguarding against renal oxidative stress, autophagy, and apoptosis. The anti-inflammatory effects of RA in conditions of lipopolysaccharide-induced neuroinflammatory injury *in vitro* and *in vivo* were recently evaluated by Wei et al. (2018). Results suggest that RA significantly reduced TLR4 and CD14 expression and NF-κB and NLRP3 inflammasome activation. A recent review provides a comprehensive perspective on the chemistry and pharmacology of RA related to their potential therapeutic applications to dementia (Alzheimer's disease and stroke) (Habtemariam 2018).

12.4.4 Other Pharmacological Activities

In addition to the biological functions above described there are also other properties described for RA, namely the chemopreventive, cardioprotective, anti-diabetic, among other. There are several studies reporting the chemopreventive properties of RA against for instance skin and colorectal cancer. Oral administration of RA prevented the formation of skin tumors during 7,12-dimethylbenz(a)anthracene (DMBA)-induced mouse skin carcinogenesis (Sharmila and Manoharan 2012). Furthermore, using a murine two-stage skin model, Osakabe et al. (2004) observed that topical application of a RA-rich extract after tumor initiation with DMBA significantly suppressed tumorigenesis. Supplementation of RA to carcinogen treated rats protected them from the deleterious effects caused by 1,2-dimethylhydrazine, a procarcinogen agent that can induce colorectal adeno-carcinoma in rats (Venkatachalam et al. 2016). More recently, Han et al. (2018) observed that this compound inhibited metastasis of colorectal cancer via activation of AMP-activated protein kinase.

Results from Chen et al. (2017) showed that RA is a potential therapeutic medication for tumor-like ocular disease (pterygium) since it inhibits cells viability through regulation of extrinsic and intrinsic apoptosis pathways, and enhanced the antioxidant system and scavenged ROS production by promoting nuclear Nrf2 protein expression and elevating HO-1, NQO1, SOD, and CAT expression levels in pterygium epithelial. Although the reports are scarce, there are also some evidences of the importance of RA on the cardiovascular system, such as the cardioprotective effects. Results from Karthik et al. (2011) with fructose-fed rats suggested that RA may be useful in reducing the cardiovascular risk associated with insulin resistance. More recently, Javidanpour et al. (2017) also demonstrated the cardioprotective effects of RA against acute myocardial infarction and arrhythmia, which may be due to its ability to enhance expression of plasma antioxidant enzymes and genes involved in Ca^{2+} homeostasis $SERCA_2$ and RyR_2. Runtuwene et al. (2016) showed that RA may be useful for treating diabetes mellitus. These authors investigated the effects of RA on glucose homeostasis and insulin regulation in rats with streptozocin (STZ)-induced type 1 diabetes or high-fat diet (HFD)-induced type 2 diabetes. Results showed that RA administration exerted a marked hypoglycemic effect on STZ-induced diabetic rats and enhanced glucose utilization and insulin sensitivity in HFD-fed diabetic rats. The capacity of RA to inhibit some metabolic enzymes including glutathione S-transferase, lactoperoxidase, acetylcholinesterase, butyrylcholinesterase and carbonic anhydrase isoenzymes, was also reported (Gülçin et al. 2016).

12.5 Food Applications

The incorporation of antioxidants in food products to avoid oxidative deterioration and maintain integral quality is a frequent practice (Caleja et al. 2018). However, if consumed excessively, the commonly used artificial antioxidants were associated with several adverse effects. This fact improves the search for natural antioxidants including from plant origin for instance to incorporate in food products to improve their preservation and also their bioactivity levels in the functionalized food products. As a powerful antioxidant and free radical scavenger, RA can be used to protect food from oxidative degradation and to develop nutraceutical products with nutritional and health benefit value. There are some recent reports describing the incorporation of RA in food products, such as dairy and bakery products. Overall, those findings indicated that the incorporation of RA or RA-rich extracts into bakery products provides advantageous functional properties, namely the antioxidant activity (Caleja et al. 2018; Ou et al. 2018).

Edible oils rich in omega-3 are very sensitive to autoxidation because of the large number of double bonds in their molecular structure. The preservation of these oils against oxidation is essential to enable a longer product lifetime and maintain their nutritional, organoleptic and sensorial properties. Guitard et al. (2016) studied several natural phenolic antioxidants in comparison to synthetic ones with regard to their ability to protect omega-3 oils from autoxidation. Results showed that three natural antioxidants including RA are more efficient than α-tocopherol and synthetic antioxidants for the preservation of omega-3 oils. Results from Saoudi et al. (2017) also suggest RA as an antioxidant agent to protect oils against oxidation.

The use of RA in food products is less investigated than its therapeutic applications, however the studies available demonstrated that this compound has a great potential to be explored in this field in the future.

12.6 Pharmacokinetic and Bioavailability

The pharmacokinetics of RA in target organs and their metabolic fate in serum, remain to be fully elucidated. Several studies conducted to understand the pharmacokinetic profile of RA using animal models have been recently revised by Nunes et al. (2017). Also, many *in vivo* studies have been reported concerning the metabolism and bioavailability of RA.

Studies from Li et al. (2007), after intravenous administration of RA in rats, showed that this compound was eliminated mainly via kidney into urine and suggest that it does not efficiently cross the blood-brain barrier. Furthermore, it was observed a more rapid distribution and elimination of RA from systemic circulation after intravenous than oral administration to rats (Lai et al. 2011). Differences were found in the metabolism of RA between humans and rats (Nakazawa and Ohsawa 2000; Baba et al. 2005). Results conducted in rats demonstrated that RA was absorbed, degraded, and/or conjugated as mhydroxyphenylpropionicacid, m-coumaric acid, and sulphated forms of caffeic acid and ferulic acid before being excreted gradually in the urine (Baba et al. 2004). Recent results demonstrated that the pharmacokinetic properties of RA also in rats after oral administration were characterized as rapid absorption, middle-speed elimination and poor absolute bioavailability (Wang et al. 2017). In humans, Baba et al. (2005) verified that RA was absorbed, conjugated, and methylated in tissues (e.g., digestive tract and liver), with a small portion being degraded into various components, and then rapidly excreted in urine. These authors suggest the involvement of microbial esterases in the RA metabolism in the digestive tract, hydrolyzing the ester linkage in RA. Results from Bel-Rhlid et al. (2009) confirmed that RA is degraded by gut microflora before its absorption and being then metabolized in various tissues, such as intestine, liver, and kidney. It has been noted that the extent of metabolism may vary considerably, due to differences in microflora and food matrices (Van Duynhoven et al. 2011).

Several studies have been conducted to evaluate the bioaccessibility and also the bioactivity of RA using *in vitro* digestion models. Gayoso et al. (2016) investigated the bioaccessibility of different phenolic compounds, including RA, after *in vitro* gastrointestinal digestion using different models and evaluate the antioxidant activity during the digestion process. Authors demonstrated that the bioaccessible fraction is strongly affected by the methodology used at the intestinal level. Moreover, despite the fact that the antioxidant activity in the bioaccessible fractions decreased significantly, a remarkable antioxidant activity could still reach the colon level. Costa et al. (2014) showed that the antioxidant and anti-cholinesterase activities of a *Lavandula viridis* L'Hér extract and of its main component, RA, is assured after *in vitro* gastrointestinal digestion. Recently, Villalva et al. (2018) studied the bioavailability of a RA-rich extract using an *in vitro* digestion/Caco-2 cell culture model as well as the anti-inflammatory and antioxidant activity of the basolateral fraction. Results demonstrated that the bioavailable fraction of the extract showed remarkable antioxidant and anti-inflammatory activities.

12.7 Delivery Systems

Considering the broad range of bioactive properties of phenolic compounds including of RA there is a growing interest on the possibility of their application in functional food, nutraceutical, and pharmaceutical products. Nevertheless, the bioavailability of RA after oral use is reduced, it is not stable and may interact easily with some matrices in which it is incorporated and therefore, a delivery system could be essential to protect it from these events, thus improving bioavailability, stability and retaining their beneficial biological properties. In this sense, several approaches have been tested namely the use of solid lipid nanoparticles (SLN) (prepared with different lipid matrices) (Camposa et al. 2014; Bhatt et al. 2015; Madureira et al. 2016a, 2016b). SLNs combine several advantages, such as physical stability, controlled release and good tolerability. Madureira et al. (2016a) evaluated the safety profile of SLNs (loaded with RA and produced with Witepsol and Carnauba waxes) *in vitro* (genotoxicity and cytotoxicity tests) and *in vivo* in rats orally treated. Authors observed that SLNs are safe when loaded with moderate concentrations of RA. The same group also evaluated the digestion effects by exposing SLNs to simulated gastrointestinal tract (GIT) conditions existing in stomach and small intestine (Madureira et al. 2016b). SLNs protected RA bioactivity (antioxidant activity) until reaching the intestine, a controlled release of RA from SLN was achieved and a significant absorption was observed at intestinal cells. Moreover, the results varyed according to the lipid matrix used with witepsol showed a higher stability than Carnauba. Yang et al. (2015) results showed that an oil solution of RA–phospholipid complex could act as an effective delivery device for RA to enhance oral bioavailability and bioefficacy. The phospholipid complex emerged as one of the most successful methods for improving the bioavailability and bioefficacy of poorly absorbed plant compounds.

The ability of RA to cross the blood–brain barrier is low, which significantly reduces its bioavailability in the cerebrum and limit the use of this compound in neurodegenerative diseases therapy. To overcome this physiological limitation, several strategies have been tested, e.g. the conjugation of RA with targeting biomolecules. The potential use of SLNs as a drug delivery system to enhance the brain-targeting efficiency of RA following intranasal administration was also investigated (Bhatt et al. 2015). RA-loaded SLNs treatment significantly attenuate 3NP-induced deficits in body weight, beam walk, locomotor and motor coordination, and are also able to significantly attenuate 3NP-induced striatal oxidative stress. Overall, the results indicated that nasal administration of RA-loaded SLNs could be a promising strategy for the Huntington's disease management. More recently, Kuo and Rajesh (2017) developed poly-acrylamide-chitosan-poly(lactide-co-glycolide) nanoparticles with surface cross-reacting material 197 and apolipoprotein E to retard the degenerative progress of Aβ-insulted neurons. Results from these authors showed that this can be an effective dosage form for brain-targeting behavior and have potential in neuronal rescue in AD treatment.

Some studies indicated that RA could be a potent inhibitor of retinal neovascularization and may be applied in the treatment of vasoproliferative retinopathies. Thus, Silva et al. (2016) investigated the potential of chitosan-based nanoparticles for RA ocular delivery. Nanoparticles showed mucoadhesive properties and no relevant cytotoxicity against ocular cell lines tested, and, therefore, may be a promising drug delivery system for ocular application in oxidative eye conditions. Microencapsulation with chitosan also proved to be an efficient strategy for topical delivery of RA (Casanova et al. 2016). This may be a good strategy to overcome the limitations associated with the application of this compound in cosmetic formulations, such as instability, discoloration, poor solubility in water and low partition coefficient.

12.8 Biotechnological Production

In the last decades considerable progress has been made concerning the production of plant secondary metabolites by using plant tissue culture techniques owing the advantages of this platform over other production systems. This appears as environmentally friendly alternative method for the production of secondary metabolites when natural supply is limited, and traditional methods are unfeasible. Some of the advantages of plant tissue culture techniques are (Murthy et al. 2014; Isah et al. 2018): allow the large-scale production of secondary metabolites in a year-round system without seasonal constraints; cultures can be established in any part of the world independently of the plant growth requisites; cultures are free of microbes and insects avoiding the use of pesticides and herbicides; provide a simple, reliable and predictable method for isolating the secondary metabolites at high efficiency within a short time when compared to the extraction from wild plant populations; and opens the opportunity to apply traditional or metabolic engineering strategies to promote the accumulation of desired compounds by *in vitro* cultures. This is an appellative method for commercial application that is inclusively endorsed by Food and Agriculture Organization as safe for the production of compounds for food application. Dias et al. (2016) recently reviewed the production of phenolic compounds by using plant tissue culture techniques. They concluded that there are numerous reports directed to the production of phenolic extracts, but there is a lack in the production of individual compounds. They also observed that elicitation strategy is often used to increase the production of phenolics. Concerning RA, Khojasteh et al. (2014) reviewed the insights in biotechnological production of this compound.

The biotechnological production of RA, by cell suspension cultures of *Coleus blumei*, was described for the first time in 1977 (Zenk et al. 1977). Afterwards the production of RA was reported in cell cultures of species from different species particularly of

Lamiaceae and Boraginaceae families (Petersen and Simmonds 2003; Matkowski 2008). Production of RA in hairy roots was also reported in several species (Bulgakov et al. 2012; Khojasteh et al. 2014, and references therein). Cell suspension cultures of *Satureja khuzistanica* Jamzad (Lamiaceae) (Khojasteh et al. 2016; Sahraroo et al. 2016) and hairy roots of *Ocimum basilicum* (Srivastava et al. 2016) were recently described as rich sources of RA.

The manipulation of growth factors (e.g., nutrients, growth regulators type and concentration, sucrose levels, etc.) and elicitation has been reported as strategies to enhance RA yields. Several elicitors, such as methyl jasmonate, salicylic acid, yeast extract, vanadyl sulphate, etc., have been tested (Bulgakov et al. 2012; Khojasteh et al. 2014). The use of permeabilizing agents, such as DMSO, cyclodextrins and coronatin, to facilitate downstream processes is another important approach used (Khojasteh et al. 2014; Khojasteh et al. 2016).

The last step in the process development of a biotechnological production system is scaling up to bioreactor level (Khojasteh et al. 2016). There are several reports describing the cultivation of cell suspensions in bioreactors for the production of RA as reviewed by Khojasteh et al. (2014). The use of bioreactor for RA production was initially described for cell suspensions of *Coleus blumei* and *Anchusa officinalis* (Ulrich et al. 1985; Su and Lei 1993). Later on, Zhong et al. (2001) showed that a large turbine impeller reactor was preferable for cell growth and RA production in *Salvia miltiorrhiza* cell cultures growing in stirred bioreactors. Pavlov et al. (2005) also used a stirred bioreactor for RA production in *Lavandula vera* cell suspensions and studied the effect of dissolved oxygen concentration and agitation. Recently, Khojasteh et al. (2016) used a wave-mixed bioreactor to produce RA in *Satureja khuzistanica* cell suspensions and showed the effectiveness of methyl jasmonate in increasing the production yield. The optimum conditions described by these authors allow a percentage of RA 14-fold higher than in the wild plants and suggest this platform suitable for the commercial RA production. Despite the efforts in the last years the large-scale production of RA in bioreactors still needs optimization. According to Khojasteh et al. (2014) single-use bioreactors, which has been successfully implemented for the production of other plant secondary metabolites, could be a promising technology for RA production in the future.

Metabolic engineering offers a new perspective to understand the expression of genes involved in the biosynthesis of secondary metabolites through overexpression studies allowing the alteration of biosynthetic pathways (O'Connor 2015; Swamy et al. 2018). Theoretically the secondary metabolites productivity of plant cell cultures can be improved trough the overexpression of genes encoding regulatory enzymes involved in their biosynthetic pathways (Verpoorte et al. 2002). The eight enzymes involved in the biosynthetic pathway of RA have been already characterized but its regulation remains unclear (Khojasteh et al. 2014). The effects of some elicitors, particularly methyl jasmonate, have been investigated on RA biosynthesis and expression of phenylpropanoid biosynthetic genes in cell cultures of several species (e.g., Kim et al. 2013; Xing et al. 2013).

Several genes related with RA biosynthesis from some plant species, particularly from Lamiaceae family, were cloned and characterized as RA biosynthetic genes in recent years. Several genes were isolated from *Salvia miltiorrhiza* Bunge., such as

SmPAL, SmC4H, Sm4CL, SmTAT, SmHPPR, SmHPPD, SmRAS and *R2R3 MYB* transcription factor gene (*SmMYB39*) (e.g., Xiao et al. 2011; Zhang et al. 2013). Transcriptomic analyses also recently revealed two genes related to RA biosynthesis in the species *Dracocephalum tanguticum* Maxim, which include a *C4H* (c76971_g1) and a *4CL* gene (c92424_g8) (Li et al. 2017). The RA contents of six plant organs were significantly correlated to the expression of these two genes. Similarly, overexpression of a *S. miltiorrhiza 4CH* gene increased RA production in hairy root cultures (Xiao et al. 2011). Also *SoHPPR*, the gene coding for HPPR (the key enzyme responsible for the RA biosynthesis), was cloned from this species and its transcript expression level monitored during cell suspension cultures, and a relationship with scavenger activity and RA yield was observed (Barberini et al. 2013).

12.9 Concluding Remarks

Many investigations have been conducted in recent years to evaluate the biological function of RA *in vitro* and also *in vivo* using several disease models. From the diverse range of health benefits attributed to this compound, its antioxidant, anti-inflammatory and neuroprotective effects are particularly evident. Nevertheless, studies also showed that the low bioaccessibility and bioavailability of this compound as well as its biochemical interaction with biological components of food matrices, can limit its therapeutic and nutraceutical applications. In this sense several strategies have been tested to increase its bioavailability and bioactivity, including the use of nanoparticles and microencapsulation. Although the results are promising novel delivery systems are still required to improve the performance of this compound in therapeutic and food applications.

Many biotechnological platforms have been established to produce RA, including shoots, cell suspensions and hairy root cultures of several species. Despite the many progresses in the biotechnological production of this compound in the last years its production in large-scale needs further optimization. The molecular understanding of its biosynthesis and the application of metabolic engineering tools are crucial to improve the biotechnological production.

ACKNOWLEDGMENT

This work received financial support from the project INTERREG - MD.Net: When Brand Meets People. S. Gonçalves acknowledge the financial support from the Foundation for Science and Technology (FCT), Portugal.

REFERENCES

Adomako-Bonsu, A.G., Chan, S.L., Pratten, M. and Fry, J.R., 2017. Antioxidant activity of rosmarinic acid and its principal metabolites in chemical and cellular systems: Importance of physico-chemical characteristics. *Toxicology In Vitro*, 40, pp. 248–255.

Andrade, M.M.A., Passos, C.S., Rubio, M.A.K., Mendonça, J.N., Lopes, N.P. and Henriques, A.T., 2016. Combining in vitro and *in silico* approaches to evaluate the multifunctional profile of rosmarinic acid from *Blechnum brasiliense* on targets related to neurodegeneration. *Chemico-Biological Interactions*, 254, pp. 135–145.

Baba, S., Osakabe, N., Natsume, M. and Terao, J., 2004. Orally administered rosmarinic acid is present as the conjugated and/ or methylated forms in plasma and is degraded and metabolized to conjugated forms of caffeic acid, ferulic acid and m-coumaric acid. *Life Sciences*, 75(2), pp. 165–178.

Baba, S., Osakabe, N., Natsume, M., Yasuda, A., Muto, Y., Hiyoshi, T., Takano, H., Yoshikawa, T. and Terao, J., 2005. Absorption, metabolism, degradation and urinary excretion of rosmarinic acid after intake of *Perilla frutescens* extract in humans. *European Journal of Nutrition*, 44(1), pp. 1–9.

Bacanlı, M., Aydın, S., Taner, G., Göktaş, H.G., Şahin, T., Başaran, A.A. and Başaran, N., 2016. Does rosmarinic acid treatment have protective role against sepsis-induced oxidative damage in Wistar Albino rats? *Human and Experimental Toxicology*, 35(8), pp. 877–886.

Barberini, S., Savona, M., Raffi, D., Leonardi, M., Pistelli, L., Stochmal, A., Vainstein, A., Pistelli, L. and Ruffoni, B., 2013. Molecular cloning of SoHPPR encoding a hydroxyphenylpyruvate reductase, and its expression in cell suspension cultures of *Salvia officinalis*. *Plant Cell, Tissue and Organ Culture (PCTOC)*, 114(1), pp. 131–138.

Bayomy, N.A., Elbakary, R.H., Ibrahim, M.A. and Abdelaziz, E.Z., 2017. Effect of lycopene and rosmarinic acid on gentamicin induced renal cortical oxidative stress, apoptosis, and autophagy in adult male albino rat. *The Anatomical Record*, 300(6), pp. 1137–1149.

Bel-Rhlid, R., Crespy, V., Page-Zoerkler, N., Nagy, K., Raab, T. and Hansen, C.E., 2009. Hydrolysis of rosmarinic acid from rosemary extract with esterases and *Lactobacillus johnsonii* in vitro and in a gastrointestinal model. *Journal of Agricultural and Food Chemistry*, 57(17), pp. 7700–7705.

Bhatt, R., Singh, D., Prakash, A. and Mishra, N., 2015. Development, characterization and nasal delivery of rosmarinic acid-loaded solid lipid nanoparticles for the effective management of Huntington's disease. *Drug Delivery*, 22(7), pp. 931–939.

Bulgakov, V.P., Inyushkina, Y.V. and Fedoreyev, S.A., 2012. Rosmarinic acid and its derivatives: Biotechnology and applications. *Critical Reviews in Biotechnology*, 32(3), pp. 203–217.

Caleja, C., Barros, L., Barreira, J.C., Ciric, A., Sokovic, M., Calhelha, R.C., Beatriz, M., Oliveira, P.P. and Ferreira, I.C., 2018. Suitability of lemon balm (*Melissa officinalis* L.) extract rich in rosmarinic acid as a potential enhancer of functional properties in cupcakes. *Food Chemistry*, 250, pp. 67–74.

Camposa, D.A., Madureira, A.R., Gomes, A.M., Sarmento, B. and Pintado, A.M., 2014. Optimization of the production of solid Witepsol nanoparticles loaded with rosmarinic acid. *Colloids and Surfaces B: Biointerfaces*, 115, pp. 109–117.

Casanova, F., Estevinho, B.N. and Santos, L., 2016. Preliminary studies of rosmarinic acid microencapsulation with chitosan and modified chitosan for topical delivery. *Powder Technology*, 297, pp. 44–49.

Chen, Y.Y., Tsai, C.F., Tsai, M.C., Hsu, Y.W. and Lu, F.J., 2017. Inhibitory effects of rosmarinic acid on pterygium epithelial cells through redox imbalance and induction of extrinsic and intrinsic apoptosis. *Experimental Eye Research*, 160, pp. 96–105.

Costa, P., Grevenstuk, T., da Costa, A.M.R., Gonçalves, S. and Romano, A., 2014. Antioxidant and anti-cholinesterase activities of *Lavandula viridis* L'Hér extracts after in vitro gastrointestinal digestion. *Industrial Crops and Products*, 55, pp. 83–89.

Dias, M.I., Sousa, M.J., Alves, R.C. and Ferreira, I.C., 2016. Exploring plant tissue culture to improve the production of phenolic compounds: A review. *Industrial Crops and Products*, 82, pp. 9–22.

Ellis, B.E. and Towers, G.H.N., 1970. Biogenesis of rosmarinic acid in *Mentha. Biochemical Journal*, 118(2), pp. 291–297.

Fonteles, A.A., de Souza, C.M., de Sousa Neves, J.C., Menezes, A.P.F., do Carmo, M.R.S., Fernandes, F.D.P., de Araújo, P.R. and de Andrade, G.M., 2016. Rosmarinic acid prevents against memory deficits in ischemic mice. *Behavioural Brain Research SreeTestContent1*, 297, pp. 91–103.

Gayoso, L., Claerbout, A.S., Calvo, M.I., Cavero, R.Y., Astiasarán, I. and Ansorena, D., 2016. Bioaccessibility of rutin, caffeic acid and rosmarinic acid: Influence of the *in vitro* gastrointestinal digestion models. *Journal of Functional Foods*, 26, pp. 428–438.

Guitard, R., Paul, J.F., Nardello-Rataj, V. and Aubry, J.M., 2016. Myricetin, rosmarinic and carnosic acids as superior natural antioxidant alternatives to α-tocopherol for the preservation of omega-3 oils. *Food Chemistry*, 213, pp. 284–295.

Gülçin, İ., Scozzafava, A., Supuran, C.T., Koksal, Z., Turkan, F., Çetinkaya, S., Bingöl, Z., Huyut, Z. and Alwasel, S.H., 2016. Rosmarinic acid inhibits some metabolic enzymes including glutathione S-transferase, lactoperoxidase, acetylcholinesterase, butyrylcholinesterase and carbonic anhydrase isoenzymes. *Journal of Enzyme Inhibition and Medicinal Chemistry*, 31(6), pp. 1698–1702.

Habtemariam, S., 2018. Molecular pharmacology of rosmarinic and salvianolic acids: Potential seeds for Alzheimer's and vascular dementia drugs. *International Journal of Molecular Sciences*, 19(2), p. 458.

Han, Y.H., Kee, J.Y. and Hong, S.H., 2018. Rosmarinic acid activates AMPK to inhibit metastasis of colorectal cancer. *Frontiers in Pharmacology*, 9, p. 68.

Hwang, E.S., Kim, H.B., Choi, G.Y., Lee, S., Lee, S.O., Kim, S. and Park, J.H., 2016. Acute rosmarinic acid treatment enhances long-term potentiation, BDNF and GluR-2 protein expression, and cell survival rate against scopolamine challenge in rat organotypic hippocampal slice cultures. *Biochemical and Biophysical Research Communications*, 475(1), pp. 44–50.

Isah, T., Umar, S., Mujib, A., Sharma, M.P., Rajasekharan, P.E., Zafar, N. and Frukh, A., 2017. Secondary metabolism of pharmaceuticals in the plant *in vitro* cultures: Strategies, approaches, and limitations to achieving higher yield. *Plant Cell, Tissue and Organ Culture (PCTOC)*, pp. 1–27.

Javidanpour, S., Dianat, M., Badavi, M. and Mard, S.A., 2017. The cardioprotective effect of rosmarinic acid on acute myocardial infarction and genes involved in Ca^{2+} homeostasis. *Free Radical Research*, 51(11–12), pp. 911–923.

Jiang, K., Ma, X., Guo, S., Zhang, T., Zhao, G., Wu, H., Wang, X. and Deng, G., 2018. Anti-inflammatory effects of rosmarinic acid in lipopolysaccharide-induced mastitis in mice. *Inflammation*, 41(2), pp. 437–448.

Jin, B.R., Chung, K.S., Cheon, S.Y., Lee, M., Hwang, S., Hwang, S.N., Rhee, K.J. and An, H.J., 2017. Rosmarinic acid suppresses colonic inflammation in dextran sulphate sodium (DSS)-induced mice via dual inhibition of NF-κB and STAT3 activation. *Scientific Reports*, 7, p. 46252.

Karthik, D., Viswanathan, P. and Anuradha, C.V., 2011. Administration of rosmarinic acid reduces cardiopathology and blood pressure through inhibition of p22phox NADPH oxidase in fructose-fed hypertensive rats. *Journal of Cardiovascular Pharmacology*, 58(5), pp. 514–521.

Khojasteh, A., Mirjalili, M.H., Hidalgo, D., Corchete, P. and Palazon, J., 2014. New trends in biotechnological production of rosmarinic acid. *Biotechnology Letters*, 36(12), pp. 2393–2406.

Khojasteh, A., Mirjalili, M.H., Palazon, J., Eibl, R. and Cusido, R.M., 2016. Methyl jasmonate enhanced production of rosmarinic acid in cell cultures of *Satureja khuzistanica* in a bioreactor. *Engineering in Life Sciences*, 16(8), pp. 740–749.

Kim, G.D., Park, Y.S., Jin, Y.H. and Park, C.S., 2015. Production and applications of rosmarinic acid and structurally related compounds. *Applied Microbiology and Biotechnology*, 99(5), pp. 2083–2092.

Kim, Y.B., Kim, J.K., Uddin, M.R., Xu, H., Park, W.T., Tuan, P.A., Li, X., Chung, E., Lee, J.H. and Park, S.U., 2013. Metabolomics analysis and biosynthesis of rosmarinic acid in *Agastache rugosa* Kuntze treated with methyl jasmonate. *PLoS One*, 8(5), p. e64199.

Kuo, Y.C. and Rajesh, R., 2017. Targeted delivery of rosmarinic acid across the blood–brain barrier for neuronal rescue using polyacrylamide-chitosan-poly (lactide-co-glycolide) nanoparticles with surface cross-reacting material 197 and apolipoprotein E. *International Journal of Pharmaceutics*, 528(1–2), pp. 228–241.

Lai, X.J., Zhang, L., Li, J.S., Liu, H.Q., Liu, X.H., Di, L.Q., Cai, B.C. and Chen, L.H., 2011. Comparative pharmacokinetic and bioavailability studies of three salvianolic acids after the administration of *Salviae miltiorrhizae* alone or with synthetical borneol in rats. *Fitoterapia*, 82(6), pp. 883–888.

Lee, A.Y., Hwang, B.R., Lee, M.H., Lee, S. and Cho, E.J., 2016. *Perilla frutescens* var. *japonica* and rosmarinic acid improve amyloid-β25-35 induced impairment of cognition and memory function. *Nutrition Research and Practice*, 10(3), pp. 274–281.

Li, X., Yu, C., Lu, Y., Gu, Y., Lu, J., Xu, W., Xuan, L. and Wang, Y., 2006. Pharmacokinetics, tissue distribution, metabolism and excretion of depside salts from *Salvia miltiorrhiza* in rats. *Drug Metabolism and Disposition*, 35, pp. 234–239.

Li, H., Fu, Y., Sun, H., Zhang, Y., and Lan, X., 2017. Transcriptomic analyses reveal biosynthetic genes related to rosmarinic acid in *Dracocephalum tanguticum. Scientific Reports*, 7, p. 74.

Liang, Z., Xu, Y., Wen, X., Nie, H., Hu, T., Yang, X., Chu, X., Yang, J., Deng, X. and He, J., 2016. Rosmarinic acid attenuates airway inflammation and hyper-responsiveness in a murine model of asthma. *Molecules*, 21(6), p. 769.

Lu, C., Zou, Y., Liu, Y. and Niu, Y., 2017. Rosmarinic acid counteracts activation of hepatic stellate cells via inhibiting the ROS-dependent MMP-2 activity: Involvement of Nrf2 antioxidant system. *Toxicology and Applied Pharmacology*, 318, pp. 69–78.

Madureira, A.R., Nunes, S., Campos, D.A. et al., 2016a. Safety profile of solid lipid nanoparticles loaded with rosmarinic acid for oral use: *In vitro* and animal approaches. *International Journal of Nanomedicine*, 11, p. 3621.

Madureira, A.R., Campos, D.A., Oliveira, A., Sarmento, B., Pintado, M.M. and Gomes, A.M., 2016b. Insights into the protective role of solid lipid nanoparticles on rosmarinic acid bioactivity during exposure to simulated gastrointestinal conditions. *Colloids and Surfaces B: Biointerfaces*, 139, pp. 277–284.

Matkowski, A., 2008. Plant *in vitro* culture for the production of antioxidants—A review. *Biotechnology Advances*, 26(6), pp. 548–560.

Murthy, H.N., Lee, E.J. and Paek, K.Y., 2014. Production of secondary metabolites from cell and organ cultures: Strategies and approaches for biomass improvement and metabolite accumulation. *Plant Cell, Tissue and Organ Culture (PCTOC)*, 118(1), pp. 1–16.

Nakazawa, T. and Ohsawa, K., 2000. Metabolites of orally administered Perilla frutescens extract in rats and humans. *Biological and Pharmaceutical Bulletin*, 23(1), pp. 122–127.

Nunes, S., Madureira, A.R., Campos, D., Sarmento, B., Gomes, A.M., Pintado, M. and Reis, F., 2017. Therapeutic and nutraceutical potential of rosmarinic acid—Cytoprotective properties and pharmacokinetic profile. *Critical Reviews in Food Science and Nutrition*, 57(9), pp. 1799–1806.

O'Connor, S.E., 2015. Engineering of secondary metabolism. *Annual Review of Genetics*, 49, pp. 71–94.

Osakabe, N., Yasuda, A., Natsume, M. and Yoshikawa, T., 2004. Rosmarinic acid inhibits epidermal inflammatory responses: Anticarcinogenic effect of *Perilla frutescens* extract in the murine two-stage skin model. *Carcinogenesis*, 25(4), pp. 549–557.

Ou, J., Teng, J., El-Nezami, H.S. and Wang, M., 2018. Impact of resveratrol, epicatechin and rosmarinic acid on fluorescent AGEs and cytotoxicity of cookies. *Journal of Functional Foods*, 40, pp. 44–50.

Park, S.U., Uddin, R., Xu, H., Kim, Y.K. and Lee, S.Y., 2008. Biotechnological applications for rosmarinic acid production in plant. *African Journal of Biotechnology*, 7(25), pp. 4959-4965.

Pavlov, A.I., Georgiev, M.I. and Ilieva, M.P., 2005. Production of rosmarinic acid by *Lavandula vera* MM cell suspension in bioreactor: Effect of dissolved oxygen concentration and agitation. *World Journal of Microbiology and Biotechnology*, 21(4), pp. 389–392.

Perez-Fons, L., GarzÓn, M.T. and Micol, V., 2009. Relationship between the antioxidant capacity and effect of rosemary (*Rosmarinus officinalis* L.) polyphenols on membrane phospholipid order. *Journal of Agricultural and Food Chemistry*, 58(1), pp. 161–171.

Petersen, M., Abdullah, Y., Benner, J. et al., 2009. Evolution of rosmarinic acid biosynthesis. *Phytochemistry*, 70(15–16), pp. 1663–1679.

Petersen, M., Häusler, E., Meinhard, J., Karwatzki, B. and Gertlowski, C., 1994. The biosynthesis of rosmarinic acid in suspension cultures of *Coleus blumei*. In *Primary and Secondary Metabolism of Plants and Cell Cultures III* (pp. 171–179). Springer, Dordrecht, the Netherlands.

Petersen, M. and Simmonds, M.S., 2003. Rosmarinic acid. *Phytochemistry*, 62(2), pp. 121–125.

Petersen, M., 2013. Rosmarinic acid: New aspects. *Phytochemistry Reviews*, 12(1), pp. 207–227.

Rahbardar, M.G., Amin, B., Mehri, S., Mirnajafi-Zadeh, S.J. and Hosseinzadeh, H., 2018. Rosmarinic acid attenuates development and existing pain in a rat model of neuropathic pain: An evidence of anti-oxidative and anti-inflammatory effects. *Phytomedicine*, 40, pp. 59–67.

Rizk, H.A., Masoud, M.A. and Maher, O.W., 2017. Prophylactic effects of ellagic acid and rosmarinic acid on doxorubicin-induced neurotoxicity in rats. *Journal of Biochemical and Molecular Toxicology*, 31(12), p. e21977.

Rocha, J., Eduardo-Figueira, M., Barateiro, A. et al., 2015. Anti-inflammatory effect of rosmarinic acid and an extract of *Rosmarinus officinalis* in rat models of local and systemic inflammation. *Basic and Clinical Pharmacology and Toxicology*, 116(5), pp. 398–413.

Runtuwene, J., Cheng, K.C., Asakawa, A., Amitani, H., Amitani, M., Morinaga, A., Takimoto, Y., Kairupan, B.H.R. and Inui, A., 2016. Rosmarinic acid ameliorates hyperglycemia and insulin sensitivity in diabetic rats, potentially by modulating the expression of PEPCK and GLUT4. *Drug Design, Development and Therapy*, 10, p. 2193.

Sahraroo, A., Mirjalili, M.H., Corchete, P., Babalar, M. and Moghadam, M.R.F., 2016. Establishment and characterization of a *Satureja khuzistanica* Jamzad (Lamiaceae) cell suspension culture: A new in vitro source of rosmarinic acid. *Cytotechnology*, 68(4), pp. 1415–1424.

Saoudi, S., Chammem, N., Sifaoui, I., Jiménez, I.A., Lorenzo-Morales, J., Piñero, J.E., Bouassida-Beji, M., Hamdi, M. and Bazzocchi, I.L., 2017. Combined effect of carnosol, rosmarinic acid and thymol on the oxidative stability of soybean oil using a simplex centroid mixture design. *Journal of the Science of Food and Agriculture*, 97(10), pp. 3300–3311.

Scarpati, M.L., and Oriente, G., 1958. Isolamento e costituzione dell' acido rosmarinico (dal *Rosmarinus off.*). *Ricerca Scientifica*, 28, pp. 2329–2333.

Sharmila, R. and Manoharan, S., 2012. Anti-tumor activity of rosmarinic acid in 7, 12-dimethylbenz (a) anthracene (DMBA) induced skin carcinogenesis in Swiss albino mice. *Indian Journal of Experimental Biology*, 50, pp. 187–194.

Silva, SB., Ferreira, D., Pintado, M. and Sarmento, B., 2016. Chitosan-based nanoparticles for rosmarinic acid ocular delivery—*In vitro* tests. *International Journal of Biological Macromolecules*, 84, pp. 112–120.

Srivastava, S., Conlan, X.A., Adholeya, A. and Cahill, D.M., 2016. Elite hairy roots of *Ocimum basilicum* as a new source of rosmarinic acid and antioxidants. *Plant Cell, Tissue and Organ Culture (PCTOC)*, 126(1), pp. 19–32.

Su, W.W. and Lei, F., 1993. Rosmarinic acid production in perfused *Anchusa officinalis* culture: Effect of inoculum size. *Biotechnology Letters*, 15(10), pp. 1035–1038.

Swamy, M.K., Sinniah, U.R. and Ghasemzadeh, A., 2018. Anticancer potential of rosmarinic acid and its improved production through biotechnological interventions and functional genomics. *Applied Microbiology and Biotechnology*, pp. 1–19.

Thammason, H., Khetkam, P., Pabuprapap, W., Suksamrarn, A. and Kunthalert, D., 2018. Ethyl rosmarinate inhibits lipopolysaccharide-induced nitric oxide and prostaglandin E2 production in alveolar macrophages. *European Journal of Pharmacology*, 824, pp. 17–23.

Ulbrich, B., Wiesner, W. and Arens, H., 1985. Large-Scale Production of Rosmarinic Acid from Plant Cell Cultures of Coleus blumei Benth. In *Primary and Secondary Metabolism of Plant Cell Cultures* (pp. 293–303). Springer, Berlin, Germany.

Van Duynhoven, J., Vaughan, E.E., Jacobs, D.M. et al., 2010. Metabolic fate of polyphenols in the human superorganism. *Proceedings of the National Academy of Sciences*, 108 (suppl 1), p. 201000098.

Venkatachalam, K., Gunasekaran, S. and Namasivayam, N., 2016. Biochemical and molecular mechanisms underlying the chemopreventive efficacy of rosmarinic acid in a rat colon cancer. *European Journal of Pharmacology*, 791, pp. 37–50.

Verpoorte, R., Contin, A. and Memelink, J., 2002. Biotechnology for the production of plant secondary metabolites. *Phytochemistry Reviews*, *1*(1), pp. 13–25.

Villalva, M., Jaime, L., Aguado, E., Nieto, J.A., Reglero, G. and Santoyo, S., 2018. Anti-inflammatory and antioxidant activities from the basolateral fraction of Caco-2 cells exposed to a rosmarinic acid enriched extract. *Journal of Agricultural and Food Chemistry*, *66*(5), pp. 1167–1174.

Wang, J., Li, G., Rui, T., Kang, A., Li, G., Fu, T., Li, J., Di, L. and Cai, B., 2017. Pharmacokinetics of rosmarinic acid in rats by LC-MS/MS: Absolute bioavailability and dose proportionality. *RSC Advances*, *7*(15), pp. 9057–9063.

Wei, Y., Chen, J., Hu, Y., Lu, W., Zhang, X., Wang, R. and Chu, K., 2018. Rosmarinic acid mitigates lipopolysaccharide-induced neuroinflammatory responses through the inhibition of TLR4 and CD14 expression and NF-κB and NLRP3 inflammasome activation. *Inflammation*, *41*(2), pp. 732–740.

Xiao, Y., Zhang, L., Gao, S., Saechao, S., Di, P., Chen, J. and Chen, W., 2011. The c4h, tat, hppr and hppd genes prompted engineering of rosmarinic acid biosynthetic pathway in *Salvia miltiorrhiza* hairy root cultures. *PLoS One*, *6*(12), p. e29713.

Xing, B.Y., Dang, X.L., Zhang, J.Y., Wang, B., Chen, Z. and Dong, J.-E. 2013. Effects of methyl jasmonate on the biosynthesis of rosmarinic acid and related enzymes in *Salvia miltiorrhiza* suspension cultures. *Plant Physiology Journal*, *49*, pp. 1326–1332.

Yang, J.H., Zhang, L., Li, J.S., Chen, L.H., Zheng, Q., Chen, T., Chen, Z.P., Fu, T.M. and Di, L.Q., 2015. Enhanced oral bioavailability and prophylactic effects on oxidative stress and hepatic damage of an oil solution containing a rosmarinic acid–phospholipid complex. *Journal of Functional Foods*, *19*, pp. 63–73.

Yoou, M.S., Park, C.L., Kim, M.H., Kim, H.M. and Jeong, H.J., 2016. Inhibition of MDM2 expression by rosmarinic acid in TSLP-stimulated mast cell. *European Journal of Pharmacology*, *771*, pp. 191–198.

Zenk, M.H., El-Shagi, H. and Ulbrich, B., 1977. Production of rosmarinic acid by cell-suspension cultures of *Coleus blumei*. *Naturwissenschaften*, *64*(11), pp. 585–586.

Zhang, S., Ma, P., Yang, D., Li, W., Liang, Z., Liu, Y. and Liu, F. 2013. Cloning and characterization of a putative R2R3 MYB transcriptional repressor of the rosmarinic acid biosynthetic pathway from *Salvia miltiorrhiza*. *PLoS One*, *8*(9), p. e73259.

Zhong, J.J., Chen, H. and Chen, F., 2001. Production of rosmarinic acid, lithospermic acid B, and tanshinones by suspension cultures of Ti-transformed *Salvia miltiorrhiza* cells in bioreactors. *Journal of Plant Biotechnology*, *3*(2), pp. 107–112.

Zhou, H., Fu, B., Xu, B., Mi, X., Li, G., Ma, C., Xie, J., Li, J. and Wang, Z., 2017. Rosmarinic acid alleviates the endothelial dysfunction induced by hydrogen peroxide in rat aortic rings via activation of AMPK. *Oxidative Medicine and Cellular Longevity*, *2017*.

Section III

Bioactive Potential of Medicinal Plants and Treatment against Diseases

13

Scientific Validation of the Usefulness of Withania somnifera Dunal. in the Prevention of Diseases

Gnanasekeran Karthikeyan, Mallappa Kumara Swamy, Madheshwar RajhaViknesh, Rajendran Shurya, and Natesan Sudhakar

CONTENTS

13.1 Introduction

In humanity through the ages, medicinal plants have occupied a very important place throughout every civilization. Plentiful references of using plants as medicaments in the common medication practices have been acquired up until now, of which, *Withania somnifera* has awesome therapeutic qualities and has been considered with ample animal and human trials examining its impact on various conditions. The *in vivo* and *in vitro* pharmacological studies published so far have revealed that *W. somnifera* is viable against many illnesses and has the capability to exhibit properties such as hypolipidemic, anti-inflammatory, anti-oxidative, anti-microbial, anti-anxiety, aphrodisiac, anti-diabetic, anti-ulcer, anti-cancer, anti-depressant, anti-bacterial and possess endocrine, cardiopulmonary, central nervous systems, immunomodulation, and cardiovascular protection activities. The scientific evaluations of the above-delineated properties of *W. somnifera* have offer hands to help our justification behind a few traditional uses of *W. somnifera* and have been examined about in detail in this chapter.

W. somnifera, commonly called as the winter cherry, is classified as a shrub, and it contains a long, tapering, brittle, light-brown root, which is internally white in color and has a peculiar pungent odor of horse urine. The common name, Ashwagandha, is a combination of the Sanskrit word "ashva," which means horse, and "gandha," which means smell. The root of *W. somnifera* is observed abundantly in dry regions of the South Asia, Africa, South Africa, and North Africa, particularly in Egypt, Morocco, Congo, Jordan, and in South Asia particularly in India, Sri Lanka, Pakistan, Afghanistan, and Bangladesh. In India

W. somnifera is cultivated commercially in Madhya Pradesh, Punjab, Uttar Pradesh, Gujarat, and Rajasthan. It is widely used in the Ayurveda since the ancient ages for treating patients with disorders as an anti-inflammatory agent, nerve-tonic, anti-cancer agent and considered to exhibit a high healing potential in many cognitive and neurological disorders and as a love potion which stimulates sexual desire (Kulkarni and Dhir 2008). Furthermore, the leaves and root of *W. somnifera* are found to exhibit narcotic properties; the latter is also considered to be a diuretic and deobstruent.

W. somnifera is a completely respected herb of the Indian Ayurvedic system of medicine. *W. somnifera* comprises a wide range of active components consisting of withaferin A, withanone and other flavonoids exhibiting robust anti-oxidant activities. *W. somnifera* has been placed in the family Solanaceae, which is regularly known as the nightshade family. The origin of name *W. somnifera* begins from the Latin name for sleep-inducing somnifera. The systematic position of *W. somnifera* is depicted below.

Class: Dicotyledons
Order: Solanales
Family: Solanaceae
Genus: *Withania*
Species: *somnifera*

13.2 Bioactive Constituents of *W. somnifera*

W. somnifera has a great medicinal value and is believed to contain over 35 chemical constituents (Rastogi and Mehrotra 1998) that are of medicinal value, of which withanolides and alkaloids are the major secondary groups characterized from *W. somnifera* and are of great medicinal interest (Tripathi et al. 1996). Detailed chemical studies have revealed the occurrence of several phytocompounds, namely anaferine, anahygrine, cuscohygrine, choline, pseudotropine, tropine, pseudo-tropine, 3-α-gloyloxytropane, isopelletierine, somniferinine, somniferiene, sitoindosides, tropanol, withanine and withananine, phytosterols (e.g., β-sisterol), polyphenol (e.g., chlorogenic acid), benzopyrone (e.g., scopoletin) and steroidal lactones (e.g., withanolides A-Y) (Elsakka et al. 1990; Bone 1996). Several types of withanolides, extracted from the roots and leaves are attributed to the medicinal properties of this plant.

Withanolides are a group of 300 naturally occurring steroids, which are built on ergostane skeleton and resembles the active constituent, ginsenoside, a glycosylated triterpene, which is one of the active compounds of Asian ginseng (*Panax ginseng*). Fifty withanolides, and more than a few sitoindosides from *W. somnifera* have been isolated so far and reported (Sharma et al. 2013). Much of *W. somnifera*'s pharmacological activity comes from the two main withanolides, namely withaferin A, and withanolide D (Figure 13.1). HPLC (High-performance liquid chromatography) investigation of various organs of *W. somnifera* shows gradual decrease in the content of withanolide A from aerial parts, i.e., from young leaves to the root (Praveen and Murthy 2013).

The specific alkaloids reported in the leaves are choline, chlorogenic acid, cuscohygrine, somniferinine, somnine, somniferine,

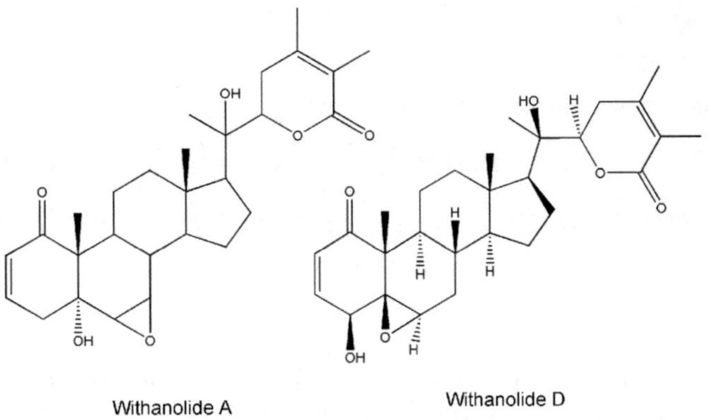

FIGURE 13.1 The structures of two main Withanolides of *Withania somnifera*, namely Withaferin A (IUPAC name: 3-rhamnopyranosyl(1-4)-glucopyranosyl-12-diacetoxy-20-hydroxywitha-5,24-dienolide) and Withanolide D (IUPAC name: 5,6:22,26-diepoxyergosta-2,24-diene-1,26-dione).

pseudo-withanine, withananine, pseudo-tropine, tropine, isopelletierine, 3-α-gloyloxytropane, anaferine, and anahydrine (Chaurasia et al. 2000). Withanolides are the primary compound present predominantly in the roots of *W. somnifera*, which are essential in its phenomenal therapeutic properties. The roots of *W. somnifera* additionally contain two acyl steryl glucoside namely sitoindoside VII and sitoindoside VIII (Ramarao et al. 1995). Shoots of *W. somnifera* consists of withaferin, a lactone belonging to the withanolide group (Mehrotra et al. 2013). Metabolic profiling by GC-MS and NMR spectroscopy reveals numerous chemically diverse metabolites encompassing sugars, tocopherols, polyols, fatty acids, organic acids, sterols, phenolic acids, semiessential proteinogenic amino acids, such as cysteine, aliphatic and aromatic acids, and withanamides. Squalene and tocopherol have been identified for the first time in the fruit, which are high anti-oxidant-containing natural compounds (Bhatia et al. 2013).

13.3 Scientifically Validated Pharmacological Activities of *W. somnifera*

13.3.1 Anti-microbial Activity

For the first time, Kurup (1956) demonstrated the antibacterial capabilities of *W. somnifera* against the pathogen, *Staphylococcus aureus*. Later, the major phytocompounds withanolides are described as having anti-microbial properties against a variety of bacteria and fungi (Mishra and Singh 2000; Owais 2005; Mehrotra et al. 2011). Numerous studies by various researchers reveal the antibiotic activity of the *W. somnifera* leaves and roots and have uncovered a significant measure of information up until this point. Ahmed et al. (2015) testified that withaferin A at a concentration of 10 μg/mL is high enough to suppress the growth of aerobic bacilli, gram-positive bacteria, acid-fast bacilli, fundal pathogen and certain viruses, like Ranikhet virus. Particularly, it was found to be potentially active against *Micrococcus pyrogens* var. *aureus*. Also, it inhibited glucose-6-phosphate dehydrogenase produced by *Bacillus subtilis*. Withaferin A possessing antibiotic activity is endorsed to the occurrence of the unsaturated lactone ring, which is stronger than penicillin. Meher et al. (2016) testified that the extract of the *W. somnifera* shrub is effective against the *Vaccinia* virus and the enteric parasite, *Entamoeba histolytica*, and additionally have protecting activity against a systemic *Aspergillus* infection. In addition, the author reveals that the activation of macrophage function was probably due to protective activity as observed from the increased phagocytosis activity and intracellular peritoneal macrophages killing, prompted by the treatment of *W. somnifera* in mice.

13.3.2 Anti-cancer Activity

Analoids of *W. somnifera* have been explored in many animal studies inspecting their effect on cancer (Singh et al. 2010). Cancer is a hyper proliferative condition of a cell that results in apoptosis, transformation, and metastasis. Millions of human beings are affected with numerous types of cancer and die every year (WHO 2007). *W. somnifera* is a pride herb of Ayurveda, which incredibly exhibits anti-cancerous activity against a variety of cancer cell lines because of the withaferin A (WFA) present in it (Mayola et al. 2011). Many studies have proved that withaferin A, D, and E have an incredible anti-tumor activity against human epidermoid carcinoma cell of nasopharynx (KB) cells. Further, *in vivo* anti-cancer activity was also evidenced against Ehrlich ascites carcinoma, sarcoma black (BL), sarcoma 180, and EO771 mammary tumors of mice. Moreover, Withaferin A and D owns growth inhibitory and radio-sensitizing activity on induced tumors in lab mouse (Devi et al. 1992; Pavan Kumar et al., 2018; Mathur et al. 2004). A study performed by Yadav et al. (2010) on five human cancer cell lines such as pc-3 and DU-145 (human prostate cancer cell lines), HCT-15 (human colon and colorectal adenocarcinoma), A-549 (adenocarcinomic human alveolar basal epithelial cell lines), and IMR (neuroblastoma cell lines) showed 0%–98% cytotoxicity activity ranging by 50% ethanol extract of leaves, stem, and root of *W. somnifera*; however, leaves showed the maximum activity.

Withaferin A and withanolide D are identified to be considerable anti-tumor and radio-sensitizing withanolides (Mathur et al. 2004). 1-Oxo-5ß, 6β-epoxy-withanolide is one of the major constituents of *W. somnifera*, which has been proved to reduce skin carcinoma induced by means of UV radiations (Davis and Kuttan 1998). Moreover, Withaferin A acts as an inhibitor of mitotic division of the human larynx carcinoma cells, particularly it affects the metaphase (Kuttan 1996). A study by Abdullah et al. (2014) reveals that mice with ovarian tumors treated with *W. somnifera* alone or in combination with an anti-cancer drug had 70%–80% reduction in tumor growth. Moreover, the remedy additionally averted metastasis. Research on animals reveals that *W. somnifera* decreases the levels of the Kappa B, which is a nuclear factor, reduces the intracellular tumor necrosis element, and complements apoptotic signaling in cancer-causing cell lines. In another study, by McKenna et al. (2015), the herb was tested for its anti-tumor activity in urethane-induced lung tumors in adult male mice. After the treatment with *W. somnifera* for 7 months, the histopathological nature of the lungs of treated animals becomes identical as the lungs of healthy animals.

Methanolic extract of *W. somnifera* has been used in stem cell proliferation. Moreover, it inhibited growth of breast, lung, CNS

and colon cancer cell lines through decreasing their viability in a dose-dependent manner and therefore proves to be a potent chemotherapeutic agent (Jayaprakasam et al. 2003). Likewise, withaferin A mediated the suppression of breast cancer cells viability by inducing apoptosis. Further, the cytoplasmic histone-related DNA fragmentation, DNA condensation, and cleavage of poly-(ADP-ribose)-polymerase were noticed (Lee et al. 2016). The chemo-preventive activity of *W. somnifera* is attributed partially to the anti-oxidant activity of the exo cytoskeleton structure alteration by means of annexin II which covalently binds (Falsey et al. 2006). Anti-tumor activity via inhibition of activity similar to proteasomal chymotrypsin (Yang et al. 2007) and the inhibition of protein kinase C or activation of caspase-3 resulting in apoptosis induction were additionally been explored by Senthil et al. (2007). This research is suggestive of anti-tumor activity in addition to enhancement of the outcomes of radiation (Bhattacharya et al. 1997).

W. somnifera leaves water extract (ASH-WEX) during *in vitro* and *in vivo* assays by Wadhwa et al. (2013) show anti-cancerous activity and triethylene glycol (TEG) has been recognized as an active anti-cancerous factor. In addition, ASH-WEX was selectively cytotoxic to most cancer cells and results in *in vivo* tumor suppression. Activation of tumor suppressor protein p53 and pRB in ASH-WEX and TEG treated cancer cell is also confirmed by means of Molecular analysis (Wadhwa et al. 2013). *W. somnifera* (leaves) flavonoid compounds display anti-cancerous activities in MCF-7, A549, and PA1 cell lines. The study confirms that *W. somnifera* leaves efficiently reduces cytotoxicity on MCF-7 (10 µg) than PA1 and A459 cancer cell lines (Nema et al. 2013).

A long-term tumor-inducing study on a Swiss albino mouse model indicates that *W. somnifera* roots inhibit the benzo(a) pyrene-induced forestomach papillomagenesis and results in 92% inhibition in tumor occurrence and its proliferation. Further, it also inhibits the 7,12-dimethylbenzanthracene-induced pores and skin papillomagenesis by 45% and 71% inhibition in tumor occurrence and its proliferation (Padmavathi et al. 2005). SiHa, squamous cervical cell line, treated with ethanolic extract of *W. somnifera* roots exhibits apoptosis (Jha et al. 2012). Further, it was found that tubocapsanolide A, a member of withanolide group present in *W. somnifera* roots, inhibits the cancer cells through suppressing the expression of oncoprotein Skp2 take part in the ubiquitin-mediated degradation (Chang et al. 2007; Sindhu and Santhi 2009).

Withaferin A complements radiation-caused apoptosis in Caki (human renal cancer) cells through inducing ROS, Bcl-2 down regulation, and Akt dephosphorylation (Yang et al. 2011). Yu et al. (2010) carried out an *in vitro* and *in vivo* study on pancreatic cancer cells to show withaferin *in vitro* anti-proliferative activity against pancreatic cancer cell strains p.c-1, MiaPaCa and BxPc3 with the IC50 value of 1.24, 2.93, and 2.78 µg respectively. In addition, Yu et al. (2010) proved that withaferin-A inhibits Hsp90, a chaperone protein that assembles or disassembles other macromolecules by binding with it via an ATP-independent mechanism, ends in protein degradation, and reveals *in vivo* anti-cancer activity towards pancreatic cancers.

Numerous studies were undergone by researchers on the anti-tumor activities of *W. somnifera* (Koduru et al. 2010; Muralikrishnan et al. 2010; Gupta et al. 2014). A novel bioactive compound withanolide sulfoxide acquired from the methanol extract of *W. somnifera* roots have been shown to inhibit nuclear transcription element-kappa-B (NFkB), cyclooxygenase, and proliferation of tumor (MCF-7) cells (Mulabagal et al. 2009). Animal cell cultures tested with *W. somnifera* reveal that the herb significantly reduces the level of NFkB, intercellular tumor necrosis component, and increases the apoptotic signaling in cancer cell lines (Patel et al. 2016). In addition, the leaf extract of *W. somnifera* produces anti-proliferative activity against MCF-7 human breast tumor cell lines (Yadav et al. 2010). Another feasible property of *W. somnifera* is its potential against most cancer by means of reducing tumor size (Ichikawa et al. 2006; Singh et al. 2010).

In a study, anti-tumor effect of *W. somnifera* in urethane-induced lung tumors in adult male mice was examined and evaluated (Agilandeswari et al. 2017). Following the treatment of *W. somnifera* for seven months continuously, the histological appearance of the lungs of animals treated with *W. somnifera* become much like the lungs of control animals (Srivastava et al. 2016). Diverse metabolites of *W. somnifera*, particularly the roots of *W. somnifera*, have proven to exhibit potent activities towards different types of cancers than the other parts of the plant. The withanolides and withaferins in addition to other metabolites inclusive of withanone (WN) and withanosides have been suggested to show potential activity against distinctive varieties of cancer cell lines (Rai et al. 2016).

13.3.3 Animal Studies on Immunomodulatory Activity

W. somnifera shows a widespread immunomodulatory activity in animal studies. *W. somnifera* administration has been suggested to be essential to prevent myelosuppression in three immunosuppressive drug treated mice namely cyclophosphamide, azathioprine, and prednisolone. Administration of *W. somnifera* in mice resulted in noticeably increased haemoglobulin, RBC count, platelet count, and body weight in *W. somnifera* treated mice (Qamar et al. 2012). Similarly, *W. somnifera* extract treatment increases the number of β-esterase producing cells in the bone marrow of CTX treated animals, than CTX alone treated animals (Naidu et al. 2006).

Treatment of *W. somnifera* extracts reduce leucopenia induced with sub-lethal dose of gamma radiation. *W. somnifera* extract experimentally increased the cell mediated immunity (CMI) in normal mice and root extract known to enhance the level of interleukin (IL-2), interferon gamma (IFN-γ), and granulocyte macrophages stimulating element (GM-CSF) in treated mice. Oral treatment with *W. somnifera* extract was effective in preventing the infection in guinea pigs by suppressing inflammatory cytokine responses (Verma et al. 2011). For increasing the phagocytic activity, it activates and mobilizes macrophages, potentiates the activity of lysosomal enzymes and acts as an anti-stress molecule and anti-inflammatory agent in mice and rat (Rasool and Varalakshmi 2006).

Ethanolic extract of *W. somnifera* induces humoral immune response in Swiss albino mice in the 7 days of administration and the results were found to be 12% as compared to control mice treated with cyclophosphamide (54%). The cell-mediated immune response was observed with enhancement in the values 19.27% in comparison with control mice treated with cyclosporine

(37.63%) (Verma and Kumar 2011). In tumor-infected mice, the effect of the methanolic extract on hematological parameter indicates an elevation in the number of RBCs as well as a decrease in WBCs than the untreated mice (Verma and Kumar 2011). Treatment with *W. somnifera* root extract (20 mg/dose/animal) in BALB/c mice showed an enhancement of the total WBC count, increase in bone marrow cell number and substantial increase in α-esterase cell quantity (Gupta et al. 2001).

Davis and Kuttan (2000) proved that management of *W. somnifera* extract in addition with sheep red blood cells (SRBC) antigen showed a remarkable increase in circulating antibody titer, the maximum range of plaque forming cells (985 pigmented cells (pc)/10 on fourth day) in spleen and inhibition of delayed type of hypersensitivity in mice. It also confirmed an elevation in phagocytic activity of peritoneal macrophages (76.5 pigmented cells/200 cells) while in comparison to control (31.5/200 cells). *W. somnifera* is an extremely good immunomodulator due to the presence of cyclophosphamide. It affects T- cell production, enhances neutrophil counts and produces enormous humoral response against sheep RBCs. The observed immunomodulatory activities may be because of withaferin A and withanolide D (Grover et al. 2010). When treated with *W. somnifera* plant extract orally, Swiss albino mice showed immunomodulatory effects on azoxymethane induced colon cancer (Muralikrishnan et al. 2010). The interferon gamma (IFNγ), interleukin-2 (IL-2) and granulocyte-macrophage colony-stimulating factor (GM-CSF) levels of cyclophosphamide treated mice become reversed to normal while administrated with *W. somnifera* extract. It also improves the immunopotent and myeloprotective effect by lowering the level of tumor necrosis factor-α (Davis and Kuttan 1999).

Withaferin A and withanolide E exhibits a precise immunosuppressive effect on both human and mice thymocytes (Kuttan 1996). Moreover, withaferin A affected both B and T lymphocytes. *W. somnifera* has a profound effect on the hematopoietic system by means of acting as an immunoregulator and chemoprotective agent.

13.3.4 Anti-Alzheimer and Anti-Parkinsonian Activity

Parkinson's disease is the loss of certain neurons of substaintia nigra pars compacta, and hence causes a severe neurogenerative disorder. Studies have proven that *W. somnifera* promotes antioxidant activity that protects nerve cells from harmful free radicals (Kuboyama et al. 2014). *W. somnifera* dietary supplements may also enhance brain function, memory response instances, and the ability to carry out responsibilities. Healthy men who were given 500 mg of the herb daily reported substantial improvements of their reaction time, task, and overall performance compared to men who received a placebo. *W. somnifera* is widely used against haloperidol or respine-induced catalepsy, hence provides an effective remedy for Parkinson's disease. Anti-Parkinsonian effect of *W. somnifera* extract has been reported because of the potent anti-oxidant, anti-peroxidative, and free radical scavenging properties (Chaurasia et al. 2007; Kumar and Kulkarni 2016). *W. somnifera* significantly reversed the catalepsy, tardive dyskinesia, and 6-hydroxydopamine elicited toxic manifestations and hence shows a new light in discovering new therapeutic drugs against Parkinson's disease (Lal and Rana, 2017).

Animal research recommends that *W. somnifera* might also reduce reminiscence and brain problems caused by injury or disease. It is investigated through the scientists of National Brain Research Centre (NBRC), New Delhi, that the extract of *W. somnifera* root administered on mice brought from Jackson Labs in the United States with Alzheimer's disease had reversed memory loss of the mice. Based on these findings scientists assumed that administration of *W. somnifera* root extract can be a powerful remedy of Alzheimer's disease in human, too. Management of haloperidol or reserpine substantially triggered catalepsy in mice. *W. somnifera* appreciably inhibited haloperidol or reserpine-triggered catalepsy and provide hope for treatment of Parkinson's disease (Kulkarni and Dhir 2008).

W. somnifera extract have been reported to have anti-Parkinsonian activity, which is aided by its anti-peroxidative, anti-oxidant, and free radical quenching activities in various scenarios. Tardive dyskinesia is a disorder resulting due to involuntary, repetitive body movements. A prolonged treatment with an indole alkaloid, reserpine on alternating days up to five days in mice noticeably encouraged tongue protrusions and vacuous chewing movements in rats. However, feeding *W. somnifera* root extract for a long duration (up to 4 weeks) to reserpine-treated animals significantly reduced the reserpine-prompted tongue protrusions and vacuous chewing movements in a dose-dependent manner (Naidu et al. 2006). Likewise, *W. somnifera* glycowithanolides (WAG) administration concomitantly with haloperidol for 28 days, inhibited the induction of the neuroleptic Tardive dyskinesia (Bhattacharya et al. 2000). Haloperidol-induced Tardive dyskinesia was also lessened by the anti-oxidant, vitamin E. But, however, it remains unaffected by means of the GABA mimetic anti-epileptic agent and sodium valproate which were administered for 28 days along with WAG. *W. somnifera* extensively reversed the catalepsy, tardive dyskinesia and 6- hydroxydopamine elicited poisonous manifestations and thus provide a novel therapeutic approach to the remedy of Parkinson's disorder (Bhattacharya et al. 2000).

Treating Parkinsonian mice models with the *W. somnifera* plant has proven to provide neuroprotection of dopaminergic neurons in substantia nigra pars compacta location of mid-brain. The prevailing literatures evaluate and enlighten the vital role of Indian ginseng to diminish neurodegenerative disorder such as Parkinson's disorder. However, vast research is needed to prove its therapeutic efficacy in neuronal disorders (Singh et al. 2015).

13.3.5 Anti-arthritic Properties

W. somnifera powder has been found effective against acute rheumatoid arthritis and reduces the discomfort associated with arthritis (Gupta and Surendra, 2014). This property has been attributed to the active primary metabolite withaferin A. The extracts of *W. somnifera* show autonomic ganglion blocking action and results in the hypotensive effect (Chaurasia et al. 2007). In a study by Kumar et al. (2015) with 90 patients who tested positive for rheumatoid factor and increased ESR level, a combination of *W. somnifera* and Sidh Makardhwaj (an Ayurvedic medicine, with herbal and mineral ingredients) treatment of Ayurveda decreased RA factor with increase in urinary mercury level. The findings of the above study suggest that the above combined Ayurvedic treatment may be used against rheumatoid arthritis (Kumar et al. 2015).

13.3.6 Anti-cortisol Activity

Cortisol, otherwise called as "stress hormone," is a hormone secreted by the adrenal gland and regulates blood sugar level. The adrenal gland releases cortisol in response to stress and when blood sugar levels get too low. Cortisol levels may additionally end up chronically elevated, which results in excessive blood sugar levels and elevated fat storage inside the stomach. Moreover, literature study reveals that *W. somnifera* administration helps in decreasing cortisol levels in chronically stressed people. In a research on the adults who were subjected to stress for a prolonged time, the group that provided with dietary supplements along with *W. somnifera* showed a significantly higher reduction in cortisol than the control group (Chandrasekhar et al. 2012).

13.3.7 Anti-depressant Activity

Studies reveal that *W. somnifera* may assist to reduce extreme depression. In a 60-day experimental study in stressed adults, people who took 600 mg/day stated a 79% reduction in extreme depression; however, at the identical time, the placebo group reported a merely 10% increase only (Bhattacharya et al. 1997).

13.3.8 Anti-diabetic Effect

The hyperglycaemic activity of Streptozotocin (STZ) is a result of decrease in pancreatic islet cell superoxide dismutase (SOD) activity essential to the accumulation of degenerative oxidative free radicals in islet-beta cells. Sarangi et al. (2013) studied the opportunities of the usage of the extracts obtained from roots and leaves of *W. somnifera* towards diabetes mellitus (DM) and also experimented their hypoglycaemic and hypolipidaemic outcomes upon rats which induced by STZ. The experimental results reveal that *W. somnifera* extract possesses hypoglycaemic and hypolipidaemic properties as a mechanism against diabetes mellitus. Moreover, Aggarwal et al. (2012) also confirms the hyperglycaemic activity of *W. sominifera*.

W. somnifera root and leaf extract exhibit hypoglycaemic and hypolipidaemic effect on alloxan drug-induced diabetic rats (Udayakumar et al. 2009). It was found that increase in hepatic metabolism, increased insulin secretion from pancreatic β-cells or insulin-sparing effect may be the key factors for the antidiabetic activity (Khatak et al. 2013). Hypoglycemic and diuretic activity of *W. somnifera* roots have also been assessed in humans. When administered, *W. somnifera* roots are found to lower the blood glucose level similar to administration of oral hypoglycemic drugs. Also elevated levels in urine sodium, urine quantity, with lowered serum cholesterol, triglycerides, and low-density lipoprotein levels were recorded by Andallu and Radhika, (2000). In another study, by Khatak et al. (2013) diabetes-induced rats exhibit tremendous anti-diabetic activity by *W. somnifera* while compared to standard diabetic drug Glibenclamide.

13.3.9 Anti-epileptic Activity

Administration of *W. somnifera* root extract to humans decreases the jerks and clonus by 70% and in animals by 10% with the dose of 100 mg/kg (Khare 2007). *W. somnifera* root extract reduced the severity of motor seizures induced by electric stimulation in right basilateral amygdaloid nuclear complex in brain by means of bipolar electrodes. The defensive impact of *W. somnifera* extract in convulsions has been stated to involve GABA-ergic mediation (Gamma-aminobutyric acid) (Singh et al. 2011).

13.3.10 Anti-inflammatory Activity

W. somnifera has been reported to increase natural killer cell activity and reduce markers of inflammation. Numerous animal studies also have proven that *W. somnifera* facilitates reducing the inflammation. It additionally decreases markers of inflammation, along with C-reactive protein (CRP) (Khan et al. 2015). Withaferin A is a prime biologically active steroid component involved in anti-inflammatory activity. Moreover, it is similar to the anti- inflammatory drug hydrocortisone sodium succinate (Giri, 2016). Primary segment of adjuvant-induced arthritis and formaldehyde-induced arthritis are models of inflammations which were subjected to distinct research to analyze the release of serum β-1 globulin during inflammation. The experiments showed exciting outcomes as Acute Phase Reactants (APR) has been stimulated in a completely lesser time but also suppressed the degree of inflammation (Gupta and Singh 2014).

W. somnifera uses inhibition of complement, delayed-type hypersensitivity, and lymphocyte proliferation to be a good anti-inflammatory agent (Rasool and Varalakshmi. 2006). The *W. somnifera* extracts have been subjected to several rheumatological conditions to exhibit its anti-inflammatory outcomes (AI Hindawi et al. 1992; Gupta and Singh. 2014). The extract was found to decrease the glycosaminoglycans content almost 100% in the granuloma tissue with the aid of oxidative phosphorylation by considerably lowering the ADP/O (phosphate/oxygen) ratio and resulted in elevating the Mg^{2+} dependent-ATPase enzyme activity and subsequent reduction in activity of succinate dehydrogenase inside the mitochondria of the granuloma tissue. The mechanism of action of *W. somnifera* has been found to be the cyclooxygenase inhibition (Balakrishnan et al. 2017). *W. somnifera* causes dose-dependent suppression of α_2-macroglobulin in rats infected with sub-plantar injection of carrageenan suspension which is an indicator for anti-inflammatory drugs while examining their serum. The doses of *W. somnifera* root powder for about 500–1200 mg/kg was given orally 3–4 h before inducing the inflammation and it was observed that highest activity (approximately 75%) reached at 1000 mg/kg (Gupta and Singh 2014).

Anti-inflammatory activity of *W. somnifera* is also due to the steroids present in it. Withaferin A is a primary component of *W. somnifera* and is much powerful as hydrocortisone sodium succinate dose, a regularly used anti-inflammatory drug (Khare 2007). Rats treated with powder of *W. somnifera* orally 1 h earlier than the injection of anti-inflammatory agent for 3 days produces anti-inflammatory responses that might be similar to hydrocortisone sodium succinate (Gupta and Singh 2014). Withaferin A has been observed to suppress the arthritic syndrome, effectively without any toxic impact. In arthritic syndrome, animals treated with hydrocortisone show weight loss however animals treated with withaferin-A showed gain in weight (Khare 2007).

13.3.11 Anti-oxidant Activity/Anti-aging Activity

Anti-oxidant protects the body against free radical damage. Many studies suggest potential phytochemical anti-oxidant because they are more secure to health and exhibits higher anti-oxidant activity than artificial anti-oxidants (Ansari et al. 2013). Different parts of *W. somnifera* extract exhibits *in vitro* anti-oxidant activity and suggests that those are the capable scavengers of radicals and protector of lipid membrane in order of leaves > fresh tubers > dry tubers. The anti-oxidant activity can be due to withanolides, glycowithanolides, and sitoindosides VII–X, so the study shows that *W. somnifera* might be proved as a natural supply of secure anti-oxidative agent (Sumathi et al. 2007).

The brain and nervous system contains high lipids as well as iron and hence are distinctly more at risk by damage due to free radical than any other tissues and promotes the generation of reactive oxygen (Halliwell and Gutteridge 2007). Nervous tissue damage by free radical may lead to neural damage in cerebral ischemia and can ably reduce the risk of getting older and affected by neurodegenerative diseases such as epilepsy, schizophrenia, Parkinson's and Alzheimer's disease (Kuboyama et al. 2005).

The active compounds isolated from *W. somnifera*, sitoindosides VII–X and withaferin A (glycowithanolides), are reported to elevate the levels of endogenous superoxide dismutase (SOD), catalase (CAT), glutathione peroxidase (GPX), and ascorbic acid, and also a remarkable decrease in lipid peroxidation by many researchers (Dhuley 2000; Bhattacharya et al. 2001; Jayaprakasam et al. 2004; Bhatnagar et al. 2005; Mirjalili et al. 2009). The reduction in the above-indicated enzyme activities results in the accumulation of oxidative free radicals and henceforth degenerative consequences occurs. *W. somnifera* acts as a powerful anti-oxidant by means of increasing the level of three obviously occurring enzymes which initiates anti-oxidant activity namely superoxide dismutase, catalases, and glutathione peroxidase in the brain of rats (Bhattacharya et al. 2001).

In a study by Jeyanthi and Subramanian (2009), GEN (gentamycin-induced nephrotoxicity)-treated rat displaying nephrocytotoxicity while administrated with *W. somnifera* (500 mg/kg) appreciably reverses the signs and symptoms of tubular necrosis. Typical results suggest that *W. somnifera*'s nephroprotective effect is a result of the improved antioxidant activity with natural anti-oxidants and scavenging free radicals found in it.

Published literature reveals that the root powder of *W. somnifera* plant prevents cadmium-induced oxidative stress in chickens and lead-induced oxidative damage in mouse (Chaurasia et al. 2000; Mahadik et al. 2008; Bharvi et al. 2010). Extensive studies performed to discover the anti-oxidant activity of *W. somnifera* extracts by using the usage of diverse *in vitro* techniques like 2, 2-diphenyl, 1-picrylhydrazyl (DPPH) radical scavenging activity, lowering the ability, compared with DMSO, hydroxy group reducing ability, estimation of overall phenol and estimation of ascorbic acid. The highest percentage of DPPH scavenging activity (83.07) was observed in polar flavonoid extract of *W. somnifera* (Aruoma and Cuppett 1997). In addition, free radical scavenging activity of ethanolic and aqueous extract of *W. somnifera* leaves studied through DPPH and Nitroblue Tetrazolium (NBT) techniques and the outcomes suggest that leaf possess anti-oxidant activity due to the presence of flavonoids and tannins inside the leaf extract (Panchawat 2011). A clinical trial carried out with 101 healthy male aging between (50–59) for twelve months to test the anti-aging property of *W. somnifera* and the research confirmed a substantial development in hemoglobin, red blood cell count, hair melanin, seated stature, decreased serum cholesterol, and preserved nail calcium in subjects (Alam et al. 2016).

13.3.12 Anti-stress, Adaptogenic, and Aphrodisiac Activity

Antistress activity associated with glycosides (sitoindosides VII and VIII) present in *W. somnifera* was reported by Bhattacharya (2000). Lohar et al. (1992) reveals that the higher concentrations of inorganic elements like Fe, Mg, K, and Ni in the roots of *W. somnifera* play a significant role in the diuretic and aphrodisiac activity. Likewise, Singh and Kumar (1998) suggest that the decoction made up of root boiled with milk and ghee is recommended for women with sterility.

Traditionally, *W. somnifera* has been orally administered to stabilize the mood of patients having behavior-related disturbances and experimentally, it is known to produce anti-depressant and anti-anxiety effects more than anti-depressant drug imipramine and anti-anxiety drug lorazepam (Archana and Namasivayam 1999). *W. somnifera* in India have been widely used as a tranquilizer, improving the reproductive and nervous system, rejuvenating body, improving vitality and recovery after chronic illness; therefore, it holds an important position similar to ginseng in China (Bhattacharya et al. 2000).

The anti-stress activity of *W. somnifera* was examined in rats for cold water swimming stress treatment and it was found that the animals that are treated reveals much stress tolerance. Similarly, a withanolide-free aqueous extract of roots shows dose-dependent anti-stress activity in mice (Khare 2007). Research on the extract of *W. somnifera* is reported to produce GABA-like activity that is responsible for their anti-anxiety effects (Mehta et al. 1991). GABA is an inhibitory neurotransmitter in the brain. It decreases the activity of neuron and inhibit nerve cells from over-firing and thus produces a calming effect.

Two acylsterylglucosides, sitoindoside VII and VIII, which are obtained from the *W. somnifera* root, show anti-stress activity and also indicate the low order of acute toxicity of compounds (Bhattacharya et al. 2000). Singh and Ramasamy (2017) revealed that a standardized *W. somnifera* root and leaf extract can recover chronically stressed humans from stress-related problems. It was found that *WS-E* treatment shows a greater dose-dependent response in parameters such as modified Hamilton anxiety (mHAM-A), serum cortisol, serum C-reactive protein, pulse rate, blood pressure and mean serum DHEAS (Dehydro epiandrosterone-serum), and hemoglobin. It also shows significantly greater responses in mean fasting blood glucose, serum lipid as compared to placebo (Auddy et al. 2008).

To study the adaptogenic activity of *W. somnifera*, chronical stress was induced in Wistar rats through mild, unpredictable foot shock stress procedure once in a day for continuously 21 days. The significant conditions like hyperglycaemia, glucose intolerance, gastric ulcerations, cognitive defects, elevated plasma cortisol levels, male sexual dysfunction, immunosuppression and mental depression induced by chronical stress were prevented by *W. somnifera* (25 and 50 mg/kg) and Panax ginseng (100 mg/kg), which were given 1h before foot shock for 21 days (Davis and Kuttan 2000).

In another study by Bhattacharya et al. (2000), adaptogenic activity of a novel withananolide free aqueous fraction from roots of *W. somnifera* was conducted on various parameters like hypoxia time, autoanalgesia, anti-fatigue effect, swimming performance time, immobilization-induced gastric ulceration, swimming induced gastric ulceration, hypothermia, and biochemical changes in the adrenal gland in dose-related manner (Singh et al. 2001).

13.3.13 Anxiolytic Effect

W. somnifera induces a soothing anxiolytic effect which is similar to the drug Lorazepam while examining the three standard anxiety tests, namely the elevated plus maze (test used to measure anxiety in laboratory animals), social interaction and the feeding latency in an unfamiliar environment. The anxiogenic agent, pentylenetetrazole increased the level of tribulin, an endocoid marker of clinical anxiety during the administration in rat brain but the combination of *W. somnifera* and Lorazepam, reduced the level of tribulin (Bhattacharya et al 2000). *W. somnifera* also exhibits an anti-depressant effect in two standard tests namely the "behavioral despair" and "learned helplessness" tests and the results were compared with the imipramine drug. The above study by Abdel-Magied et al. (2001) supports the use of *W. somnifera* as a good anxiolytic in nature.

13.3.14 Cardioprotective Activity

Myocardial infarction is one of the deadly cardiovascular disease. It is one of the most critical subjects of intense research by scientists (Bolli 1994). Nowadays there is an extended realization that herbs can maintain the stability of body and may have an impact on coronary heart diseases and its treatment by providing dietary substances (Dhar et al. 1968; Kushwaha et al. 2012). Even though the therapeutic properties of W. somnifera like immunomodulatory, adaptogenic, anti-oxidant, hypoglycemic and anti-cancerous are widely known (Lavie et al. 1965); very few studies are available on its cardioprotective activity. Recently, *W. somnifera* was confirmed to be a cardioprotective agent that provides a scientific reason for the rationale of the use of this medicinal plant in Ayurveda as Maharasayana (Gupta et al. 2004; Mohanty et al. 2004; Sehgal et al. 2012).

Mohanty et al. (2004) examined vitamin E (a well-known cardioprotective anti-oxidant) and hydroalcoholic extract of *W. somnifera* for their cardioprotective impact using rats affected with myocardial necrosis induced by isoprenaline (isoproterenol). The basic parameters tested were hemodynamic, histopathological, and biochemical and in comparison both the drugs naturally decrease the malondialdehyde levels to restore the myocardial anti-oxidant reputation and hold membrane integrity. Cardioprotective impact of both *W. somnifera* and vitamin E was histopathologically examined and *W. somnifera* indicates maximum cardioprotective effect at 50 mg/kg dose than vitamin E (Mohanty et al. 2004). Moreover, the extract of *W. somnifera* was evaluated for cardioprotective activity in the cardiovascular and respiratory systems in dogs and frogs. The alkaloids of *W. somnifera* show hypotensive, cardiac, and respiratory stimulant action in dogs (Malhotra et al. 1981). The cardio-inhibitory action in puppies seemed to be because of ganglion blocking and direct cardio-depressant actions (Mishra et al. 2000). In another study, left ventricular heart rate dysfunction has been decreased, left ventricular rate of peak positive and negative pressure exchange was decreased and the left ventricular end-diastolic pressure has been elevated within the animal models increasing the overall cardiopulmonary function (Mohanty et al. 2004). In addition, *W. somnifera* is also used as a general tonic, because of its beneficial outcomes on the cardiopulmonary system.

13.3.15 Effect on Nervous System

W. somnifera is reported to have the sedative rather than simulative action on the central nervous system, making it a superior medicine in exhaustion with nervous irritability. *W. somnifera* alters the concentration of neurotransmitters that are known to play an important role in brain processes such as memory. The effect on nervous system are associated with Ashwagandholine (root extracts), which potentiates barbiturate, ethanol and urethane-induced hypnosis in mice. In addition, it causes relaxant and antispasmodic effects in intestinal, uterine, tracheal, and vascular muscles against smooth muscle contractions producing agents. The study reports also suggest that the cortical and basal forebrain cholinergic-signal transduction cascade events were potentially influenced by the bioactive compounds of *W. somnifera* (Malhotra et al. 1965). *W. somnifera* extracts exhibits the cognition and memory enhancing effects and this was explained by cortical muscarinic acetylcholine receptor capacity while induced by the drug (Schliebs et al. 1997). In general, *W. somnifera* is traditionally used as a tonic and nootropic agent and also eliminates the scopolamine-induced memory deficits in mice. *W. somnifera* extracts also show an anti-Parkinsonian effect on neuroleptic-induced catalepsy by inhibiting haloperidol or reserpine-induced catalepsy resulting in properties such as potent anti-oxidant, anti-peroxidative, and free-radical quenching activities (Ahmad et al. 2005).

A variety of neural disorders can be treated with the traditional medicine made of *W. somnifera*. Chronic neurodegenerative conditions can also be treated with *W. somnifera* extract (Saykally 2017). *W. somnifera* is treated against the neurodegenerative disorders like Alzheimer, Parkinson, and Huntington at any stage of disease. Roots of *W. somnifera* possess potential activity against neurotic atrophy, synaptic loss, along with GABA mimetic effect and also support the formation of dendrites. This potential activity was due to glycowithanolides withaferin A VII–X therapeutic activity (Abbas et al. 2004, 2005 and Ahmad et al. 2005). Scientists of the institute of Natural Medicine at the Toyama Medical and Pharmaceutical University of Japan, by using the valid model of damaged nerve cell and impaired nerve signaling pathway, showed that the *W. somnifera* supports significant regeneration of the axon and dendrites of nerve cells. *W. somnifera* plant extracts supports the reconstruction of synapses or networks of the nervous system and it may act as potential treatment of Alzheimer and Parkinsons, which are two of the primary neurodegenerative diseases (Kuboyama et al. 2005). In another study at the same institute, Tohda et al. (2000)

found that *W. somnifera* help in growth of nerve cell dendrites which suggests that *W. somnifera* help in healing the brain tissue changes that accompany dementia.

W. somnifera extract treatment to rats at concentration 100, 200 and 300 mg/kg dose for 3 weeks significantly reverse all the parameters such as glutathione content, activities of glutathione-S-transferase, glutathione reductase, glutathione peroxidase, etc., to normal in dose dependent manner. This demonstrates that *W. somnifera* extract can be treated against Parkinson's disease by neuronal injury protection (Ahmad et al. 2005). *W. somnifera* root extract increases the capacity of cortical muscarinic acetylcholine receptor by affecting mainly the cortical and basal forebrain cholinergic signal transduction cascade which leads to the cognition-enhancing and memory-improving effect in animals and humans (Schliebs et al. 1997). The pre-treatment of *W. somnifera* results in attenuation of cerebral ischemia-reperfusion and long-term hypoperfusion-induced alterations in rats and hence reveals that *W. somnifera* exhibits neuroprotective activity (Trigunayat et al. 2007). In another study by Jain et al. (2001), female albino rats were subjected to immobilization stress for 14 h and then treated with the *W. somnifera* root extract which shows a notable reduction (80%) in the number of hippocampal sub-region degenerative cells of experimental animals Sankar et al. (2007) studied the influence of *W. somnifera* root extract on Parkinsonism in mice intoxicated with 1-methyl-4-phenyl-1,2,3,6-tetrahydropyridine (MPTP). The results show a remarkable improvement in behavior, anti-oxidant status and also a considerable decrease in level of lipid peroxidation.

13.3.16 Hypolipidemic Effect

W. somnifera root powder is used to treat defects in total lipids, cholesterol, and triglycerides in hypercholesteremic animals. In addition, notably elevated plasma HDL-cholesterol levels and bile acid content of liver can also be treated with *W. somnifera*. A massive lowering of lipid-peroxidation level lowering took place in *W. somnifera* comparably in administered hypercholesteremic animals than the placebo animals (Bhattacharya et al. 2001). In another study with aqueous extract of the *Withania coagulans* in hyperlipidemic rats, a weight-loss plan for 7 weeks resulted in lowering of serum cholesterol, triglycerides and lipoprotein levels substantially. In addition, *W. somnifera* also exhibits hypolipidemic activity in triton-induced hypercholesterolemia. Degenerative adjustments were found to be relatively lesser than hyperlipidemic controls when the liver tissues of handled hyperlipidemic rats were subjected to histopathological examination (Visavadiya and Narasimhacharya 2006).

In another study, roots of *W. somnifera* were subjected for hypoglycemic, diuretic, and hypocholesterolemic effects in human trails. Six models of each moderate noninsulin-dependent diabetes mellitus (NIDDM) and moderate hypercholesterolemic subjects have been administered with the powder of roots of *W. somnifera* continuously for 30 days. The blood and urine samples of the subjects were also studied at the side of nutritional pattern earlier and on the end of administration. Blood glucose level was found to be lower, which is similar to an effect on oral hypoglycemic drug streptozotocin (Sarangi et al. 2013).

13.3.17 Insomnia

Insomnia is a common sleeping disorder which results in difficult to falling asleep or maintaining it. The available drugs currently for insomnia develop adverse effects and hence alternative choice of treatment for insomnia relies on natural therapies. The plant extract of *W. somnifera* induces sleep and was widely used in Indian system of traditional home medicine and also in Ayurveda. A study clearly demonstrates that the triethylene glycol extracted from the leaves of *W. somnifera* is an active sleep-inducing component and could potentially be useful for insomnia therapy (Kaushik et al. 2017).

13.3.18 Nootropic Effect

Outcomes of sitoindosides VII–X and withaferin obtained with the administration of aqueous methanol extract of roots of *W. somnifera* have been researched in rats. The acetylcholinesterase (ache) activity inside the lateral septum and globus pallidus was improved noticeably by *W. somnifera* compounds and ache activity in the vertical diagonal band was decreased. In addition, enhanced M1-muscarinic-cholinergic receptor binding sites have been increased as the number of cortical regions including cingulate, frontal, parietal, and retrospinal cortex were increased and these results in the Nootrophic activity of *W. somnifera* (Dhuley 2001).

Oral administration of withanoside IV shows that sominone, an aglycone of withanoside IV, is the primary metabolite in nootropic activity. Rat cortical neurons damaged by ABETA (amyloid-β), when injected with sominone-induced axonal and dendritic regeneration and synaptic reconstruction considerably (Kuboyama et al. 2006).

In another study, neurite outgrowth of rat was induced when inoculated with withanoside IV in cortical neurons. Beside, memory deficits in ABETA-injected mice were substantially improved and loss of axons, dendrites, and synapses were averted by oral administration of withanoside IV extensively. Withanolide IV additionally eliminates the neuronal dysfunction in Alzheimer's disease with metabolism in somninone (Zhao et al. 2002).

Chronic *W. somnifera* administration considerably reverses reserpine-induced retention deficits (Kumar and Kulkarni 2016). In a distinct study *W. somnifera* root extract improved retention of a passive avoidance task in a step-down paradigm in mice. Additionally, the scopolamine-induced disruption of acquisition and retention were reversed and amnesia that was induced through acute treatment with electroconvulsive shock (ECS) was reverted by *W. somnifera* (Dhuley 2001). Chronic treatment in mice receiving ECS and received a disrupted memory consolidation drastically progressed memory consolidation on day 7 after subjected for 6 successive days at 24 h intervals. On the basis of these results, it's far recommended that *W. somnifera* reveals a nootropic impact in mice (Dhuley 2000).

13.3.19 Rejuvenating Effect

W. somnifera was reported to possess growth-promoting effect when administered alone in a powder form or in combination with other drugs. The growth promoting activity is attributed to withanolides (Budhiraja et al. 2000). The study conducted in both children and old age people registered a remarkable change in parameters like hemoglobin, packed cell volume (PCV), mean cell volume (MCV), serum iron, body weight, hand grip, and total proteins. In addition, serum cholesterol level was lowered, the nail calcium was preserved in adults and ESR level was lowered considerably. Among the subjects, 71.4% reported to have increased health vigorously (Kuppurajan et al. 1980). These studies prove that *W. somnifera* can be used as a general health tonic and possess rejuvenating effect in both younger and older populations. Similar results were depicted by Sandhu et al. (2010).

13.3.20 Sexual Behavior

W. somnifera is also used as a tonic in the treatment of spermatopathia, impotence and seminal depletion (Khanna et al. 2006) and the men who used the herb enjoyed higher vigor performance (Boone 1998). *W. somnifera* enables increased testosterone levels and considerably increases sperm quality and fertility in men. In a study of 75 infertile men, the group that has been treated with *W. somnifera* had an increase of sperm count and motility (Chandrasekhar et al. 2012).

In animal studies, orally administered methanolic root extract of *W. somnifera* in rats at the dose of 3,000 mg/kg/day for seven days triggered a remarkable impairment in libido, sexual performance, sexual vigor, and penile erectile dysfunction. The sexual behavior of rats were evaluated 7 days prior to treatment, on day 3, 7, 14, and 30 post-treatment by pairing every male with a receptive female, which were partly reversible on cessation of treatment. This anti-masculine impact has no connection with the testosterone levels however attributed to hyperprolactinemic, GABA-ergic, serotonergic, or sedative activities of *W. somnifera* root extract; however, Andallu and Radhika (2000) reveal that *W. somnifera* roots may be adverse to male sexual competence.

13.4 Conclusion

Since ancient times, natural herbal products have been used to treat different type of diseases in numerous ways. The *W. somnifera* contains different type of phytoconstituents showing different pharmacological activities. Herbal medicines show very fewer side effects and hence they were used to treat various ailments. Traditional uses of *W. somnifera* keep on being a vital common solution for different sicknesses and researchers have confirmed them throughout the years. In spite of the fact that roots, leaves and fruits of *W. somnifera* show a tremendous therapeutic potential, root of *W. somnifera* is turned out to be very potential. Multiple health benefits of this herb make it as an ideal rejuvenator of physical and mental wellbeing, in treatment of diabetes, to treat pressure, elevated cholesterol and triglyceride level, malignancy, neurodegenerative disease, nervousness, different bacterial and fungal infections. In conclusion, *W. somnifera* is one among potential herbs in the world. Further studies are required to find many more activities of this plant. In short, the drug can be a broad-spectrum medicine for the treatment of various disorders and also can be used by healthy individual for maintaining a positive health.

REFERENCES

Abbas, S.S., Bhalla, M. and Singh, N., 2005. A clinical study of Organic Ashwagandha in some cases of uterine tumors (fibroids) and dermatofibrosarcoma. In *Proceedings of the Workshop on Essential Medicines, Adverse Drug Reactions and Therapeutic Drug Monitoring* (pp. 143–144).

Abbas, S.S., Singh, V., Bhalla, M. and Singh, N., 2004. Clinical study of organic Ashwagandha in cases of Parkinsonism, neuropathy, paralysis and uterine tumours (fibroids and other tumours) including Cutaneous Endodermal carcinoma. In *Proc of the National Seminar on "Eco-friendly Herbs of Ayurveda in Healthcare of Mankind: A Strategy for Scientific Evaluation an Uniform Standardization"- Lucknow* (Vol. 81).

Abdel-Magied, E.M., Abdel-Rahman, H.A. and Harraz, F.M., 2001. The effect of aqueous extracts of Cynomorium coccineum and *Withania somnifera* on testicular development in immature Wistar rats. *Journal of Ethnopharmacology*, 75(1), pp. 1–4.

Achar, G.S., Prabhakar, B.T., Rao, S., George, T., Abraham, S, Sequeiram, N. and Baliga, M.S., 2018. Scientific validation of the usefulness of *Withania somnifera* Dunal in the prevention and treatment of cancer. In: Akhtar M., Swamy M. (Eds.), *Anticancer Plants: Properties and Application.* Singapore, Springer.

Aggarwal, R., Diwanay, S., Patki, P. and Patwardhan, B. 2012. Studies on immunomodulatory activity of *Withania somnifera* (Ashwagandha). *Journal of applied Pharmaceutical Science*, 2(1), pp. 170–175.

Agilandeswari, D. and Mohan Maruga Raja, M.K., 2017. Antimicrobial studies on rasam: A south Indian traditional functional food. *World Journal of Pharmaceutical Research*, 6(5), pp. 766–774.

Ahmad, M., Saleem, S., Ahmad, A.S., et al. 2005. Neuroprotective effects of *Withania somnifera* on 6-hydroxydopamine induced Parkinsonism in rats. *Human & Experimental Toxicology*, 24(3), pp.137–147.

Ahmed Abdullah, W., Mohamed, A., Nasser, E.A. and Doaa, E. 2014. Potential toxicity of Egyptian Ashwagandha: Significance for their therapeutic bioactivity and anti-cancer properties, *International Journal of Science and Research*, 4(2), pp. 2170–2176.

Al-Hindawi, M.K., Al-Khafaji, S.H. and Abdul-Nabi, M.H., 1992. Anti-granuloma activity of Iraqi *Withania somnifera*. *Journal of Ethnopharmacology*, 37(2), pp. 113–116.

Andallu, B. and Radhika, B., 2000. Hypoglycemic, diuretic and hypocholesterolemic effect of winter cherry (*Withania somnifera*, Dunal) root. *Indian Journal of Experimental Biology*, 38(6), pp. 607–609.

Ansari, A.Q., Ahmed, S.A., Waheed, M.A. and Juned, S., 2013. Extraction and determination of antioxidant activity of *Withania somnifera* Dunal. *European Journal of Experimental Biology*, 3(5), pp. 502–507.

Archana, R. and Namasivayam, A., 1998. Antistressor effect of *Withania somnifera*. *Journal of Ethnopharmacology*, 64(1), pp. 91–93.

Aruoma, O.I. and Cuppett, S.L. Eds., 1997. *Antioxidant Methodology: In Vivo and in Vitro in Vitro Concepts.* The American Oil Chemists Society. pp. 2–29.

Auddy, B., Hazra, J., Mitra, A., Abedon, A. and Ghosal, S.A., 2008. Standardized withanis somnifera extract significantly reduces stress-related parameters in chronically stressed humans: A double-blind randomized placebo-controlled study. *Journal of the American Nutraceutical Association*, 11(1), pp. 50–56.

Balakrishnan., A.S., Nathan, A.A., Kumar, M., Ramamoorthy, S. and Mothilal, S.K.R., 2017. *Withania somnifera* targets interleukin-8 and cyclooxygenase-2 in human prostate cancer progression. *Prostate International Journal*, 5(2): 75–83.

Begum, V.H. and Sadique, J., 1988. Long term effect of herbal drug *Withania somnifera* on adjuvant induced arthritis in rats. *Indian Journal of Experimental Biology*, 26(11), pp. 877–882.

Bharavi, K., Reddy, A.G., Rao, G.S., Reddy, A.R. and Rao, S.R., 2010. Reversal of cadmium-induced oxidative stress in chicken by herbal adaptogens *Withania somnifera* and Ocimum sanctum. *Toxicology International*, 17(2), p. 59.

Bhatia, A., Bharti, S.K., Tewari, S.K., Sidhu, O.P. and Roy, R., 2013. Metabolic profiling for studying chemotype variations in *Withania somnifera* (L.) Dunal fruits using GC–MS and NMR spectroscopy. *Phytochemistry*, 93, pp. 105–115.

Bhatnagar, M., Sisodia, S.S. and Bhatnagar, R., 2005. Antiulcer and antioxidant activity of Asparagus racemosus Willd and *Withania somnifera* Dunal in rats. *Annals of the New York Academy of Sciences*, 1056(1), pp. 261–278.

Bhattacharya, A., Ghosal, S. and Bhattacharya, S.K., 2001. Anti-oxidant effect of *Withania somnifera* glycowithanolides in chronic foot-shock stress-induced perturbations of oxidative free radical scavenging enzymes and lipid peroxidation in rat frontal cortex and striatum. *Journal of Ethnopharmacology*, 74(1), pp. 1–6.

Bhattacharya, S.K., Bhattacharya, A., Sairam, K. and Ghosal, S., 2000. Anxiolytic-antidepressant activity of *Withania somnifera* glycowithanolides: an experimental study. *Phytomedicine*, 7(6), pp. 463–469.

Bhattacharya, S.K., Satyan, K.S. and Chakrabarti, A., 1997. Effect of Trasina, an Ayurvedic herbal formulation, on pancreatic islet superoxide dismutase activity in hyperglycaemic rats. *Indian Journal of Experimental Biology*, 35(3), pp. 297–299.

Bolli, R., 1994. Myocardial ischaemia: Metabolic disorders leading to cell death. *Revista portuguesa de cardiologia: orgao oficial da Sociedade Portuguesa de Cardiologia= Portuguese Journal of Cardiology: An Official Journal of the Portuguese Society of Cardiology*, 13(9), pp. 649–653.

Bone, K. and Morgan, M., 1996. *Clinical Applications of Ayurvedic and Chinese Herbs: Monographs For the Western Herbal Practitioner.* Phytotherapy Press.

Boone, K., 1998. Withania: The Indian ginseng and anti-aging adaptogen. *Nutrition and Healing*, 5(6), pp. 5–7.

Budhiraja, R.D., Krishan, P. and Sudhir, S., 2000. Biological activity of withanolides. *Journal of Scientific and Industrial Research*, 59(11), pp. 904–911.

Chandrasekhar, K., Kapoor, J. and Anishetty, S., 2012. A prospective, randomized double-blind, placebo-controlled study of safety and efficacy of a high-concentration full-spectrum extract of ashwagandha root in reducing stress and anxiety in adults. *Indian Journal of Psychological Medicine*, 34(3), p. 255.

Chang, H.C., Chang, F.R., Wang, Y.C., et al. 2007. A bioactive withanolide Tubocapsanolide A inhibits proliferation of human lung cancer cells via repressing Skp2 expression. *Molecular Cancer Therapeutics*, 6(5), pp. 1572–1578.

Chaurasia, S.S., Panda, S. and Kar, A., 2000. *Withania somnifera* root extract in the regulation of lead-induced oxidative damage in male mouse. *Pharmacological Research*, 41(6), pp. 663–666.

Chaurasiya, N.D., Gupta, V.K. and Sangwan, R.S., 2007. Leaf ontogenic phase-related dynamics of withaferin a and withanone biogenesis in ashwagandha (*Withania somnifera* Dunal.): An important medicinal herb. *Journal of Plant Biology*, 50(4), p. 508.

Davis, L. and Kuttan, G., 1998. Suppressive effect of cyclophosphamide-induced toxicity by *Withania somnifera* extract in mice. *Journal of Ethnopharmacology*, 62(3), pp. 209–214.

Davis, L. and Kuttan, G., 1999. Effect of *Withania somnifera* on cytokine production in nol and cyclophosphamide treated mice. *Immunopharmacology and Immunotoxicology*, 21(4), pp. 695–703.

Davis, L. and Kuttan, G., 2000. Immunomodulatory activity of *Withania somnifera*. *Journal of Ethnopharmacology*, 71(1–2), pp. 193–200.

Devi, P.U., Sharada, A.C., Solomon, F.E. and Kamath, M.S., 1992. *In vivo* growth inhibitory effect of *Withania somnifera* (Ashwagandha) on a transplantable mouse tumor, Sarcoma 180. *Indian Journal of Experimental Biology*, 30(3), pp. 169–172.

Dhar, M.L., Dhar, M.M., Dhawan, B.N., Mehrotra, B.N. and Ray, C., 1968. Screening of Indian plants for biological activity: I. *Indian Journal of Experimental Biology*, 6(4), pp. 232–247.

Dhuley, J.N., 2000. Adaptogenic and cardioprotective action of ashwagandha in rats and frogs. *Journal of Ethnopharmacology*, 70(1), pp. 57–63.

Dhuley, J.N., 2001. Retracted: Nootropic-like effect of ashwagandha (*Withania somnifera* L.) in mice. *Phytotherapy Research*, 15(6), pp. 524–528.

Elsakka, M., Grigorescu, E., Stănescu, U. and Dorneanu, V., 1990. New data referring to chemistry of *Withania somnifera* species. *Revista medico-chirurgicala a Societatii de Medici si Naturalisti din Iasi*, 94(2), pp. 385–387.

Falsey, R.R., Marron, M.T., Gunaherath, G.K.B., et al. 2006. Actin microfilament aggregation induced by withaferin A is mediated by annexin II. *Nature Chemical Biology*, 2(1), p. 33.

Giri, D.K.R., 2016. Comparative study of anti-inflammatory activity of *Withania somnifera* (Ashwagandha) with hydrocortisone in experimental animals (Albino rats). *Journal of Medicinal Plants Studies*, 4(1), pp. 78–83.

Grover, A., Shandilya, A., Punetha, A., Bisaria, V.S. and Sundar, D., 2010. Inhibition of the NEMO/IKKβ association complex formation, a novel mechanism associated with the NF-κB activation suppression by *Withania somnifera*'s key metabolite withaferin A. In *BMC Genomics* 11(4), pp. S25.

Gupta, A. and Singh, S., 2014. Evaluation of anti-inflammatory effect of Withania somnifera root on collagen-induced arthritis in rats. *Pharmaceutical Biology*, 52(3), pp. 308–320.

Gupta, G.L. and Rana, A.C., 2007. PHCOG MAG.: Plant review *Withania somnifera* (Ashwagandha): A review. *Pharmacognosy Reviews*, 1(1), pp. 129–136.

Gupta, M.L., Misra, H.O., Kalra, A. and Khanuja, S.P.S., 2004. Root-rot and wilt: A new disease of ashwagandha (*Withania somnifera*) caused by *Fusarium solani*. *Journal of Medicinal and Aromatic Plant Sciences*, 26(2), pp. 285–287.

Halliwell, B. and Gutteridge, J.M., 2015. *Free Radicals in Biology and Medicine*. New York, Oxford University Press.

Ichikawa, H., Takada, Y., Shishodia, S., Jayaprakasam, B., Nair, M.G. and Aggarwal, B.B., 2006. Withanolides potentiate apoptosis, inhibit invasion, and abolish osteoclastogenesis through suppression of nuclear factor-κB (NF-κB) activation and NF-κB–regulated gene expression. *Molecular Cancer Therapeutics*, 5(6), pp. 1434–1445.

Ilayperuma, I., Ratnasooriya, W.D. and Weerasooriya, T.R., 2002. Effect of *Withania somnifera* root extract on the sexual behaviour of male rats. *Asian Journal of Andrology*, 4(4), pp. 295–298.

Jain, S., Shukla, S.D., Sharma, K. and Bhatnagar, M., 2001. Neuroprotective effects of *Withania somnifera* Dunn. in Hippocampal Sub-regions of Female Albino Rat. *Phytotherapy Research*, 15(6), pp. 544–548.

Jayaprakasam, B., Strasburg, G.A. and Nair, M.G., 2004. Potent lipid peroxidation inhibitors from *Withania somnifera* fruits. *Tetrahedron*, 60(13), pp. 3109–3121.

Jayaprakasam, B., Zhang, Y., Seeram, N.P. and Nair, M.G., 2003. Growth inhibition of human tumor cell lines by withanolides from *Withania somnifera* leaves. *Life Sciences*, 74(1), pp. 125–132.

Jeyanthi, T. and Subramanian, P., 2009. Nephroprotective effect of *Withania somnifera*: A dose-dependent study. *Renal Failure*, 31(9), pp. 814–821.

Jha, A.K., Jha, M. and Kaur, J., 2012. Ethanolic Extracts of *Ocimum sanctum, Azadirachta indica* and *Withania somnifera* cause apoptosis in SiHa cells. *Research Journal of Pharmaceutical, Biological and Chemical Sciences*, 3(2), pp. 557–562.

Kaushik, M.K., Kaul, S.C., Wadhwa, R., Yanagisawa, M. and Urade, Y., 2017. Triethylene glycol, an active component of Ashwagandha (*Withania somnifera*) leaves, is responsible for sleep induction. *PLoS One*, 12(2), p. e0172508.

Khan, M.A., Subramaneyaan, M., Arora, V.K., Banerjee, B.D. and Ahmed, R.S., 2015. Effect of *Withania somnifera* (Ashwagandha) root extract on amelioration of oxidative stress and autoantibodies production in collagen-induced arthritic rats. *Journal of Complementary and Integrative Medicine*, 12(2), pp. 117–125.

Khanna, P.K., Kumar, A., Ahuja, A. and Kaul, M.K., 2006. Multipurpose efficacious medicinal plant *Withania somnifera* (L.) dunal (Ashwagandha). *Journal of Plant Biology*, 33(3), p. 185.

Khare, C.P., 2007. *Indian Medicinal Plants: An Illustrated Dictionary*. 1st Indian Reprint. New Delhi, India, Springer, p. 28.

Khatak, M., Sehrawat, R. and Khatak, S., 2013. A comparative study: Homoeopathic medicine and a medicinal plant *Withania somnifera* for antidiabetic activity. *Journal of Pharmacognosy and Phytochemistry*, 2(3), pp. 109–112.

Koduru, S., Kumar, R., Srinivasan, S., Evers, M.B. and Damodaran, C., 2010. Notch-1 inhibition by Withaferin-A: A therapeutic target against colon carcinogenesis. *Molecular Cancer Therapeutics*, 9(1), pp. 202–210.

Kuboyama, T., Tohda, C. and Komatsu, K., 2005. Neuritic regeneration and synaptic reconstruction induced by withanolide A. *British Journal of Pharmacology*, 144(7), pp. 961–971.

Kuboyama, T., Tohda, C. and Komatsu, K., 2006. Withanoside IV and its active metabolite, sominone, attenuate Aβ (25–35)-induced neurodegeneration. *European Journal of Neuroscience*, 23(6), pp. 1417–1426.

Kuboyama, T., Tohda, C. and Komatsu, K., 2014. Effects of Ashwagandha (roots of *Withania somnifera*) on neurodegenerative diseases. *Biological and Pharmaceutical Bulletin*, 37(6), pp. 892–897.

Kulkarni, S.K. and Dhir, A., 2008. *Withania somnifera*: An Indian ginseng. *Progress in Neuro-Psychopharmacology and Biological Psychiatry*, 32(5), pp. 1093–1105.

Kumar, G., Srivastava, A., Sharma, S.K., Rao, T.D. and Gupta, Y.K., 2015. Efficacy & safety evaluation of Ayurvedic treatment (Ashwagandha powder & Sidh Makardhwaj) in rheumatoid arthritis patients: A pilot prospective study. *The Indian Journal of Medical Research*, 141(1), p. 100.

Kuppurajan, K. and Bone, K., 1980. As cited in Bone K. *Clinical Applications of Ayurvedic and Chinese Herbs. Monographs for the Western Herbal Practitioner*. Phytotherapy Press, pp. 137–141.

Kurup, U.S., 1956. The Cure. Antibiotic principals of the leaves of *W.sonmnifera*. *Current Science*, 25, pp. 57–60.

Kushwaha, S., Betsy, A. and Chawla, P., 2012. Effect of *Ashwagandha* (*Withania somnifera*) root powder supplementation in treatment of hypertension. *Studies on Ethno- Medicine*, 6(2), pp. 111–115.

Kuttan, G., 1996. Use of *Withania somnifera* Dunal as an adjuvant during radiation therapy. *Indian Journal of Experimental Biology*, 34(9), pp. 854–856.

Lavie, D., Glotter, E., Shvo, Y. 1965. Constituents of *Withania somnifera* Dun. Part IV The structure of withaferin-A. *Journal of the American Chemical Society*, 30, pp. 7517–7531.

Lavie, D., Kirson, I. and Glotter, E., 1968. Constituents of *Withania somnifera* Dun. Part X. The Structure of Withanolide D. *Israel Journal of Chemistry*, 6(5), pp. 671–678.

Lohar, D.R., Chaturvedi, D. and Varma, P.N., 1991. Mineral elements of a few medicinally important plants. *Indian Drugs*, 29(6), pp. 271–273.

Mahadik, K.R., Gopu, C.L., Gilda, S.S. and Paradkar, A.R., 2008. Comparative evaluation of antioxidant potential of Ashwagandha arishta and self-generated alcoholic preparation of *Withania somnifera* Dunal. *Planta Medica*, 74(09), p. PA288.

Malhotra, A., Penpargkul, S., Fein, F.S., Sonnenblick, E.H. and Scheuer, J., 1981. The effect of streptozotocin-induced diabetes in rats on cardiac contractile proteins. *Circulation Research*, 49(6), pp. 1243–1250.

Malhotra, C.L., Das, P.K., Dhalla, N.S. and Prasad, K. 1981. Studies on Withania ashwagandha, Kaul. III. The effect of total alkaloids on the cardiovascular system and respiration. *Indian Journal of Medical Sciences*, 49, pp. 448–460.

Malhotra, C.L., Mehta, V.L., Das, P.K. and Dhalla, N.S., 1965. Studies on Withania-ashwagandha, Kaul. V. The effect of total alkaloids (ashwagandholine) on the central nervous system. *Indian Journal of Physiology and Pharmacology*, 9(3), pp. 127–136.

Mathur, S., Kaur, P., Sharma, M., et al. 2004. The treatment of skin carcinoma, induced by UV B radiation, using 1-oxo-5β, 6β-epoxy-witha-2-enolide, isolated from the roots of *Withania somnifera*, in a rat model. *Phytomedicine*, 1(5), pp. 452–460.

Mayola, E., Gallerne, C., Degli Esposti, D., et al. 2011. Withaferin A induces apoptosis in human melanoma cells through generation of reactive oxygen species and down-regulation of Bcl-2. *Apoptosis*, 16(10), p. 1014.

McKenna, M.K., Gachuki, B.W., Alhakeem, S.S., et al. 2015. Anticancer activity of withaferin A in B-cell lymphoma. *Cancer Biology & Therapy*, 16(7), pp. 1088–1098.

Meher, S.K., Das, B., Panda, P., Bhuyan, G.C. and Rao, M.M., 2016. Uses of *Withania somnifera* (Linn) Dunal (Ashwagandha) in Ayurveda and its Pharmacological Evidences. *Research Journal of Pharmacology and Pharmacodynamics*, 8(1), p. 23.

Mehta, A.K., Binkley, P., Gandhi, S.S. and Ticku, M.K. 1991. Pharmacological effect of W. somnifera roots extracts on GABA receptor complex. *Indian Journal of Medical Research*, 94, pp. 312–315.

Mirjalili, M.H., Moyano, E., Bonfill, M., Cusido, R.M. and Palazón, J., 2009. Steroidal lactones from *Withania somnifera*, an ancient plant for novel medicine. *Molecules*, 14(7), pp. 2373–2393.

Mishra, L.C., Singh, B.B. and Dagenais, S., 2000. Scientific basis for the therapeutic use of *Withania somnifera* (ashwagandha): A review. *Alternative Medicine Review*, 5(4), pp. 334–346.

Mohanty, I., Arya, D.S., Dinda, A., et al. 2004a. Mechanisms of cardioprotective effect of *Withania somnifera* in experimentally induced myocardial infarction. *Basic & Clinical Pharmacology & Toxicology*, 94(4), pp. 184–190.

Mohanty, I., Gupta, S.K., Talwar, K.K., et al. 2004b. Cardioprotection from ischemia and reperfusion injury by *Withania somnifera*: A hemodynamic, biochemical and histopathological assessment. *Molecular and Cellular Biochemistry*, 260(1), pp. 39–47.

Mulabagal, V., Subbaraju, G.V., Rao, C.V., et al. 2009. Withanolide sulfoxide from Aswagandha roots inhibits nuclear transcription factor-kappa-B, cyclooxygenase and tumor cell proliferation. *Phytotherapy Research*, 23(7), pp. 987–992.

Muralikrishnan, G., Dinda, A.K. and Shakeel, F., 2010. Immunomodulatory effects of *Withania somnifera* on azoxymethane induced experimental colon cancer in mice. *Immunological Investigations*, 39(7), pp. 688–698.

Naidu, P.S., Singh, A. and Kulkarni, S.K., 2006. Effect of *Withania somnifera* root extract on reserpine-induced orofacial dyskinesia and cognitive dysfunction. *Phytotherapy Research*, 20(2), pp. 140–146.

Narinderpal, K., Junaid, N. and Raman, B., 2013. A review on pharmacological profile of *Withania somnifera* (Ashwagandha). *Research & Reviews: Journal of Botanical Sciences*, 2, pp. 6–14.

Nema, R., Khare, S., Jain, P. and Pradhan, A., 2013. Anticancer activity of *Withania somnifera* (leaves) flavonoids compound. *International Journal of Pharmaceutical Sciences Review and Research*, 19(1), pp. 103–106.

Owais, M., Sharad, K.S., Shehbaz, A. and Saleemuddin, M., 2005. Antibacterial efficacy of *Withania somnifera* (ashwagandha) an indigenous medicinal plant against experimental murine salmonellosis. *Phytomedicine*, 12(3), pp. 229–235.

Padmavathi, B., Rath, P.C., Rao, A.R. and Singh, R.P., 2005. Roots of *Withania somnifera* inhibit forestomach and skin carcinogenesis in mice. *Evidence-Based Complementary and Alternative Medicine*, 2(1), pp. 99–105.

Panchawat, S., 2011. *In vitro* free radical scavenging activity of leaves extracts of *Withania somnifera*. *Recent Research in Science and Technology*, 3(11).

Patel, S.B., Rao, N.J. and Hingorani, L.L., 2016. Safety assessment of *Withania somnifera* extract standardized for Withaferin A: Acute and sub-acute toxicity study. *Journal of Ayurveda and Integrative Medicine*, 7(1), pp. 30–37.

Praveen, N. and Murthy, H.N., 2013. Withanolide A production from *Withania somnifera* hairy root cultures with improved growth by altering the concentrations of macro elements and nitrogen source in the medium. *Acta Physiologiae Plantarum*, 35(3), pp. 811–816.

Rai, M., Jogee, P.S., Agarkar, G. and Santos, C.A.D., 2016. Anticancer activities of *Withania somnifera*: Current research, formulations, and future perspectives. *Pharmaceutical Biology*, 54(2), pp. 189–197.

Ramarao, P., Rao, K.T., Srivastava, R.S. and Ghosal, S., 1995. Effects of glycowithanolides from *Withania somnifera* on morphine-induced inhibition of intestinal motility and tolerance to analgesia in mice. *Phytotherapy Research*, 9(1), pp. 66–68.

Rasool, M. and Varalakshmi, P., 2006. Immunomodulatory role of *Withania somnifera* root powder on experimental induced inflammation: An *in vivo* and *in vitro* study. *Vascular Pharmacology*, 44(6), pp. 406–410.

Rastogi, R., 1998. *Compendium of Indian Medicinal Plants*, Vol. V. CDRI Lucknow and NISC New Delhi, 80, pp. 144–174.

Rastogi, R.P. and Mehrotra, B.N., 1998. *Compendium of Indian Medicinal Plants*. 2nd Reprint. Central Drug Research Institute, Lucknow and National Institute of Science Communication, Council of Scientific and Industrial Research, New Delhi, 1, pp. 434–436.

Sandhu, J.S., Shah, B., Shenoy, S., Chauhan, S., Lavekar, G.S. and Padhi, M.M., 2010. Effects of Withania somnifera (Ashwagandha) and Terminalia arjuna (Arjuna) on physical performance and cardiorespiratory endurance in healthy young adults. *International Journal of Ayurveda Research*, 1(3), p. 144.

Sankar, S.R., Manivasagam, T., Krishnamurti, A. and Ramanathan, M., 2007. The neuroprotective effect of *Withania somnifera* root extract in MPTP-intoxicated mice: An analysis of behavioral and biochemical variables. *Cellular & Molecular Biology Letters*, 12(4), p. 473.

Sarangi, A., Jena, S., Sarangi, A.K. and Swain, B., 2013. Antidiabetic effects of *Withania somnifera* root and leaf extracts on streptozotocin induced diabetic rats. *Journal of Cell and Tissue Research*, 13(1), p. 3597.

Saykally, J.N., Hatic, H., Keeley, K.L., Jain, S.C., Ravindranath, V. and Citron, B.A., 2017. Withania somnifera extract protects model neurons from in vitro traumatic injury. *Cell Transplantation*, 26(7), pp. 1193–1201.

Schliebs, R., Liebmann, A., Bhattacharya, S.K., et al. 1997. Systemic administration of defined extracts from *Withania somnifera* (Indian Ginseng) and Shilajit differentially affects cholinergic but not glutamatergic and GABAergic markers in rat brain. *Neurochemistry International*, 30(2), pp. 181–190.

Sehgal, N., Gupta, A., Valli, R.K., et al. 2012. *Withania somnifera* reverses Alzheimer's disease pathology by enhancing low-density lipoprotein receptor-related protein in liver. *Proceedings of the National Academy of Sciences*, 109(9), pp. 3510–3515.

Senthil, V., Ramadevi, S., Venkatakrishnan, P., et al. 2007. Withanolide induces apoptosis in HL-60 leukemia cells via mitochondria mediated cytochrome c release and caspase activation. *Chemico-Biological Interactions*, 167(1), pp. 19–30.

Sharma, R.A., Goswami, M. and Yadav, A., 2013. GC-MS screening of alkaloids of *Withania somnifera* L. *in vivo* and in vitro. *Indian Journal of Applied Research*, 3, pp. 63–66.

Sindhu, K. and Santhi, N. 2009. Molecular Interaction of withanolides from *W. somnifera* against the oncoprotein Skp2 using glide. *Advanced Biotech*, 9(6), pp. 28–33.

Singh, A., Malhotra, S. and Subban, R., 2008. Anti-inflammatory and analgesic agents from Indian medicinal plants. *International Journal of Integrative Biology*, 3(1), pp. 57–72.

Singh, B., Saxena, A.K., Chandan, B.K., et al. 2001. Adaptogenic activity of a novel, withanolide-free aqueous fraction from the roots of *Withania somnifera* Dun. *Phytotherapy Research*, 15(4), pp. 311–318.

Singh, N., Rai, S.N., Singh, D. and Singh, S.P., 2015. *Withania somnifera* shows ability to counter Parkinson's Disease: An Update.

Singh, S. and Kumar, S., 1998. *Withania somnifera. The Indian Ginseng Ashwagandha*. Lucknow.

Srivastava, A.N., Ahmad, R. and Khan, M.A., 2016. Evaluation and comparison of the in vitro cytotoxic activity of Withania somnifera Methanolic and ethanolic extracts against MDA-MB-231 and Vero cell lines. *Scientia Pharmaceutica*, 84(1), p. 41.

Sumathi, S., Padma, P.R., Gathampari, S. and Vidhya, S., 2007. Free radical scavenging activity of different parts of *Withania somnifera. Ancient Science of Life*, 26(3), p. 30.

Tohda, C., Kuboyama, T. and Komatsu, K., 2000. Dendrite extension by methanol extract of Ashwagandha (roots of *Withania somnifera*) in SK-N-SH cells. *Neuroreport*, 11(9), pp. 1981–1985.

Trigunayat, A., Raghavendra, M., Singh, R.K., Bhattacharya, A.K. and Acharya, S.B., 2007. Neuroprotective effect of *Withania somnifera* (WS) in cerebral ischemia-reperfusion and long-term hypoperfusion induced alterations in rats. *Journal of Natural Remedies*, 7(2), pp. 234–246.

Tripathi, A.K., Shukla, Y.N. and Kumar, S., 1996. Ashwagandha *Withania somnifera* Dunal (Solanaceae): A status report. *Journal of Medicinal and Aromatic Plant Sciences*, 8, pp. 46–62.

Udayakumar, R., Kasthurirengan, S., Mariashibu, T.S., et al. 2009. Hypoglycaemic and hypolipidaemic effects of *Withania somnifera* root and leaf extracts on alloxan-induced diabetic rats. *International Journal of Molecular Sciences*, 10(5), pp. 2367–2382.

Verma, S.K. and Kumar, A., 2011. Therapeutic uses of *Withania somnifera* (ashwagandha) with a note on withanolides and its pharmacological actions. *Asian Journal of Pharmaceutical and Clinical Research*, 4(1), pp. 1–4.

Visavadiya, N.P. and Narasimhacharya, A.V.R.L., 2007. Hypocholesteremic and antioxidant effects of *Withania somnifera* (Dunal) in hypercholesteremic rats. *Phytomedicine*, 14(2–3), pp. 136–142.

Wadhwa, R., Singh, R., Gao, R., et al. 2013. Water extract of Ashwagandha leaves has anticancer activity: Identification of an active component and its mechanism of action. *PLoS One*, 8(10), p. e77189.

World Health Organization, 2007. WHO calls for prevention of cancer through healthy work places. In WHO calls for prevention of cancer through healthy workplaces.

Yadav, B., Bajaj, A., Saxena, M. and Saxena, A.K., 2010. *In vitro* anticancer activity of the root, stem and leaves of *Withania somnifera* against various human cancer cell lines. *Indian Journal of Pharmaceutical Sciences*, 72(5), p. 659.

Yang, E.S., Choi, M.J., Kim, J.H., Choi, K.S. and Kwon, T.K., 2011. Withaferin A enhances radiation-induced apoptosis in Caki cells through induction of reactive oxygen species, Bcl-2 downregulation and Akt inhibition. *Chemico-Biological Interactions*, 190(1), pp. 9–15.

Yang, H., Shi, G. and Dou, Q.P., 2007. The tumor proteasome is a primary target for the natural anticancer compound Withaferin A isolated from "Indian winter cherry". *Molecular Pharmacology*, 71(2), pp. 426–437.

Yu, Y., Hamza, A., Zhang, T., et al. 2010. Withaferin A targets heat shock protein 90 in pancreatic cancer cells. *Biochemical Pharmacology*, 79(4), pp. 542–551.

Zhao, J., Nakamura, N., Hattori, M., Kuboyama, T., Tohda, C. and Komatsu, K., 2002. Withanolide derivatives from the roots of *Withania somnifera* and their neurite outgrowth activities. *Chemical and Pharmaceutical Bulletin*, 50(6), pp. 760–765.

14

Plant Essential Oils and Anticancer Properties: An Update

Rajat Nath, Priyanka Saha, Deepa Nath, Manabendra Dutta Choudhury, and Anupam Das Talukdar

CONTENTS

14.1 Introduction

Cancer is one among the key health issue globally because it is the second most-notable reason for mortality next to cardiovascular disease (Bhalla et al. 2013). Cancer is generally characterized by uncontrolled cell division and cell proliferation (Roy et al. 2017). Nearly 1/6th of the death occurs due to various forms of cancer. According to the recent report of World Health Organisation (WHO) in 2015, cancer accounted for 8.8 billion deaths worldwide. Amongst them, 14 million cases were diagnosed [WHO fact sheet February 2018]. Carcinoma appraisal takes place in altered food habits and irregular lifestyles (Roy et al. 2017). Tobacco smoking accounts for almost 35% of cancer deaths (Russo et al. 2012). Another 30%–35% deaths occur due to irregular lifestyle like poor diet, obesity, excessive drinking of alcohol and lack of physical activity (Allott and Hursting 2015, Jayasekara et al. 2016). The remaining factors include environmental pollution, ionizing radiation exposure to sunlight, and certain infections (Anand et al. 2008). The most widely recognized kinds of malignancy in males are prostate cancer, colorectal cancer, stomach cancer and lung cancer (McGuire 2016). On the other hand, it is seen that breast cancer, lung cancer, colorectal cancer, and cervical cancer is prominent in females (World Cancer Report 2014).

In children, brain tumors and acute lymphoblastic leukemia are most common (World Cancer Report 2017). On the basis of the category and the stages of cancer, there are many kinds of treatments available. It may be a single treatment or combination of more than one treatment. Various types of cancer treatments are as follows:

1. *Surgery:* In this kind of therapy, the cancer is removed from the body by the surgical operation.
2. *Radiation Therapy:* High doses of radiation are applied to the cancerous cell and shrink the tumor in that kind of therapy.
3. *Chemotherapy:* It is the process killing the cancer cell by applying chemical drugs.
4. *Immunotherapy:* It helps the immune system to improve the fighting capacity to kill the cancerous cell.
5. *Targeted Therapy:* It is another sort of tumor treatment that focuses on the adjustments in growth cells that assist development, partition, and spread.
6. *Hormone Therapy:* This kind of therapy is mainly associated with breast and prostate cancer. It is the therapy that stops or slows down the growth of cancer cells.

7. *Stem Cell Transplant:* Where the high dose of chemotherapy and radiotherapy destroyed the blood-forming stem cells, stem cell transplant is the process of restoring them.

8. *Precision Medicine:* By understanding the genetic basis of the disease, precision medicines are used to treat the cancer (https://www.cancer.gov/about-cancer/treatment/types).

Although a number of cancer therapies are available today, the main drawback of such kinds of therapy is their excessive or high rate of side effects. The side effects start when the healthy tissues and organs are affected by the treatment. The common side effects of available cancer therapy are appetite loss, constipation, anemia, bleeding and bruising (thrombocytopenia), edema (swelling), delirium, fatigue, diarrhea, fertility issues in men and women, lymphedema, infection and neutropenia, hair loss (alopecia), nausea and vomiting, memory or concentration problem, pain, mouth and throat problem, nerve problem (peripheral neuropathy), sexual health issues, sleep problem, skin and nail changes, urinary and bladder problems, and so on (www.cancer.gov/about-cancer/treatment/side-effects). These problems may be minimized by the therapy using natural products and phytochemicals in particular. Among the herbal candidates, essential oils holds a promising position as anticancer agents as it is a potential source for bioactive natural molecules and contains all the potentialities to act as natural therapeutics. Though it may not be the substitute for established therapies such as radiotherapy or chemotherapy, but can at least minimize the deleterious effect of present therapies. This is usually accomplished by acting in combination with the routine methods. This essential oil therapy, though not able to replace the established regime yet, focuses a ray of hope that essential oil ingredients are bestowed with potential anticancer properties.

14.2　Need for Herbal Medicine

Presently in the market there are no enormously effective medicines available for cancer treatment. Therefore, there is a vital need for anti-cancer drugs with least antagonistic side effects, higher efficiency and cost effective (Cai et al. 2004). Plants offer a vast pool of valuable agents to treat diseases, which can be the best preference. Opting for plant-based treatment could bring a paradigm shift to the existing treatment regime because of their minimum toxicity and economical affordability (Efferth et al. 2007). In *in vitro* and *in vivo* studies it is seen that various plant-derived products, viz. saponins, alkaloids, glycosides, triterpenes and polyphenols among others, are promising agents of anticancer therapy (Mukherjee 2003). Thousands of plants are reported having noteworthy anticancer properties (Mukherjee 2003). Some of the plant-derived compounds being widely used are colchicine, vincristine, ellipticine, vinblastine, lepachol, irinotecan, taxol, camptothecin, podophyllotoxin, paclitaxel, and etoposide (Cragg and Newman 2005). Besides the plant-obtained compounds, it is reported that essential oils (EOs) derived from aromatic plants also retain anticancer properties (Yu et al. 2011, Mitoshi et al. 2012). Several analysts over the last two decades have found anticancer properties in essential oils and their components *in vivo*

and *in vitro* models (Bhalla et al. 2013). Essential oils can be utilized as a part of a blend with the standard treatment utilized for malignancy to counter the negative impacts of the treatment and can enhance the state of patients as these are the font of novel anticancer compounds (Bhalla et al. 2013).

14.3　Chemical Classifications of Essential Oils and Their Constituents

Essential oils are volatile, concentrated liquids that are hydrophobic in nature and derived from aromatic plants (Gautam et al. 2014). These plant secondary metabolites are also known as ethereal or volatile oil and exist in very little amount in different plants parts. The biological properties of EOs depend upon various components in it. On the basis of the structural components, EOs are classified (Özkan and Erdoğan 2011). A general classification on the basis of chemical structures is enlisted in Table 14.1.

14.4　Essential Oils and Their Mechanism on Cancer Therapy

The main target of cancer therapy is to focus on apoptosis and arresting the cell cycle of the transformed cells at various developmental stages of pathology. Natural EOs having reported apoptotic properties and are a good sources of anticancer drug as it act bidirectionally, both by chemoprevention as well as in cancer suppression (Gautam et al. 2014).

Some of the important properties of EOs against cancer cells are discussed below.

14.4.1　Essential Oils as Antioxidant Agent

Free radicals are responsible for causing various types of diseases such as cancer, Alzheimer's disease, liver disease, inflammation, arthritis, aging, Parkinson's disease, inflammation, atherosclerosis, diabetes, AIDS, and so on (Bhalla et al. 2013). Damages due to oxidative stress can be prevented by the use of natural antioxidants and especially essential oils (EOs) (Moon and Shibamoto 2009). Hence, antioxidant action of EOs is the most important studied. In the recent investigation with EOs by the several researchers, it was found that EOs are the ideal antioxidant sources (Stadler et al. 1995). In eukaryotes, mitochondrial DNA is damaged by the hydroxyl radicals, which are produced by hydrogen peroxide and superoxide anions. That damaged DNA inhibits electron transport protein expression which leads the amassing of Reactive Oxygen Species (ROS) (Van Houten et al. 2006). The free radicals generated by damaged mitochondrial layers when joined with essential oil deliver receptive phenoxy radicals that consolidate with ROS and avert additionally harm (Sakihama et al. 2002, Azmi et al. 2006). Many aromatic compounds, such as terpenoids and tarpenes, including α-terpinene, β-terpinene, β-terpinolene, linalool, eugenol, thymol, α-tocopherol, citronella, isomethone, and methone of EOs, show high antioxidant activities (Bhalla et al. 2013) (Figure 14.1).

TABLE 14.1

General Classification of EOs in the Basis of Their Chemical Structure

Types	Component	General Structure	General Formula	References
Terpene hydrocarbons	Monoterpene		$C_{10}H_{16}$	Gautam et al. (2014)
	Sesquiterpenes		$C_{15}H_{24}$	
	Diterpene		$C_{20}H_{32}$	
Oxygenated terpenes	Oxygenated monoterpene		$C_{10}H_{16}O$	
	Oxygenated sesquiterpenes		$C_{15}H_{24}O$	
	Simple alcohols	R-OH	ROH	
	Monoterpene alcohols		$C_{10}H_{17}OH$	
	Sesquiterpenes alcohols		$C_{15}H_{25}OH$	
Other oxygenated terpenes	Phenols		RC_6H_5OH	
	Ketones	R′ R	$RC(=O)R′$	
	Esters	R′O R	$RCO_2R′$	
	Lactones		$C_3H_6O_3$	
	Coumarins		$C_9H_6O_2$	

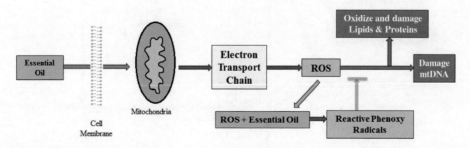

FIGURE 14.1 Antioxidant mechanism of essential oils.

14.4.2 Essential Oil as Anti-proliferative Agent

Various reports have stated essential oil as anti-proliferative entity. Work has been done in cancer cell line such as human erythroleukemic K562 cells where α-pinene, α-terpineol, β-terpenene, caryophyllene, and 4-terpineol showed impressive anti-proliferative activity with IC50 values of 329–398 μm mL^{-1} (Lampronti et al. 2006). Another work was done by Patil et al. in colon cancer cells of human (SW-480) where he reported that volatile components of *Citrus aurantifolia*, at a concentration of 100 μg mL^{-1} showed 78% apoptosis facilitated proliferation inhibition after 48 h. Lime volatile oil showed the capability of fragmentation of DNA and caspase-3 induction up to 1.8–2-fold after 24 and 48 h correspondingly in chronic sarcoma condition (Patil et al. 2009). In cell line of metastatic breast cancer (MDA-MB231), carvacrol, a phenolic monoterpenes also shows the antiproliferative activity by activating the mitochondrial pathway of apoptotic response which act by reducing the membrane potential in mitochondria and increasing the production of cytochrome c; which in turn increases the caspases activity and poly ADP-ribose polymerase. This takes place mainly by down regulating the prime apoptotic markers, Bcl-2/Bax ratio and also by DNA fragmentation in such cases (Arunasree 2010).

14.4.3 Essential Oil as Antimutagenic Agent

Preventing error-prone repair of DNA or upholding error-prone DNA repair are the key rule of antimutagenic components (Kada and Shimoi 1987). EOs shows the antimutagenic activities by directly scavenging the mutagens and inactivating them; prevention of the infiltration of mutagens into the cell, restraint of metabolic modification by P450 of promutagens to mutagens, capture of mutagen formed free radicals by activation of antioxidant enzymes of cell (De and Ramel 1988, Sharma et al. 2001). α-Terpinene, citral α-terpineol, D-limonene, citronellal, and 1,8-cineole camphor are EOs that show the antimutagenic activity (De-Oliveira et al. 1997). Some current report showed EOs also act upon UV-light-derived mutations also (Bakkali et al. 2006, 2008).

14.5 Mode of Action of Essential Oils

Essential oils have various target areas by which they acts on a malignant cell. Different kinds of essential oils follow different pathways to act upon a diseased cell. Some mode of actions of essential oils is described below.

14.5.1 Essential Oil Act by Enrichment of Immunity

Aromatherapy is a well-known process by which EOs are used into different medias to enhance the immune function. Aromatherapy improves the immune functions by releasing of stress, regulating adrenal gland mediated hormones, fortifying the immune response by supporting the expulsion of toxins from the lymph, and invigorating immune-boosting cells creation and hurtful micro-organisms obliteration (Bhalla et al. 2013). Studies shows that the zedoarondiol, a sub-type of EOs, is capable of inhibiting the NF-κB, prompted by lipopolysaccharide (LPS)-stimulated murine macrophages by inhibiting COX-2, iNOX, and pro-inflammatory cytokines (Cho et al. 2009). NF-κB also involved in transcription of cytokines, which are liable for acute inflammatory response generation. Such cytokines are IL-1, TNF-α, IL-8, and IL-6 (Yoon et al. 2009). The ability of essential oils to regulate these pro-inflammatory proteins in inflammatory response makes them potent natural therapeutics (Figure 14.2).

14.5.2 Essential Oils Act on Cell Cycle

Uncontrolled cell division is the key sign of malignancy that is happening because of the negative regulation of cell cycle movement due to various noxious agents present in the surrounding (Diaz-Moralli et al. 2013). A nascent cell experiences different cell cycle stages, for example, G1, S, G2, and metaphase. So, targeting any of these phases of a cancerous cell may affect the entire cell cycle processes, which lead to inhibition of the cell division and cell growth (Gabrielli et al. 2012). In colorectal cancer cell, it

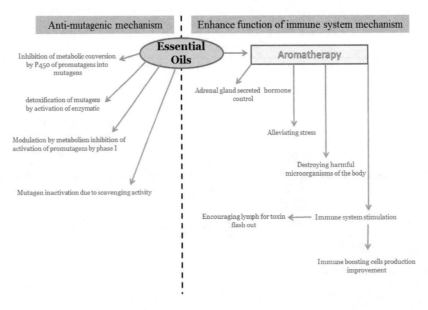

FIGURE 14.2 Mechanism of anti-mutagenic and enhance immune system function of EOs.

was reported that increasing dose of EOs can suppress the expression of CDK4 and Cyclin D1 along with up regulating P21, which is a negative regulator of G1 phase transition (Jeong et al. 2013). Garaniol, carvaccrol, thymol such component of EOs inhibits different stages of cell cycle such as S/G2, G2/M, transition (Gautam et al. 2014). Cell cycle arresting by altering the expression of IL8, DDIT3 and CDKNIA genes are also reported by monoterpenes class of compounds (Frank et al. 2009).

14.5.3 Essential Oil Act by Enzyme Detoxification

Altering the metabolism of procarcinogen, carcinogenesis can be inhibited by sulphur-containing agents mainly by detoxification, by increasing phase II enzyme level such as UDP-glucoronyltransferase (UGT), glutathione S-transferase (GST), and quinonereductase (QR), and by reducing phase I enzymes level such as cytochrome P_{450} (Guengerich 1992, Wattenberg 1993). The underlying mechanism of EOs may be due to its capability of diminishing early stage of transformation by metabolizing the toxic elements into dormant form of intermediate metabolites. This process is generally accompanied by conjugation reaction, thereby detoxifying the reactive intermediates (Singletary 1996). Various important components of EOs such as 1,8-cineol, camphor, borneol, and α-terpeniolare are capable of inducing the cytochrome P_{450} (CYP) family, particularly CYP2B by which such aberrations may revert back into normal (Debersac et al. 2001). Monoterpenes such as *trans*-citral is reported as a good GST class π (GSTPI) inducer, which acts synergistically to increase the activity of the total content of GST (Debersac et al. 2001). This condition is important in respect to enriching the antioxidant system. So, these components of EOs are good chemopreventives towards carcinogenesis related to inflammation (Mulder et al. 1995).

14.5.4 Essential Oil Act by Modulation of DNA Damage and Repair Signaling

Various degree of DNA damage occurs due to increasing levels of ROS in the cell, which eventually leads to cell death. EOs are capable of increasing the ROS level they especially act on immortal cells, ultimately leading to their death. Because of their action specificity, normal cells remain untouched (Mantha 2013). Considering this unique properties of EOs, an effective treatment method is established targeting DNA repair pathways, which inhibit the unlimited and uncontrolled cell proliferation (Kelley et al. 2001). Various research works in this related field has showed that potential genotoxin like hydrogen peroxide causes DNA damage that can be minimized by periodic pre-treatment with Myrcene, linalool and eucalyptol (Mitić-Ćulafić et al. 2009). Monoterpene, like thujone and camphor, are also reported as a DNA repair process mediator in various cases of induced cell-toxicity model system (Nikolić et al. 2011). On the contrary, some DNA repair system viz; HDAC4 and H2AFX are found to be liable for DNA repair and cell cycle progression. This features helps in mitigating cancer specifically gall bladder cancer. A number of reports has depicted this characterizing property of essential oils where they prove to be promising natural therapeutics that target various DNA repair pathways. Other oils like Frankincense also show such bioactive properties that inhibit or rather checks the abnormal cell proliferation (Frank et al. 2009).

14.5.5 Essential Oil Act by Inducing Apoptosis

Programmed cell death, i.e., apoptosis, is an important programme for a cell which helps maintain its cellularity and integrity. This scheduled event depends on various cellular signaling pathways, genetic materials and many more cellular events. Cardile et al. reported that EOs are capable of damaging DNA of transformed cells, which then undergoes apoptosis (Cardile et al. 2009). Genetic modification by the EOs also serves the purpose of inducing cell apoptosis. For bladder cancer cells, the genes involved in apoptosis like DEDD2, CDKN1A, IER3, SGK, IL6, NUDT2, GAD45B, and TNFAIP3 are reported. These marker genes could be modulated by oil such as frankincense oil (Frank et al. 2009) under many circumstances. The apoptosis process also takes place by formation of caspase-3 which is generally formed due to activation of caspase-9. Consequently, cytochrome C releases its content into cytoplasm of the transformed cells mainly due to the overexpression of Bcl-2 and Bax, the two prime apoptotic marker enzyme by the activity of EOs (Cha, Kim, and Kim 2010). The stimulated caspase-3 inturn increases the c-jun N-terminal kinase, extra cellular phosphorylation (ERK),and p38 MAPK (Cha et al. 2010) aiding the process. EOs are also very active to cleave the poly (ADP-ribose) polymerase-1 (PARP) which also leads to apoptosis (Soldani and Scovassi 2002).

14.5.6 Essential Oils and Oxidative Stress

Oxidative stress is a leading cause of many chronicle and imperative diseases. It is one of the major causes of cell mutations and developing malignancy in a cell. Essential oils act as an antioxidant agent to fight against oxidative stress. Different EOs follow different physiological pathways to counteract on the family of reactive oxygen species. In MAPK pathway, MAPKs such as ERK, JNK, and p38 kinase signaling molecules become active in response to oxidative stress in the cell. MAPKs are responsible for the cancer cell apoptosis process (Ki et al. 2013). Essential oil from *Neolitsea sericea* can mediate its apoptosis activity by phosphorylating of MAPK of the cells (Martindale and Holbrook 2002).

EOs derived from *Artemisia vessels* instigate apoptosis in mouth cancer cells. It is because of expanding DNA binding activity of AP-1. AP-1 (Activator protein-1) that assumes an imperative part in transformation, multiplication, differentiation, and apoptosis. Ap-1 movement likewise managed by MAP kinase protein (Cha et al. 2009).

Tumor suppressor protein p53 is regulated by protein Akt. It is reported that some EOs derived from *Boswellia sacra* inhibit pPDK1 and mTOR leading to dephosphorylation of Akt protein which in turn activates caspase-3 and caspase-9. All such processes ultimately result in apoptosis in lung cancerous cells (Seal et al. 2012).

NF-κB, a nuclear transcription factor, is activated in the tumor cell, which is an important anticancer drug target as it helps to show anticancer activities of the natural compounds. α-Terpineol, component of EOs is reported a target of NF-κB and can down regulate the responsible genes like *IL-1β, IFNG, IL1R1, ITK*, and *EGFR* (Hassan et al. 2010). All this action leads to a comparatively stable state of recovery towards varieties of carcinoma (Figure 14.3, Tables 14.2 and 14.3).

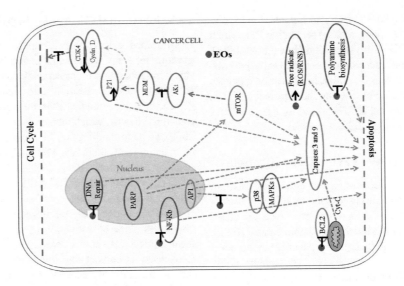

FIGURE 14.3   EOs and their constituent target multiple pathways in cancer cell.

TABLE 14.2

List of EOs Constituents with Their Mode of Actions

Sl. No.	Mode of Action	Constituents	Derived from Plant	References
1.	Antioxidant	Santolina alcohol, borneol, sabinol	*Achillea ligustica*	Tuberoso et al. (2005)
		Thymol, isopropyl, isothiocyanate	*Capparis spinosa*	Kulisic-Bilusic et al. (2012)
		α-Pinene, globulol, aromadendrene	*Egyptian corn silk*	El-Ghorab et al. (2007)
		2-Methyl-5-hexenenitrile,3-butenyl isothiocyanate	*Coriandrum sativum*	Alenzi et al. (2010)
		1,8-Cineole	*Foeniculum vulgare*	Mohamad et al. (2011)
		α-BisabololoxideA, (*E*)-β-farnesene	*Matricaria recutita*	McKay and Blumberg (2006a)
		α-Thujone	*Melaleuca alternifolia*	Kim et al. (2004)
		Methyl cinnamate	*Nigella sativa* and *Nigella damascene*	Edris (2009)
		10-Epi-eudesmol, pinene	*Pelargonium graveolens*	Fayed (2009)
		1,8-Cineole, α-pinene	*Pituranthos tortuosus*	Abdallah and Ezzat (2011a)
		Eucalyptol	*Rosmarinus officinalis*	Xue et al. (2012)
		Carvacrol, thymol, thymoquinone	*Satureja montana*	Grosso et al. (2009)
		β-Caryophyllene	*Syzygium aromaticum*	Prashar et al. (2006)
		Viridiflorene, β-terpineol	*Tanacetum parthenium*	Mohsenzadeh et al. (2011)
2	Antiproliferative	Methyl chavicol	*Agastache rugosa*	Kim et al. (2001)
		α-Thujone, terpenyl acetate	*Anisomeles malabarica*	Preethy et al. (2012)
		Tetradecanoic acid, (Z, Z)- 9,12-octadecadienoic acid, Hexadecanoic acid	*Calamintha origanifolia*	Formisano et al. (2007)
		γ –terpinene, Thymol	*Carum copticum*	Yin et al. (2012)
		β-caryophyllene, Eugenol	*Citrus aurantifolia*	Patil et al. (2009)
		p-Cymene, ascaridole	*Crambe orientalis*	Razavi and Nejad-Ebrahimi (2009)
		Limonene	*Eucalyptus benthamii*	Döll-Boscardin et al. (2012)
		(E)-β-Ionone, hexahydrofarnesylacetone	*Genista sessilifolia* and *Genista tinctoria*	Rigano et al. (2009)
		α-BisabololoxideA, (E)-β-farnesene	*Matricaria recutita*	McKay and Blumberg (2006b)
		Carvacrol, γ -terpinene, p-cymene	*Oplopanax horridus*	Inui et al. (2010)
		Bornyl acetate, camphene	*Satureja hortensis*	Lampronti et al. (2006)
		Carvacrol, thymol, thymoquinone	*Satureja montana*	Grosso et al. (2009)
3.	Antimutagenic	Methyl chavicol	*Agastache rugosa*	Kim et al. (2001)
		Citronellol, *trans*-geraniol	*Origanum onites*	Bostancıoğlu et al. (2012)

(Continued)

TABLE 14.2 (*Continued*)

List of EOs Constituents with Their Mode of Actions

Sl. No.	Mode of Action	Constituents	Derived from Plant	References
4.	Apoptosis induction	α-Thujone, terpenyl acetate	*Anisomeles malabarica*	Preethy et al. (2012)
		α-Thujene, α-pinene, myrcene	*Boswellia sacra*	Suhail et al. (2011)
		trans-Chrysanthenyl acetate	*Centipeda minima*	Su et al. (2010)
		2-Decenoic acid	*Commiphora gileadensis*	Amiel et al. (2012)
		Carvacrol	*Santalum album*	Bommareddy et al. (2012)
		Thujones, fenchone	*Thuja occidentalis*	Biswas et al. (2011)
5.	Cytotoxic	Sabinene, α-pinene, α-thujene	*Astrodaucus orientalis*	Zahri and Ghasemian (2011)
		α-Humulene, 7-epi-α-eudesmol	*Callicarpa americana*	Jones et al. (2007)
		(E)-Cinnamaldehyde	*Cinnamomum zeylanicum*	Unlu et al. (2010)
		Epicurzerenone, curdione	*Croton regelianus*	Bezerra et al. (2009)
		syn-7-Hydroxy-7-anisylnorbornene, pregeijerene	*Cyperus kyllingia*	Khamsan et al. (2011)
		cis-α-Terpineol, 6,11-oxidoacor-4-ene, citronellol	*Dictamnus dasycarpus*	Lei et al. (2008)
		Limonene	*Eucalyptus benthamii*	Döll-Boscardin et al. (2012)
		(E)-β-Ionone, hexahydrofarnesylacetone	*Genista sessilifolia* and *Genista tinctoria*	Rigano et al. (2009)
		β-Caryophyllene, sabinene	*Helichrysum gymnocephalum*	Rigano et al. (2009)
		(Z)-Citral, (E)-citral	*Lavandula officinalis*	Zamanian-Azodi et al. (2012)
		Eucalyptol	*Lindera strychnifolia*	Yan et al. (2009)
		1,8-Cineole	*Mangifera indica*	Simionatto et al. (2010)
		Menthol	*Melaleuca armillaris*	Chabir et al. (2011)
		α-Phellandrene, p-cymene	*Mentha piperita* and *Mentha spicata*	Hussain et al. (2010b)
		β-Elemene	*Myrica gale*	Sylvestre et al. (2005)
		β-myrcene, sabinene, γ-terpinene, Terpinen-4-ol	*Photinia serrulata*	Hou et al. (2007)
6.	Antitumor	α-Cadinol, caryophyllene oxide, α-muurolol	*Curcuma zedoaria*	Chen et al. (2011)
		Myrcene, limonene	*Moringa oleifera*	Bose (2007)
7.	Antineoplastic	1-p-menthen-8-ethyl acetate, 1,8-Cineole	*Lantana camara*	Badakhshan et al. (2009)
8.	Synergistic	α-Gurjunene, β-selinene	*Malus domestica*	Walia et al. (2012)
9.	Anti-inflammatory	Santalol	*Salvia libanotica*	Itani et al. (2008)

TABLE 14.3

List of EO Bearing Plants or Compounds Having Anticancer Properties

Sl. No.	Type of Cancer	Essential Oil-Bearing Plant or Compound Tested	Model Used	References
1.	Brain cancer	*Croton regelianus* and ascaridole compound	SF-295 (IC50 = 48.0 and 8.4 μg/mL respectively)	Bezerra et al. (2009)
		Afrostyrax lepidophyllus and *Scorodophloeus zenkeri*	T98G (IC50 = 15.4 and 12.4 μg/mL respectively)	Bayala et al. (2014a)
		α-Bisabolol	T67 and C6: 50% of cell death after 24 h treatment with 2.5 μM	Cavalieri et al. (2004)
		Casearia sylvestris	U87 (IC50 = 27.1 μg/mL)	Bou et al. (2013)
		L. multiflora, O. basilicum, Z. officiniale and *A. conizoides*	SF-767 (IC$_{50}$ = 0.31, 0.30, 0.43 and 0.48 mg/mL respectively) and SF-763 (IC$_{50}$ = 0.47, 0.43, 0.44 and 0.38 respectively)	Bayala et al. (2014b)
		Malus domestica	C-6 (1 mg/mL = 58.5% inhibition)	Walia et al. (2012)

(Continued)

TABLE 14.3 (*Continued*)

List of EO Bearing Plants or Compounds Having Anticancer Properties

Sl. No.	Type of Cancer	Essential Oil-Bearing Plant or Compound Tested	Model Used	References
2.	Breast cancer	*Scorodophloeus zenkeri* and *Afrostyrax lepidophyllus*	MDA-MB 231 (IC50 = 8.0 and 10.9 µg/mL respectively)	Fogang et al. (2014)
		Satureja khuzistanica	MCF7 (IC50 = 125 µg/mL)	Yousefzadi et al. (2014)
		Casearia sylvestris	MCF-7 (IC50 = 42.2 µg/mL)	Bou et al. (2013)
		Cedrelopsis grevei	MCF-7 (IC50 = 21.5 mg/L)	Afoulous et al. (2013)
		Solanium spirale	MCF-7 (IC50 = 19.69 µg/mL)	Keawsa-Ard et al. (2012)
		Carbazole alkaloids	MCF-7 (IC50 = 2.12 µg/mL)	Nagappan et al. (2011)
		Helichrysum gymnocephalum	MCF-7 (IC50 = 16 µg/mL)	Afoulous et al. (2011)
		Pituranthos tortuosus	MCF-7 (IC50 = 3.38 µg/mL)	Abdallah and Ezzat (2011b)
		Melaleuca armillaris	MCF-7 (IC50 = 12 µg/mL)	Chabir et al. 2011)
		Rosmarinus officinalis	MCF-7 (IC50 = 190.1 µg/mL)	Hussain et al. (2010a)
		Schinus terebinthifolius and *Schinus molle*	MCF-7 (IC50 = 47, and 54 mg/mL respectively)	Bendaoud et al. (2010)
		Erigeron acris	MCF-7 (IC50 = 14.5 µg/mL)	Nazaruk et al. (2010)
		Aquilaria sinensis	MCF-7 (99.6% inhibition at 500 µg/mL)	Xu et al. (2010)
		Thymus vulgaris	MCF-7 (IC50 = 0.030% (v/v))	Zu et al. (2010)
		Aristolochia mollissima	MCF-7 (IC50 = 20.6 and 21.1 µg/mL respectively) and MDA-MB- 435S (IC50 = 22.1 and 20.3 µg/mL respectively)	Yu et al. (2007)
		Schefflera heptaphylla	MDA-MB-231	Kim et al. (2014)
		β-Caryophyllene oxide	Crl (IC50 = 47.7; 91.8 and 63.6 µg/mL respectively)	Sigurdsson et al. (2005)
3.	Cervix cancer	*Ocimum basilicum*	HeLa (IC50 = 90.5 µg/mL)	Kathirvel and Ravi (2012)
		Carbazole alkaloids	HeLa (IC50 = 1.98 µg/mL)	Nagappan et al. (2011)
		Aristolochia mollissima	HeLa (IC50 = 38.6 µg/mL)	Yu et al. (2007)
4.	Ovary cancer	D-Limonene	V79	Mauro et al. (2013)
		Malus domestica	CHOK1 (1000 µg/mL = 68.3% inhibition)	Walia et al. (2012)
5.	Colon cancer	*Kadsura longipedunculata*	SW-480 (IC50 = 136.62 µg/mL)	Mulyaningsih et al. (2010)
		Comptonia peregrina	DLD-1 (IC50 = 46 µg/mL)	Sylvestre et al. (2007)
		Satureja khuzistanica	SW480 (IC50 = 62.5 µg/mL)	Yousefzadi et al. (2014)
		1,8-Cineol	HCT116 and RKO	Murata et al. (2013)
		Artemisia indica	Caco-2 (IC50 = 19.5 µg/mL)	Rashid et al. (2013)
		Pituranthos tortuosus	HCT116 (IC50 = 1.34 µg/mL)	Abdallah and Ezzat (2011b)
		Ascaridole compound and *Croton regelianus*	HCT-8 (IC50 = 18.4 and40.0 µg/mL respectively)	Bezerra et al. (2009)
		Cymbopogon flexuosus	502713 (IC50 = 4.2 µg/mL)	Sharma et al. (2009)
		Eugenol	SNU-C5 (IC50 = 129.4 µM)	Yoo et al. (2005)
		Geraniol and 5-fluorouracil	Caco-2 (IC50 = 250 and 0.4 µM respectively) and SW620 (IC50 = 330 and 2.0 µM respectively)	Carnesecchi et al. (2004)
		Afrostyrax lepidophyllus and *Scorodophloeus zenkeri*	HCT116 (IC50 = 12.4 and 8.5 µg/mL respectively)	Fogang et al. (2014)
		Athanasia brownii	HCT 116 (IC50 = 29.53 µg/mL)	Rasoanaivo et al. (2013)
6.	Kidney cancer	*Prangos asperula* and *Platycladus orientalis*	ACHN (IC50 = 139.17 and 121.93 µg/mL respectively)	Loizzo et al. (2008)
		Laurus nobilis	ACHN (IC50 = 78.24 µg/mL)	Loizzo et al. (2007)
		Aristolochia mollissima	ACHN (IC50 = 22.3 and 33.8 µg/mL respectively)	Yu et al. (2007)
		Satureja khuzistanica	Vero (IC50 = 31.26 µg/mL)	Yousefzadi et al. (2014)

(Continued)

TABLE 14.3 (*Continued*)

List of EO Bearing Plants or Compounds Having Anticancer Properties

Sl. No.	Type of Cancer	Essential Oil-Bearing Plant or Compound Tested	Model Used	References
7.	Leukaemia	*Isointermedeol and Cymbopogon flexuosus*	HL-60 (IC50 = 20 and 30 µg/mL, respectively)	Yousefzadi et al. (2014)
		Casearia sylvestris	HL-60 (IC50 = 29 µg/mL)	Bou et al. (2013)
		Artemisia indica	THP-1 (IC50 = 10 µg/mL)	Rashid et al. (2013)
		Malus domestica	THP-1 (1000 µg/mL = 68.3% inhibition)	Walia et al. (2012)
		Carbazole alkaloids	P388 (IC50 = 5.00 µg/mL)	Nagappan et al. (2011)
		Ascaridole compound and Croton regelianus	HL-60 (IC50 = 6.32 and 22.2 µg/mL respectively)	Bezerra et al. (2009)
		Eugenol	HL-60 (IC50 = 23.7 µM)	Yoo et al.(2005)
		Ocimum basilicum	P388 (IC50 = 0.0362 mg/mL)	Manosroi et al. (2006)
8.	Liver cancer	*Schefflera heptaphylla*	HepG2 (IC50 = 6.9 µg/mL)	Li et al. (2009)
		Curcuma wenyujin	HepG2 (IC50 = 70 µg/mL)	Xiao et al. (2008)
		Curcuma zedoaria	SMMC-7721 (IC50 = 30.7 µg/mL)	Chen et al. (2011)
		Patrinia scabra	Bel-7402 (IC50 = 16 µg/mL)	Sun et al. (2005)
		Eugenol	HepG2 (IC50 = 118.6 µM) and U-937 (IC50 = 39.4 µM)	Yoo et al. (2005)
		Thymus citriodorus	HepG2 (IC50 = 0.34% v/v)	Wu et al. (2013)
		Artemisia indica	HEP-2 (IC50 = 15.5 µg/mL)	Rashid et al. (2013)
		Pituranthos tortuosus	HEPG2 (IC50 = 1.67 µg/mL)	Abdallah and Ezzat (2011b)
		Kadsura longipedunculata	HepG2 (IC50 = 136.96 µg/mL)	Mulyaningsih et al. (2010)
		Aristolochia mollissima	Bel-7402 (IC50 = 33.1 and 49.5 µg/mL respectively) and Hep G2 (IC50 = 33.2 and 40.7 µg/mL respectively)	Yu et al. (2007)
9.	Lung Cancer	*Artemisia indica*	A-549 (IC50 = 25 µg/mL)	Rashid et al. (2013)
		Tridax procumbens	B16F-10 in vitro (70.2% of inhibition for 50 µg) and in vivo	Manjamalai et al. (2012)
		Solanium spirale	NCI-H187 (IC50 = 24.02 µg/mL)	Keawsa-Ard et al. (2012)
		Malus domestica	A549 (1000 µg/mL = 60.7% inhibition)	Walia et al. (2012)
		Thymus vulgaris	A549 (IC50 = 0.011% (v/v))	Zu et al. (2010)
		Comptonia peregrina	A-549 (IC50 = 66 µg/mL)	Sylvestre et al. (2007)
		Xylopia frutescens	NCI-H358M (IC50 = 24.6 µg/mL)	Ferraz et al. (2013)
10.	Mouth epidermal carcinoma	*Psidium guajava*	KB (IC50 = 0.0379 mg/mL)	Manosroi et al. (2006)
11.	Multiple myeloma	β-Caryophyllene oxide	U266 and MM1.S	Kim et al. (2014)
12.	Nasopharyngeal cancer	*Centipeda minima*	CNE (IC50 = 5.2 µg/mL after 72 h)	Su et al. (2010)
13.	Neuroblastoma	*Cymbopogon flexuosus*	IMR-32 (IC50 = 4.7 µg/mL)	Sharma et al. (2009)
14.	Oral cancer	*Solanium spirale*	KB (IC50 = 26.42 µg/mL)	Keawsa-Ard et al. (2012)
		Salvia officinalis	UMSSC1 (IC50 = 135 µg/mL)	Sertel et al. (2011a)
		Levisticum officinale	HNSCC (IC50 = 292.6 µg/mL)	Sertel et al. (2011b)
15.	Ovary	*Patrinia scabra*	HO-8910 (IC50 = 21 µg/mL)	Sun et al. (2005)
		Cymbopogon citratus	Chinese Hamster Ovary (CHO)	Kpoviessi et al. (2014)
16.	Pancreas	*Angelica archangelica*	PANC-1 (IC50 = 58.4)	Sigurdsson et al. (2005)
		Kadsura longipedunculata	MIA PaCa-2 (IC50 = 133.5 µg/mL)	Mulyaningsih et al. (2010)

(Continued)

TABLE 14.3 (*Continued*)

List of EO Bearing Plants or Compounds Having Anticancer Properties

Sl. No.	Type of Cancer	Essential Oil-Bearing Plant or Compound Tested	Model Used	References
17.	Prostate	*Xylopia frutescens*	PC-3M (IC50 = 40 µg/mL) and in vivo at 37.5% of inhibition	Ferraz et al. (2013)
		Nagami kumquats	LNCaP (200 ppm = 55, 61 and 63.4 % inhibition at 24, 48, 72 h	Jayaprakasha et al. (2012)
		Rosmarinus officinalis	LNCaP (IC50 = 180.9 µg/mL)	Hussain et al. (2010a)
		a-Humulene	LNCaP (IC50 = 11.24 µg/mL)	Loizzo et al. (2007)
		β-Caryophyllene oxide	DU145	Kim et al. (2014)
		Thymus vulgari	PC-3 (IC50 = 0.010% (v/v))	Zu et al. (2010)
		O. basilicum, L. multiflora, A. conizoides, and *Z. officiniale*	LNCaP (0.46, 0.58, 0.35 and 0.38 mg/mL respectively) and PC3 (0.45, 0.30, 0.49 and 0.42 respectively)	Bayala et al. (2014b)
18.	Skin (melanoma)	*Athanasia brownii*	A375 (IC50 = 19.85 µg/mL)	Rasoanaivo et al. (2013)
		Scorodophloeus zenkeri and *Afrostyrax lepidophyllus*	A375 (IC50 = 17.7 and 20.6 µg/mL respectively)	Fogang et al. (2014)
		Casearia sylvestris	A2058 (IC50 = 41.1 µg/mL)	Bou et al. (2013)
		Curcuma zedoaria	B16BL6 (IC50 = 41.8 µg/mL)	Chen et al. (2011)
		Ascaridole compound and *Croton regelianus*	MDA-MB-435 (IC50 = 10.5and 47.3 µg/mL respectively)	Bezerra et al. (2009)
		Schefflera heptaphylla	A375 (IC50 = 7.5 µg/mL)	Li et al. (2009)
		Cupressus sempervirens ssp. *pyramidalis*	C32 (IC50 = 104.90 µg/mL)	Loizzo et al. (2008)
		Laurus nobilis	C32 (IC50 = 75.45 µg/mL)	Loizzo et al. (2007)
19.	Stomach cancer	*Nigella sativa* seeds	SCL, SCL-37'6, SCL-6, Kato-3 and NUGC-4 (IC50 = 155.02; 120.40; 185.77; 286.83 and 384.53 respectively)	Islam et al. (2004)

14.6 Multidrug Resistance in Cancer

Multidrug-resistance (MDR) is one of the major problems of therapeutic drugs of various diseases including cancer. Proper chemotherapeutic drugs have become useless by the development of multidrug resistance. Occurrence of tumors periodically decreased the potency of intracellular drug along with impairment of signaling cascades of apoptosis by increasing efflux of drug by energy-dependent drug transporter, which belongs to a superfamily of ATP-binding cassette (ABC) (Gottesman and Pastan 1993). It is reported that linalool a monoterpene is able to reverse the doxorubicin resistance in wild type MCF7 and MCF7 adriamycine-resistance (ADR) adenocarcinoma cell lines (Ravizza et al. 2008). There are not many illustrative reports (Calcabrini et al. 2004, Effenberger-Neidnicht and Schobert 2011) that prove the combinational therapy of EOs' capability can overcome the multidrug resistance.

14.7 Conclusion and Future Prospects

The fundamental focal point of this chapter is to feature the remedial and chemo preventive properties of essential oils with throwing its light in their mode of actions. As the available cancer therapy regime is costly yet plated with many lethal side effects, so natural therapeutics is a vital need in cancer treatment. EOs have many efficient properties towards anticancer activities that are ultimately an effective substitute for modern cancer medications. In spite of the fact that some of EOs are still in various phases of clinical trial for anticancer treatment, it might be prescribed as a part of complementation to natural therapeutics in near future.

ACKNOWLEDGEMENT

The authors are thankful to DBT (Govt. of India) Sponsored Bioinformatics Infrastructure Facility (BIF) of Assam University and DelCON's e-Journal Access Facility.

CONFLICT OF INTEREST

The authors declare that there is no conflict of interest regarding the publication of this book chapter.

REFERENCES

Abdallah, H. M. & Ezzat, S. M. 2011a. Effect of the method of preparation on the composition and cytotoxic activity of the essential oil of Pituranthos tortuosus. *Z Naturforsch C,* 66, 143–148.

Abdallah, H. M. & Ezzat, S. M. 2011b. Effect of the method of preparation on the composition and cytotoxic activity of the essential oil of Pituranthos tortuosus. *Zeitschrift für Naturforschung C,* 66, 143–148.

Afoulous, S., Ferhout, H., Raoelison, E. G., Valentin, A., Moukarzel, B., Couderc, F. & Bouajila, J. 2011. Helichrysum gymnocephalum essential oil: chemical composition and cytotoxic, antimalarial and antioxidant activities, attribution of the activity origin by correlations. *Molecules*, 16, 8273–8291.

Afoulous, S., Ferhout, H., Raoelison, E. G., Valentin, A., Moukarzel, B., Couderc, F. & Bouajila, J. 2013. Chemical composition and anticancer, antiinflammatory, antioxidant and antimalarial activities of leaves essential oil of Cedrelopsis grevei. *Food and Chemical Toxicology*, 56, 352–362.

Alenzi, F., El-Bolkiny, Y. E.-S. & Salem, M. 2010. Protective effects of Nigella sativa oil and thymoquinone against toxicity induced by the anticancer drug cyclophosphamide. *British Journal of Biomedical Science*, 67, 20–28.

Allott, E. H. & Hursting, S. D. 2015. Obesity and cancer: Mechanistic insights from transdisciplinary studies. *Endocrine Related Cancer*, 22, R 365–86.

Amiel, E., Ofir, R., Dudai, N., Soloway, E., Rabinsky, T. & Rachmilevitch, S. 2012. β-Caryophyllene, a compound isolated from the biblical balm of gilead (Commiphora gileadensis), is a selective apoptosis inducer for tumor cell lines. *Evidence-based Complementary and Alternative Medicine*, 2012.

Anand, P., Kunnumakkara, A. B., Sundaram, C., Harikumar, K. B., Tharakan, S. T., Lai, O. S., Sung, B. & Aggarwal, B. B. 2008. Cancer is a preventable disease that requires major lifestyle changes. *Pharmaceutical Research*, 25, 2097–116.

Arunasree, K. 2010. Anti-proliferative effects of carvacrol on a human metastatic breast cancer cell line, MDA-MB 231. *Phytomedicine*, 17, 581–588.

Azmi, A. S., Bhat, S. H., Hanif, S. & Hadi, S. 2006. Plant polyphenols mobilize endogenous copper in human peripheral lymphocytes leading to oxidative DNA breakage: A putative mechanism for anticancer properties. *FEBS Letters*, 580, 533–538.

Badakhshan, M. P., Sreenivasan, S., Jegathambigai, R. N. & Surash, R. 2009. Anti-leukemia activity of methanolic extracts of Lantana camara. *Pharmacognosy Research*, 1, 274.

Bakkali, F., Averbeck, S., Averbeck, D. & Idaomar, M. 2008. Biological effects of essential oils–a review. *Food and Chemical Toxicology*, 46, 446–475.

Bakkali, F., Averbeck, S., Averbeck, D., Zhiri, A., Baudoux, D. & Idaomar, M. 2006. Antigenotoxic effects of three essential oils in diploid yeast (Saccharomyces cerevisiae) after treatments with UVC radiation, 8-MOP plus UVA and MMS. *Mutation Research/Genetic Toxicology and Environmental Mutagenesis*, 606, 27–38.

Bayala, B., Bassole, I. H. N., Gnoula, C., Nebie, R., Yonli, A., Morel, L., Figueredo, G., Nikiema, J.-B., Lobaccaro, J.-M. A. & Simpore, J. 2014b. Chemical composition, antioxidant, anti-inflammatory and anti-proliferative activities of essential oils of plants from Burkina Faso. *PLoS One*, 9, e92122.

Bayala, B., Bassole, I. H., Scifo, R., Gnoula, C., Morel, L., Lobaccaro, J.-M. A. & Simpore, J. 2014a. Anticancer activity of essential oils and their chemical components: A review. *American Journal of Cancer Research*, 4, 591.

Bendaoud, H., Romdhane, M., Souchard, J. P., Cazaux, S. & Bouajila, J. 2010. Chemical composition and anticancer and antioxidant activities of Schinus molle L. and Schinus terebinthifolius Raddi berries essential oils. *Journal of Food Science*, 75.

Bezerra, D. P., José Filho, D., Alves, A. P. N., Pessoa, C., De Moraes, M. O., Pessoa, O. D. L., Torres, M. C. M., Silveira, E. R., Viana, F. A. & Costa-Lotufo, L. V. 2009. Antitumor activity of the essential oil from the leaves of Croton regelianus and its component ascaridole. *Chemistry & Biodiversity*, 6, 1224–1231.

Bhalla, Y., Gupta, V. K. & Jaitak, V. 2013. Anticancer activity of essential oils: A review. *Journal of the Science of Food and Agriculture*, 93, 3643–3653.

Biswas, R., Mandal, S. K., Dutta, S., Bhattacharyya, S. S., Boujedaini, N. & Khuda-Bukhsh, A. R. 2011. Thujone-rich fraction of Thuja occidentalis demonstrates major anti-cancer potentials: evidences from in vitro studies on A375 cells. *Evidence-Based Complementary and Alternative Medicine*, 2011.

Bommareddy, A., Rule, B., Vanwert, A. L., Santha, S. & Dwivedi, C. 2012. α-Santalol, a derivative of sandalwood oil, induces apoptosis in human prostate cancer cells by causing caspase-3 activation. *Phytomedicine*, 19, 804–811.

Bose, C. K. 2007. Possible role of Moringa oleifera Lam. root in epithelial ovarian cancer. *Medscape General Medicine*, 9, 26.

Bostancıoğlu, R. B., Kürkçüoğlu, M., Başer, K. H. C. & Koparal, A. T. 2012. Assessment of anti-angiogenic and anti-tumoral potentials of Origanum onites L. essential oil. *Food and Chemical Toxicology*, 50, 2002–2008.

Bou, D. D., Lago, J. H. G., Figueiredo, C. R., Matsuo, A. L., Guadagnin, R. C., Soares, M. G. & Sartorelli, P. 2013. Chemical composition and cytotoxicity evaluation of essential oil from leaves of Casearia sylvestris, its main compound α-zingiberene and derivatives. *Molecules*, 18, 9477–9487.

Cai, Y., Luo, Q., Sun, M. & Corke, H. 2004. Antioxidant activity and phenolic compounds of 112 traditional Chinese medicinal plants associated with anticancer. *Life Sciences*, 74, 2157–2184.

Calcabrini, A., Stringaro, A., Toccacieli, L., Meschini, S., Marra, M., Colone, M., Arancia, G., Molinari, A., Salvatore, G. & Mondello, F. 2004. Terpinen-4-ol, the main component of Melaleuca alternifolia (tea tree) oil inhibits the in vitro growth of human melanoma cells. *Journal of Investigative Dermatology*, 122, 349–360.

Cardile, V., Russo, A., Formisano, C., Rigano, D., Senatore, F., Arnold, N. A. & Piozzi, F. 2009. Essential oils of Salvia bracteata and Salvia rubifolia from Lebanon: Chemical composition, antimicrobial activity and inhibitory effect on human melanoma cells. *Journal of Ethnopharmacology*, 126, 265–272.

Carnesecchi, S., Bras-Gonçalves, R., Bradaia, A., Zeisel, M., Gossé, F., Poupon, M.-F. & Raul, F. 2004. Geraniol, a component of plant essential oils, modulates DNA synthesis and potentiates 5-fluorouracil efficacy on human colon tumor xenografts. *Cancer Letters*, 215, 53–59.

Cavalieri, E., Mariotto, S., Fabrizi, C., De Prati, A. C., Gottardo, R., Leone, S., Berra, L. V., Lauro, G. M., Ciampa, A. R. & Suzuki, H. 2004. α-Bisabolol, a nontoxic natural compound, strongly induces apoptosis in glioma cells. *Biochemical and Biophysical Research Communications*, 315, 589–594.

Cha, J.-D., Kim, Y.-H. & Kim, J.-Y. 2010. Essential oil and 1, 8-cineole from Artemisia lavandulaefolia induces apoptosis in KB Cells via mitochondrial stress and caspase activation. *Food Science and Biotechnology*, 19, 185–191.

Cha, J. D., Moon, S. E., Kim, H. Y., Cha, I. H. & Lee, K. Y. 2009. Essential oil of artemisia capillaris induces apoptosis in KB Cells via mitochondrial stress and caspase activation mediated by MAPK-stimulated signaling pathway. *Journal of Food Science*, 74.

Chabir, N., Romdhane, M., Valentin, A., Moukarzel, B., Marzoug, H. N. B., Brahim, N. B., Mars, M. & Bouajila, J. 2011. Chemical study and antimalarial, antioxidant, and anticancer activities of Melaleuca armillaris (Sol Ex Gateau) Sm essential oil. *Journal of Medicinal Food,* 14, 1383–1388.

Chen, W., Lu, Y., Gao, M., Wu, J., Wang, A. & Shi, R. 2011. Anti-angiogenesis effect of essential oil from Curcuma zedoaria in vitro and in vivo. *Journal of Ethnopharmacology,* 133, 220–226.

Cho, W., Nam, J.-W., Kang, H.-J., Windono, T., Seo, E.-K. & Lee, K.-T. 2009. Zedoarondiol isolated from the rhizoma of Curcuma heyneana is involved in the inhibition of iNOS, COX-2 and pro-inflammatory cytokines via the downregulation of NF-κB pathway in LPS-stimulated murine macrophages. *International Immunopharmacology,* 9, 1049–1057.

Cragg, G. M. & Newman, D. J. 2005. Plants as a source of anticancer agents. *Journal of Ethnopharmacology,* 100, 72–79.

De, S. F. & Ramel, C. 1988. Mechanisms of inhibitors of mutagenesis and carcinogenesis: Classification and overview. *Mutation Research,* 202, 285–306.

Debersac, P., Heydel, J.-M., Amiot, M.-J., Goudonnet, H., Artur, Y., Suschetet, M. & Siess, M.-H. 2001. Induction of cytochrome P450 and/or detoxication enzymes by various extracts of rosemary: Description of specific patterns. *Food and Chemical Toxicology,* 39, 907–918.

De-Oliveira, A. C., Ribeiro-Pinto, L. F. & Paumgartten, F. J. 1997. In vitro inhibition of CYP2B1 monooxygenase by β-myrcene and other monoterpenoid compounds. *Toxicology Letters,* 92, 39–46.

Diaz-Moralli, S., Tarrado-Castellarnau, M., Miranda, A. & Cascante, M. 2013. Targeting cell cycle regulation in cancer therapy. *Pharmacology & Therapeutics,* 138, 255–271.

Döll-Boscardin, P. M., Sartoratto, A., Maia, S., De Noronha, B. H. L., Padilha De Paula, J., Nakashima, T., Farago, P. V. & Kanunfre, C. C. 2012. In vitro cytotoxic potential of essential oils of Eucalyptus benthamii and its related terpenes on tumor cell lines. *Evidence-Based Complementary and Alternative Medicine,* 2012.

Edris, A. E. 2009. Anti-cancer properties of Nigella spp. essential oils and their major constituents, thymoquinone and beta-elemene. *Current Clinical Pharmacology,* 4, 43–46.

Effenberger-Neidnicht, K. & Schobert, R. 2011. Combinatorial effects of thymoquinone on the anti-cancer activity of doxorubicin. *Cancer Chemotherapy and Pharmacology,* 67, 867–874.

Efferth, T., Li, P. C., Konkimalla, V. S. & Kaina, B. 2007. From traditional Chinese medicine to rational cancer therapy. *Trends in Molecular Medicine,* 13, 353–361.

El-Ghorab, A., El-Massry, K. F. & Shibamoto, T. 2007. Chemical composition of the volatile extract and antioxidant activities of the volatile and nonvolatile extracts of Egyptian corn silk (Zea mays L.). *Journal of Agricultural and Food Chemistry,* 55, 9124–9127.

Fayed, S. A. 2009. Antioxidant and anticancer activities of Citrus reticulate (Petitgrain Mandarin) and Pelargonium graveolens (Geranium) essential oils. *Research Journal of Agriculture and Biological Sciences,* 5, 740–747.

Ferraz, R. P., Cardoso, G. M., Da Silva, T. B., Fontes, J. E. D. N., Prata, A. P. D. N., Carvalho, A. A., Moraes, M. O., Pessoa, C., Costa, E. V. & Bezerra, D. P. 2013. Antitumour properties of the leaf essential oil of Xylopia frutescens Aubl. (Annonaceae). *Food Chemistry,* 141, 196–200.

Fogang, H. P., Maggi, F., Tapondjou, L. A., Womeni, H. M., Papa, F., Quassinti, L., Bramucci, M., Vitali, L. A., Petrelli, D. & Lupidi, G. 2014. In vitro biological activities of seed essential oils from the cameroonian spices afrostyrax lepidophyllus mildbr. and scorodophloeus zenkeri harms rich in sulfur-containing compounds. *Chemistry & Biodiversity,* 11, 161–169.

Formisano, C., Rigano, D., Napolitano, F., Senatore, F., Arnold, N., Piozzi, F. & Rosselli, S. 2007. Volatile constituents of Calamintha origanifolia Boiss: Growing wild in Lebanon. *Natural Product Communications,* 2, 1253–1256.

Frank, M. B., Yang, Q., Osban, J., Azzarello, J. T., Saban, M. R., Saban, R., Ashley, R. A., Welter, J. C., Fung, K.-M. & LIN, H.-K. 2009. Frankincense oil derived from Boswellia carteri induces tumor cell specific cytotoxicity. *BMC Complementary and Alternative Medicine,* 9, 6.

Gabrielli, B., Brooks, K. & Pavey, S. 2012. Defective cell cycle checkpoints as targets for anti-cancer therapies. *Frontiers in Pharmacology,* 3, 9.

Gautam, N., Mantha, A. K. & Mittal, S. 2014. Essential oils and their constituents as anticancer agents: A mechanistic view. *BioMed Research International,* 2014.

Gottesman, M. M. & Pastan, I. 1993. Biochemistry of multidrug resistance mediated by the multidrug transporter. *Annual Review of Biochemistry,* 62, 385–427.

Grosso, C., Oliveira, A., Mainar, A., Urieta, J., Barroso, J. & Palavra, A. 2009. Antioxidant activities of the supercritical and conventional Satureja montana extracts. *Journal of Food Science,* 74.

Guengerich, F. P. 1992. Metabolic activation of carcinogens. *Pharmacology and Therapeutics,* 54, 17–61.

Hassan, S. B., Gali-Muhtasib, H., Göransson, H. & Larsson, R. 2010. Alpha terpineol: A potential anticancer agent which acts through suppressing NF-κB signalling. *Anticancer Research,* 30, 1911–1919.

Hou, J., Sun, T., Hu, J., Chen, S., Cai, X. & Zou, G. 2007. Chemical composition, cytotoxic and antioxidant activity of the leaf essential oil of Photinia serrulata. *Food Chemistry,* 103, 355–358.

https://www.cancer.gov/about-cancer/treatment/side-effects.

https://www.cancer.gov/about-cancer/treatment/types.

Hussain, A. I., Anwar, F., Chatha, S. A. S., Jabbar, A., Mahboob, S. & Nigam, P. S. 2010a. Rosmarinus officinalis essential oil: Antiproliferative, antioxidant and antibacterial activities. *Brazilian Journal of Microbiology,* 41, 1070–1078.

Hussain, A. I., Anwar, F., Nigam, P. S., Ashraf, M. & Gilani, A. H. 2010b. Seasonal variation in content, chemical composition and antimicrobial and cytotoxic activities of essential oils from four Mentha species. *Journal of the Science of Food and Agriculture,* 90, 1827–1836.

Inui, T., Wang, Y., Nikolic, D., Smith, D. C., Franzblau, S. G. & Pauli, G. F. 2010. Sesquiterpenes from Oplopanax horridus. *Journal of Natural Products,* 73, 563–567.

Islam, N., Begum, P., Ahsan, T., Huque, S. & Ahsan, M. 2004. Immunosuppressive and cytotoxic properties of Nigella sativa. *Phytotherapy Research,* 18, 395–398.

Itani, W. S., El-Banna, S. H., Hassan, S. B., Larsson, R. L., Bazarbachi, A. & Gali-Muhtasib, H. U. 2008. Anti colon cancer components from lebanese sage (Salvia libanotica) essential oil: Mechanism basis. *Cancer Biology & Therapy,* 7, 1765–1773.

Jayaprakasha, G. K., Murthy, K. N. C., Demarais, R. & Patil, B. S. 2012. Inhibition of prostate cancer (LNCaP) cell proliferation by volatile components from Nagami kumquats. *Planta Medica,* 78, 974–980.

Jayasekara, H., Macinnis, R. J., Room, R. & English, D. R. 2016. Long-term alcohol consumption and breast, upper aero-digestive tract and colorectal cancer risk: A systematic review and meta-analysis. *Alcohol Alcohol,* 51, 315–330.

Jeong, J. B., Choi, J., Lou, Z., Jiang, X. & Lee, S.-H. 2013. Patchouli alcohol, an essential oil of Pogostemon cablin, exhibits anti-tumorigenic activity in human colorectal cancer cells. *International Immunopharmacology,* 16, 184–190.

Jones, W. P., Lobo-Echeverri, T., Mi, Q., Chai, H.-B., Soejarto, D. D., Cordell, G. A., Swanson, S. M. & Kinghorn, A. D. 2007. Cytotoxic constituents from the fruiting branches of Callicarpa americana collected in Southern Florida, 1. *Journal of Natural Products,* 70, 372–377.

Kada, T. & Shimoi, K. 1987. Desmutagens and bio-antimutagens–their modes of action. *Bioessays,* 7, 113–116.

Kathirvel, P. & Ravi, S. 2012. Chemical composition of the essential oil from basil (Ocimum basilicum Linn.) and its in vitro cytotoxicity against HeLa and HEp-2 human cancer cell lines and NIH 3T3 mouse embryonic fibroblasts. *Natural Product Research,* 26, 1112–1118.

Keawsa-Ard, S., Liawruangrath, B., Liawruangrath, S., Teerawutgulrag, A. & Pyne, S. G. 2012. Chemical constituents and antioxidant and biological activities of the essential oil from leaves of Solanum spirale. *Natural Product Communications: An International Journal for Communications and Reviews,* 7(7), 955–958.

Kelley, M. R., Cheng, L., Foster, R., Tritt, R., Jiang, J., Broshears, J. & Koch, M. 2001. Elevated and altered expression of the multifunctional DNA base excision repair and redox enzyme Ape1/ref-1 in prostate cancer. *Clinical Cancer Research,* 7, 824–830.

Khamsan, S., Liawruangrath, B., Liawruangrath, S., Teerawutkulrag, A., Pyne, S. G. & Garson, M. J. 2011. Antimalarial, anticancer, antimicrobial activities and chemical constituents of essential oil from the aerial parts of cyperus kyllingia endl. *Records of Natural Products,* 5, 324.

Ki, Y.-W., Park, J. H., Lee, J. E., Shin, I. C. & Koh, H. C. 2013. JNK and p38 MAPK regulate oxidative stress and the inflammatory response in chlorpyrifos-induced apoptosis. *Toxicology Letters,* 218, 235–245.

Kim, C., Cho, S. K., Kapoor, S., Kumar, A., Vali, S., Abbasi, T., Kim, S. H., Sethi, G. & Ahn, K. S. 2014. β-caryophyllene oxide inhibits constitutive and inducible STAT3 signaling pathway through induction of the SHP-1 protein tyrosine phosphatase. *Molecular Carcinogenesis,* 53, 793–806.

Kim, H. J., Chen, F., Wu, C., Wang, X., Chung, H. Y. & Jin, Z. 2004. Evaluation of antioxidant activity of Australian tea tree (Melaleuca alternifolia) oil and its components. *Journal of Agricultural and Food Chemistry,* 52, 2849–2854.

Kim, M. H., Chung, W. T., Kim, Y. K., Lee, J. H., Lee, H. Y., Hwang, B., Park, Y. S., Hwang, S. J. & Kim, J. H. 2001. The effect of the oil of Agastache rugosa O. Kuntze and three of its components on human cancer cell lines. *Journal of Essential Oil Research,* 13, 214–218.

Kpoviessi, S., Bero, J., Agbani, P., Gbaguidi, F., Kpadonou-Kpoviessi, B., Sinsin, B., Accrombessi, G., Frédérich, M., Moudachirou, M. & Quetin-Leclercq, J. 2014. Chemical composition, cytotoxicity and in vitro antitrypanosomal and antiplasmodial activity of the essential oils of four Cymbopogon species from Benin. *Journal of Ethnopharmacology,* 151, 652–659.

Kulisic-Bilusic, T., Schmöller, I., Schnäbele, K., Siracusa, L. & Ruberto, G. 2012. The anticarcinogenic potential of essential oil and aqueous infusion from caper (Capparis spinosa L.). *Food Chemistry,* 132, 261–267.

Lampronti, I., Saab, A. M. & Gambari, R. 2006. Antiproliferative activity of essential oils derived from plants belonging to the Magnoliophyta division. *International Journal of Oncology,* 29, 989–995.

Lei, J., Yu, J., Yu, H. & Liao, Z. 2008. Composition, cytotoxicity and antimicrobial activity of essential oil from Dictamnus dasycarpus. *Food Chemistry,* 107, 1205–1209.

Li, Y. L., Yeung, C. M., Chiu, L., Cen, Y. Z. & Ooi, V. E. 2009. Chemical composition and antiproliferative activity of essential oil from the leaves of a medicinal herb, Schefflera heptaphylla. *Phytotherapy Research,* 23, 140–142.

Loizzo, M. R., Tundis, R., Menichini, F., Saab, A. M., Statti, G. A. & Menichini, F. 2007. Cytotoxic activity of essential oils from Labiatae and Lauraceae families against in vitro human tumor models. *Anticancer Research,* 27, 3293–3299.

Loizzo, M., Tundis, R., Menichini, F., Saab, A., Statti, G. & Menichini, F. 2008. Antiproliferative effects of essential oils and their major constituents in human renal adenocarcinoma and amelanotic melanoma cells. *Cell Proliferation,* 41, 1002–1012.

Manjamalai, A., Kumar, M. & Grace, V. 2012. Essential oil of Tridax procumbens L induces apoptosis and suppresses angiogenesis and lung metastasis of the B16F-10 cell line in C57BL/6 mice. *Asian Pacific Journal of Cancer Prevention,* 13, 5887–5895.

Manosroi, J., Dhumtanom, P. & Manosroi, A. 2006. Antiproliferative activity of essential oil extracted from Thai medicinal plants on KB and P388 cell lines. *Cancer Letters,* 235, 114–120.

Mantha, A. 2013. APE1: A molecule of focus with neuroprotective and anti-cancer properties. *Journal of Biotechnology & Biomaterials,* 3.

Martindale, J. L. & Holbrook, N. J. 2002. Cellular response to oxidative stress: Signaling for suicide and survival. *Journal of Cellular Physiology,* 192, 1–15.

Mauro, M., Catanzaro, I., Naselli, F., Sciandrello, G. & Caradonna, F. 2013. Abnormal mitotic spindle assembly and cytokinesis induced by D-Limonene in cultured mammalian cells. *Mutagenesis,* 28, 631–635.

Mcguire, S. 2016. World Cancer Report 2014. Geneva, Switzerland: World Health Organization, International Agency for Research on Cancer, WHO Press, 2015. *Advances in Nutrition,* 7, 418–419.

McKay, D. L. & Blumberg, J. B. 2006a. A review of the bioactivity and potential health benefits of chamomile tea (Matricaria recutita L.). *Phytotherapy Research,* 20, 519–530.

McKay, D. L. & Blumberg, J. B. 2006b. A review of the bioactivity and potential health benefits of peppermint tea (Mentha piperita L.). *Phytotherapy Research,* 20, 619–633.

Mitić-Ćulafić, D., Žegura, B., Nikolić, B., Vuković-Gačić, B., Knežević-Vukčević, J. & Filipič, M. 2009. Protective effect of linalool, myrcene and eucalyptol against t-butyl hydroperoxide induced genotoxicity in bacteria and cultured human cells. *Food and Chemical Toxicology,* 47, 260–266.

Mitoshi, M., Kuriyama, I., Nakayama, H., Miyazato, H., Sugimoto, K., Kobayashi, Y., Jippo, T., Kanazawa, K., Yoshida, H. & Mizushina, Y. 2012. Effects of essential oils from herbal plants and citrus fruits on DNA polymerase inhibitory, cancer cell growth inhibitory, antiallergic, and antioxidant activities. *Journal of Agricultural and Food Chemistry,* 60, 11343–11350.

Mohamad, R. H., El-Bastawesy, A. M., Abdel-Monem, M. G., Noor, A. M., Al-Mehdar, H. A. R., Sharawy, S. M. & El-Merzabani, M. M. 2011. Antioxidant and anticarcinogenic effects of methanolic extract and volatile oil of fennel seeds (Foeniculum vulgare). *Journal of Mmedicinal Food,* 14, 986–1001.

Mohsenzadeh, F., Chehregani, A. & Amiri, H. 2011. Chemical composition, antibacterial activity and cytotoxicity of essential oils of Tanacetum parthenium in different developmental stages. *Pharmceutical Biology,* 49, 920–926.

Moon, J.-K. & Shibamoto, T. 2009. Antioxidant assays for plant and food components. *Journal of Agricultural and Food Chemistry,* 57, 1655–1666.

Mukherjee, P. K. 2003. Exploring botanicals in Indian system of medicine: Regulatory perspectives. *Clinical Research and Regulatory Affairs,* 20, 249–264.

Mulder, T. P., Verspaget, H. W., Sier, C. F., Roelofs, H. M., Ganesh, S., Griffioen, G. & Peters, W. H. 1995. Glutathione S-transferase π in colorectal tumors is predictive for overall survival. *Cancer Research,* 55, 2696–2702.

Mulyaningsih, S., Youns, M., El-Readi, M. Z., Ashour, M. L., Nibret, E., Sporer, F., Herrmann, F., Reichling, J. & Wink, M. 2010. Biological activity of the essential oil of Kadsura longipedunculata (Schisandraceae) and its major components. *Journal of Pharmacy and Pharmacology,* 62, 1037–1044.

Murata, S., Shiragami, R., Kosugi, C., Tezuka, T., Yamazaki, M., Hirano, A., Yoshimura, Y., Suzuki, M., Shuto, K. & Ohkohchi, N. 2013. Antitumor effect of 1, 8-cineole against colon cancer. *Oncology Reports,* 30, 2647–2652.

Nagappan, T., Ramasamy, P., Wahid, M. E. A., Segaran, T. C. & Vairappan, C. S. 2011. Biological activity of carbazole alkaloids and essential oil of Murraya koenigii against antibiotic resistant microbes and cancer cell lines. *Molecules,* 16, 9651–9664.

Nazaruk, J., Karna, E., Wieczorek, P., Sacha, P. & Tryniszewska, E. 2010. In vitro antiproliferative and antifungal activity of essential oils from Erigeron acris L. and Erigeron annuus (L.) Pers. *Zeitschrift für Naturforschung C,* 65, 642–646.

Nikolić, B., Mitić-Ćulafić, D., Vuković-Gačić, B. & Knežević-Vukčević, J. 2011. Modulation of genotoxicity and DNA repair by plant monoterpenes camphor, eucalyptol and thujone in Escherichia coli and mammalian cells. *Food and Chemical Toxicology,* 49, 2035–2045.

Özkan, A. & Erdoğan, A. 2011. A comparative evaluation of antioxidant and anticancer activity of essential oil from Origanum onites (Lamiaceae) and its two major phenolic components. *Turkish Journal of Biology,* 35, 735–742.

Patil, J. R., Jayaprakasha, G., Murthy, K. C., Tichy, S. E., Chetti, M. B. & Patil, B. S. 2009. Apoptosis-mediated proliferation inhibition of human colon cancer cells by volatile principles of Citrus aurantifolia. *Food Chemistry,* 114, 1351–1358.

Prashar, A., Locke, I. C. & Evans, C. S. 2006. Cytotoxicity of clove (Syzygium aromaticum) oil and its major components to human skin cells. *Cell Proliferation,* 39, 241–248.

Preethy, C. P., Padmapriya, R., Periasamy, V. S., Riyasdeen, A., Srinag, S., Krishnamurthy, H., Alshatwi, A. A. & Akbarsha, M. A. 2012. Antiproliferative property of n-hexane and chloroform extracts of Anisomeles malabarica (L). R. Br. in HPV16-positive human cervical cancer cells. *Journal of Pharmacology & Pharmacotherapeutics,* 3, 26.

Rashid, S., Rather, M. A., Shah, W. A. & Bhat, B. A. 2013. Chemical composition, antimicrobial, cytotoxic and antioxidant activities of the essential oil of Artemisia indica Willd. *Food Chemistry,* 138, 693–700.

Rasoanaivo, P., Fortuné Randriana, R., Maggi, F., Nicoletti, M., Quassinti, L., Bramucci, M., Lupidi, G., Petrelli, D., Vitali, L. A. & Papa, F. 2013. Chemical composition and biological activities of the essential oil of Athanasia brownii Hochr. (Asteraceae) endemic to Madagascar. *Chemistry & Biodiversity,* 10, 1876–1886.

Ravizza, R., Gariboldi, M. B., Molteni, R. & Monti, E. 2008. Linalool, a plant-derived monoterpene alcohol, reverses doxorubicin resistance in human breast adenocarcinoma cells. *Oncology Reports,* 20, 625–630.

Razavi, S. M. & Nejad-Ebrahimi, S. 2009. Chemical composition, allelopatic and cytotoxic effects of essential oils of flowering tops and leaves of Crambe orientalis L. from Iran. *Natural Product Research,* 23, 1492–1498.

Rigano, D., Cardile, V., Formisano, C., Maldini, M. T., Piacente, S., Bevilacqua, J., Russo, A. & Senatore, F. 2009. Genista sessilifolia DC. and Genista tinctoria L. inhibit UV light and nitric oxide-induced DNA damage and human melanoma cell growth. *Chemico-Biological Interactions,* 180, 211–219.

Roy, A., Ahuja, S. & Bharadvaja, N. 2017. A review on medicinal plants against cancer. *Journal of Plant Sciences and Agricultural Research,* 1, 008.

Russo, P., Cardinale, A., Margaritora, S. & Cesario, A. 2012. Nicotinic receptor and tobacco-related cancer. *Life Sciences,* 91, 1087–1092.

Sakihama, Y., Cohen, M. F., Grace, S. C. & Yamasaki, H. 2002. Plant phenolic antioxidant and prooxidant activities: Phenolics-induced oxidative damage mediated by metals in plants. *Toxicology,* 177, 67–80.

Seal, S., Chatterjee, P., Bhattacharya, S., Pal, D., Dasgupta, S., Kundu, R., Mukherjee, S., Bhattacharya, S., Bhuyan, M. & Bhattacharyya, P. R. 2012. Vapor of volatile oils from Litsea cubeba seed induces apoptosis and causes cell cycle arrest in lung cancer cells. *PLoS One,* 7, e47014.

Sertel, S., Eichhorn, T., Plinkert, P. & Efferth, T. 2011a. Anticancer activity of Salvia officinalis essential oil against HNSCC cell line (UMSCC1). *HNO,* 59, 1203–1208.

Sertel, S., Eichhorn, T., Plinkert, P. K. & Efferth, T. 2011b. Chemical composition and antiproliferative activity of essential oil from the leaves of a medicinal herb, Levisticum officinale, against UMSCC1 head and neck squamous carcinoma cells. *Anticancer Research,* 31, 185–191.

Sharma, N., Trikha, P., Athar, M. & Raisuddin, S. 2001. Inhibition of benzo [a] pyrene-and cyclophoshamide-induced mutagenicity by Cinnamomum cassia. *Mutation Research/Fundamental and Molecular Mechanisms of Mutagenesis,* 480, 179–188.

Sharma, P. R., Mondhe, D. M., Muthiah, S., Pal, H. C., Shahi, A. K., Saxena, A. K. & Qazi, G. N. 2009. Anticancer activity of an essential oil from Cymbopogon flexuosus. *Chemico-Biological Interactions,* 179, 160–168.

Sigurdsson, S., Ögmundsdóttir, H. M. & Gudbjarnason, S. 2005. The cytotoxic effect of two chemotypes of essential oils from the fruits of Angelica archangelica L. *Anticancer Research,* 25, 1877–1880.

Simionatto, E., Peres, M., Hess, S., Da Silva, C., Chagas, M., Poppi, N., Prates, C., Matos, M. D. F., Santos, E. & De Carvalho, J. 2010. Chemical composition and cytotoxic activity of leaves essential oil from Mangifera indica var. coquinho (Anacardiaceae). *Journal of Essential Oil Research,* 22, 596–599.

Singletary, K. W. 1996. Rosemary extract and carnosol stimulate rat liver glutathione-S-transferase and quinone reductase activities. *Cancer Letters,* 100, 139–144.

Soldani, C. & Scovassi, A. 2002. Poly (ADP-ribose) polymerase-1 cleavage during apoptosis: An update. *Apoptosis,* 7, 321–328.

Stadler, R. H., Markovic, J. & Turesky, R. J. 1995. In vitro anti-and pro-oxidative effects of natural polyphenols. *Biological Trace Element Research,* 47, 299–305.

Su, M., Wu, P., Li, Y. & Chung, H. Y. 2010. Antiproliferative effects of volatile oils from Centipeda minima on human nasopharyngeal cancer CNE cells. *Natural Product Communications,* 5, 151–156.

Suhail, M. M., Wu, W., Cao, A., Mondalek, F. G., Fung, K.-M., Shih, P.-T., Fang, Y.-T., Woolley, C., Young, G. & Lin, H.-K. 2011. Boswellia sacra essential oil induces tumor cell-specific apoptosis and suppresses tumor aggressiveness in cultured human breast cancer cells. *BMC Complementary and Alternative Medicine,* 11, 129.

Sun, H., Sun, C. & Pan, Y. 2005. Cytotoxic activity and constituents of the volatile oil from the roots of Patrinia scabra Bunge. *Chemistry & Biodiversity,* 2, 1351–1357.

Sylvestre, M., Legault, J., Dufour, D. & Pichette, A. 2005. Chemical composition and anticancer activity of leaf essential oil of Myrica gale L. *Phytomedicine,* 12, 299–304.

Sylvestre, M., Pichette, A., Lavoie, S., Longtin, A. & Legault, J. 2007. Composition and cytotoxic activity of the leaf essential oil of Comptonia peregrina (L.) Coulter. *Phytotherapy Research,* 21, 536–540.

Tuberoso, C. I., Kowalczyk, A., Coroneo, V., Russo, M. T., Dessì, S. & Cabras, P. 2005. Chemical composition and antioxidant, antimicrobial, and antifungal activities of the essential oil of Achillea ligustica All. *Journal of Agricultural and Food Chemistry,* 53, 10148–10153.

Unlu, M., Ergene, E., Unlu, G. V., Zeytinoglu, H. S. & Vural, N. 2010. Composition, antimicrobial activity and in vitro cytotoxicity of essential oil from Cinnamomum zeylanicum Blume (Lauraceae). *Food and Chemical Toxicology,* 48, 3274–3280.

Van Houten, B., Woshner, V. & Santos, J. H. 2006. Role of mitochondrial DNA in toxic responses to oxidative stress. *DNA Repair,* 5, 145–152.

Walia, M., Mann, T. S., Kumar, D., Agnihotri, V. K. & Singh, B. 2012. Chemical composition and in vitro cytotoxic activity of essential oil of leaves of Malus domestica growing in Western Himalaya (India). *Evidence-Based Complementary and Alternative Medicine,* 2012.

Wattenberg, L. 1993. Chemoprevention of carcinogenesis by minor nonnutrient constituents of the diet. *Food, Nutrition and Chemical Toxicity,* 31, 287–300.

World Cancer Report [Online]. World Health Organization. 2014.

World Cancer Report [Online]. World Health Organization. 2017.

Wu, S., Wei, F., Li, H., Liu, X., Zhang, J. & Liu, J. 2013. Chemical composition of essential oil from Thymus citriodorus and its toxic effect on liver cancer cells. *Zhong yao cai= Zhongyaocai= Journal of Chinese Medicinal Materials,* 36, 756–759.

Xiao, Y., Yang, F.-Q., Li, S.-P., Hu, G., Lee, S. M.-Y. & Wang, Y.-T. 2008. Essential oil of Curcuma wenyujin induces apoptosis in human hepatoma cells. *World Journal of Gastroenterology: WJG,* 14, 4309.

Xu, W., Gao, X., Guo, X., Chen, Y., Zhang, W. & Luo, Y. 2010. Study on volatile components from peel of Aquilaria sinensis and the anti-tumor activity. *Zhong yao cai= Zhongyaocai= Journal of Chinese Medicinal Materials,* 33, 1736–1740.

Xue, Y., Jin-Gang, W., Tao, Y., Shan-Shan, L., Hong-Wei, R., Guo-Hua, F. & Shu-Fang, G. 2012. Establishment of rapid propagation system for Saponaria officinalis L. progeny after EMS mutation. *Journal of Northeast Agricultural University (English Edition),* 19, 21–26.

Yan, R., Yang, Y., Zeng, Y. & Zou, G. 2009. Cytotoxicity and antibacterial activity of Lindera strychnifolia essential oils and extracts. *Journal of Ethnopharmacology,* 121, 451–455.

Yin, Q.-H., Yan, F.-X., Zu, X.-Y., Wu, Y.-H., Wu, X.-P., Liao, M.-C., Deng, S.-W., Yin, L.-L. & Zhuang, Y.-Z. 2012. Antiproliferative and pro-apoptotic effect of carvacrol on human hepatocellular carcinoma cell line HepG-2. *Cytotechnology,* 64, 43–51.

Yoo, C.-B., Han, K.-T., Cho, K.-S., Ha, J., Park, H.-J., Nam, J.-H., Kil, U.-H. & Lee, K.-T. 2005. Eugenol isolated from the essential oil of Eugenia caryophyllata induces a reactive oxygen species-mediated apoptosis in HL-60 human promyelocytic leukemia cells. *Cancer Letters,* 225, 41–52.

Yoon, W.-J., Kim, S.-S., Oh, T.-H., Lee, N. H. & Hyun, C.-G. 2009. Cryptomeria japonica essential oil inhibits the growth of drug-resistant skin pathogens and LPS-induced nitric oxide and pro-inflammatory cytokine production. *Polish Journal of Microbiology,* 58, 61–68.

Yousefzadi, M., Riahi-Madvar, A., Hadian, J., Rezaee, F., Rafiee, R. & Biniaz, M. 2014. Toxicity of essential oil of Satureja khuzistanica: in vitro cytotoxicity and anti-microbial activity. *Journal of Immunotoxicology,* 11, 50–55.

Yu, J.-Q., Lei, J.-C., Zhang, X.-Q., Yu, H.-D., Tian, D.-Z., Liao, Z.-X. & Zou, G.-L. 2011. Anticancer, antioxidant and antimicrobial activities of the essential oil of Lycopus lucidus Turcz. var. hirtus Regel. *Food Chemistry,* 126, 1593–1598.

Yu, J. Q., Liao, Z. X., Cai, X. Q., Lei, J. C. & Zou, G. L. 2007. Composition, antimicrobial activity and cytotoxicity of essential oils from Aristolochia mollissima. *Environmental Toxicology and Pharmacology,* 23, 162–167.

Zahri, S. & Ghasemian, A. 2011. Phytochemical prospection and biological activity of Astrodaucus orientalis (L.) Drude growing wild in Iran. *Pharmacologia,* 2(10), 299–303.

Zamanian-Azodi, M., Rezaie-Tavirani, M., Heydari-Kashal, S., Kalantari, S., Dailian, S. & Zali, H. 2012. Proteomics analysis of MKN45 cell line before and after treatment with Lavender aqueous extract. *Gastroenterology and Hepatology from Bed to Bench,* 5, 35.

Zu, Y., Yu, H., Liang, L., Fu, Y., Efferth, T., Liu, X. & Wu, N. 2010. Activities of ten essential oils towards Propionibacterium acnes and PC-3, A-549 and MCF-7 cancer cells. *Molecules,* 15, 3200–3210.

15

Isolation, Extraction, Preclinical and Clinical Studies on Major Anticancer Compounds of Natural Origin

Somorita Baishya, Subrata Das, Manabendra Dutta Choudhury, and Anupam Das Talukdar

CONTENTS

15.1 Introduction: Cancer and Chemotherapy

Cancer is one of the primary causes of death worldwide (Kumar et al., 2016, Bhanot et al., 2011, Rayan et al., 2017). Cancer mainly occurs due to environmental mutations, genetic predispositions and incorrect life style (Bhanot et al., 2011). From a molecular perspective, cancer occurs when the mutations strike on either or both proto-oncogenes and tumor suppressor genes, and they start proliferating rapidly (Rayan et al., 2017). Activation of oncogenes transforms normal cells to cancerous cells whereby they defy apoptotic pathways i.e. programmed cell death by ignoring the death signals and proliferate abnormally to form tumor cells, which can be benign or malignant. The malignant tumors have the capacity to invade and spread to healthy cells, unlike benign tumors, by a process called metastasis. Generally, the malignant tumors are also regarded as cancerous cells and are divided into three categories viz. sarcomas, carcinomas, and lymphomas or leukemia (Višnjić and Crljen-Manestar, 2004, Rayan et al., 2017).

Chemotherapy, radiotherapy and surgery have been the preference of cure against cancer since a long time (DeVita and Chu, 2008, MacDonald, 2009). However, until 1960, radiotherapy and surgery dominated the cancer therapy regimens. Later it was discovered that conjugating radiotherapy or surgery with chemotherapy gave better treatment results (DeVita and Chu, 2008). Although the choice of therapy to be subjected on the patients depends on different parameters such as type and grade of tumor, patient's stage and tolerance of the patient towards side-effects of treatment (MacDonald, 2009).

Development and standardization of model systems for testing a large variety of chemicals came as a major step forward in cancer chemotherapeutic research (DeVita and Chu, 2008). Positive effects of the use of estrogen on breast cancer and prostate cancer patients evoked excitement in oncology research (Beatson, 1896, Huggins and Hodges, 2002). In the 1940s, use of nitrogen mustard for treating lymphomas had created elation amongst oncologists; however, remissions resulted in the back stepping of its use (DeVita and Chu, 2008).

In the meantime nutritional researchers identified that folic acid antagonists like aminopterin and amethopterin could treat lymphomas; however, remissions were very common (Farber et al., 1948). Although discovery of the activities of aminopterin and amethopterin gave impetus to the synthesis of newer drugs like 6-thioquanine and 6-mercaptopurine that played a great role in the treatment of acute leukemia (Farber et al., 1948). In 1950s and 1960s, antibiotics like penicillin, actinomycin D also gained popularity as antitumor medicines (DeVita and Chu, 2008).

Focus on chemotherapy gained more attention in the mid-1950s with the success of fluorouracil (5-FU) that had broad spectrum antitumor activities (DeVita and Chu, 2008). Discovery of anti-leukemic activity of *Vincarosea* alkaloids gave oncologic research a great boost (Johnson et al., 1963). Following this finding, oncologists like DeVita, Moxley, and Frei began various programs, which initially received a lot of backlash from the NIH Clinical Center, but by the end of 1960 these programs proved that chemotherapy could cure cancer (DeVita and Chu, 2008).

With the establishment of the pivotal role that chemotherapeutics played in curing cancer, oncologists started giving thought on use of chemotherapeutics as adjuvants. However, the choice was difficult as exposing patients to such therapies who have or were undergoing localized treatment could unnecessarily subject their healthy cells to various side effects. But researchers tried to give adjuvant chemotherapy a shot because of the success of combination chemotherapy. And with this began the era of adjuvant chemotherapy. Oncologists started using certain drugs as adjuvant to surgery in breast cancer, colorectal cancer, acute leukemia, lymphomas and metastatic testicular cancer, and they were getting positive results. Those results changed today's chemotherapeutics regimens and also the increased the quality of patient's life (DeVita and Chu, 2008, MacDonald, 2009).

15.2 Cons of Chemotherapy

Undoubtedly chemotherapy has emerged as a boon for cancer patients (DeVita and Chu, 2008, MacDonald, 2009). Most of these treatments have less complications (Vail and Withrow, 2007). However, such treatments do impart a lot of side effects like multi-system toxicities, healthy cell death along with the cancerous cells, bone marrow suppression, immune system suppression, alopecia, neutropenia and gastrointestinal problems (MacDonald, 2009, Rayan et al., 2017). Along with these general side effects there are chances that cancer might recur again with drug resistance, thereby degrading the patients' quality of life (Rayan et al., 2017). These major disadvantages brought about a paradigm shift in the anti-cancer drug research. Researchers shifted their focus on documenting and developing anti-cancer drugs from natural origin (Rayan et al., 2017). Success of antibiotics as anti-tumor medicines and anecdotal evidences of anti-cancer and anti-tumor activities of plant secondary metabolites gave momentum to this research (Newman and Cragg, 2005).

15.3 Chemotherapeutics from Natural Origin

Natural products from plants, marine organisms, and microorganisms have been a rich source of medicines from time immemorial. Natural compounds like alkaloids, terpenoids, flavonoids, saponins, polysaccharides, and others possess various types of bioactivity. Many of these natural products possess anti-cancer activity. More or less than 60% of the anti-cancer drugs that are presently used for curing cancer, are from natural origin. Isolation of anti-cancer drugs from natural origin are popular because natural products offer less toxicity, large diversity in structure and less adverse reaction in addition to providing improved efficacy towards drug target, thus improving treatment regimens (Gordaliza, 2007, Rayan et al., 2017, Newman et al., 2003, Saklani and Kutty, 2008, Bhanot et al., 2011). Moreover, modern techniques developed for screening, separation, structural elucidation, and combinatorial synthesis have also given a new lease to new drug research from natural products (Saklani and Kutty, 2008). Therefore, most of the anti-cancer drugs that are in clinical trials are obtained from plants, microorganisms, and marine sources (Butler, 2008).

With the discovery of anti-cancer potentials of vinca alkaloids, vinblastine and vincristine from plant sources and development of anti-cancer agents from them in 1950 marked the beginning of search of anti-cancer agents from plant sources. Receiving impetus from this, the United States National Cancer Institute (NCI) had set up a huge plant collection program in the temperate regions from 1960 to 1982. Within this period many novel chemotypes like taxanes and camptothecins came up. Many of those anti-cancer agents have reached pre-clinical stages (Cragg and Newman, 2005) (Figure 15.1).

Since 2002–2007, many anti-cancer agents from plant origin, marine origin and microbial origin have been approved (Bhanot et al., 2011). List of anti-cancer drugs obtained from natural sources like plants, marine organisms and microorganisms that are in clinical and preclinical use are shown in Table 15.1.

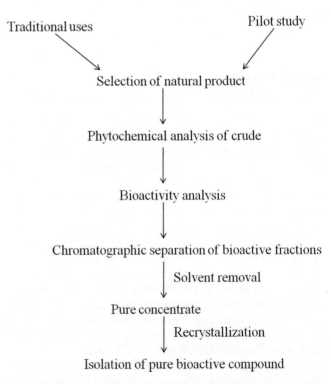

FIGURE 15.1 Scheme of plan for extraction and isolation of bioactive natural products.

TABLE 15.1

List of a Few Anti-cancer Compounds Obtained from Natural Sources That Are in Clinical and Pre-clinical Trials

Sl. No.	Organism	Compound	Used in Cancer	References
1.	*Catharanthusroseus*	Vinblastine	Neoplasias, chonioepithelioma, breast cancer and bronchus cancer	Johnson et al. (1960), Svoboda et al. (1959)
2.	*Catharanthusroseus*	Vincristine	Childhood leukemia	Johnson et al. (1960), Svoboda et al. (1959)
3.	*Taxusbrevifolia*	Taxol	Breast, ovarian, and non-small cell lung cancer	Nikolic et al. (2011)
4.	*Cephalotaxusharringtonia* var. *drupacea*	Homoharringtonine	Leukemia	Powell (1974)
5.	*Camptothecaacuminata*	Camptothecin	Ovarian and small cell lung cancers and colorectal cancers	Buta and Novak (1978)
6.	*Nothapodytesfoetida*	Camptothecin	Ovarian and small cell lung cancers and colorectal cancers	Srivastava (2005)
7.	*Bleekeriavitensis, Ochrosiaelliptica*	Ellipticine	Breast cancer	Iqbal et al. (2017)
8.	*Podophyllum* sp.	Epipodophyllotoxin	Skin cancers	Sultan et al. (2010)
9.	*Euphorbia peplus*	Ingenol mebutate	Non-melanoma skin cancer	Seca and Pinto (2018)
10.	*Amoorarohituka, Dysoxylumbinecteriferum*	Flavopiridol	Chronic lymphocytic leukemia	Saklani and Kutty (2008)
11.	*Colchicum autumnale*	Colchicine	Hypopharyngeal cancer	Schwartsmann et al. (2000)
12.	*Combretumcaffrum*	Combrestatin	Myeloid leukemia	Pettit et al. (1982)
13.	*Curcuma longa*	Curcumin	Gastrointestinal mucosal cancers	Seca and Pinto (2018)
14.	*Platanusacerifolia*	Betulinic acid	Lung, colon, breast, prostate, hepatocellular, bladder, head and neck, stomach, pancreatic, ovarian and cervical carcinoma, glioblastoma, chronic myeloid leukemia cells and human melanoma	Seca and Pinto (2018)
15.	*Streptomyces peucetius* var. *caesius*	Doxorubicin	Breast cancer	Arcamone et al. (1969)
16.	*Streptomyces pneuceticus*	Epirubicin	Breast cancer	Bhanot et al. (2011)
17.	*Streptomyces pneuceticus*	Idarubicin	Breast cancer and leukemia	Bhanot et al. (2011)
18.	*Streptomyces caespitosus*	Mitomycin C	Gastric, colorectal, anal and lung cancer	Bhanot (2011)
19.	*Streptomyces hygroscopicus*	Rapamycin	Melanoma	Schwartsmann et al. (2000)
20.	*Streptomyces verticillus*	Bleomycin	Germ-cell, cervix and head and neck cancer	Blum (1973)
21.	Sponge	Cytarabine (Ara-C)	Leukemia	Ruiz-Torres et al. (2017)
22.	Sponge	Gemcitabine (GEM) (Gemzar)		Ruiz-Torres et al. (2017)
23.	Tunicate	Trabectedin (E7389)	Advanced liposarcoma and leiomyosarcoma	Ruiz-Torres et al. (2017)
24.	Tunicate	Plitidepsin	Myeloma	Ruiz-Torres et al. (2017)
25.	Sponge	Eribulinmesylate (E7389)	Breast cancer	Ruiz-Torres et al. (2017)
26.	Mollusca	Brentuximabvedotin	Lymphoma	Ruiz-Torres et al. (2017)
27.	Mollusk	Glembatumumabvedotin	Breast cancer	Ruiz-Torres et al. (2017)
28.	Mollusk	Elisidepsin	Antineoplastic	Ruiz-Torres et al. (2017)
29.	Soft coral	Pseudopterosins	Breast cancer	Ruiz-Torres et al. (2017)
30.	Sponge	Contignasterol derivative	Carcinoma	Ruiz-Torres et al. (2017)
31.	Sponge	Tetrahydroisoquinoline alkaloid (Zalypsis®)	Lymphoma, Ewing's sarcoma, endometrial cancer, uterine cervical cancer	Ruiz-Torres et al. (2017)
32.	Sponge	Discodermolide	Colon, ovarian, and breast cancers	Ruiz-Torres et al. (2017)
33.	Bryozoa	Bryostatin-1	Colorectal cancer, leukemia, lymphoma, ovarian cancer, melanoma	Ruiz-Torres et al. (2017)
34.	Mollusk	Pinatuzumabvedotin	Non-Hodgkin lymphoma, leukemia	Ruiz-Torres et al. (2017)
35.	Mollusk	Tisotumabvedotin	Ovarian, endometrium, cervix and prostate cancer	Ruiz-Torres et al. (2017)
36.	Sponge	Hemiasterlin	Breast cancer	Ruiz-Torres et al. (2017)
37.	Tunicate derivatives	Didemnin B, aplidine and ecteinascidin	Non-Hodgkin's lymphoma	Schwartsmann et al. (2000)
38.	Sponge	Psammaplin derivative	Non small cell lung cancer	Ruiz-Torres et al. (2017)

(Continued)

TABLE 15.1 (*Continued*)

List of a Few Anti-cancer Compounds Obtained from Natural Sources That Are in Clinical and Pre-clinical Trials

Sl. No.	Organism	Compound	Used in Cancer	References
39.	Sponge	Bengamide B derivative	Breast cancer	Haag-Richter (2009), Ruiz-Torres et al. (2017)
40.	Mollusk	Dolastatins	Leukaemia, lymphoma	Pettit et al. (1987), Schwartsmann et al. (2000)
41.	Mushroom	6-Hydroxymethylacylfulvene	Breast carcinoma, lungadenocarcinoma, prostate colon carcinoma. myeloid leukaemia	Schwartsmann et al. (2000)
42.	*Chromobacterium violaceum*	Depsipeptide	Stomach, breast and colon adenocarcinomas	Schwartsmann et al. (2000)
43.	*Saccharothrix aerocolingenes*	NSC D655649	Melanomas	Schwartsmann et al. (2000)
44.	*Omphalotusolearius*	MGI-114	Breast cancer, lung cancer, prostrate cancer	Schwartsmann et al. (2000)
45.	*Clavulariaviridis*	Clavulone II	Promyelocytic leukemia	Bhanot et al. (2011)

15.4 Plants as a Source of Anti-cancer Drugs

Plants have been used as medicine since time immemorial. Ancient literatures of Old World record the use of plants and their various parts as medicine (Saklani and Kutty 2008). Owing to their medicinal properties, structural diversity, and less toxicity, they have been a choice of source for the pharmaceutical industries (Harvey 2008). Therefore, resorting to plants for development of anti-cancer drugs was obvious. Many anti-cancer drugs from plant origin have even reached clinical trials (Bhanot et al. 2011, Rayan et al. 2017, Gordaliza 2007, Harvey 2008, Huggins and Hodges 2002). Some of them are discussed below.

15.4.1 Vinca Alkaloids

Vinca alkaloids, vinblastine, and vincristine isolated from *Catharanthus roseus* G. Don. of the Apocynaceae family was one of the first isolated agents into clinical use (Bhanot et al., 2011, Cragg and Newman, 2005, Gordaliza, 2007, Johnson et al., 1963, Noble, 1990). It is commonly known as Madagascar periwinkle or rose/rosy periwinkle. It is an ornamental evergreen herb or sub-shrub that is widely cultivated in gardens around the world although the plant is endemic to Madagascar. The plant was earlier named as *Vinca rosea* although phytochemical researches on the plant demonstrated that *C. roseus* is the appropriate name (Johnson et al., 1960).

In some parts of Brazil the folklore used the leaf infusion of this plant as mouthwash to treat tooth ache and scurvy. The infusion was also used to treat hemorrhage and chronic wounds (Johnson et al., 1963). Traditionally this plant's extract was used against diabetes. Such therapeutics aspects of *C. roseus*, especially anti-diabetic property of the plant, instigated researcher teams of University of Western Ontario and Lilly Research Laboratories to carry out research on this plant (Johnson et al., 1963). However, water-based extracts of this (Noble, 1990) plant could not lower blood sugar level in diabetic rats and rabbits, thereby had no effect on the disease. But the case became curious when intra-peritoneal administration of the extracts lead to the development of several abscesses

and ultimately death of the rats. Post-mortem analysis of those rats showed that crude extract of the plant had reduced the WBC count of the rats and destroyed the bone marrow of those animals. This gave hint to the researchers that extract of *C. roseus* might possess some compounds that could be useful in treating cancer of the hematopoietic system like lymphomas and leukemias (Noble, 1990). Since the crude aqueous extract constituted a mixture of many substances; therefore, the researchers thought to isolate the active compounds from that plant (Noble, 1990).

Phytochemical analysis of the crude extract of *C. roseus* revealed that alkaloids were the most bioactive constituents and leaves of *C. roseus* contained most of the alkaloids (Johnson et al., 1960, 1963). Therefore, leaves of this plant were used for extraction of the alkaloids (Johnson et al., 1963). For the extraction process, 9 kg of the whole plant was stirred with two portions of 45 l of Skelly B for removal of fat. This extract was Mayer's test positive. 3 l of the concentration was treated with 2 N HCl and then again subjected to treatment with NH₄OH and ether. This fraction gave 3.1 gm alkaloids and was marked as group E by the researchers.

The defatted drug was tartrated with of 2% tartaric acid and extracted with benzene. The extracted product was concentrated to 9 l in vacuum. This extract was again extracted with ethylene dichloride at 3.2 pH and ultimately 8.86 gm was yielded. This group was marked A1.

For the third group (A), the acidic solution was made alkaline by treating with NH₄OH and extracted with ethylene dichloride, which yielded 20.07 gm product after drying and evaporation.

The fourth group had two sub-groups-B, B1. The acid extract was bubbled with NH₃ till the extract became alkaline. The alkaline extract was again treated with benzene and concentrated to give yield of 6.13 gm. and 21.87 gm respectively.

For the next group i.e. F, the alkaline solution was treated with pellets of sodium hydroxide at pH 11 followed by extraction with chloroform and subsequent drying and evaporation of the alkaloid (Svoboda et al., 1959).

Of all these fractions, chromatographic purification of fraction A on alumina yielded many inactive alkaloids along with vinleurosine and vincaleukoblastine (VLB) that had anti-tumor activity.

Amongst the isolated alkaloids, some post-VLB fractions were the most bioactive. These alkaloids were then subjected to gradient partitioning between benzene and buffers at pH 2.8–7.5. The materials were crystallized and ultimately pure form of vincristine and vinrosidine were obtained. Two more pure compounds viz. vindolidine and locherinine were isolated from fraction E. All the four isolated alkaloids had different nature and showed different bioactivities. Therefore, the physical, chemical and biological properties of the alkaloids were analyzed. Physical analysis revealed the structural properties of the alkaloids, while the chemical analysis showed that the alkaloids had indole and dihydnoindole moieties and had almost similar structures. Biological activity showed that vinblastine and vincristine were active against human neoplasms. Vinblastine could treat neoplasias of various types, Hodgkin's disease, chonioepithelioma, breast cancer and bronchus cancer. Vincristine, on the other hand, could remission childhood leukemia. Thus, from a group of vinca alkaloids, two alkaloids were isolated that underwent clinical trial as an anti-cancer drug and is still used widely (Johnson et al., 1963, Svoboda et al., 1959).

15.4.2 Taxanes

Taxanes obtained from *Taxus brevifolia* Nutt. of the Taxaceae family contributed a new generation of anti-cancer drug from plant origin. *T. brevifolia*, commonly known as Pacific yew, is a small to medium ever-blooming conifer native to northwestern Pacific (Cragg and Newman, 2005, Newman et al., 2003). The folklore of Native America used various parts of *Taxus* plants as medicine for treatment of diseases other than cancer, while in Ayurvedic literature it is mentioned that the leaves of *T. baccata* were used against cancer (Cragg and Newman, 2005).

The bark extract of *Taxus* was first analyzed by NCI during their plant collection program of 1963 and the active compound of the extract, paclitaxel, was first identified in 1971. Studies on the bioactivities of the extract revealed that the extract could inhibit growth and multiplication of cell, i.e., the extract was cytostatic in nature (Wani et al., 1971). However, the development of paclitaxel as drug was slow because the yield of the plant was very less. But, with the discovery of its mechanism of action, interest of the researchers grew in this regard. It was found that paclitaxel could be obtained naturally and semi-synthetically from *T. yannanensis* and *T. baccata*. In fact, a group of researchers found that fungus of *Taxomyces andreanae* could also produce paclitaxel, although the yield was very little. This finding hinted the researchers that endophytic fungus might have a role in paclitaxel production and later it was observed that endophytes from *T. chinensis* var. *mairei* and *T. yunnanensis* like BT2 could produce paclitaxel. Moreover, this finding also suggested that paclitaxel could be alternatively produced by paclitaxel-producing microorganisms. This approach of paclitaxel production was followed in spite of very less yield as compared to yews because fungus had higher growth rate and required shorter production time (Nikolic et al., 2011).

The isolation of paclitaxel from the plant is a mainly a five-step process (Bui-Khac and Potier, 2004):

1. Washing the raw material to remove impurities: The plant material is washed with deionized water at 20°–25° for 3 hours. The water containing all the impurities is discarded.

2. Extraction using organic solvents: The wet raw material is extracted using three different solvent systems- methanol, acetone, and 1:1 methanol-acetone mixture. The extract is then filtered and transferred to a double walled water tank where the temperature is maintained at 65°C–70°C. Then distillation is done to remove the organic solvents and again the dilute extract is transferred to another tank.

3. Biomass isolation from the extraction: To precipitate the biomass, the diluted extract is further diluted with methanol and water and rigorously stirred with high concentration of sodium chloride. The formed wet biomass is then subjected to filtration or centrifugation and dried either by lyophilization or ventilation.

4. Removal of organic pigments like raisins: Treating the dried biomass with acetone-hexane mixture followed by addition of pure hexane to make 1:4 ratio of the solvent mixture to remove raisins and pigments in a better way. Water is then added to make an oily phase mixture of taxanes, which is then transferred to a decanting tank where the oily phase mixture settles at the bottom.

5. Purification of the solution by chromatographic technique: The oily phase mixture of taxanes contains paclitaxel which is isolated by first mixing the oily phase mixture of taxanes with silica gel and dried in a vent. This mixture of oil phase and silica gel is loaded on a silica gel column and run in a 65% hexane: 35% acetone system. The fractions containing paclitaxel are collected, dried and kept in acetone, to which 3/4 volumes of hexane is added for crystallization.

Paclitaxel thus obtained might contain some of its analogues. Therefore, in order to obtain pure fraction reverse-phase liquid chromatography preferably on C_{18} column is done. However, to increase the amount of product in a short period of time different procedures may be followed. One such process is the Accelerated solvent extraction (ASE). Obviously, the temperature, pressure and solvent conditions will be different (Nikolic et al., 2011).

15.4.3 Camptothecin

Camptotheca acuminata Decne from the Nyssaceae family gave another anti-cancer drug—camptothecin—to the health care armamentarium. Camptothecin was originally isolated from the stems of *C. acuminata*, an ornamental deciduous tree native to

China and Tibet; however, the yield was very less. Camptothecin was also isolated from *Nothapodytes foetida* (Wight) Sleumer of Icacinaceae family. *N. foetida* is tree found in the Western Ghats of India (Srivastava et al., 2005). The alkaloid showed inhibitory actions on cell division and anti-tumor activities by blocking DNA Topoisomerase enzyme activity (Buta and Novak, 1978, Wall et al., 1966).

For extraction and isolation of camptothecin from *C. acuminate* the stem was air-dried and ground. About 69 gm of the ground chips were extracted for 24 hours in Soxhlet extractor with 95% ethanol. The extract was vacuum-evaporated and partitioned with dichloromethane and water. The water-based extracts were extracted twice with dichloromethane and then all the dichloromethane extracts were combined and evaporated until a concentrate of about 500 mg was obtained. This concentrate was then dissolved in at least 4:1 ethyl acetate-dichloromethane and then filtered. This filtrate was then subjected to size exclusion chromatography with 4:1 ethyl acetate-dichloromethane solvent system. Here a gel permeation column which was packed with biobeads of S-X1 type was used. Ultraviolet lamp was used to visualize the alkaloids. Camptothecin was observed as a fluorescence bright blue band. Camptothecin and its analogs were then eluted out in order to obtain pure fractions. The collected fractions were subjected to thin layer chromatography (TLC) in C_6H_6-ethyl acetate-methanol (10: 10: 1 v/v) solvent system. Although the band indicated that only camptothecin was present, a few fractions contained a combination of camptothecin and 10-methoxycamptothecin. In the next step, solvent was removed from the fractions and methanol was used to recrystallize the obtained camptothecin alkaloids. Now to obtain pure fraction of camptothecin, melting temperature, UV spectra and TLC was compared and column chromatography was performed on silica gel (Buta and Novak, 1978).

For extraction and isolation of campothecin from *N. foetida*, leaves, bark and fruits were collected in between late winter to early spring, air dried and powdered. The dry powder was then extracted with petroleum ether at 60°C–80°C in Soxhlet extractor. Chloroform was used to extract the remaining tissues. This portion was then reduced under pressure. In the next step, two solvent system, i.e., dichloromethane, chloroform, ethyl acetate, ether, acetone, methanol, ethanol and acetonitrile were used sequentially and successively for extraction. Vacuum evaporation of the extracts was done to remove the solvents. This dry extract is then precipitated. The precipitate is subjected to column chromatography and TLC to obtain pure campothecin (Srivastava et al., 2005).

Although campothecin had reached clinical trials by the dint of its shown anti-cancer properties, its use discontinued due to some toxicity issues. However, the sodium salt derivative of campothecin—Topotecan and Irinotecan—are still used for treatment cancers, respectively (Cragg and Newman, 2005).

15.4.4 Homoharringtonine

Homoharringtonine is another plant-based anti-cancer drug that is in clinical use. It is isolated from *Cephalotaxus harringtonia*

var. *drupacea*, commonly known as Japanese Plum Yew. This plant is an evergreen shrub that is mostly found between East Asia to Japan and blooms during spring (Cragg and Newman, 2005, Powell et al., 1974). A pilot study on the plant revealed that the alkaloids of this shrub had effect on leukemic mice. The alkaloids of this plant viz. harringtonine, homoharringtonine, isoharringtonine, and deoxyharringtonine are found mainly in the seeds and in leaves, stems, and roots in a lesser quantity. However, the extraction of the alkaloids from the seeds is tough since large number of seeds for such processing is not available. Extraction from leaves pose similar troubles because leaves contain emulsions of lipids and other ethanol-soluble materials (Powell et al., 1974).

Pharmacological investigations on the alkaloids of this plant by NCI gave green signal to harringtonine and homoharringtonine to be used as anti-cancer drugs. Further investigations supported the use of homoharringtonine over harringtonine as the former was found in more concentration than the later and also a lower dose of the former was more effective than the latter (Powell et al., 1974).

For extraction and isolation of the alkaloids, the following steps were employed (Powell et al., 1974):

Collection of plant material: *C. harringtonia* trees were collected in January and air-dried at room temperature for about 6 months. For extraction purpose the branches and roots of the trees were used. Those air-dried branches and roots were then ground.

Extraction: Extraction began in two batches, first comprising of upper and lower branches and the second batch containing remaining of the lower branches and roots. Both the batches were then separately soaked in 95% ethanol for 48 hours. This extract was transferred to another tank and filtered. This step was repeated twice. The solid residues are discarded. Both the extracts were combined and concentrated.

Crude alkaloid isolation: The extract was then treated with 2.5% tartaric acid at pH 2 and rigorously stirred so that emulsification cannot occur. The extract is then treated with chloroform to remove lipids and other non-polar residues. The extract now contained only basic alkaloids which were recovered NH_4OH at pH 9 and then extracted with chloroform. The chloroform extracts were then filtered and concentrated by evaporation.

Alkaloid separation: The aqueous portion is discarded and the concentrated extract is subjected to countercurrent distribution with chloroform and McIlvaine's buffer at pH 5. The active alkaloids are recovered with chloroform and concentrated by column chromatography.

Isolation of pure forms: The pure forms of the alkaloids were isolated by countercurrent distribution.

15.4.5 Ellipticine

Ellipticine is an anti-cancer drug obtained from natural sources. It is isolated from *Bleekeria vitensis* and *Ochrosia elliptica*

Labill. of the Apocynaceae family. Ellipticine was first isolated from *O. elliptica*, a flowering plant native to Australia, while *B. vitensis* is a medicinal plant from Fiji (Cragg and Newman, 2005, Iqbal et al., 2017, Kizek et al., 2012). This alkaloid showed topoisomerase-inhibiting activity, DNA intercalating activity, efficacy towards many types of cancer and lacked toxicity, and thus attracted attention of researchers (Iqbal et al., 2017, Kizek et al., 2012).

For isolation of ellipticine, leaves and stems of *O. elliptica* were collected. The plant materials were air-dried and powdered. The ground material was mixed with 95% ethanol and stirred at 55°C. This ethanolic extract was filtered and again treated with ethanol at 55°C. In the next step the alcoholic extract was concentrated to about 2 l in vacuum until it became black colored and syrupy. This concentrate was added to 15 l of water that contained 2 kg of Supercel and was stirred rigorously. Glacial acetic acid was added to maintain pH of 5–4. The mixture was filtered and the residue was washed with very little water and again dried in vacuum 32°C. The dry mixture was extracted in a Soxhlet apparatus with ether and alcohol. The filtrate was again added to 7 l of water that contained glacial acetic acid. Next the filtrate was treated with chloroform and sodium carbonate. The chloroform based filtrates were dried and concentrated to 2.5 l using magnesium sulfate. Column chromatography of the chloroform-based filtrates was done on acid washed alumina. Twenty-four fractions were collected, of which following fractions showed ellipticine:

1. 14th fraction showed ellipticine on paper chromatogram
2. 15–16th fraction gave 25 mg ellipticine
3. 17–20th fraction gave ellipticine contaminated with methoxyellipticine

Although crystals of ellipticine were not obtained from the fractions, crystallized ellipticine were made by treating with methanol. These crystals appeared as bright lemon-yellow colored needles. However, removal of methanol from the crystals was tough. Ethyl acetate could also recrystallize ellipticine that looked like yellow or orange rosettes or orange rods (Goodwin et al., 1959).

15.4.6 Podophyllotoxin

Podophyllotoxin, a lignan from *Podophyllum peltatum* L. and *P. hexandnum* Royle (syn. *P. emodii* Wallich) posseses broad range of medicinal properties. *P. hexandnum* commonly known as Himalayan mayapple is a perennial herb found in India and is known to possess many ayurvedic properties. Traditionally the plant was used to treat constipation, septic wounds, inflammation, mental disorder, genital warts, leukemia, etc. *P. peltatum*, also known as May Apple, was used in homeopathic medicinal system to cure acute diarrhea, gastrointestinal disorders, and gall stones (Canel et al., 2000, Cragg and Newman, 2005).

Initial works on *Podophyllum* extracts were not satisfactory because of its bitter taste and toxic nature. Even crystals of podophyllotoxin showed toxicity similar to colchicines. Based on these findings the American Pharmacopoeia in 1942 eliminated *Podophyllum* extract from their list. However, interest in podophyllin rekindled when its topical application on venereal warts was successful. Discovery of anti-mitotic activity of podophyllin also gave air to the zeal. Based on the success from these findings, researchers focused more on the bioactivity of secondary metabolites of *Podophyllum* and came up with the break through discovery of anti-cancer property of podophyilin and podophyllotoxin. Their findings established that podophyllotoxin could kill cancer cells more than podophyilin, while quercetin and picropodophyllotoxin lacked any activity. Podophyllotoxin could inhibit assembly of microtubule while its derivatives like etoposide and teniposide had topoisomerase inhibiting activity (Imbert, 1998).

For extraction and isolation of *Podophyllum* secondary metabolites from *P. hexandrum*, roots were collected, air-dried, ground and stored at 40°C. The powdered material was extracted in 100% methanol in Soxhlet apparatus. The extract was filtered and the process was repeated thrice. Methanol was removed at 50°C in a Rotavapor and the extract was vacuum evaporated. The extracts were then subjected to column chromatography and TLC respectively with methanolic solvent system at seven different pH (1–7). LC-MS was done to identify the compounds like podophyllotoxins, glycosides, quercitin, and so on. The extract was again dissolved in HPLC grade methanol to make the volume 1 mL. 5 µL of the sample is injected into the column. Conditions like 1 mL/minute flow rate were maintained at 30°C at 283 nm wavelength. The extracts were again evaporated in vacuum and podophyllotoxins were isolated by recrystallization (Sultan et al., 2010).

Glycosidic derivatives of podophyllotoxins gave birth to etoposide and teniposide that were more dynamic and water soluble than their parent molecule (Imbert, 1998).

15.4.7 Combrestatin

Combrestatin is isolated from *Combretum caffrum* (Eckl. & Zeyh.) Kuntze, a bushwillow tree, which is found in South Africa. The plant has many medicinal properties. Traditionally this plant was used for its anti-cancer properties. The bark, leaves and the roots of this plant were used as general tonic, against conjunctivitis and body pain reliever respectively (Cunningham, 1990). Due its anti-cancer properties, NCI took interest in this plant (Pettit et al., 1982a).

The root bark of the plant had anti-cancer properties. Methylene chloride—methanol ethanol was used for extraction. The crude was diluted with water and methylene chloride was removed by using methanol water with ligroin, carbon tetrachloride and methylene chloride in 9:1, 4:1 and 3:2 ratios, respectively. Sephadex LH-20 and silica gel were used to separate the methylene chloride fraction and pure combretastatin was obtained. The pure fraction was recrystallized as needles of combretastatin using acetone-hexane at 130°C–131°C (Pettit et al., 1982a) (Table 15.2).

TABLE 15.2

Plant-Based Anti-cancer Drugs at Various Stages of Clinical Trials

Sl. No.	Compound	Name	Clinical Status	References
1.	Homoharringtonine	Ceflatonin®	Phase II	Saklani and Kutty (2008)
2.	Ellipticine	Elliptinium	Phase II	Newman and Cragg (2005)
3.	Podophyllotoxin	NK-611	Phase I	Saklani and Kutty (2008)
		Tafluposide 105		
4.	Combrestatin	AVE-8062 (AC-7700)	Phase I	
		AVE-8064		
		AVE-8063		
		CA4PO4 (combrestatin A-4 phosphate)	Phase II	
5.	Vinca alkaloids	Hydravin™	Phase II	Saklani and Kutty (2008)
		Venorelbine	Phase III	
		Vincristine sulfate TCS (OncoTCS)/Marqibo		
		Javlor®		
6.	Paclitaxel	MST-997 (TL-909)	Phase I	Saklani and Kutty (2008)
		PNU-166945		
		BMS-275183		
		BMS-188797	Phase II	
		BMS-184476		
		DJ-927		
		MAC-321 (TL-00139)		
		TPI-287		
		Ortataxel		
		TXD-258 (XRP-6258, RPR-116258A)	Phase IIa	
		XRP_9881 (RPR-109881 A)	Phase III	
		ABI-007 (suspension)		
		DHA-paclitaxel		
		Paclitaxel poliglumex		
		RPR-116258A		
7.	Campothecin	BN-80927	Phase I	Saklani and Kutty (2008)
		Karenitecin®	Phase I/II	
		LE-SN38		
		Diflomotecan (BN-80915) 100	Phase II	
		DRF-1042		
		Gimatecan (ST-1481)		
		Irinotecan (Hycamp)		
		Lurtotecan		
		NK012		
		PG-Camptothecin		
		Rubitecan (9-nitro camptothecin)	Phase III	
		Oral topotecan		
		Exatecanmesilate		
		9-amino camptothecin		

15.5 Microorganisms as a Source of Anti-cancer Drugs

Microorganisms are a rich source of secondary metabolites and can be produced in large scale by culturing. Many of those antibiotics have anti-tumor properties. Therefore, microorganisms have become a choice of anti-cancer drugs. Many drugs isolated from microorganisms are in clinical use (Bhanot et al., 2011). Extraction and isolation of a few of such drugs are discussed below:

15.5.1 Doxorubicin

Doxorubicin, produced by *Streptomyces peucetius* ATCC 27952, belongs to anthracycline class. Enzyme inhibition activity and intercalating activity of doxorubicin leads to DNA disruption and thus, cause cell death. Hence, doxorubicin is used to treat leukemia, breast cancer, sarcoma, and myeloma (Niraula et al., 2010).

For production of doxorubicin, 300 mL of media containing dextrose, yeast, agar, and salts was sterilized and shaken on a rotary shaker and incubated at 28°C. Inoculum was prepared in a round-bottomed flask and kept in rotary shaker. Various analytical methods were employed to determine reducing sugar, dry weight, and estimation of total pigments. Paper chromatography and TLC was done to separate doxorubicin from the mixture of pigments using oxalic acid, chloroform, and methanol solvent system. The bacteria were allowed to grow for 10 days at 28°C. The bloomed mycelium was homogenized and suspended in sterile water. This suspension was again incubated for 48 hours. The incubated suspension was used for production of doxorubicin. Since the mycelium of the bacteria mainly produced doxorubicin, it was extracted using acetone and aqueous sulfuric acid at 4:1 ratio. The residue was filtered out and the extract was vacuum evaporated. Lipids and other lipid soluble particles were removed using chloroform first at acidic pH and again with chloroform and methanol 9:1 ratio. Methanolic hydrogen chloride and ether were used to precipitate the organic extract. Using partition chromatography the extract was purified with cellulose as stationary phase and using paper chromatography and ultraviolet spectrum it was analyzed in n-butanol system at pH 5. The antibiotic was again eluted and transferred to water at acidic pH and again extracted with chloroform. The concentrate was crystallized as hydrochloride salt on addition of methanolic hydrogen chloride and recrystallized with ethanol or methanol-propanol mixture (Arcamone et al., 1969).

15.5.2 Bleomycins

Bleomycins are a glycopeptides class that from *Streptornyces verticillus*. This antibiotic is of A and B type and is effective against Ehrlich carcinoma, sarcoma. It was seen that bleomycins can inhibit DNA synthesis. Blenoxane is drug produced from bleomycin A and B (Blum et al., 1973, Boger et al., 1992, Pittillo et al., 1971).

Column chromatography on silica gel of blenoxane was done after dissolving it in 0.2 mL of water. Methanol, ammonium acetate, and ammonium solution were used as eluents and retention factor was checked. The fraction bearing the desired components were vacuum dried first. The concentrate was again diluted and run on Amberlite XAD-2 column. The fraction was desalted with water and eluted again with methanol. Bleomycin was obtained as white solids. Earlier ion-exchange chromatography was employed for purification of bleomycin (Boger et al., 1992).

15.6 Marine Organisms as a Source of Anti-cancer Drugs

Marine organisms are a rich source of secondary metabolites, which are still unexplored. These metabolites are produced in response to competition or harsh marine environmental conditions. Since the marine world is very diverse and has gotten very little attention, researchers have focused on marine pharmacology for the last two decades. Since inception of this newly field, many molecules possessing anti-cancer properties

have been identified. Some of these molecules have even reached clinical trials (Ruiz-Torres et al., 2017). A few anticancer molecules isolated from marine origin are discussed below:

15.6.1 Dolastatins

Dolstatins are peptides obtained from a mollusk, *Dolabella auricularia*, found in Indian Ocean. Preclinical studies on this mollusk showed that it could hinder assembly of microtubules, binding of GTP, and arrest cells at metaphase stage. The peptide showed positive in vitro activity against lymphoma, leukemia, and solid tumor cell lines. These pilot studies initiated the want of development of anti-cancer drug from this peptide (Schwartsmann et al., 2000).

For extraction of dolstatins wet *D. auricularia* is extracted with mixture of ethanol and 2-propanol. The crude extract was then subjected to series of partitioning with different solvent system and a methylene chloride concentrate was obtained. Column chromatography and HPLC of the concentrate using different solvent system gave colorless and amorphous pure dolstatin (Pettit et al., 1987).

15.6.2 Bryostatins

Bugula neritina, a member of Bryozoa phylum, produces a lactone called bryostatin which is reported to show in vitro cytostatic activity in leukemia, melanoma, lymphoma, and ovarian cancer. This lactone probably works by downrgulating protein kinase C (PKC) (Schwartsmann et al., 2000).

For isolation of bryostatins, wet *B. neritina* was extracted with methylene chloride. This crude extract was further partitioned in solvent systems like methanol-water with ligroin and carbon tetrachloride. Column chromatography was done to purify the fraction containing carbon tetrachloride. Recrystallization of the methanolic fraction gave crystals of bryostatins (Pettit et al., 1982b).

15.7 Limitations of Anti-cancer Drug Discovery from Natural Products

Natural products are the first choice for developing drugs owing to their diverse medicinal properties. Therefore, production of anti-cancer drugs from natural sources is quite obvious. In fact, 60% of the marketed anti-cancer drugs are from natural origin (Bhanot et al., 2011, Rayan et al., 2017). However, there are some limitations with development of anti-cancer drugs from natural sources. Firstly, collection of huge amounts of raw material without affecting biodiversity is not possible (Harvey, 2008). Secondly, ensuring that the natural product of interest will affect the effector organ or product in the desired way; for this mechanism of action of the natural product has to be elucidated (Koehn and Carter, 2005). But technical drawbacks like limitations with characterization, purification, and structure elucidation procedures of natural products are always associated with these drug development methods. Moreover, issues concerning IPR also contribute to the decreased interest of researchers in developing drugs from natural origin (Harvey, 2008).

15.8 Conclusion

Cancer is a deadly disease, killing many people around the world. Although advances in the fields of surgery, radiotherapy, chemotherapy, immunotherapy, and hormonal therapy could provide a momentary support system against this lethal disease, however complete relieve from cancer has not been achieved. This situation is aggravating with the development of resistant cancer types and heterogenous nature of the cancer cell population. Therefore, development of more anti-cancer drugs is essential (Ruiz-Torres et al., 2017). In this line, natural products play a promising role. This is because they have less toxicity, less adverse side effects and more diverse structures and reports suggest that about 60% of the anti-cancer drugs that are available in the market are of natural origin (Bhanot et al., 2011, Gordaliza, 2007, Newman and Cragg, 2005, Rayan et al., 2017, Ruiz-Torres et al., 2017, Saklani and Kutty, 2008). But, inadequate supply of raw materials, slow rate of product development following traditional extraction methods and concerns with intellectual property rights (IPR) has become a bummer in drug discovery (Cheuka et al., 2016, Seca and Pinto, 2018). So, combining traditional methods with high throughput screening (HTS) methods and new biotechnological advances with may give efficient products in less time. This approach might cure cancer in the future (Harvey, 2008, Seca and Pinto, 2018).

REFERENCES

Arcamone, F., Cassinelli, G., Fantini, G., Grein, A., Orezzi, P., Pol, C. & Spalla, C. 1969. Adriamycin, 14-hydroxydaimomycin, a new antitumor antibiotic from S. Peucetius var. caesius. *Biotechnology and Bioengineering*, 11, pp. 1101–1110.

Beatson, G. 1896. On the treatment of inoperable cases of carcinoma of the mamma: Suggestions for a new method of treatment, with illustrative cases. *The Lancet*, 148, pp. 162–165.

Bhanot, A., Sharma, R. & Noolvi, M. N. 2011. Natural sources as potential anti-cancer agents: A review. *International Journal of Phytomedicine*, 3, p. 9.

Blum, R. H., Carter, S. K. & Agre, K. 1973. A clinical review of bleomycin—A new antineoplastic agent. *Cancer*, 31, pp. 903–914.

Boger, D. L., Menezes, R. F. & Yang, W. 1992. A simple method for the purification and isolation of bleomycin A2. *Bioorganic & Medicinal Chemistry Letters*, 2, pp. 959–962.

Bui-Khac, T. & Potier, M. 2004. Process for isolation and purification of paclitaxel from natural sources. Google Patents.

Buta, J. G. & Novak, M. J. 1978. Isolation of camptothecin and 10-methoxycamptothecin from Camptotheca acuminata by gel permeation chromatography. *Industrial & Engineering Chemistry Product Research and Development*, 17, pp. 160–161.

Butler, M. S. 2008. Natural products to drugs: Natural product-derived compounds in clinical trials. *Natural Product Reports*, 25, 475–516.

Canel, C., Moraes, R. M., Dayan, F. E. & Ferreira, D. 2000. Podophyllotoxin. *Phytochemistry*, 54, pp. 115–120.

Cheuka, P. M., Mayoka, G., Mutai, P. & Chibale, K. 2016. The role of natural products in drug discovery and development against neglected tropical diseases. *Molecules*, 22, p. 58.

Cragg, G. M. & Newman, D. J. 2005. Plants as a source of anti-cancer agents. *Journal of Ethnopharmacology*, 100, pp. 72–79.

Cunningham, A. 1990. People and medicines: The exploitation and conservation of traditional Zulu medicinal plants. *Mitteilungen aus dem Institut für Allgemeine Botanik Hamburg*, pp. 979–990.

Devita, V. T. & Chu, E. 2008. A history of cancer chemotherapy. *Cancer Research*, 68, pp. 8643–8653.

Farber, S., Diamond, L. K., Mercer, R. D., Sylvester JR, R. F. & Wolff, J. A. 1948. Temporary remissions in acute leukemia in children produced by folic acid antagonist, 4-aminopteroyl-glutamic acid (aminopterin). *New England Journal of Medicine*, 238, pp. 787–793.

Goodwin, S., Smith, A. & Horning, E. 1959. Alkaloids of Ochrosia elliptica Labill. *Journal of the American Chemical Society*, 81, pp. 1903–1908.

Gordaliza, M. 2007. Natural products as leads to anticancer drugs. *Clinical and Translational Oncology*, 9, pp. 767–776.

Haag-Richter, S. 2009. Bengamide derivatives, process for preparing them, and their use. Google Patents.

Harvey, A. L. 2008. Natural products in drug discovery. *Drug Discovery Today*, 13, pp. 894–901.

Huggins, C. & Hodges, C. V. 2002. Studies on prostatic cancer: I. The effect of castration, of estrogen and of androgen injection on serum phosphatases in metastatic carcinoma of the prostate. *The Journal of Urology*, 168, pp. 9–12.

Imbert, T. 1998. Discovery of podophyllotoxins. *Biochimie*, 80, pp. 207–222.

Iqbal, J., Abbasi, B. A., Mahmood, T., Kanwal, S., Ali, B. & Khalil, A. T. 2017. Plant-derived anticancer agents: A green anticancer approach. *Asian Pacific Journal of Tropical Biomedicine*, 7, pp. 1129–1150.

Johnson, I. S., Armstrong, J. G., Gorman, M. & Burnett, J. P. 1963. The vinca alkaloids: A new class of oncolytic agents. *Cancer Research*, 23, pp. 1390–427.

Johnson, I. S., Wright, H. F., Svoboda, G. H. & Vlantis, J. 1960. Antitumor principles derived from Vinca rosea L. I. Vincaleukoblastine and leurosine. *Cancer Research*, 20, pp. 1016–1022.

Kizek, R., Adam, V., Hrabeta, J., Eckschlager, T., Smutny, S., Burda, J. V., Frei, E. & Stiborova, M. 2012. Anthracyclines and ellipticines as DNA-damaging anticancer drugs: Recent advances. *Pharmacology & Therapeutics*, 133, pp. 26–39.

Koehn, F. E. & Carter, G. T. 2005. The evolving role of natural products in drug discovery. *Nature Reviews Drug Discovery*, 4, p. 206.

Kumar, S., Bajaj, S. & Bodla, R. B. 2016. Preclinical screening methods in cancer. *Indian Journal of Pharmacology*, 48, pp. 481.

Macdonald, V. 2009. Chemotherapy: managing side effects and safe handling. *The Canadian Veterinary Journal*, 50, pp. 665.

Newman, D. J. & Cragg, G. M. 2005. The discovery of anticancer drugs from natural sources. *Natural Products*. Springer.

Newman, D. J., Cragg, G. M. & Snader, K. M. 2003. Natural products as sources of new drugs over the period 1981–2002. *Journal of Natural Products*, 66, pp. 1022–1037.

Nikolic, V. D., Savic, I. M., Savic, I. M., Nikolic, L. B., Stankovic, M. Z. & Marinkovic, V. D. 2011. Paclitaxel as an anticancer agent: Isolation, activity, synthesis and stability. *Central European Journal of Medicine*, 6, p. 527.

Niraula, N. P., Kim, S.-H., Sohng, J. K. & Kim, E.-S. 2010. Biotechnological doxorubicin production: Pathway and regulation engineering of strains for enhanced production. *Applied Microbiology and Biotechnology*, 87, pp. 1187–1194.

Noble, R. L. 1990. The discovery of the vinca alkaloids—chemotherapeutic agents against cancer. *Biochemistry and Cell Biology*, 68, pp. 1344–1351.

Pettit, G. R., Cragg, G. M., Herald, D. L., Schmidt, J. M. & Lohavanijaya, P. 1982a. Isolation and structure of combretastatin. *Canadian Journal of Chemistry*, 60, pp. 1374–1376.

Pettit, G. R., Herald, C. L., Doubek, D. L., Herald, D. L., Arnold, E. & Clardy, J. 1982b. Isolation and structure of bryostatin 1. *Journal of the American Chemical Society*, 104, pp. 6846–6848.

Pettit, G. R., Kamano, Y., Herald, C. L., Tuinman, A. A., Boettner, F. E., Kizu, H., Schmidt, J. M., Baczynskyj, L., Tomer, K. B. & Bontems, R. J. 1987. The isolation and structure of a remarkable marine animal antineoplastic constituent: Dolastatin 10. *Journal of the American Chemical Society*, 109, pp. 6883–6885.

Pittillo, R. F., Woolley, C. & Rice, L. S. 1971. Bleomycin, an antitumor antibiotic: Improved microbiological assay and tissue distribution studies in normal mice. *Applied Microbiology*, 22, pp. 564–566.

Powell, R. G., Rogovin, S. P. & Smith JR, C. R. 1974. Isolation of antitumor alkaloids from Cephalotaxus harringtonia. *Industrial & Engineering Chemistry Product Research and Development*, 13, pp. 129–132.

Rayan, A., Raiyn, J. & Falah, M. 2017. Nature is the best source of anticancer drugs: Indexing natural products for their anticancer bioactivity. *PLoS One*, 12, p. e0187925.

Ruiz-Torres, V., Encinar, J. A., Herranz-López, M., Pérez-Sánchez, A., Galiano, V., Barrajón-Catalán, E. & Micol, V. 2017. An updated review on marine anticancer compounds: The use of virtual screening for the discovery of small-molecule cancer drugs. *Molecules*, 22, p. 1037.

Saklani, A. & Kutty, S. K. 2008. Plant-derived compounds in clinical trials. *Drug Discovery Today*, 13, pp. 161–171.

Schwartsmann, G., Brondani, A., Berlinck, R. & Jimeno, J. 2000. Marine organisms and other novel natural sources of new cancer drugs. *Annals of Oncology*, 11, pp. 235–243.

Seca, A. M. & Pinto, D. C. 2018. Plant secondary metabolites as anticancer agents: Successes in clinical trials and therapeutic application. *International Journal of Molecular Sciences*, 19, p. 263.

Srivastava, S. K., Khan, M. & Khanuja, S. P. S. 2005. Process for isolation of anticancer agent camptothecin from Nothapodytes foetida. Google Patents.

Sultan, P., Shawl, A., Abdellah, A. & Ramteke, P. 2010. Isolation, characterization and comparative study on podophyllotoxin and related glycosides of podophyllum heaxandrum. *Current Research Journal of Biological Sciences*, 2, pp. 345–351.

Svoboda, G. H., Neuss, N. & Gorman, M. 1959. Alkaloids of Vinca rosea L. (Catharanthus roseus G. Don.) V. Preparation and characterization of alkaloids. *Journal of Pharmaceutical Sciences*, 48, pp. 659–666.

Vail, D. M. & Withrow, S. J. 2007. Tumors of the skin and subcutaneous tissues. *Small Animal Clinical Oncology. 4a ed. Saunders. United States*, pp. 375–401.

Visnjic, D. & Crljen-Manestar, V. 2004. *The Cell: A Molecular Approach* (Geoffrey M Cooper and Robert E Hausman) Chapter 13: Cell Signaling.

Wall, M. E., Wani, M. C., Cook, C., Palmer, K. H., Mcphail, A. A. & Sim, G. 1966. Plant antitumor agents. I. The isolation and structure of camptothecin: A novel alkaloidal leukemia and tumor inhibitor from camptotheca acuminata1, 2. *Journal of the American Chemical Society*, 88, pp. 3888–3890.

Wani, M. C., Taylor, H. L., Wall, M. E., Coggon, P. & Mcphail, A. T. 1971. Plant antitumor agents. VI. Isolation and structure of taxol, a novel antileukemic and antitumor agent from Taxus brevifolia. *Journal of the American Chemical Society*, 93, pp. 2325–2327.

16

Anticancer Properties of Medicinal Plants Listed in the Herbal Pharmacopoeia of the United Mexican States

Erick P. Gutiérrez-Grijalva, Laura A. Contreras-Angulo, Alexis Emus-Medina, Gabriela Vázquez-Olivo, and J. Basilio Heredia

CONTENTS

16.1 Introduction

Several plant species have been traditionally used in the folk medicine to alleviate a wide variety of ailments. In this sense, plants listed in the Herbal Pharmacopoeia of the United Mexican States have been traditionally used in Mexico as herbal remedies against a wide variety of diseases, such as kidney stones, urinary infections, fever, diabetes, inflammation, etc (Secretaría de Salud, 2013). Furthermore, studies regarding the anticancer properties of these plant species are also on the rise due to the global increasing incidences of cancer. Cancer is one of the main causes of death in Mexico, and according to the World Health Organization around 5.6% of Mexicans die due to one or the other types of cancers (Globocan 2014; World Health Organization, 2018). Hence, the most recent findings on the anticancer properties of the Farmacopea Herbolaria de los Estados Unidos Mexicanos (FHEUM) are summarized in this chapter.

This chapter focuses on the anticancer properties of medicinal plants from the Herbal Pharmacopoeia of the United Mexican States. The data is compiled from the recent scientific literature obtained from the Scopus, Web of Science, and Science Direct databases.

16.2 Major Anticancer Medicinal Plants

16.2.1 *Allium sativum* L. (Garlic)

Garlic (*Allium sativum* L.) is a plant belonging to the genus, *Allinum*, a native ofthe central Asia. Since ancient times, it has been traditionally used to treat respiratory infections, bacterial infections, wounds, skin diseases, etc. (Petrovska and Cekovska 2010; Lanzotti et al. 2014). Garlic is characterized by its high content of phenolic compounds, saponins, sapogenins, minerals, vitamins, amides, nitrogen oxides, and sulfur compounds, which are known to have proved medicinal properties (Petropoulos et al. 2018; Szychowski et al. 2018).

Garlic is currently used in several medications, and its anticancer property has been proved scientifically by *in vitro* studies. For example, Kim et al. (2012) analyzed the effect of hexane extracts of garlic cloves at different concentrations (12.5, 25, 50, 100 µg/mL) on the human hepatocarcinoma cells, Hep3B cells. They observed decreased cell viability with the increased concentration of the extract. At 50 and 100 µg mL^{-1} concentrations, 70% and 58% cell viability was observed, respectively. Also, the authors observed an increase in the concentration of cells in the sub-G1 phase, indicating the inhibition of Hep3B proliferation which is attributed to the induction of apoptosis (caspase-dependent) and mediated by reactive oxygen species (ROS). Further, Ban et al. (2009) investigated the anti-proliferative effect of the combination of thiacremonone sulfur compound obtained from garlic (50 μg L^{-1}) and docataxel (5 nM) (a chemotherapeutic agent). This combination was found to have a significant anti-proliferative effect on prostate cancer cells (PC3 and DU145) by inhibiting the nuclear factor NF-κB, which is involved in counteracting apoptosis and inhibiting cell growth. Similarly, Gruhlke et al. (2017) analyzed the anti-proliferative effect of allicin, a well-known distinctive compound of garlic against different cancer cell lines, such as human lung epithelium carcinoma (A549), mouse fibroblast (3T3), human umbilical vein endothelial cell (HUVEC), human colon carcinoma (HT29) and human breast cancer (MCF7). The authors have reported the concentration dependent activity of cell growth inhibition, cell proliferation and apoptosis. Further, it was observed that all cells were inhibited at the concentration between 0.0375 and 0.15 mM. While, in 3T3 and MCF7 cell lines, higher degrees of apoptosis was observed at the concentration of 0.15 mM. Wang et al. (2012c) evaluated allicin at 1, 4, and 8 μg mL^{-1} on T-lymphocytes EL-4 cells incubated for 24, 36, 48, and 60 h. The results showed that allicin at 4 and 8 μg mL^{-1} after 24 and 48 h of incubation produced a significant increase in the activity of caspase-3, -12, and cyt C. Also, the ratio of Bax:BCL-2 was found to be higher. These parameters indicates the induction of apoptosis by mitochondrial pathway. Likewise, another compound of garlic that has attracted interest in medications is the diallyltrisulfide (DATS). This compound has been tested in basal cell carcinoma (BCC) and the results have shown a dose-dependent cytotoxic effect due to apoptosis via mitochondrial pathways, endoplasmic reticulum stress, and the induction of ROS production. The authors also reported the increase in the levels of GPRP78/BiP, CHOP/GADD153, and Caspase-4, a decrease in the levels of antiapoptotic proteins, Bcl-2 and Bcl-xl, and the increase of Bax and phospo-p53 proteins (Wang et al. 2012a, 2012b). In the same way, Shin et al. (2014) studied the apototic effect of DATS (0, 20, 40, 60, 80, and 100 μM) in T24 human bladder cancer cells, where they observed an anti-proliferative effect mainly mediated by increased levels of Bax, decrease of Bcl-2, Bcl-xl and the inhibition of XIAP, cIAP-1 and cIAP2. Furthermore, along with the activation of caspase-8 and -9 enzymes, increased depolymerization of the mitochondrial membrane was observed. This suggests that DATS induces apoptosis by activating c-Jun N-terminal kinase (JNK) and inactivation of PI3K/Akt.

16.2.2 *Cynara scolymus* L. (Artichoke)

Cynara scolymus L., which is also known as artichoke, is a perennial herbaceous plant belonging to the family, *Asteraceae*. It is mainly originated from the Mediterranean (Sękara et al. 2015). Its edible part corresponds to the internal bracts and receptacles that represent around 35%–55% of its fresh weight (Ceccarelli et al. 2010). This plant has been studied extensively due to its antimicrobial, antiallergic, anti-inflammatory, and anti-cancer effects. The plant possesses several bioactive compounds, such as phenolic acids, flavonoids, etc. (Joy and Haber 2007; Miraj and Kiani 2016; Tang et al. 2017). Some

investigations have related the occurrence of artichoke phytochemical compounds to their possible medicinal applications. For instance, Miccadei et al. (2008) reported that main phenolic compounds in artichoke extract corresponded to chlorogenic acid, di-caffeoylquinic acid 4, di-caffeoylquinic acid 5, luteolin-glycoside, and apigenin-glycoside. This extract was tested at different concentrations (400, 800, and 1200 μM) in HepG2 cells, observing a time-dose dependent behavior, showing a reduction in cell viability up to 80% in the highest extract concentrations. Similarly, Mileo et al. (2015) found that artichoke extracts containing monocaffeoylquinic acids, dicaffeoylquinic acids, luteolin, and apigenin glycoside showed an anti-proliferative effect in human breast cancer cell (MDA-MB231) in chronic doses and low at concentrations (2.5, 10, 30, and 60 μM), there was no increase in caspase-9 (caspases-independent mechanism) and a significant increase in senescence-associated with β-galactosidase (SA-β-gal), p21$^{Cip1/Waf1}$ and p16^{INK4a} was observed, suggesting that the extracts of artichoke induced senescence in MDA-MB231 cells. It has been reported that the main constituents of artichoke leaf extracts are 3,5-dicaffeoyl quinic acid, 4-caffeoyl quinic acid, luteolin-7-o-glucoside, and quinic acid are able to inhibit the activity of NF-κB and the enzyme, AKR1B1 (Aldo-keto reductase family 1, member B1) in human leukemic cell (THP-1) (Miláčková et al. 2017). Similarly, aqueous and alcoholic extracts of artichoke showed a cytotoxic effect against T-47D and ZR-75-human breast cancer cell lines, where the alcoholic extracts showed a greater effect (Vígh et al. 2016). On the other hand, Gezer et al. (2015) evaluated the effect of cyanarin (1,3-O-dicaffoylquinic acid), a typical compound found in artichoke against three skin cell lines, namely fibroblasts (FSF-1), telomerase-immortalized mesenchymal stem cell (hTERT-MSC), and cervical cancer cell (HeLa). They observed a dose-dependent relationship in the cytotoxic effect. FSF-1 and hTERT-MSC cells showed a greater tolerance to cyanarin, however cell proliferation was inhibited at a higher concentration of 500 μM. While, HeLa cells were more susceptible to cyanidin and exhibited a marked inhibition at 75 μM concentration. In a study by Ramadan et al. (2014), adult male Wistar albino rats treated with N-nitrosodiethylamine (NDA) and carbon tetrachloride (CCL4) were challenged with the extracts of artichoke (0.75 and 1.5 g/kg). They found that the extracts significantly limited the increase of aspartate aminotransferase (AST), serum alkaline phosphatase (ALP), alanine monostransferase (ALT), and bilirubin levels. Also, the examination of tissues showed that there was a significant serum albumin increase and an improvement in the hepatocytes compared to the control.

16.2.3 *Cinnamomum verum* (Cinnamon)

Cinnamon is a plant that belongs to the genus, *Cinnamomum* (Family, Lauraceae). It is native to Asia and is characterized by being an aromatic in nature of the inner bark of the trees (Vangalapati et al. 2012). It is traditionally used in the herbal medicine preparations in China and India (Qin et al. 2003). It is having antimicrobial, antioxidant, antitumor, anti-inflammatory, hypoglycemic, and gastro-protective properties (Bandara et al. 2012). The most known compounds of cinnamon are cinnamaldehyde, cinnamate, and cinnamic acid, which are responsible for its characteristic smell and taste (Mollazadeh and Hosseinzadeh 2016).

Many investigations have been carried out to correlate the relationship of cinnamon's phyto-chemical contents and their medicinal properties. For example, Kwon et al. (2009) analyzed the effect of cinnamon extracts and its main compounds, i.e., trans-cinnamic acid and cinnamic aldehyde against cancer cells. Also, the *in vivo* study showed that cinnamon extracts inhibited tumor growth in mice (injected subcutaneously with B16F10 cells) treated with doses of 400 μg g⁻¹ mouse weight. Further, cinnamon extracts regulated the expression of growth factors. A compound of interest from cinnamon is the 2-methoxycinnamaldehyde (2-MCA), which was shown to exhibit anti-proliferative effect against human hepatocellular carcinoma cells, SK-Hep-1 through the induction of apoptosis, activating caspases 3 and 9 and inhibiting topoisomerase I and II enzymes (Perng et al. 2016b). They also reported a significant suppression of tumor growth (70% reduction) using *in vivo* mice model studies. Similar results were observed by Liu et al. (2017), where human lung squamous cell carcinoma NCI-H520 cells proliferation was inhibited by 2-MCA by inducing apoptosis pathways. Also, 2-MCA suppressed the proliferation, induced apoptosis and inhibited the growth of tumors in rats, human colorectal adenocarcinoma COLO 2015 cells (Perng et al. 2016a; Tsai et al. 2016a) and human lung adenocarcinoma A549 cells (Wong et al. 2016). Cuminaldehyde is another compound obtained from cinnamon that was reported to have an anti-proliferative activity and suppresses the tumor growth as witnessed in Human Lung Squamous Cell Carcinoma NCI-H520 (Yang et al. 2016) and Human Colorectal Adenocarcinoma COLO 2015 cells (Tsai et al. 2016b).

16.2.4 *Aloe vera* L. Burm (Aloe)

Aloe vera is a succulent plant with green fleshy leaves (gelatinous viscous inside). It belongs to the family of *Xanthorrhoeaceae* and the genus, *Aloe* and has been used in traditional medicine by Egyptians, Chinese, Japanese, and Greeks, to name a few. It is characterized by its content of carbohydrates, amino acids, vitamins, organic acids, phenolic compounds, anthraquinones, phytosterols, and lignans (Guo and Mei 2016). It is used against some diseases such as liver problems, diabetes, ulcers, inflammation, wounds, and cancer, among others (Baruah et al. 2016).

Ghazanfari et al. (2013) analyzed different concentration (0.01, 0.02, 0.05, 0.1, 0.2, 1, 2, and 5 mg mL⁻¹) of aqueous extracts of *aloe vera* in adenocarcinoma gastric cell line (AGS) and did not observe a cytotoxic effect. On the other hand, extracts of *aloe vera* showed an induction of apoptosis and aloe emodin (1,8-dihydroxy-3-[hydroximethyl]-anthraquinone) induced necrosis in B16F10 cells (Çandöken et al. 2017). In another study it was found that aloe emodin induced apoptosis in breast cancer cell line MCF-7 in a concentrated and time-dependent manner on, increasing up to 37% the number of young apoptotic cells and induced the expression of Fas (CD95) (Fakhari et al. 2014). Similarly, Jose et al. (2014) reported that isolated flavonoids of *aloe vera* showed cytotoxic activity against cell line MCF-7 (IC$_{50}$ = 54 μg mL⁻¹). Similar results were observed by Srihari et al. (2015) when reporting a cytotoxic effect of methanol extracts of *Aloe vera* in MCF-7 with an IC50 = 74.33 μg mL⁻¹. Likewise Hussain et al. (2015) demonstrated that extracts of *Aloe vera*

caused an increase in cell death in MCF-7 and HeLa cell lines, by inducing apoptosis in a time-concentration-dependent manner, in addition to not observing a cytotoxic effect against lymphocytes in the same concentrations. *In vivo* studies have shown that orally supplied aloe vera and honey (640 μLkg⁻¹) for 20 days in rats with tumors (Walker 256 carcinoma cell), decreased the tumor growth as compared to a control, also an increase in Bax and a decrease in Ki67-LI and Bcl-2 were observed (Tomasin and Cintra Gomes-Marcondes 2011). It has also been reported that extracts of *Aloe vera* (300 mg kg⁻¹) and training for six weeks in rats induced with breast cancer decreased the expression of COX-2 and VEGF related to the process of tumor angiogenesis (Shirali et al. 2017). Similarly, Im et al. (2016) conducted a study on the effect of oral administration of *Aloe vera* processed gel (200 and 400 mg/kg/day) in rats given colon cancer promoters (azoxymethane and dextran sodium sulfate), and observed a decrease in the development of adenomas and adenocarcinomas, together with an inhibition of COX-2 and NF-κB.

16.2.5 *Elettaria cardamomum* L. Maton (Cardamom)

Cardamom belongs to the genus *Ellatteria* in the family *Zingiberaceae* (Sengottuvelu, 2011; Parthasarathy and Prasath 2012). Cardamom seeds are the part of the plant most used, it is low in fat and high in protein, iron, and vitamins B and C. Cardamom essential oils are rich in terpenoids, these secondary metabolites are used in the pharmaceutical and food industry for their medicinal and flavor properties, currently various investigations direct their efforts in testing the anticancer properties. The most abundant essential oil in cardamom are α-terpinyl acetate and 1,8-cineole, also small amounts of α-terpineol, linalool, borneol, camphor, and α-pinene have been identified (Gadekar and Jadhav 2017).

There are several *in vitro* and *in vivo* cardamom anti-proliferative studies. For instance, Elguindy et al. (2016) evaluated the chemoprotective effect of cardamom in hepatocellular carcinoma induced by diethylnitrosamine (DENA) in rats. Before inducing liver cancer, cardamom essential oil obtained by steam distillation was orally administered to rats in concentrations of 100 and 200 mg/kg for one week and after induction with DENA, the essential oil was continuously supplied daily until the end of the experiment. The DENA treatment showed an increase in the levels of the cytokines TNF-α (tumor necrosis factor α) and IL-1β (interleukin-1β), while cardamom-treated cells significantly decreased the levels of TNF-α and IL-1β due to its ability to inactivate NF-κB factor. Cardamom-reduced liver injury, the activity of ornithine decarboxylase, the hepatic malondialdehyde (hepatic lipid peroxidation end product) formation, and increased the activity of antioxidant enzymes such as catalase, superoxide dismutase, glutathione peroxidase, glutathione reductase, and glutathione-S-transferase in the liver of rats. This study demonstrated that the cardamom treatment in rats was effective in preventing hepatocellular carcinoma, by reducing reactive oxygen species, decreasing oxidative stress, and reducing pro-inflammatory cytokines and NF-κB. This effect was attributed to the synergistic effect of its phytocompounds such as tannic acid, gallic acid, caffeic acid, and 4,5-dicaffeoyl quinic acid.

Another study by Das et al. (2012) reported the efficacy of cardamom against 7,12-dimethylbenz[a]anthracene (DMBA)-induced skin papillomatogenesis (similar to non-melanoma skin cancer in humans) in Swiss albino mice. To carry out the experiment, the treatments applied were DMBA (application dorsal skin) and cardamom (Oral-500 mg/kg), and the evaluation period was 12 weeks. The administration of cardamom showed elevated levels of the antioxidant enzymes glutathione peroxidase, glutathione reductase, reduced glutathione, glutathione-S-transferase, catalase, and superoxide dismutase, which are known to increase the cellular defense levels against different stresses. On the other hand, the ingestion of cardamom inhibited the loss of expression of the nuclear erythroid factor 2-related factor 2 (Nrf2 protein), and Kelch ECH associating protein 1 (Keap1), both proteins are of great importance because they are related to the production of genes involved in the production of antioxidant enzymes (Kansanen et al. 2013). The expression of COX-2 (Cyclo-oxygenase-2), considered as one of the markers of cancer progression, is also reduced by 37%, indicating an inhibition of neoplasia in mouse skin due to the ingestion of cardamom. With this study, they demonstrated that cardamom has a chemopreventive action in the progression of skin cancer in mice, and this may be due to the presence of limonene, cineole, linalool, pinene, and borneol, which act as antioxidants (Acharya et al. 2010). A similar study was carried out by Qiblawi et al. (2012) to evaluate the effect of aqueous suspension of cardamom (100 µL per mouse per day) on DMBA-induced skin papillomatogenesis (100 µg in 50 µL acetone per animal). The treatment with cardamom was applied orally prior to the induction of carcinogenesis by DMBA. Qiblawi et al. (2012) showed a 50% reduction in the incidence of tumors and a diminishing appearance of tumors (decreased diameter and weight), GSH (reduced glutathione enzyme) levels rise as lipid peroxidation is reduced.

Cardamom was evaluated in pan masalainduced lung cancer in mice by Kumari and Dutta (2013). Pan masala is a mixture of several species of seeds, such as areca nut and catechu of which some have cytotoxic, mutagenic, genotoxic, and carcinogenic properties. The experiment was carried out in Swiss mice that were fed with pan masala at 2% and pan masala plus 2 and 0.2% cardamom, respectively. Mice were fed daily and after nine months, the dose of cardamom was continued for three more months in order to observe the protective and improvement effect. Histological evaluations were performed to observe the damage induced by the pan masala treatment, as well as the possible protective effect of cardamom, finding that the 0.2% dose of cardamom was effective in reducing the damaging in lung (leakage edema, fibrin, and erythrocytes). The continuous application of cardamom after nine months and then three months with only cardamom showed a protective effect against the loss of membrane integrity, high levels of edema, and emphysematous alteration. On the other hand, the enzymatic activity of acid phosphatase, alkaline phosphatase, and lactate dehydrogenase decreased significantly after three months, where only cardamom was added to the diet indicating a decrease in lung damage. All those effects can be attributed to the phenolic compounds present in cardamom (Kumari and Dutta, 2013). Moreover, Majdalawieh and Carr (2010) evaluated the *in vitro* effect of cardamom extract on YAC-1 tumor cells (mouse lymphoma cells) at doses of 1, 10, 50, and 100 µg of cardamom/mL, and did not find a direct toxic effect on the cancer cell line; however the extracts improved the cytotoxic activity

of the NK cells (natural killer cells against cancer cells), which is an indicative of a potential immuno-stimulatory effect of cardamom. This effect might be caused by the eugenol content in cardamom extract (Park et al. 2007).

16.2.6 *Aesculus hippocastanum* L. (Seeds of Horse Chestnut)

This plant is originated from the south part of Balkan Peninsula and Caucasian regions in Europe and is widely distributed all over the world (Čukanović et al. 2011; Felipe et al. 2013). It belongs to the sub-family of *Hippocastanaceae*, genus *Aesculus* and the family Sapindaceae, commonly known as "horse chestnut." This plant has been used as herbal medicine in humans and animals, and also in the pharmaceutical and cosmetic industry. The most used parts of the plant are leaves and seeds as cardiotonics and anti-inflammatories, also barks and twigs as febrifuges and dermatitis (Ćalić-Dragosavac et al. 2011; Foca et al. 2011). Each part of the plant contains different classes of bioactive compounds: aescin in seeds, essential oils in leaves and flowers, coumarins in bark (Kapusta et al. 2007). The main active compound of the seed horse chestnut is β-escin or aescin which is a mixture of triterpenesaponin, and its components include glycosides of protoaescigenin, barringtogenol C, allantoin, sterols, leucocyanidin, leucodelphinidin, and tannins (Ćalić-Dragosavac et al. 2011; Felipe et al. 2013; Abudayeh et al. 2015). The anti-inflammatory, antiedamatous, antiexudative, and antioxidant activity of seeds horse chestnuts have been mainly related to the aescin compound. For example, Patlolla et al. (2013) studied the *in vivo* effect of β-escin, a saponin isolated from horse chestnut seeds, in mice using the tobacco carcinogen 4-(methyl-nitrosamino)-1-(3-pyridyl)-1-butanone (NNK) to induce lung adenoma and adenocarcinoma, at the same time β-escin was tested against H460 human lung cancer cells to determinate the mode of action. The results obtained established the chemopreventive effect of β-escin in the inhibition of induced lung adenocarcinoma and inhibition of the high activity of Aldehyde dehydrogenase (ALDH1A1) in human lung H460 cancer cells. Jiang et al. (2011) investigated the protective effect of aescin in mice with lipopolysaccharide-induced liver injury. The LPS (40 mg/kg), escin (3.6 mg/kg), and LPS plus escin (0.9, 1.8, or 3.6 mg/kg) were applied, and after 6 h the liver was removed from the mice for further evaluations. The results showed that escin significantly attenuated liver damage in LPS-induced endotoxemic mice, also increased the GSH levels and the activities of SOD and GSH-Px; furthermore, the pretreatment of escin decreased liver NO (Nitric Oxide) content and inhibited the production of TNF-α, IL-1β.

The cytotoxic effect of horse chestnuts against human breast cancer cells (MCF-7) was studied by Gulcin Sagdicoglu Celep et al. (2012). They used different parts of the plant such as flowers, leaves, seeds and bark and evaluated the antioxidant capacity (DPPH radical scavenging activity), the lipid peroxidation inhibition activity and total of polyphenols, in order to determine which plant organs should be used for the cell assay. The results indicated that the bark extract was more effective as inhibitor of lipid peroxidation and had more antioxidant capacity (DPPH), and a high amount of phenolic compounds. The bark extracts (0.01 mg/mL) decreased the viability of MCF-7 cancer cells at around 70%, an effect that was mainly attributed to the polyphenolic compounds. Güney et al. (2013) used aescin from horse chestnut at 10, 20, 30, 50, and 50 μM/mL for 24, 48, and 72 h to determine its anti-proliferative effect on H-R as 5RP7 cells. They found that the aescin anti-proliferative effect was concentration and time-dependent, and the apoptosis increased 36.9% in cells treated for 72 h. Rimmon et al. (2013) evaluated the effect of escin (2–30 μM for 72 h), alone and combined with chemotherapy on pancreatic cancer cells (Panc-1, p34, COLO 357, and MIA-Paca); in accordance with the aforementioned studies, the effect of escin was concentration-dependent, with a decreased growth of human pancreatic carcinoma cell, inhibition of NF-κB signaling, and sensitized pancreatic cancer cell to the cytotoxic effect of chemotherapy. Another study that used escin from *Aesculus hippocastanum* at 3.5, 7, 14, and 21 μg/mL against A549 human lung adenocarcinoma cell line showed that the increase in the concentration of escin on A549 cells induced apoptosis (Çiftçi et al. 2015).

16.2.7 *Echinacea angustifolia, E. pallida and E. purpurea* (Equinacea)

This plant is also known as purple coneflower, echinacea, snakeroot, black sampson, scurvy root, Indian head among others; it is widely distributed in North America. It belongs to the *Asteraceae* family, and there are nine known species, from which the most known are *E. angustifolia, E. pallida,* and *E. purpurea* (Zhai et al. 2009; Kindscher 2016; Kindscher and Riggs, 2016). Chemical composition of *Echinacea* includes alkamides, caffeic acid derivates, polysaccharides, glycoproteins, polyacetylenes, polyenes, flavonoids, and terpenoids; and these compounds may be responsible for the immunostimulatory activity of this plant. However its content depends of the plant part used, including roots, flowers, and entire aerial (Cao and Kindscher 2016).

Echinacea, used traditionally for a variety of ailments by Native American tribes, has been a subject of significant scientific studies (Cao and Kindscher 2016). Spelman et al. (2009) demonstrated that undeca-2E-ene-8, 10-dynoic acid derived from *Echinacea angustifolia* have affinity for the cannabinoid-2 (CB2) receptor and inhibits IL-2 secretion in Jurkat T (human E6.1) cells through PPARγ (peroxisome proliferator activated receptor gamma) activity. This receptor have role in different diseases including diabetes, atherosclerosis, inflammation, cancer, and autoimmune disorder. Another study by Hou et al. (2010) used different parts from three different species (*E. purpurea, pallida* and *angustifolia*), such as aerial and roots lyophilized, and were then extracted by supercritical fluids and used in murine macrophage cells (RAW 264.7) and human breast adenocarcinoma cells (MCF-7). The latter were used to correlate the metabolites of the extracts with their potential anti-inflammatory, to determine if the luciferase reporter gene activity was employed. Both *E. purpurea* and *E. angustifolia* had high content of alkamides (74.06% and 90.73%, respectively), which is related to the suppression of the 12-O-tetradecanoylphorbol-13-acetate (TPA)-induced COX-2 promoter activity in MCF-7 cancer cells. Further immunohistochemical analysis in mouse skin showed that *E. angustifolia* extract suppressed COX-2 promoter activity in MCF-7 cells. COX-2 promoters in transfected MCF-7 cells were not affected by treatment with *E. pallida* roots extracts, which was mainly attributed to the content of alkamides in each *Echinacea* species.

In murine macrophages, it was demonstrated that alkamides significantly inhibited COX-2 activity and the LPS-induced expression of COX-2, inducible NO synthase (NOS2), and specific cytokines or chemokines.

Yaglıoglu et al. (2013) studied the antiproliferative effect of methanol extract, pentadeca-(8E, 13Z)-dein-11-yn-2-one, and (E)-1,8-pentadecadiene, isolated from *E. pallida* roots against C6 cells (rat brain tumor cells) and HeLa cells (human uterus carcinoma). The cells were treated with 5, 10, 20, 30, 40, 50, 75, and 100 μg/mL for 24 h. The results showed that the extract and pentadeca-(8E, 13Z)-dein-11-yn-2-one inhibited the proliferation of C6 cells at 75 μg/mL, but against HeLa cells they were effective at 5, 10, 20 30, and 40 μg/mL, indicating a cell specific activity. On the other hand (E)-1,8-pentadecadiene had no antiproliferative activity against C6 cells and HeLa cells. These results suggested the anticancer properties of the *E. pallida*.

Equinacea purpurea flowers extract and cichoric acids were investigated to know their cytotoxic effect on human colon cancer cells (CaCo-2 and HCT-116). Dry aqueous ethanol extracts obtained of *E. purpurea* (0–2000 μg/mL) and commercial cichoric acid (0–200 μg/mL) were used against the colon cancer cells. Cichoric acid is a caffeic acid derivate phytochemical found in *Equinacea*. The results indicated that the aqueous ethanol extract reduced the viability of both colon cancer cells when treated for 48 h, which means that the response was dose-dependent. From the cichoric acid effect after incubation for 24 h, it was observed that the cell viability was decreased at 150 or 200 mg/mL, and after 48 h the viability of Caco-2 and HCT-116 were decreased. The cichoric acid treatment decreased the telomerase activity (a reverse transcriptase enzyme) in HCT-116 and induced apoptosis in colon cancer cells. The authors suggest that cichoric acid represses telomerase activity, thus, exerting its cytotoxic effect on colon cancer cells (Tsai et al. 2012).

Echinacea angustifolia was investigated against human cervix adecarcinoma (HeLa) by Cenić-Milošević et al. (2013). They obtained dry ethanolic extract and it was applied at 12.5, 25, 50, 100, and 200 μg/mL to the cancer cells. The inhibitory concentration on HeLa cells was 43.52 μg/mL, exhibiting good cytotoxic activity. This effect could be due to its chemical constituents composed of alkylamides and polysaccharides. Flavonoids were found in *E. purpurea* (Yildiz et al. 2014), and the extracts of the plants were investigated for their cytotoxic effect on cancer cells human colon adenocarcinoma (CaCo-2), human breast adenocarcinoma (MCF-7), human glioblastoma-astrocytoma (A549), epithelial-like cell line (U87MG), human cervical cancer cells (HeLa), and kidney epithelial cells from African Green monkey (VERO). To carry out the experiment the plant was ground to made extractions by different methods (supercritical CO_2 extraction, subcritical water extraction, microwave-assisted, and Soxhlet extraction). The extracts were used against cancer cells at 0.5, 5.0, and 50 μg/mL. The extracts obtained did not showed inhibition in cell proliferation of cancer cells at 50 μg/mL concentration of the extracts. Pomari et al. (2014), showed the effect anti-inflammatory effect of different extracts of plants among them *E. angustifolia*. The inflammatory process was induced by H_2O_2 in macrophages, and then treated with *E. angustifolia*. The results obtained indicated that the extracts showed an anti-inflammatory activity, down-regulating inflammatory genes. Another study was developed by Aarland et al. (2017) to determine the *in vivo* and *in vitro* effect of hydroalcoholic extracts of *E. purpurea* and *angustifolia* obtained from different parts of the plant (roots, leaves, flowers, and seeds). In the *in vivo* assay the inflammation was induced in Winstar rats of 84 days age by a sub-plantar injection of carrageenan 1% (w/v). The rats were previously treated with Echinacea extracts (daily dose 0.4 mL/kg). In the *in vitro* MCF-7, HeLa and HCT-15 cells were treated with five different concentrations of Echinacea. Root, leaves, flowers, and seeds from *E. purpurea* showed a robust anti-inflammatory effect in rats, while other two extracts did not have a significant effect. This result may be due to the presence of saponins content. On the other hand, the antiproliferative effect showed by the extract of *E. purpurea* plus *E. angustifolia* (mixture of aerial parts and roots of both species) was much more toxic for HeLa cell, and moderately for HCT15 and MCF-7 cell lines, being an attributable effect due to the higher alkylamides content. Likewise, Cichello et al. (2016) used a blend of *E. purpurea* and *E. angustifolia* to determine the proliferative effect on various cancer cervical and bile duct cell lines. The extracts (12.5, 25, 50, 100, or 200 μg/mL) were applied to cervical cancer cell (HeLa), cholangiocarcinoma (QBC-939), lung/bronchial epithelial (Beas-2b), and human vein epithelial cells (HUVEC). The inhibitory effect (inhibition up to 100%) in HeLa cell with Echinacea-blend treatments showed a concentration-dependent from 12.5 to 25 μg/mL, however it was reduced to 90% at 50, 100, and 200 μg/mL. However, the HUVEC cell line showed an eight-fold increase over HeLa cells. The extracts promoted the growth of cancer cell proliferation, and may interfere with cancer treatment, but could have a role in the regeneration of epithelial tissue.

16.2.8 *Citrus aurantium* L

Citrus aurantium, also known as bitter orange, sour orange, or marmalade orange, is a citrus tree that belongs to the *Rutaceae* family and genus *Citrus*, and is widely distributed in tropical and subtropical southeast regions of the world (Han et al. 2012). *C. aurantium* peels of fruits are commonly used in folk herbal medicine. Some recent studies have shown the potential of *C. aurantium* as anti-proliferative agents.

For instance, Han et al. (2012) showed that the methanolic extracts from *C. aurantium* had anti-proliferative effect by induction of apoptosis on U937 human leukemia cells; this effect was concentration-dependent. Moreover, Huang et al. (2012) showed that flavonoid extracts from *C. aurantium* rich in hesperetin and naringenin are able to enhance melanin synthesis through the stimulation of MAPKs and induction of the accumulation of B-catenin. Furthermore, Karimi et al. (2012) evaluated the extracts from *Citrus aurantium* bloom, rich in gallic acid, pyrogallol, syringic acid, caffeic acid, rutin, quercetin, and naring; which were attributed for the anti-proliferative potential in breast cancer cell lines (MCF-7; MDA-MB-231), human colon adenocarcinoma (HT-29), and Chang cell, as a normal human hepatocyte. Some flavonoids from *C. aurantium* have been identified as naringin, hesperidin, poncirin, isosiennstein, hexamethoxyflavone, sineesytin, hexamethoxyflavone, tetramrthnl-o-isoscutellaeein, nobiletin, heptamethoxyflavone, 3-hydoxynobiletin, tangeretin, hydroxypentamethoxyflavone, and hexamethoxyflavone. These have also shown to inhibit cell cycle progression in the G2/M

phase and decrease in the expression level of cyclin B1, cdc2, cdc 25c in AGS cells. The anti-proliferative effect of these flavonoids on AGS cells with IC50 value of 99 μg/mL (Lee et al. 2012). Moreover, flavonoid-rich extracts from *C. aurantium* peel (naringenin, hesperidin, nobiletin, heptamethoxyflavone, poncirin, and tangeretin) also showed anti-proliferative properties on human hepatoblastoma cells (HepG2) in a concentration-dependent manner. The suggested mode of action of flavonoid extracts from *C. aurantium* was proposed through the increased expression of Bax/Bcl-xL ratio in HepG2 treated cells (Lee et al. 2015).

Furthermore, essential oils from the blossoms of *C. aurantium* have shown antiproliferative properties in LO2, MCF-7, and 3t3-L1 cells through the inhibition of NO, IL-6, TNF-α, and IL-1β. Simultaneously, the proliferation inhibitory rate of the essential oil on MCF-7 cells was far less than that of 5-fluoroacil. The major constituents of *C. aurantium* essential oil were characterized as linalool, α-terpineol, limonene, and linalyl acetate (Shen et al. 2017a). Likewise, Park et al. (2014) described that flavonoids isolated from *C. aurantium* have also shown induced cancer cell apoptosis in A549 cells through the regulation of the apoptosis related protein, cleaved caspase-3 and p-p53. The identified flavonoids were naringin, hesperidin, poncirin, isosinnesetin, hexamethoxyflavone, sinestin, hexamethoxyflavone, tetramethyl-o-isoscutellaein, nobiletin, heptamethoxyflavone, 3-hydroxynobiletin, tangeretin, hydroxypentamethoxyflavone, and hexamethoxyflavone. In addition, crude polysaccharides from *C. aurantium* have also shown cytotoxicity, in a concentration-dependent manner, against MCF-7 human breast cancer cells, as well as the HCC827 lung cancer cells. Interestingly, the inhibitory rate was higher than that of the positive control 5-fluoracil at 62.5 μg/mL (Shen et al. 2017b).

16.2.9 *Origanum vulgare* L.

Origanum vulgare, also known as European oregano is a flowering perennial plant belonging to the *Lamiaceae* family, native to Eurasia and the Mediterranean region (USDA, 2018b). *O. vulgare* L. subsp. *hirtum* essential oils inhibit DNA damage by preventing its oxidation, as evaluated by Aybastıer et al. (2018) in malignant lung cells (A549). It is suggested that this effect is due to the carvacrol, thymol, protocatechuic acid, caffeic acid, epicatechin and rosmarinic acid content in water and water-methanol samples.

Begnini et al. (2014) showed that EO from *O vulgare* had antiproliferative effect against human breast adenocarcinoma (MCF-7), and human colon adenocarcinoma (HT-29). The EO was composed mainly of 4-terpineol; and EO had a concentration-dependent manner and was most effective against HT-29 cell line. The antiproliferative effect of *O vulgare* subsp. *hirtum* essential oil and its main constituents (carvacrol, thymol, citral and limonene) has also shown anti-proliferative activity against hepatocarcinoma HepG2 cells (Elshafie et al., 2017). The crude EO extract has significantly reduced the cell viability of the hepatocarcinoma cell line in dose-dependent mode (25–800 μg/μL). Elshafie et al. (2017) also showed that carvacrol and citral might be the responsible constituents for the effect as evaluated by the reduction in cell viability of hepatocarcinoma cells in the range from 0.01 to 0.25 μg/μL.

Essential oils from *O vulgare* and *O majorana* were evaluated for their cytotoxic effect. In the MTT assay, *O majorana* essential oil was more cytotoxic than *O vulgare* essential oil against different cancer cell types, such as MCF-7, LNCaP and NIH-3T3, with IC_{50} values of 70, 85.3, and 300.5 µg/mL, respectively (Hussain et al. 2011). *O. vulgare* ethanol extracts have shown around 70% cytotoxicity against breast cancer cells (MCF-7) in a concentration-dependent manner at concentrations of 100, 200, 300, 400, and 500 µg/mL (Nile et al. 2017). Marrelli et al. (2015) tested the hydroalcoholic extracts of *O vulgare* L. subsp. *viridilum* against breast cancer (MCF-7), hepatic cancer (HepG2), and colorectal cancer (LoVo). The study showed a selective anti-proliferative effect on HepG2 cells with IC_{50} value of 32.59 µg/mL. According to the criteria of the American National Cancer Institute, the $IC50$ values of less than 20, 20100 and more than 100 µg/mL are regarded as active (A), moderately active (MA) and inactive /IA), respectively.

Isolated compounds from *O vulgare* such as β-caryophyllene oxide, a sesquiterpene, have also shown anti-proliferative effect in human prostate and breast cancer cell through the inhibition of the constitutive activation PI3K/AKT/mTOR/S6K1 and MAPK activation pathways, while activating ERK, JNK, and p38MAPK. The isolated sesquiterpeneβ-caryophyllene oxide also potentiated the apoptotic effect by down-regulating several proliferation-related genes (cyclyn D1, Bcl-2, Bcl-xL, IAP-1, and IAP-2) (Park et al. 2011).

Savini et al. (2009) showed that *O. vulgare* ethanol extracts led to growth arrest and cell death of Caco-2 cells in a dose and time-dependent manner, which can be caused by changes in glutathione content and increase of glutathione oxidized form. The antioxidant status of *O. vulgare* aqueous extracts was evaluated in 1,2-dimethylhydrzine (DMH)-induced rat colon carcinogenesis. *O. vulgare* extracts managed to reverse the decreased levels of antioxidant enzymes such as superoxide dismutase, catalase, reduced glutathione, glutathione reductase, glutathione peroxidase and glutathione-S-transferase caused by DMH. Oregano extracts were rich in rosmarinic and caffeic acid. The activity of oregano was thought to be due to its phenolic constituents such as rosmarinic acid, caffeic acid and certain volatile compounds, namely linalool, alcohol, phenol, terpenes, and some flavonoids (Srihari et al. 2008). Didymin, a flavonoid isolated from *O. vulgare* has shown strong inhibition of the viability, clonogenicity and migration in a concentration-dependent manner of hepatocarcinoma human cells (HepG2). The mode of action of didymin was assessed by Wei et al. (2017), and reported that didymin induced apoptosis and cell cycle arrest at G2/M phase by regulation of cyclin B1, cyclin D1 and CDK4; didymin also inhibited the ERK/MAPK and PI3K/Akt pathways by increasing the level of Raf kinase inhibitor protein (RKIP). The IC_{50} values were found to be 66.3, 48.6, and 31.2 µM after incubation for 24, 48 and 72 h, respectively.

16.2.10 *Urtica dioica* L.

Urtica dioica is an herbaceous perennial flowering plant that belongs to the *Urticaceae* family; and native of Europe, Asia, northern Africa, and North America (USDA, 2018b). *U. dioica* is also known as common nettle, European nettle, giant nettle, urtiga, ortiga, etc (USDA, 2018a, 2018b). *U. dioica* extracts have a preventive role against the toxicity caused by cisplatin, a hepatotoxic and nephrotoxic agent used against cancer. *U. dioica* extracts managed to prevent cisplatin toxicity in Erhlich ascites tumor-bearing mice by decreasing lipid and protein peroxidation

and increasing the activity of antioxidant enzymes such as superoxide dismutase, catalase, and glutathione peroxidase (Özkol et al. 2012). Moreover, *U. dioica* dichloromethane extracts have cytotoxic effect on breast cancer cells in a BALB/c mouse model. Mice treated with the extracts showed a decreased tumor size through suppression of tumor growth cells (Mohammadi et al. 2016b). Mohammadi et al. (2016a) showed that the combination of paclitaxel and *U. dioica* extract had a synergistic effect on MDA-MB-468 breast cancer cells. This was shown when the IC_{50} dose for *U. dioica* extract and paclitaxel in 24 h were 29.46 µg/mL and 6.73 µM respectively; while at 48 h results were found at 15.54 µg/mL and 3.51 µM, respectively. While the combination of *U. dioica* extract and paclitaxel resulted in decreased IC_{50} dose of paclitaxel. A study by Fattahi et al. (2013) showed that *U. dioica* aqueous extract had a dose-dependent anti-proliferative effect on MCF-7 cells after 72 h with an IC_{50} value of 2 mg/mL. This activity was associated with an increase in the apoptosis confirmed by DNA fragmentation, the appearance of apoptotic cells and increased amounts of calpain 1, calpastatin, caspase 3, caspase 9, Bax, and Bcl-2.

16.2.11 *Rosmarinus officinalis* L.

Rosmarinus officinalis L., also known as rosemary, is a perennial flowering herb belonging to the *Lamiaceae* family and the genus *Rosmarinus*; it is native to Northern Africa, Western Asia, Southeastern, and Southwestern Europe, and the Mediterranean region (USDA 2018a, 2018b). Yesil-Celiktas et al. (2010) evaluated the anti-proliferative effect of methanolic and supercritical CO_2 extracts and their active compounds such as carnosic and rosmarinic acid. And found that supercritical CO_2 extracts showed higher antiproliferative activity against several cancer cell lines (NCI-82, DU-145, Hep-3B, K562, MCF-7, PC-3, and MDA-MB-231), and the effect is mainly attributed to its carnosic acid content.

The anticancer activity of rosemary extract and carnosic acid was evaluated in a nude mouse model by Yan et al. (2015) and the mode of action investigated in colon cancer cell lines. Rosemary extract and carnosic acid increased apoptosis and decreased viability in a concentration-dependent manner in colon cancer cell lines. Moreover, rosemary extract and carnosic acid upregulated the expression of Nrf2 in colon cells and inhibited HCT116 xenograft tumor formation in mice.

The 1,8-cineole, α-pinene, and β-pinene are among the main components of rosemary essential oil. In this regard, Wang et al. (2012b) reported that *Rosmarinus officinalis* essential oil extracts exhibited strong cytotoxicity towards three human cancer cells. It's IC_{50} values on SK-OV-3, HO-8910, and Bel-7420 were 0.025%, 0.076%, and 0.13% (v/v), respectively. As it can be observed by the lower IC_{50} value, rosemary essential oils showed a higher anti-proliferative effect against SK-OV-3 cell line. Moreover, at a concentration of 0.0625% (v/v), the cell viability treated by *R. officina*lis L. essential oil, α-pinene, β-pinene and 1,8-cineole for SK-OV-3 were 36.13%, 45.85%, 67.77%, and 93.03%, respectively.

Polyphenol-enriched rosemary extract has also shown anti-proliferative effect. And the mode of action was assessed by Valdés et al. (2016), who showed that rosemary extract alleviates cellular stress through the activation of the transcription factor

Nrf2. The effect of carnosic acid on neuroblastoma IMR-32 cells was reported by Tsai et al. (2011) showed that there is a relationship between the reactive oxygen species and mitogen-activated protein kinase associated with carcinogenesis. Carnosic acid decreased IMR-32 cells viability in a concentration-dependent manner by inducing apoptosis, downregulating the apoptotic Bcl-2 protein, increasing the activation of p38, and decreasing the activation of extracellular signal-regulated kinase. In addition, Tong et al. (2017) showed that in HepG2 cells, rosemary extracts composed of 23.2% carnosic acid, and 12.4% carnosol, induced the expression of Sestrin2 and MRP2 which suggest an upregulation of the Nrf2 protein. Tai et al. (2012) reported that rosemary extracts have anti-proliferative activity on human ovarian cancer A2780 and its cisplatin resistant daughter cell line A2780CP70. Interestingly, carnosol and rosmarinic acid showed synergistic antiproliferation effect with cisplatin-resistant and A2780 cells. Similarly, Shrestha et al. (2016) showed that co-treatment of sageone, an icetexanediterpenoid from rosemary, with a subtoxic dose of cisplatin had synergistic effect on apoptosis induction in stomach cancer cells (SNU-1).

16.2.12 *Zingiber officinale* Roscoe (Ginger)

Ginger (*Zingiber officinale* Roscoe) is a plant that belongs to the *Zingiberaceae* family, genus *Zingiber* Mill (Integrated Taxonomic Information System, 2018b). It is indigenous to Southeast Asia (Zancan et al. 2002). Ginger is an underground root or rhizome borne from the herbaceous perennial, monocotyledonous *Z. officinale* Roscoe plant. *Z. officinale* Roscoe grows in most tropical regions including China, Haiti, India, Nigeria, Jamaica, and Australia (Balladin et al. 1998; Bailey-Shaw et al. 2008). Ginger rhizome has been used as a medicine in Asian, Indian, and Arabic herbal traditions, and its main active ingredients are gingerols, shogaols, and essential oils (Shukla and Singh 2007; Bailey-Shaw et al. 2008). Both the extracts and isolated compounds of ginger have been reported to have anticancer properties against different cancer cell lines and has been demonstrated in *in vivo* and *in vitro* experiments. For instance, Brown et al. (2009) reported that 6-gingerol, one of the main pharmacological active compound of ginger extract suppresses the growth of YYT cells and showed anti-angiogenesis activity. The ethyl acetate fraction of ginger rhizome extract reduces the telomerase activity and protein production by means of down regulation of hTERT and c-Myc expression in human lung carcinoma cell line A549 (Tuntiwechapikul et al. 2010).

The ginger extract regulates cell-cycle and apoptosis regulatory molecules, impaired reproductive capacity, perturbed cell-cycle progression, and induced apoptosis in several prostate cancer cell lines (PC-3, LNCaP, C4-2, C4-2B, and DU145). Additionally, the ginger extract suppressed tumor tissue in mice (Karna et al. 2011). The anticancer property of ethanol extract of ginger was evaluated against cholangiocarcinoma (CCA). The ginger rhizome extract was well tolerated at a dose of 5,000 mg/Kg body weight in an *in vivo* assay. In the same study, the authors demonstrated that the extract had *in vitro* cytotoxic and apoptotic activity against human CCA cell line CL-6, additionally, the extract induced the expression of drug resistant genes which needs further molecular studies to understand the mechanisms of action of ginger extract and its major components (Plengsuriyakarn et al. 2012a). In a

similar study, the rhizome extract of ginger showed anticancer activity against CCA in nude mouse xenograft model. The authors suggested that the anticancer properties of this extract may be attributed to the phytochemicals present in ginger extract such as 6-gingerol and 6-paradol (Plengsuriyakarn et al. 2012b).

In another work, 6-shogaol showed strong cytotoxic activity against human promyelocytic leukemia HL-60 and human oral epidermal carcinoma KB cell lines. The authors attributed this activity to the enone group to 6-shogaol (Peng et al. 2012). Additionally, in human (LNCaP, DU145, and PC3) and mouse (HMVP2) prostate cancer cell lines, 6-shogaol, reduced the survival and induced apoptosis. Moreover, 6-shogaol was more effective than 6-gingerol and 6-paradol. The authors suggested that the effect of 6-shogaol was due to the ability of this compound to inhibit STAT3 and NF-κB signaling, as well as multiple signaling pathways that help in the induction of apoptosis (Saha et al. 2014). Rastogi et al. (2015) reported that 6-gingerol have antiproliferative activity against human cervical cancer cells (HeLa, CaSki and SiHa), through proteasome inhibition mediated p53 reactivation, increasing of oxidative stress, induction of DNA damage associated G2/M cell cycle arrest and also it induces apoptosis, besides, 6-gingerol demonstrated antitumor activity *in vivo* (Rastogi et al. 2015). Likewise, the methanolic extract of ginger rhizome also inhibited the proliferation of human cervical cancer HeLa cells and breast cancer MDA-MB-231 cells (Ghasemzadeh et al. 2012; Ansari et al. 2016).

Akimoto et al. (2015) reported that ginger extract induced the cell death of human pancreatic cancer cell lines Panc-1 by "autosis," a new type of cell death. The authors suggested that autosis could be caused due to reactive oxygen species production (Akimoto et al. 2015). The ethanol extract of ginger root presented antiproliferative capacity against murine melanoma cell line B164A5 as well as proapoptotic effect (Danciu et al. 2015). Ginger rhizome extract induced apoptosis on human leukemic cell lines K562 and MOLT-4. The apoptotic effect may be via mitochondrial mediated apoptotic pathway. The authors also suggested that the phytochemicals (6-gingerol, α-gingeberin) present in ginger rhizome extract may be responsible for the anticancer activity of ginger rhizome extract (Bhargava et al. 2015).

The anticancer property of shogaol metabolites has also been evaluated. For instance, 5-cysteinyl-conjugated 6-shogaol exerted strong antiproliferative activity against HCT-116 and HT-29 human colon cancer cells. The authors suggested that 5-cysteinyl-conjugated 6-shogaol regulate the expression of p53 though reactive oxygen species (ROS) generation and showed apoptosis induction activity through mitochondrial pathway (Fu et al. 2014). Ginger essential oils (GEO) has also been studied, and has demonstrated a strong *in vitro* cytotoxic activity against Dalton's Lymphoma Ascites and Ehrlich Ascites Carcinoma cell line, and, to a lesser extent it was cytotoxic to L929 cancer cell line, furthermore, GEO showed antitumor activity *in vivo* (Jeena et al. 2015). Similarly, GEO of ginger showed cytotoxic activity against HT29-19A non-muco secreting and HT29-muco secreting cell lines (Al-Tamimi et al. 2016).

Further research is needed for the elucidation of the anticancer activity of each isolated phytochemical of ginger extract as well as the mechanism of actions (Brown et al., 2009; Plengsuriyakarn et al. 2012a; Bhargava et al. 2015). Nonetheless, ginger extract has been recommended as a complementary agent in cancer treatment (Tuntiwechapikul et al. 2010).

16.2.13 *Hypericum perforatum* L. (St John's-wort)

Hypericum perforatum L. is a well-known medicinal plant belonging to the *Hypericaceae* family, genus *Hypericum* L. and it is commonly called St. John's wort (USDA 2012). It is an herbaceous perennial plant, indigenous to Europe, western Asia, and northern Africa (Grieve 2018; USDA 2012). The main phytochemicals of *H. perforatum* are synthesized and stored in glands and secretory pockets in its aerial parts (Velingkar et al. 2017). *H. perforatum* has been characterized for containing phytochemicals such as essential oils and polyphenols (Pavlović et al. 2006; Rusalepp et al. 2017). Besides, the genus *Hypericum* has been found to contain a particular chemical class of compounds: the naphthodianthrones (i.e., hypericin) and phloroglucinols (i.e., hyperforin) (Sarrou et al. 2018).

Hyperforin has been reported to suppress lymphangiogenesis which is related to the metastasis of tumor cells (Rothley et al. 2009). Moreover, herbal, leaf, and flower extracts from *H. perforatum* have shown cytotoxic activity against Caco-2 intestinal cancer cell cultures, at the highest concentration tested (100 µg/mL), and, this effect was associated with the content of rutin and hyperforin as this compounds showed a strong, positive, statistical correlation (rutin ($r = 0.865$, $p \leq 0.026$) and hyperforin ($r = 0.522$, $p \leq 0.026$) (Sarrou et al. 2018). Additionally, in HT29 cells, *H. perforatum* extract demonstrated a protective effect from oxidative DNA damage. The authors suggested that the phenolic content and their antioxidant activity may have contributing effect to this cancer preventing activity (Ramos et al. 2013). On the other hand, *H. perforatum* did not exhibit cytotoxic activity on HeLa cell line (Cenić-Milošević et al. 2014).

In another study, the antiproliferative and apoptotic effect was enhanced with a combination of treatment with photoactivated hypericin and Manumycin A on colon adenocarcinoma cells HT-29 cells in comparison to single treatments (Sackova et al. 2011). Hypericin alone showed cytotoxicity against MCF-7 human breast cancer cell line through apoptosis induction (Mirmalek et al. 2016). Additionally, it has been found that cytotoxicity of *H. perforatum* extract is comparable to the one exerted by hyperforin alone against three the hepatic cell lines (HepG2, HepaRG, and WRL-68) (Martinho et al. 2016). In an *in vitro* study, the synthesized hyperforin derivatives: hyperforin acetate and *N*,*N*-dicyclohexylamine salt of hyperforin displayed antiproliferative activity against HeLa, HepG2, MCF-7, K562, A549, K562/ADR, and A375 cell lines (Sun et al. 2011). *H. perforatum* extract is also rich in hyperoside. The anti-cancer activity of this compound was assessed *in vitro* against A549 cells and demonstrated apoptosis-inducing activity. This effect may be mediated via activation of the p38 mitogen-activated protein kinase (MAPK) and c-Jun N-terminal kinase (JNK) pathway (Yang et al. 2017).

The hydrophilic extract of *H. perforatum* induced a proapototic effect which was comparable to that caused by oxaplatin after 4 h of treatment against HT-29 colon cancer cells, although this effect was lost after 8 h of treatment (Cinci et al. 2017). This extract was rich in phenolic acids, which showed to effectively reduce the presence of ROS. However, the authors did not show a correlation analysis between this phytochemical and the proapoptotic effect. Although, there is an increasing interest in the research of anticancer potential of *H. perforatum*, there is a need of further studies to demonstrate the correlation between the bioactivity and the phytochemical content of *H. perforatum* and the mechanisms involved, besides, a comparison of the bioactivity between the different parts of the plant (Sarrou et al. 2018).

16.2.14 *Humulus lupulus* L. (Hop)

Humulus lupulus L., commonly known as hop, is a perennial vine plant belonging to the *Cannabaceae* family, genus *Humulus* (Integrated Taxonomic Information System 2018a), and it is originally to Europe, western Asia, and North America, although it is widely distributed throughout the temperate regions of the world (Zanoli and Zavatti 2008; Ravindran 2017). *H. lupulus* L. plant can grow up to around 19–29 feet in length. The leaves are darkgreen colored, long petiolate, heart shaped with 3–5 lobes and a rough surface (Zanoli and Zavatti 2008). The female inflorescences consist of "bracts" and "bracteoles" around a central axis (Almaguer et al. 2014). Hop flowers have been used traditionally for medicinal purposes. *H. lupulus* extracts have been evaluated for its anticancer activity and showed antiproliferative activity against human hepatoma (Hep3B) and colon cancer (HT-29) cells, besides, the antiproliferative activity was different depending on the solvent used for the extraction (Comert Onder et al. 2016).

Prenylated flavonoids are the main bioactive constituents reported to be responsible for its medicinal applications. The most studied prenylated flavonoids of hops are isoxanthohumol (IX) and xanthohumol (XN). The XN has demonstrated anticancer properties through antiproliferative and apoptosis-induction activity against different types of cancer cell lines such as HepG2 and Huh7 (Dorn et al. 2010). The XN reduced the proliferation of human RK33 and RK45 larynx cancer cell lines; cell cycle arrest might be one of the mechanisms involved in the anticancer activity of XN, also, the induction of proapoptotic effectors (Sławińska-Brych et al. 2015). Also, cell cycle arrest was shown by XN in pancreatic cancer celllines (Jiang et al. 2015). Both, IX and XN demonstrated to inhibit the proliferation of colon cancer cell lines HT-29 and SW620 (Hudcová et al. 2014). The XN also inhibited the growth of human medullary thyroid cancer cells. This effect was associated with the induction of phosphorylated ERK1/2 (Cook et al. 2010). The XN stimulated acid sphingomyelinase-derived ceramide for XN-mediated apoptosis of on mouse bone marrow-derived dendritic cells (Xuan et al. 2010). XN-induced apoptosis in glioblastoma cells.

Oxidative stress appeared to be involved in the induction of apoptosis, also activation of p53 has been reported to be involved in apoptosis induction (Festa et al., 2011; Zajc et al. 2012). In another work, XN showed apoptotic activity in Ca Ski cervical cancer cell line. The authors suggested that this effect may be associated to the S phase arrest in cell cycle triggered by XN. Also, this study showed an activation of caspase-3, caspase-8, and caspase-9, PARP cleavage, upregulation of p53 proteins and downregulation of XIAP (WaiKuan Yong and AbdMalek 2015).

A caspase-dependent apoptotic pathway was also reported for the induction of glioblastoma cell death by XN (Chen et al. 2016).

Sławińska-Brych et al. (2016) reported that XN have anti-proliferative and pro-apoptotic activity against A549 lung adenocarcinoma cells (Yong et al. 2015). This effect may be due to the inhibition that XN exert in the extracellular-signal-regulated kinase (ERK1/2) pathway on A549 cells. Additionally, XN induced parapoptosis of HL-600 leukemia cells, and the p38 mitogen-activated protein kinases (MAPK) signaling pathway may be implicated in the mechanism (Mi et al. 2017). Furthermore, XN apoptosis-inducing activity in leukemic cells is associated with the inhibition of NFκB, through the modification of cysteine residues of the IκBα kinase and NFκB by XN (Harikumar et al. 2009). The XN can induce autophagy in U7 glioma cells (Lu et al. 2015). Furthermore, XN has demonstrated antiproliferative activity via modulation of the *Notch 1* signaling pathway, against breast (Sun et al. 2018), pancreatic (Kunnimalaiyaan et al. 2015b) and ovarian (Drenzek et al. 2011) cancer cells as well as in hepatocellular carcinoma (Kunnimalaiyaan et al. 2015a), also, XN has shown antitumor effect *in vivo* (Sun et al., 2018). Furthermore, XN has demonstrated to inhibit angiogenesis by means of NF-B activity suppression in pancreatic cancer cells (Saito et al. 2018). Additionally, it has been reported that XN inhibit the activity of complex I in mitochondria which could be the main mechanisms of this compound to induce apoptosis in cancer cells (Zhang et al. 2015). Moreover, XN at high doses (5 μM) was shown to increase ROS production in MCF-7 breast cancer cell lines (Blanquer-Rossello et al. 2013).

On the other hand, IX can be converted by the colonic microbiota to 8-prenylnaringenin (8-PN). This compound has also shown anticancer activity. Allsopp et al. (2013) evaluated the effect of IX and 8-PN in key stages of colon carcinogenesis. The authors showed that IX and 8-PN reduces Caco-2 cell viability. IX and 8-PN mechanisms for this effect were increased in the G2/M and sub-G1 cell cycle fractions and an elevated G0/G1 and an increased sub-G1 cell-cycle fraction, respectively. Also, these compounds inhibited human HT-115 colon carcinoma cells invasiveness. PC-3 prostate cancer and UO.31 renal carcinoma cells growth was inhibited by 8-PN and 6-prenylnaringenin (Busch et al. 2015). In another study, the anti-proliferative activity of 8-PN, IX, and 6-prenylnaringenin from hop on prostate cancer cell lines PC-3 and DU145 was described. All compounds displayed inhibitory effects on the growth of prostate cancer cells. This may be due to an autophagic cell death induced by the compounds (Delmulle et al. 2007). The 8-PN and naringenin demonstrated cytotoxic activity in a glioblastoma cell line (U-118 MG), although, 8-PN had a higher cytotoxic activity. Additionally, the anticancer potential of 8-PN was correlated to the accumulation of 8-PN in U-118 MG cells. The authors suggested that the prenyl group increases cellular uptake of 8-PN and in consequence, its anticancer potential (Stompor et al. 2017b). Fungal metabolites of XN showed antiproliferative activity against the human HT-29 cell line. The authors suggested that higher anticancer property was related to the methylation of the glucose hydroxyl group (Tronina et al. 2013b). Tronina et al. (2013a) evaluated three hop flavonoids and six xanthohumol metabolites obtained from the biotransformation of XN. All compounds evaluated displayed antiproliferative effects against breast cancer (MCF-7), prostate cancer (PC-3) and colon cancer (HT-29) cell lines. Moreover, compounds XN and α,β-dihydroxanthohumol showed higher cytotoxic activity

than the positive control Cisplatin. The authors suggested that the cytotoxic effect is enhanced by the reduction of the α,β-double bound while it is reduced due to the prenyl moiety modification, and isomerization of chalcone to flavanone. Other phytochemicals evaluated in hop plants are proanthocyanidins and lupulone. For instance, proanthocyanidins showed cytotoxic activity against human colon cancer HT-29 cells. This effect was associated with the increased levels of reactive oxygen species induced by proanthocyanidins, as well as an induction of the formation of protein carbonyls in HT-29 cells (Chung et al. 2009). On the other hand, synthetic compounds form 8-PN, XN, and IX have shown antiproliferative activity (Anioł et al. 2012; Stompor et al. 2017a). For instance, Tyrrell et al. (2012) synthesized unnatural lupulone derivatives from hop plants which demonstrated to be more toxic to MDA-MB-231 breast cancer cell line than the parent lupulone itself. The bitter acids, α- and β-acids and iso-α-acids, showed to inhibit the proliferation of Caco-2 cells (Machado et al. 2017). Also, synthesized acyl derivatives of xanthohumol showed antiproliferative activity against HT-29 cell line, although this activity was lower than the parent compound (XN) (Żołnierczyk et al. 2017).

16.3 Conclusions

Extracts and isolated compounds from medicinal plants listed in the Herbal Pharmacopoeia of the Mexican United States exert their anticancer effect through several mechanisms, some of which have been assessed by numerous authors. Moreover, studies regarding the anti-proliferative and anticancer effect of medicinal plants are on the rise as a result of the unwanted secondary effects or resistance of the most used drugs against different cancer types. However, pre-clinical and clinical studies are still needed to assess the pharmacokinetics and *in vivo* potential anticancer activity of these plants.

REFERENCES

Aarland, R. C., Bañuelos-Hernández, A. E., Fragoso-Serrano, M., Sierra-Palacios, E. D. C., Díaz De León-Sánchez, F., Pérez-Flores, L. J., Rivera-Cabrera, F. & Mendoza-Espinoza, J. A. 2017. Studies on phytochemical, antioxidant, anti-inflammatory, hypoglycaemic and antiproliferative activities of *Echinacea purpurea* and *Echinacea angustifolia* extracts. *Pharmaceutical Biology*, 55, pp. 649–656.

Abudayeh, Z. H. M., Al Azzam, K. M., Naddaf, A., Karpiuk, U. V. & Kislichenko, V. S. 2015. Determination of four major saponins in skin and endosperm of seeds of horse chestnut (*Aesculus hippocastanum* L.) using high performance liquid chromatography with positive confirmation by thin layer chromatography. *Advanced Pharmaceutical Bulletin*, 5, pp. 587.

Acharya, A., Das, I., Singh, S. & Saha, T. 2010. Chemopreventive properties of indole-3-carbinol, diindolylmethane and other constituents of cardamom against carcinogenesis. *Recent Patents on Food, Nutrition and Agriculture*, 2, pp. 166–177.

Akimoto, M., Iizuka, M., Kanematsu, R., Yoshida, M. & Takenaga, K. 2015. Anticancer effect of ginger extract against pancreatic cancer cells mainly through reactive oxygen species-mediated autotic cell death. *PLoS One*, 10, pp. e0126605.

Allsopp, P., Possemiers, S., Campbell, D., Gill, C. & Rowland, I. 2013. A comparison of the anticancer properties of isoxanthohumol and 8-prenylnaringenin using *in vitro* models of colon cancer. *Biofactors,* 39, pp. 441–447.

Almaguer, C., Schönberger, C., Gastl, M., Arendt, E. K. & Becker, T. 2014. *Humulus lupulus*–a story that begs to be told. A review. *Journal of the Institute of Brewing,* 120, pp. 289–314.

Al-Tamimi, M. A., Rastall, B. & Abu-Reidah, I. M. 2016. Chemical Composition, Cytotoxic, Apoptotic and antioxidant activities of main commercial essential oils in Palestine: A comparative study. *Medicines,* 3, pp. 27.

Anioł, M., Świderska, A., Stompor, M. & Żołnierczyk, A. K. 2012. Antiproliferative activity and synthesis of 8-prenylnaringenin derivatives by demethylation of 7-O-and 4′-O-substituted isoxanthohumols. *Medicinal Chemistry Research,* 21, pp. 4230–4238.

Ansari, J., Ahmad, M., Khan, A., Fatima, N., Khan, H. J., Rastogi, N., Mishra, D. P. & Mahdi, A. A. 2016. Anticancer and Antioxidant activity of *Zingiber officinale Roscoe* rhizome. *Indian Journal of Experimental Biology,* 54, pp. 767–773.

Aybastıer, Ö., Dawbaa, S., Demir, C., Akgün, O., Ulukaya, E. & Arı, F. 2018. Quantification of DNA damage products by gas chromatography tandem mass spectrometry in lung cell lines and prevention effect of thyme antioxidants on oxidative induced DNA damage. *Mutation Research/Fundamental and Molecular Mechanisms of Mutagenesis,* 808, pp. 1–9.

Bailey-Shaw, Y. A., Williams, L. A. D., Junor, G.-A. O., Green, C. E., Hibbert, S. L., Salmon, C. N. A. & Smith, A. M. 2008. Changes in the contents of oleoresin and pungent bioactive principles of Jamaican ginger (*Zingiber officinale* Roscoe.) during maturation. *Journal of Agricultural and Food Chemistry,* 56, pp. 5564–5571.

Balladin, D. A., Headley, O., Chang-Yen, I. & Mcgaw, D. R. 1998. High pressure liquid chromatographic analysis of the main pungent principles of solar dried West Indian ginger (*Zingiber officinale* Roscoe). *Renewable Energy,* 13, pp. 531–536.

Ban, J. O., Cho, J. S., Hwang, I. G., Noh, J. W., Kim, W. J., Lee, U. S. et al. 2009. Anti-cancer effect of the combination of thiacremonone and docetaxel by inactivation of NF-κB in human cancer cells. *The Korean Society of Applied Pharmacology,* 17, pp. 403–411.

Bandara, T., Uluwaduge, I. & Jansz, E. 2012. Bioactivity of cinnamon with special emphasis on diabetes mellitus: A review. *International Journal of Food Sciences and Nutrition,* 63, pp. 380–386.

Baruah, A., Bordoloi, M. & Baruah, H. P. D. 2016. *Aloe vera*: A multipurpose industrial crop. *Industrial Crops and Products,* 94, pp. 951–963.

Begnini, K. R., Nedel, F., Lund, R. G., Carvalho, P. H. D. A., Rodrigues, M. R. A., Beira, F. T. A. & Del-Pino, F. A. B. 2014. Composition and antiproliferative effect of essential oil of *Origanum vulgare* against tumor cell Lines. *Journal of Medicinal Food,* 17, pp. 1129–1133.

Bhargava, S., Malhotra, H., Rathore, O. S., Malhotra, B., Sharma, P., Batra, A., Sharma, A. & Chiplunkar, S. V. 2015. Anti-leukemic activities of alcoholic extracts of two traditional Indian medicinal plants. *Leukemia & Lymphoma,* 56, pp. 3168–3182.

Blanquer-Rossello, M. M., Oliver, J., Valle, A. & Roca, P. 2013. Effect of xanthohumol and 8-prenylnaringenin on MCF-7 breast cancer cells oxidative stress and mitochondrial complexes expression. *Journal of Cellular Biochemistry,* 114, 2785–2794.

Brown, A. C., Shah, C., Liu, J., Pham, J. T., Zhang, J. G. & Jadus, M. R. 2009. Ginger's (*Zingiber officinale* Roscoe) inhibition of rat colonic adenocarcinoma cells proliferation and angiogenesis *in vitro. Phytotherapy Research,* 23, pp. 640–645.

Busch, C., Noor, S., Leischner, C., Burkard, M., Lauer, U. M. & Venturelli, S. 2015. Anti-proliferative activity of hop-derived prenylflavonoids against human cancer cell lines. *Wiener Medizinische Wochenschrift,* 165, pp. 258–261.

Ćalić-Dragosavac, D., Stevović, S., Zdravković-Korać, S., Milojević, J., Cingel, A. & Vinterhalter, B. 2011. Secondary metabolite of horse chestnut *in vitro* culture. *Advanced Environmental Biology,* 5, pp. 267–270.

Çandöken, E., Kuruca, S. E. & Akev, N. 2017. Evaluation of anticancer effects of Aloe vera and aloe emodin on B16F10 murine melanoma and NIH3T3 mouse embryogenic fibroblast cells. *Istanbul Journal of Pharmacy,* 47, pp. 1–1.

Cao, C. and Kindscher, K., 2016. The Medicinal Chemistry of Echinacea Species. In *Echinacea* (pp. 127–145). Springer, Cham, Switzerland.

Ceccarelli, N., Curadi, M., Picciarelli, P., Martelloni, L., Sbrana, C. & Giovannetti, M. 2010. Globe artichoke as a functional food. *Mediterranean Journal of Nutrition and Metabolism,* 3, pp. 197–201.

Cenić-Milošević, D., Tambur, Z., Bokonjić, D., Ivančajić, S., Stanojković, T., Grozdanić, N. & Juranić, Z. 2013. Antiproliferative effects of some medicinal plants on Hela cells. *Archives of Biological Sciences,* 65, pp. 65–70.

Cenić-Milošević, D., Tambur, Z., Ivančajić, S., Stanojković, T., Grozdanić, N., Kulišić, Z. & Juranić, Z. 2014. Antiproliferative effects of *Tanacetipartheni, Hypericum perforatum* and propolis on HeLa cells. *Archives of Biological Sciences,* 66, pp. 705–712.

Chen, P. H., Chang, C. K., Shih, C. M., Cheng, C. H., Lin, C. W., Lee, C. C., Liu, A. J., Ho, K. H. & Chen, K. C. 2016. The miR-204-3p-targeted IGFBP2 pathway is involved in xanthohumol-induced glioma cell apoptotic death. *Neuropharmacology,* 110, pp. 362–375.

Chung, W.-G., Miranda, C. L., Stevens, J. F. & Maier, C. S. 2009. Hop proanthocyanidins induce apoptosis, protein carbonylation, and cytoskeleton disorganization in human colorectal adenocarcinoma cells via reactive oxygen species. *Food and Chemical Toxicology,* 47, pp. 827–836.

Cichello, S. A., Yao, Q. & He, X. Q. 2016. Proliferative activity of a blend of Echinacea angustifolia and *Echinacea purpurea* root extracts in human vein epithelial, HeLa, and QBC-939 cell lines, but not in Beas-2b cell lines. *Journal of Traditional and Complementary Medicine,* 6, pp. 193–197.

Çiftçi, G. A., İşcan, A. & Kutlu, M. 2015. Escin reduces cell proliferation and induces apoptosis on glioma and lung adenocarcinoma cell lines. *Cytotechnology,* 67, pp. 893–904.

Cinci, L., Di Cesare Mannelli, L., Maidecchi, A., Mattoli, L. & Ghelardini, C. 2017. Effects of *Hypericum perforatum* extract on oxaliplatin-induced neurotoxicity: *In vitro* evaluations. *Zeitschrift für Naturforschung C: A Journal of Biosciences,* 72, pp. 219–226.

Comert Onder, F., Ay, M., Aydogan Turkoglu, S., Tura Kockar, F. & Celik, A. 2016. Antiproliferative activity of *Humulus lupulus* extracts on human hepatoma (Hep3B), colon (HT-29) cancer cells and proteases, tyrosinase, beta-lactamase enzyme inhibition studies. *The Journal of Enzyme Inhibition and Medicinal Chemistry,* 31, pp. 90–98.

Cook, M. R., Luo, J., Ndiaye, M., Chen, H. & Kunnimalaiyaan, M. 2010. Xanthohumol inhibits the neuroendocrine transcription factor achaete-scute complex-like 1, suppresses proliferation, and induces phosphorylated ERK1/2 in medullary thyroid cancer. *The American Journal of Surgery,* 199, pp. 315–318.

Čukanović, J., Ninić-Todorović, J., Ognjanov, V., Mladenović, E., Ljubojević, M. & Kurjakov, A. 2011. Biochemical composition of the horse chestnut seed (*Aesculus hippocastanum* L.). *Archives of Biological Sciences,* 63, pp. 345–351.

Danciu, C., Vlaia, L., Fetea, F., Hancianu, M., Coricovac, D. E., Ciurlea, S. A., Soica, C. M. et al. 2015. Evaluation of phenolic profile, antioxidant and anticancer potential of two main representatives of Zingiberaceae family against B164A5 murine melanoma cells. *Biological Research,* 48, p. 1.

Das, I., Acharya, A., Berry, D. L., Sen, S., Williams, E., Permaul, E., Sengupta, A., Bhattacharya, S. & Saha, T. 2012. Antioxidative effects of the spice cardamom against non-melanoma skin cancer by modulating nuclear factor erythroid-2-related factor 2 and NF-κB signalling pathways. *British Journal of Nutrition,* 108, pp. 984–997.

Delmulle, L., Berghe, T. V., Keukeleire, D. D. & Vandenabeele, P. 2007. Treatment of PC-3 and DU145 prostate cancer cells by prenylflavonoids from hop (*Humulus lupulus* L.) induces a caspase-independent form of cell death. *Phytotherapy Research,* 22, pp. 197–203.

Dorn, C., Weiss, T. S., Heilmann, J. & Hellerbrand, C. 2010. Xanthohumol, a prenylated chalcone derived from hops, inhibits proliferation, migration and interleukin-8 expression of hepatocellular carcinoma cells. *International Journal of Oncology,* 36, pp. 435–441.

Drenzek, J. G., Seiler, N. L., Jaskula-Sztul, R., Rausch, M. M. & Rose, S. L. 2011. Xanthohumol decreases Notch1 expression and cell growth by cell cycle arrest and induction of apoptosis in epithelial ovarian cancer cell lines. *Gynecologic Oncology,* 122, pp. 396–401.

Elguindy, N. M., Yacout, G. A., El Azab, E. F. & Maghraby, H. K. 2016. Chemoprotective Effect of *Elettaria Cardamomum* against chemically induced hepatocellular carcinoma in rats by inhibiting NF-κB, oxidative stress, and activity of ornithine decarboxylase. *South African Journal of Botany,* 105, pp. 251–258.

Elshafie, H., Armentano, M., Carmosino, M., Bufo, S., De Feo, V. & Camele, I. 2017. Cytotoxic activity of *Origanum vulgare* L. on hepatocellular carcinoma cell Line HepG2 and evaluation of its biological activity. *Molecules,* 22, pp. 1435.

Fakhari, S., Mahmoodi, M., Hosseini, J., Hosseini Zijoud, S.-M., Khoshdel, A., Tahamtan, M., Ahmadi, A., Menbari, M. N., Gharib, A. & Hakhamaneshi, M. S. 2014. *Aloe emodin* induces apoptosis through the up-regulation of fas in the human breast cancer cell line MCF-7. *Life Science Journal,* 11, pp. 47–53.

Fattahi, S., Ardekani, A. M., Zabihi, E., Abedian, Z., Mostafazadeh, A., Pourbagher, R. & Akhavan-Niaki, H. 2013. Antioxidant and apoptotic effects of an aqueous extract of *Urticadioica* on the MCF-7 human breast cancer cell line. *Asian Pacific Journal of Cancer Prevention,* 14, pp. 5317–5323.

Felipe, M. B. M., De Carvalho, F. M., Félix-Silva, J., Fernandes-Pedrosa, M. F., Scortecci, K. C., Agnez-Lima, L. F. & De Medeiros, S. R. B. 2013. Evaluation of genotoxic and antioxidant activity of an *Aesculus hippocastanum* L. (Sapindaceae) phytotherapeutic agent. *Biomedicine & Preventive Nutrition,* 3, pp. 261–266.

Festa, M., Capasso, A., D'acunto, C. W., Masullo, M., Rossi, A. G., Pizza, C. & Piacente, S. 2011. Xanthohumol induces apoptosis in human malignant glioblastoma cells by increasing reactive oxygen species and activating MAPK pathways. *Journal of Natural Products,* 74, pp. 2505–2513.

Foca, G., Ulrici, A., Cocchi, M., Durante, C., Vigni, M. L., Marchetti, A., Sighinolfi, S. & Tassi, L. 2011. Seeds of horse chestnut (*Aesculus hippocastanum* L.) and their possible utilization for human consumption. *Nuts and Seeds in Health and Disease Prevention.* Elsevier, Amsterdam, the Netherlands.

Fu, J., Chen, H., Soroka, D. N., Warin, R. F. & Sang, S. 2014. Cysteine-conjugated metabolites of ginger components, shogaols, induce apoptosis through oxidative stress-mediated p53 pathway in human colon cancer cells. *Journal of Agricultural and Food Chemistry,* 62, pp. 4632–4642.

Gadekar, V. S. J. & Jadhav, G. V. A. U. 2017. Review of spices as a medicine. *World Journal of Pharmacy and Pharmaceutical Sciences,* 6(8), pp. 413–425.

Gezer, C., Yücecan, S. & Rattan, S. I. S. 2015. Artichoke compound cynarin differentially affects the survival, growth, and stress response of normal, immortalized, and cancerous human cells. *Turkish Journal of Biology,* 39, pp. 299–305.

Ghasemzadeh, A., Jaafar, H. Z. & Karimi, E. 2012. Involvement of salicylic acid on antioxidant and anticancer properties, anthocyanin production and chalcone synthase activity in ginger (*Zingiber officinale* Roscoe) varieties. *International Journal of Molecular Sciences,* 13, pp. 14828–14844.

Ghazanfari, T., Yaraee, R., Shams, J., Rahmati, B., Radjabian, T. & Hakimzadeh, H. 2013. Cytotoxic effect of four herbal medicines on gastric cancer (AGS) cell line. *Food and Agricultural Immunology,* 24, pp. 1–7.

Globocan, W. 2014. Population Fact Sheet. International Agency for Research on Cancer, World Health Organization. Available: http://globocan.iarc.fr/Pages/fact_sheets_population.aspx [Accessed September 27, 2017].

Grieve, M. 2018. St. John's Wort. https://botanical.com/botanical/mgmh/s/sajohn06.html [Accessed May 4, 2018].

Gruhlke, M. C. H., Nicco, C., Batteux, F. & Slusarenko, A. J. 2017. The effects of allicin, a reactive sulfur species from garlic, on a selection of mammalian cell lines. *Antioxidants,* 6.

Gulcin Sagdicoglu Celep, A., Yilmaz, S. & Coruh, N. 2012. Antioxidant capacity and cytotoxicity of *Aesculus hippocastanum* on breast cancer MCF-7 cells. *Journal of Food & Drug Analysis,* 20(3), pp. 692–698.

Güney, G., Kutlu, H. M. & Işcan, A. 2013. The apoptotic effects of escin in the H-ras transformed 5RP7 cell line. *Phytotherapy Research,* 27, pp. 900–905.

Guo, X. & Mei, N. 2016. Aloe vera: A review of toxicity and adverse clinical effects. *Journal of Environmental Science and Health, Part C,* 34, pp. 77–96.

Han, M. H., Lee, W. S., Lu, J. N., Kim, G., Jung, J. M., Ryu, C. H., Kim, G. I. Y., Hwang, H. J., Kwon, T. K. & Choi, Y. H. 2012. *Citrus aurantium* L. exhibits apoptotic effects on U937 human leukemia cells partly through inhibition of Akt. *International Journal of Oncology,* 40, pp. 2090–2096.

Harikumar, K. B., Kunnumakkara, A. B., Ahn, K. S., Anand, P., Krishnan, S., Guha, S. & Aggarwal, B. B. 2009. Modification of the cysteine residues in IkappaBalpha kinase and NF-κB (p65) by xanthohumol leads to suppression of NF-κB-regulated gene products and potentiation of apoptosis in leukemia cells. *Blood,* 113, pp. 2003–2013.

Hou, C. C., Chen, C. H., Yang, N. S., Chen, Y. P., Lo, C. P., Wang, S. Y., Tien, Y. J., Tsai, P. W. & Shyur, L. F. 2010. Comparative metabolomics approach coupled with cell- and gene-based assays for species classification and anti-inflammatory bio-activity validation of Echinacea plants. *Journal of Nutritional Biochemistry,* 21, pp. 1045–1059.

Huang, Y.-C., Liu, K.-C. & Chiou, Y.-L. 2012. Melanogenesis of murine melanoma cells induced by hesperetin: A Citrus hydro-lysate-derived flavonoid. *Food and Chemical Toxicology,* 50, pp. 653–659.

Hudcová, T., Bryndová, J., Fialová, K., Fiala, J., Karabín, M., Jelínek, L. & Dostálek, P. 2014. Antiproliferative effects of prenylflavonoids from hops on human colon cancer cell lines. *Journal of the Institute of Brewing,* 120, pp. 225–230.

Hussain, A. I., Anwar, F., Rasheed, S., Nigam, P. S., Janneh, O. & Sarker, S. D. 2011. Composition, antioxidant and che-motherapeutic properties of the essential oils from two Origanum species growing in Pakistan. *Revista Brasileira de Farmacognosia,* 21, pp. 943–952.

Hussain, A., Sharma, C., Khan, S., Shah, K. & Haque, S. 2015. Aloe vera inhibits proliferation of human breast and cervical can-cer cells and acts synergistically with cisplatin. *Asian Pacific Journal of Cancer Prevention,* 16, pp. 2939–2946.

Im, S.-A., Kim, J.-W., Kim, H.-S., Park, C.-S., Shin, E., Do, S.-G., Park, Y. I. & Lee, C.-K. 2016. Prevention of azoxy-methane/dextran sodium sulfate-induced mouse colon carcinogenesis by processed *Aloe vera* gel. *International Immunopharmacology,* 40, pp. 428–435.

Integrated Taxonomic Information System. 2018a. *Humulus lupu-lus* L. Available: https://www.itis.gov/servlet/SingleRpt/SingleRpt?search_topic=TSN&search_value=19160#null [Accessed April 2, 2018].

Integrated Taxonomic Information System. 2018b. *Zingiberofficinale Roscoe.* Available: https://www.itis.gov/servlet/SingleRpt/SingleRpt?search_topic=TSN&search_value=42402#null [Accessed April 2, 2018].

Jeena, K., Liju, V. B. & Kuttan, R. 2015. Antitumor and cytotoxic activity of ginger essential oil (*Zingiber officinale* Roscoe). *International Journal of Pharmacy and Pharmaceutical Sciences,* 2015, pp. 4.

Jiang, N., Xin, W., Wang, T., Zhang, L., Fan, H., Du, Y., Li, C. & Fu, F. 2011. Protective effect of aescin from the seeds of *Aesculus hippocastanum* on liver injury induced by endotoxin in mice. *Phytomedicine,* 18, pp. 1276–1284.

Jiang, W., Zhao, S., Xu, L., Lu, Y., Lu, Z., Chen, C., Ni, J., Wan, R. & Yang, L. 2015. The inhibitory effects of xanthohumol, a prenylated chalcone derived from hops, on cell growth and tumorigenesis in human pancreatic cancer. *Biomedicine & Pharmacotherapy,* 73, pp. 40–47.

Jose, J., Sudhakaran, S., Kumar, S., Jayaraman, S. & Variyar, E. J. 2014. A comparative evaluation of anticancer activities of flavonoids isolated from *Mimosa pudica, Aloe vera* and *Phyllanthusniruri* against human breast carcinoma cell line (MCF-7) using MTT assay. *International Journal of Pharmacy and Pharmaceutical Sciences,* 6, pp. 319–322.

Joy, J. F. & Haber, S. L. 2007. Clinical uses of artichoke leaf extract. *American Journal of Health-System Pharmacy,* 64, pp. 1904–1909.

Kansanen, E., Kuosmanen, S. M., Leinonen, H. & Levonen, A.-L. 2013. The Keap1-Nrf2 pathway: Mechanisms of activation and dysregulation in cancer. *Redox Biology,* 1, pp. 45–49.

Kapusta, I., Janda, B., Szajwaj, B., Stochmal, A., Piacente, S., Pizza, C., Franceschi, F., Franz, C. & Oleszek, W. 2007. Flavonoids in horse chestnut (*Aesculus hippocastanum*) seeds and pow-dered waste water byproducts. *Journal of Agricultural and Food Chemistry,* 55, pp. 8485–8490.

Karimi, E., Oskoueian, E., Hendra, R., Oskoueian, A. & Jaafar, H. Z. E. 2012. Phenolic compounds characterization and bio-logical activities of *Citrus aurantium* Bloom. *Molecules,* 17, pp. 1203.

Karna, P., Chagani, S., Gundala, S. R., Rida, P. C. G., Asif, G., Sharma, V., Gupta, M. V. & Aneja, R. 2011. Benefits of ginger extract in prostate cancer. *British Journal of Nutrition,* 107, pp. 473–484.

Kim, H. J., Han, M. H., Kim, G. Y., Choi, Y. W. & Choi, Y. H. 2012. Hexane extracts of garlic cloves induce apoptosis through the generation of reactive oxygen species in Hep3B human hepatocarcinoma cells. *Oncology Reports,* 28, pp. 1757–1763.

Kindscher, K. & Riggs, M. 2016. Cultivation of *Echinacea angusti-folia* and *Echinacea purpurea. Echinacea: Herbal Medicine with a Wild History.* Springer, Cham, Switzerland.

Kindscher, K. 2016. The biology and ecology of Echinacea species. *Echinacea: Herbal Medicine with a Wild History.* Springer, Cham, Switzerland.

Kumari, S. & Dutta, A. 2013. Protective effect of *Eleteriacardamomum* (L.) maton against *Pan masala* induced damage in lung of male Swiss mice. *Asian Pacific Journal of Tropical Medicine,* 6, pp. 525–531.

Kunnimalaiyaan, S., Sokolowski, K. M., Balamurugan, M., Gamblin, T. C. & Kunnimalaiyaan, M. 2015a. Xanthohumol inhibits Notch signaling and induces apoptosis in hepatocel-lular carcinoma. *PLoS One,* 10, pp. e0127464.

Kunnimalaiyaan, S., Trevino, J., Tsai, S., Gamblin, T. C. & Kunnimalaiyaan, M. 2015b. Xanthohumol-mediated suppres-sion of Notch1 signaling is associated with antitumor activity in human pancreatic cancer cells. *Molecular Cancer Therapy,* 14, pp. 1395–403.

Kwon, H.-K., Jeon, W. K., Hwang, J.-S., Lee, C.-G., So, J.-S., Park, J.-A., Ko, B. S. & Im, S.-H. 2009. Cinnamon extract sup-presses tumor progression by modulating angiogenesis and the effector function of CD8+ T cells. *Cancer Letters,* 278, pp. 174–182.

Lanzotti, V., Scala, F. & Bonanomi, G. 2014. Compounds from Allium species with cytotoxic and antimicrobial activity. *Phytochemistry Reviews,* 13, pp. 769–791.

Lee, D.-H., Park, K.-I., Park, H.-S., Kang, S.-R., Nagappan, A., Kim, J.-A., Kim, E.-H. et al. 2012. Flavonoids isolated from Korea *Citrus aurantium* L. Induce G2/M Phase arrest and apoptosis in human gastric cancer AGS cells. *Evidence-Based Complementary and Alternative Medicine,* 2012, pp. 11.

Lee, S. H., Yumnam, S., Hong, G. E., Raha, S., Venkatarame Gowda Saralamma, V., Lee, H. J., Heo, J. D. et al. 2015. Flavonoids of Korean *Citrus aurantium* L. induce apoptosis via intrinsic pathway in human hepatoblastoma HepG2 Cells. *Phytotherapy Research,* 29, pp. 1940–1949.

Liu, Y.-H., Tsai, K.-D., Yang, S.-M., Wong, H.-Y., Chen, T.-W., Cherng, J. & Cherng, J.-M. 2017. *Cinnamomum verum* ingre-dient 2-methoxycinnamaldehyde: A new antiproliferative drug targeting topoisomerase I and II in human lung squa-mous cell carcinoma NCI-H520 cells. *European Journal of Cancer Prevention,* 26, pp. 314–323.

Lu, W.-J., Chang, C.-C., Lien, L.-M., Yen, T.-L., Chiu, H.-C., Huang, S.-Y., Sheu, J.-R. & Lin, K.-H. 2015. Xanthohumol from *Humulus lupulus* L. induces glioma cell autophagy via inhibiting Akt/mTOR/S6K pathway. *Journal of Functional Foods,* 18, pp. 538–549.

Machado, J. C., Faria, M. A., Melo, A. & Ferreira, I. M. P. L. V. O. 2017. Antiproliferative effect of beer and hop compounds against human colorectal adenocarcinome Caco-2 cells. *Journal of Functional Foods,* 36, pp. 255–261.

Majdalawieh, A. F. & Carr, R. I. 2010. *In vitro* investigation of the potential immunomodulatory and anti-cancer activities of black pepper (*Piper nigrum*) and cardamom (*Elettaria cardamomum*). *Journal of Medicinal Food,* 13, pp. 371–381.

Marrelli, M., Cristaldi, B., Menichini, F. & Conforti, F. 2015. Inhibitory effects of wild dietary plants on lipid peroxidation and on the proliferation of human cancer cells. *Food and Chemical Toxicology,* 86, pp. 16–24.

Martinho, A., Silva, S. M., Garcia, S., Moreno, I., Granadeiro, L. B., Alves, G., Duarte, A. P., Domingues, F., Silvestre, S. & Gallardo, E. 2016. Effects of *Hypericum perforatum* hydroalcoholic extract, hypericin, and hyperforin on cytotoxicity and CYP3A4 mRNA expression in hepatic cell lines: A comparative study. *Medicinal Chemistry Research,* 25, pp. 2999–3010.

Mi, X., Wang, C., Sun, C., Chen, X., Huo, X., Zhang, Y., Li, G. et al. 2017. Xanthohumol induces paraptosis of leukemia cells through p38 mitogen activated protein kinase signaling pathway. *Oncotarget,* 8, pp. 31297–31304.

Miccadei, S., Di Venere, D., Cardinali, A., Romano, F., Durazzo, A., Foddai, M. S., Fraioli, R., Mobarhan, S. & Maiani, G. 2008. Antioxidative and apoptotic properties of polyphenolic extracts from edible part of artichoke (*Cynara scolymus* L.) on cultured rat hepatocytes and on human hepatoma cells. *Nutrition and Cancer,* 60, pp. 276–283.

Miláčková, I., Kapustová, K., Mučaji, P. & Hošek, J. 2017. Artichoke leaf extract inhibits AKR1B1 and reduces NF-κB activity in human leukemic cells. *Phytotherapy Research,* 31, pp. 488–496.

Mileo, A. M., Di Venere, D., Abbruzzese, C. & Miccadei, S. 2015. Long-term exposure to polyphenols of artichoke (*Cynara scolymus* L.) exerts induction of senescence driven growth arrest in the MDA-MB231 human breast cancer cell line. *Oxidative Medicine and Cellular Longevity,* 2015, Article ID 363827, 11 pages. doi:10.1155/2015/363827.

Miraj, S. & Kiani, S. 2016. Study of therapeutic effects of Cynara scolymus L.: A review. *Der Pharmacia Lettre,* 8, pp. 168–173.

Mirmalek, S. A., Azizi, M. A., Jangholi, E., Yadollah-Damavandi, S., Javidi, M. A., Parsa, Y., Parsa, T., Salimi-Tabatabaee, S. A., Ghasemzadeh Kolagar, H. & Alizadeh-Navaei, R. 2016. Cytotoxic and apoptogenic effect of hypericin, the bioactive component of *Hypericum perforatum* on the MCF-7 human breast cancer cell line. *Cancer Cell International,* 16, pp. 3.

Mohammadi, A., Mansoori, B., Aghapour, M., Shirjang, S., Nami, S. & Baradaran, B. 2016a. The *Urticadioica* extract enhances sensitivity of paclitaxel drug to MDA-MB-468 breast cancer cells. *Biomedicine & Pharmacotherapy,* 83, pp. 835–842.

Mohammadi, A., Mansoori, B., Baradaran, P. C., Khaze, V., Aghapour, M., Farhadi, M. & Baradaran, B. 2016b. *Urticadioica* extract inhibits proliferation and induces apoptosis and related gene expression of breast cancer cells *in vitro* and *in vivo*. *Clinical Breast Cancer,* 17, pp. 463–470.

Mollazadeh, H. & Hosseinzadeh, H. 2016. Cinnamon effects on metabolic syndrome: A review based on its mechanisms. *Iranian Journal of Basic Medical Sciences,* 19, pp. 1258.

Nile, S. H., Nile, A. S. & Keum, Y.-S. 2017. Total phenolics, antioxidant, antitumor, and enzyme inhibitory activity of Indian medicinal and aromatic plants extracted with different extraction methods. *3 Biotech,* 7, pp. 76.

Özkol, H., Musa, D., Tuluce, Y. & Koyuncu, I. 2012. Ameliorative influence of Urtica dioica L against cisplatin-induced toxicity in mice bearing Ehrlich ascites carcinoma. *Drug and Chemical Toxicology,* 35, pp. 251–257.

Park, K.-I., Park, H.-S., Kim, M.-K., Hong, G.-E., Nagappan, A., Lee, H.-J., Yumnam, S. et al. 2014. Flavonoids identified from Korean *Citrus aurantium* L. inhibit Non-small cell lung cancer growth *in vivo* and *in vitro*. *Journal of Functional Foods,* 7, pp. 287–297.

Park, K.-R., Lee, J.-H., Choi, C., Liu, K.-H., Seog, D.-H., Kim, Y.-H., Kim, D.-E., Yun, C.-H. & Yea, S. S. 2007. Suppression of interleukin-2 gene expression by isoeugenol is mediated through down-regulation of NF-AT and NF-κB. *International Immunopharmacology,* 7, pp. 1251–1258.

Park, K.-R., Nam, D., Yun, H.-M., Lee, S.-G., Jang, H.-J., Sethi, G., Cho, S. K. & Ahn, K. S. 2011. β-Caryophyllene oxide inhibits growth and induces apoptosis through the suppression of PI3K/AKT/mTOR/S6K1 pathways and ROS-mediated MAPKs activation. *Cancer Letters,* 312, pp. 178–188.

Parthasarathy, V. A. & Prasath, D. 2012. 8-Cardamom A2-Peter, K.V. *Handbook of Herbs and Spices (Second Edition).* Woodhead Publishing, Cambridge, UK.

Patlolla, J. M., Qian, L., Biddick, L., Zhang, Y., Desai, D., Amin, S., Lightfoot, S. & Rao, C. V. 2013. β-Escin inhibits NNK-induced lung adenocarcinoma and ALDH1A1 and RhoA/Rock expression in A/J mice and growth of H460 human lung cancer cells. *Cancer Prevention Research,* 6, pp. 1140–1149.

Pavlović, M., Tzakou, O., Petrakis, P. V. & Couladis, M. 2006. The essential oil of *Hypericum perforatum* L., *Hypericum tetrapterum* Fries and *Hypericum olympicum* L. growing in Greece. *Flavour and Fragrance Journal,* 21, pp. 84–87.

Peng, F., Tao, Q., Wu, X., Dou, H., Spencer, S., Mang, C., Xu, L. et al. 2012. Cytotoxic, cytoprotective and antioxidant effects of isolated phenolic compounds from fresh ginger. *Fitoterapia,* 83, pp. 568–585.

Perng, D.-S., Tsai, Y.-H., Cherng, J., Kuo, C.-W., Shiao, C.-C. & Cherng, J.-M. 2016a. Discovery of a novel anti-cancer agent targeting both topoisomerase I and II in hepatocellular carcinoma Hep 3B cells *in vitro* and *in vivo*: *Cinnamomum verum* component 2-methoxycinnamaldehyde. *Journal of Drug Targeting,* 24, pp. 624–634.

Perng, D.-S., Tsai, Y.-H., Cherng, J., Wang, J.-S., Chou, K.-S., Shih, C.-W. & Cherng, J.-M. 2016b. Discovery of a novel anticancer agent with both anti-topoisomerase i and ii activities in hepatocellular carcinoma sK-hep-1 cells *in vitro* and *in vivo*: Cinnamomum verum component 2-methoxycinnamaldehyde. *Drug Design, Development and Therapy,* 10, pp. 141.

Petropoulos, S. A., Fernandes, Â., Ntatsi, G., Petrotos, K., Barros, L. & Ferreira, I. C. 2018. Nutritional value, chemical characterization and bulb morphology of Greek garlic landraces. *Molecules,* 23, pp. 319.

Petrovska, B. B. & Cekovska, S. 2010. Extracts from the history and medical properties of garlic. *Pharmacognosy Reviews,* 4, pp. 106.

Plengsuriyakarn, T., Viyanant, V., Eursitthichai, V., Picha, P., Kupradinun, P., Itharat, A. & Na-Bangchang, K. 2012a. Anticancer activities against cholangiocarcinoma, toxicity and pharmacological activities of Thai medicinal plants in animal models. *BMC Complementary and Alternative Medicine*, 12, pp. 23–23.

Plengsuriyakarn, T., Viyanant, V., Eursitthichai, V., Tesana, S., Chaijaroenkul, W., Itharat, A. & Na-Bangchang, K. 2012b. Cytotoxicity, toxicity, and anticancer activity of *Zingiber officinaleRoscoe* against cholangiocarcinoma. *Asian Pacific Journal of Cancer Prevention*, 13, pp. 4597–606.

Pomari, E., Stefanon, B. & Colitti, M. 2014. Effect of plant extracts on H_2O_2-induced inflammatory gene expression in macrophages. *Journal of Inflammation Research*, 7, pp. 103–112.

Qiblawi, S., Al-Hazimi, A., Al-Mogbel, M., Hossain, A. & Bagchi, D. 2012. Chemopreventive effects of cardamom (*Elettaria cardamomum* L.) on chemically induced skin carcinogenesis in Swiss albino mice. *Journal of Medicinal Food*, 15, pp. 576–580.

Qin, B., Nagasaki, M., Ren, M., Bajotto, G., Oshida, Y. & Sato, Y. 2003. Cinnamon extract (traditional herb) potentiates *in vivo* insulin-regulated glucose utilization via enhancing insulin signaling in rats. *Diabetes Research and Clinical Practice*, 62, pp. 139–148.

Ramadan, A., Afifi, N. A., Yassin, N. Z., Abdel-Rahman, R. F., Hassan, A. H. & Hany, M. 2014. Hepatoprotective effect of artichoke extract against pre-cancerous lesion of experimentally induced hepatocellular carcinoma in rats. *Life Science Journal*, 11.

Ramos, A. A., Marques, F., Fernandes-Ferreira, M. & Pereira-Wilson, C. 2013. Water extracts of tree Hypericum sps. protect DNA from oxidative and alkylating damage and enhance DNA repair in colon cells. *Food Chemistry and Toxicology*, 51, pp. 80–86.

Rastogi, N., Duggal, S., Singh, S. K., Porwal, K., Srivastava, V. K., Maurya, R., Bhatt, M. L. B. & Mishra, D. P. 2015. Proteasome inhibition mediates p53 reactivation and anti-cancer activity of 6-Gingerol in cervical cancer cells. *Oncotarget*, 6, pp. 43310–43325.

Ravindran, P. N. 2017. *The Encyclopedia of Herbs and Spices*. CABI, Wallingford, UK.

Rimmon, A., Vexler, A., Berkovich, L., Earon, G., Ron, I. & Lev-Ari, S. 2013. Escin chemosensitizes human pancreatic cancer cells and inhibits the nuclear factor-kappaB signaling pathway. *Biochemistry Research International*, 2013, Article ID 251752, 9 pages. doi:10.1155/2013/251752.

Rothley, M., Schmid, A., Thiele, W., Schacht, V., Plaumann, D., Gartner, M., Yektaoglu, A. et al. 2009. Hyperforin and aristoforin inhibit lymphatic endothelial cell proliferation *in vitro* and suppress tumor-induced lymphangiogenesis *in vivo*. *International Journal of Cancer*, 125, pp. 34–42.

Rusalepp, L., Raal, A., Püssa, T. & Mäeorg, U. 2017. Comparison of chemical composition of *Hypericum perforatum* and *H.maculatum* in Estonia. *Biochemical Systematics and Ecology*, 73, pp. 41–46.

Sackova, V., Kulikova, L., Kello, M., Uhrinova, I. & Fedorocko, P. 2011. Enhanced antiproliferative and apoptotic response of HT-29 adenocarcinoma cells to combination of photoactivated hypericin and farnesyltransferase inhibitor manumycin A. *International Journal Molecular Science*, 12, pp. 8388–8405.

Saha, A., Blando, J., Silver, E., Beltran, L., Sessler, J. & Digiovanni, J. 2014. 6-Shogaol from dried ginger inhibits growth of prostate cancer cells both *in vitro* and *in vivo* through inhibition of STAT3 and NF-κB signaling. *Cancer Prevention Research*, 7, pp. 627–38.

Saito, K., Matsuo, Y., Imafuji, H., Okubo, T., Maeda, Y., Sato, T., Shamoto, T. et al. 2018. Xanthohumol inhibits angiogenesis by suppressing nuclear factor-κB activation in pancreatic cancer. *Cancer Science*, 109, pp. 132–140.

Sarrou, E., Giassafaki, L. P., Masuero, D., Perenzoni, D., Vizirianakis, I. S., Irakli, M., Chatzopoulou, P. & Martens, S. 2018. Metabolomics assisted fingerprint of *Hypericum perforatum* chemotypes and assessment of their cytotoxic activity. *Food Chemistry and Toxicology*, 114, pp. 325–333.

Savini, I., Arnone, R., Catani, M. V. & Avigliano, L. 2009. *Origanum vulgare* induces apoptosis in human colon cancer Caco-2 Cells. *Nutrition and Cancer*, 61, pp. 381–389.

Secretaría De Salud 2013. *Farmacopea Herbolaria de los Estados Unidos Mexicanos*, D.F., México, Secretaría de Salud.

Sękara, A., Kalisz, A., Gruszecki, R., Grabowska, A. & Kunicki, E. 2015. Globe artichoke–a vegetable, herb and ornamental of value in central Europe. *The Journal of Horticultural Science and Biotechnology*, 90, pp. 365–374.

Sengottuvelu, S. 2011. Chapter 34-Cardamom (*Elettaria cardamomum* Linn. Maton) seeds in health A2-preedy, Victor R. In: Watson, R. R. & Patel, V. B. (Eds.) *Nuts and Seeds in Health and Disease Prevention*. Academic Press, San Diego, CA.

Shen, C.-Y., Jiang, J.-G., Zhu, W. & Ou-Yang, Q. 2017a. Anti-inflammatory effect of essential oil from *Citrus aurantium* L. var. amara Engl. *Journal of Agricultural and Food Chemistry*, 65, pp. 8586–8594.

Shen, C.-Y., Yang, L., Jiang, J.-G., Zheng, C.-Y. & Zhu, W. 2017b. Immune enhancement effects and extraction optimization of polysaccharides from *Citrus aurantium* L. var. amara Engl. *Food & Function*, 8, pp. 796–807.

Shin, D. Y., Kim, G. Y., Hwang, H. J., Kim, W. J. & Choi, Y. H. 2014. Diallyl trisulfide-induced apoptosis of bladder cancer cells is caspase-dependent and regulated by PI3K/Akt and JNK pathways. *Environmental Toxicology and Pharmacology*, 37, pp. 74–83.

Shirali, S., Barari, A., Hosseini, S. A. & Khodadi, E. 2017. Effects of six weeks endurance training and *Aloe vera* supplementation on COX-2 and VEGF levels in mice with breast cancer. *Asian Pacific Journal of Cancer Prevention*, 18, pp. 31.

Shrestha, S., Song, Y. W., Kim, H., Lee, D. S. & Cho, S. K. 2016. Sageone, a diterpene from *Rosmarinus officinalis*, synergizes with cisplatin cytotoxicity in SNU-1 human gastric cancer cells. *Phytomedicine*, 23, pp. 1671–1679.

Shukla, Y. & Singh, M. 2007. Cancer preventive properties of ginger: A brief review. *Food and Chemical Toxicology*, 45, pp. 683–690.

Sławińska-Brych, A., Król, S. K., Dmoszyńska-Graniczka, M., Zdzisińska, B., Stepulak, A. & Gagoś, M. 2015. Xanthohumol inhibits cell cycle progression and proliferation oflarynx cancer cells *in vitro*. *Chemico-Biological Interactions*, 240, pp. 110–118.

Sławińska-Brych, A., Zdzisińska, B., Dmoszyńska-Graniczka, M., Jeleniewicz, W., Kurzepa, J., Gagoś, M. & Stepulak, A. 2016. Xanthohumol inhibits the extracellular signal regulated kinase (ERK) signalling pathway and suppresses cell growth of lung adenocarcinoma cells. *Toxicology*, 357–358, pp. 65–73.

Spelman, K., Iiams-Hauser, K., Cech, N. B., Taylor, E. W., Smirnoff, N. & Wenner, C. A. 2009. Role for PPARγ in IL-2 inhibition in T cells by Echinacea-derived undeca-2E-ene-8,10-diynoic acid isobutylamide. *International Immunopharmacology,* 9, pp. 1260–1264.

Srihari, R., Surendranath, A. R., Kasturacharya, N., Shivappa, K. C., Sivasitambaram, N. D. & Dhananjaya, B. 2015. Evaluating the cytotoxic potential of methonolic leaf extract of *Aloe vera* on MCF-7 breast cancer cell lines. *International Journal Pharmacy and Pharmaceutical Sciences,* 7, pp. 81–83.

Srihari, T., Sengottuvelan, M. & Nalini, N. 2008. Dose-dependent effect of oregano (*Origanum vulgare* L.) on lipid peroxidation and antioxidant status in 1,2-dimethylhydrazine-induced rat colon carcinogenesis. *Journal of Pharmacy and Pharmacology,* 60, pp. 787–794.

Stompor, M., Switalska, M., Podgorski, R., Uram, L., Aebisher, D. & Wietrzyk, J. 2017a. Synthesis and biological evaluation of 4′-O-acetyl-isoxanthohumol and its analogues as antioxidant and antiproliferative agents. *Acta Biochimica Polonica,* 64, pp. 577–583.

Stompor, M., Uram, Ł. & Podgórski, R. 2017b. *In vitro* effect of 8-prenylnaringenin and naringenin on fibroblasts and glioblastoma cells-cellular accumulation and cytotoxicity. *Molecules,* 22, p. 1092.

Sun, F., Liu, J. Y., He, F., Liu, Z., Wang, R., Wang, D. M., Wang, Y. F. & Yang, D. P. 2011. *In-vitro* antitumor activity evaluation of hyperforin derivatives. *Journal of Asian Natural Product Research,* 13, pp. 688–699.

Sun, Z., Zhou, C., Liu, F., Zhang, W., Chen, J., Pan, Y., Ma, L. et al. 2018. Inhibition of breast cancer cell survival by Xanthohumol via modulation of the Notch signaling pathway *in vivo* and *in vitro. Oncology Letters,* 15, pp. 908–916.

Szychowski, K. A., Rybczyńska-Tkaczyk, K., Gaweł-Bęben, K., Ašwieca, M., Kara, M., Jakubczyk, A., Matysiak, M., Binduga, U. E. & Gmiński, J. 2018. Characterization of active compounds of different garlic (*Allium sativum* L.) cultivars. *Polish Journal of Food and Nutrition Sciences,* 68, pp. 73–81.

Tai, J., Cheung, S., Wu, M. & Hasman, D. 2012. Antiproliferation effect of Rosemary (*Rosmarinus officinalis*) on human ovarian cancer cells *in vitro. Phytomedicine,* 19, pp. 436–443.

Tang, X., Wei, R., Deng, A. & Lei, T. 2017. Protective effects of ethanolic extracts from artichoke, an edible herbal medicine, against acute alcohol-induced liver injury in mice. *Nutrients,* 9, p. 1000.

Tomasin, R. & Cintra Gomes-Marcondes, M. C. 2011. Oral administration of Aloe vera and honey reduces walker tumour growth by decreasing cell proliferation and increasing apoptosis in tumour tissue. *Phytotherapy Research,* 25, pp. 619–623.

Tong, X.-P., Ma, Y.-X., Quan, D.-N., Zhang, L., Yan, M. & Fan, X.-R. 2017. Rosemary extracts upregulate Nrf2, Sestrin2, and MRP2 protein level in human hepatoma HepG2 Cells. *Evidence-Based Complementary and Alternative Medicine,* 2017, p. 7.

Tronina, T., Bartmańska, A., Filip-Psurska, B., Wietrzyk, J., Popłoński, J. & Huszcza, E. 2013a. Fungal metabolites of xanthohumol with potent antiproliferative activity on human cancer cell lines *in vitro. Bioorganic & Medicinal Chemistry,* 21, pp. 2001–2006.

Tronina, T., Bartmańska, A., Milczarek, M., Wietrzyk, J., Popłoński, J., Rój, E. & Huszcza, E. 2013b. Antioxidant and antiproliferative activity of glycosides obtained by biotransformation of xanthohumol. *Bioorganic & Medicinal Chemistry Letters,* 23, pp. 1957–1960.

Tsai, C.-W., Lin, C.-Y., Lin, H.-H. & Chen, J.-H. 2011. Carnosic acid, a rosemary phenolic compound, induces apoptosis through reactive oxygen species-mediated p38 activation in human neuroblastoma IMR-32 cells. *Neurochemical Research,* 36, pp. 2442.

Tsai, K.-D., Cherng, J., Liu, Y.-H., Chen, T.-W., Wong, H.-Y., Yang, S.-M., Chou, K.-S. & Cherng, J.-M. 2016a. Cinnamomum verum component 2-methoxycinnamaldehyde: A novel antiproliferative drug inducing cell death through targeting both topoisomerase I and II in human colorectal adenocarcinoma COLO 205 cells. *Food & Nutrition Research,* 60, pp. 31607.

Tsai, K.-D., Liu, Y.-H., Chen, T.-W., Yang, S.-M., Wong, H.-Y., Cherng, J., Chou, K.-S. & Cherng, J.-M. 2016b. Cuminaldehyde from *Cinnamomum verum* induces cell death through targeting topoisomerase 1 and 2 in human colorectal adenocarcinoma COLO 205 cells. *Nutrients,* 8, pp. 318.

Tsai, Y.-L., Chiu, C.-C., Chen, J. Y.-F., Chan, K.-C. & Lin, S.-D. 2012. Cytotoxic effects of Echinacea purpurea flower extracts and cichoric acid on human colon cancer cells through induction of apoptosis. *Journal of Ethnopharmacology,* 143, pp. 914–919.

Tuntiwechapikul, W., Taka, T., Songsomboon, C., Kaewtunjai, N., Imsumran, A., Makonkawkeyoon, L., Pompimon, W. & Lee, T. R. 2010. Ginger extract inhibits human telomerase reverse transcriptase and c-Myc expression in A549 lung cancer cells. *Journal of Medicinal Food,* 13, pp. 1347–1354.

Tyrrell, E., Archer, R., Tucknott, M., Colston, K., Pirianov, G., Ramanthan, D., Dhillon, R., Sinclair, A. & Skinner, G. A. 2012. The synthesis and anticancer effects of a range of natural and unnatural hop β-acids on breast cancer cells. *Phytochemistry Letters,* 5, pp. 144–149.

USDA 2012. Plant guide for common St. Johnswort *(Hypericum perforatum)*. Plant Materials Center, Cape May Court House, NJ.

USDA. 2018a. National Plant Germplasm System. Beltsville, MD: National Germplasm Resources Laboratory. Available: https://npgsweb.ars-grin.gov/gringlobal/taxonomydetail.aspx [Accessed May 23, 2018].

USDA. 2018b. *The Plants Database.* Greensboro, NC 27401-4901: National Plant Data Team. Available: http://plants.usda.gov [Accessed May 23, 2018].

Valdés, A., Artemenko, K. A., Bergquist, J., García-Cañas, V. & Cifuentes, A. 2016. Comprehensive proteomic study of the antiproliferative activity of a polyphenol-enriched rosemary extract on colon cancer cells using nanoliquid chromatography–orbitrap MS/MS. *Journal of Proteome Research,* 15, pp. 1971–1985.

Vangalapati, M., Satya, S., Prakash, S. & Avanigadda, S. 2012. A review on pharmacological activities and clinical effects of cinnamon species. *Research Journal of Pharmaceutical, Biological and Chemical Sciences,* 3, pp. 653–663.

Velingkar, V. S., Gupta, G. L. & Hegde, N. B. 2017. A current update on phytochemistry, pharmacology and herb–drug interactions of *Hypericum perforatum. Phytochemistry Reviews,* 16, pp. 725–744.

Vígh, S., Zsvér-Vadas, Z., Pribac, C., Moş, L., Cziáky, Z., Czapár, M., Mihali, C. V., Turcuş, V. & Máthé, E. 2016. Artichoke (*Cynara scolymus* L.) Extracts are showing concentration-dependent hormetic and cytotoxic effects on breast cancer cell lines. *Studia Universitatis Vasile Goldis Seria Stiintele Vietii (Life Sciences Series)*, 26, pp. 423.

Wang, H.-C., Hsieh, S.-C., Yang, J.-H., Lin, S.-Y. & Sheen, L.-Y. 2012a. Diallyl trisulfide induces apoptosis of human basal cell carcinoma cells via endoplasmic reticulum stress and the mitochondrial pathway. *Nutrition and Cancer*, 64, pp. 770–780.

Wang, W., Li, N., Luo, M., Zu, Y. & Efferth, T. 2012b. Antibacterial activity and anticancer activity of *Rosmarinus officinalis* L. essential oil compared to that of its main components. *Molecules*, 17, p. 2704.

Wang, Z., Liu, Z., Cao, Z. & Li, L. 2012c. Allicin induces apoptosis in EL-4 cells *in vitro* by activation of expression of caspase-3 and -12 and up-regulation of the ratio of Bax/Bcl-2. *Natural Product Research*, 26, pp. 1033–1037.

Wei, J., Huang, Q., Bai, F., Lin, J., Nie, J., Lu, S., Lu, C., Huang, R., Lu, Z. & Lin, X. 2017. Didymin induces apoptosis through mitochondrial dysfunction and up-regulation of RKIP in human hepatoma cells. *Chemico-Biological Interactions*, 261, pp. 118–126.

Wong, H. Y., Tsai, K. D., Liu, Y. H., Yang, S. M., Chen, T. W., Cherng, J., Chou, K. S., Chang, C. M., Yao, B. T. & Cherng, J. M. 2016. Cinnamomum verum component 2-methoxycin-namaldehyde: A novel anticancer agent with both anti-topoi-somerase I and II activities in human lung adenocarcinoma A549 cells *in vitro* and *in vivo*. *Phytotherapy Research*, 30, pp. 331–340.

World Health Organization. 2018. *Global Health Observatory Data*. Available: https://www.who.int/gho/en/ [Accessed January 27, 2018].

Xuan, N. T., Shumilina, E., Gulbins, E., Gu, S., Götz, F. & Lang, F. 2010. Triggering of dendritic cell apoptosis by xanthohumol. *Molecular Nutrition & Food Research*, 54, pp. S214–S224.

Yaglıoglu, A. S., Akdulum, B., Erenler, R., Demirtas, I., Telci, I. & Tekin, S. 2013. Antiproliferative activity of pentadeca-(8E, 13Z) dien-11-yn-2-one and (E)-1,8-pentadecadiene from *Echinacea pallida* (Nutt.) Nutt. roots. *Medicinal Chemistry Research*, 22, pp. 2946–2953.

Yan, M., Li, G., Petiwala, S. M., Householter, E. & Johnson, J. J. 2015. Standardized rosemary (*Rosmarinus officinalis*) extract induces Nrf2/sestrin-2 pathway in colon cancer cells. *Journal of Functional Foods*, 13, pp. 137–147.

Yang, S.-M., Tsai, K.-D., Wong, H.-Y., Liu, Y.-H., Chen, T.-W., Cherng, J., Hsu, K.-C., Ang, Y.-U. & Cherng, J.-M. 2016. Molecular mechanism of *Cinnamomum verum* component cuminaldehyde inhibits cell growth and induces cell death in human lung squamous cell carcinoma NCI-H520 cells in vitro and *in vivo*. *Journal of Cancer*, 7, p. 251.

Yang, Y., Tantai, J., Sun, Y., Zhong, C. & Li, Z. 2017. Effect of hyperoside on the apoptosis of A549 human non-small cell lung cancer cells and the underlying mechanism. *Molecular Medicine Reports*, 16, pp. 6483–6488.

Yesil-Celiktas, O., Sevimli, C., Bedir, E. & Vardar-Sukan, F. 2010. Inhibitory effects of rosemary extracts, carnosic acid and ros-marinic acid on the growth of various human cancer cell lines. *Plant Foods for Human Nutrition*, 65, pp. 158–163.

Yildiz, E., Karabulut, D. & Yesil-Celiktas, O. 2014. A bioactivity based comparison of *Echinacea purpurea* extracts obtained by various processes. *The Journal of Supercritical Fluids*, 89, pp. 8–15.

Yong, W. K. & Abd Malek, S. N. 2015. Xanthohumol induces growth inhibition and apoptosis in Ca Ski human cervical cancer cells. *Evidence-Based Complementary and Alternative Medicine*, 2015, p. 10.

Yong, W. K., Ho, Y. F. & Malek, S. N. 2015. Xanthohumol induces apoptosis and S phase cell cycle arrest in A549 non-small cell lung cancer cells. *Pharmacognosy Magazine*, 11, pp. S275–S283.

Zajc, I., Filipic, M. & Lah, T. T. 2012. Xanthohumol induces different cytotoxicity and apoptotic pathways in malignant and normal astrocytes. *Phototherapy Research*, 26, pp. 1709–1713.

Zancan, K. C., Marques, M. O. M., Petenate, A. J. & Meireles, M. A. A. 2002. Extraction of ginger (*Zingiber officinale* Roscoe) oleoresin with CO_2 and co-solvents: A study of the antioxidant action of the extracts. *The Journal of Supercritical Fluids*, 24, pp. 57–76.

Zanoli, P. & Zavatti, M. 2008. Pharmacognostic and pharmacological profile of *Humulus lupulus* L. *Journal of Ethnopharmacology*, 116, pp. 383–396.

Zhai, Z., Solco, A., Wu, L., Wurtele, E. S., Kohut, M. L., Murphy, P. A. & Cunnick, J. E. 2009. Echinacea increases arginase activity and has anti-inflammatory properties in RAW 264.7 macrophage cells, indicative of alternative macrophage activa-tion. *Journal of Ethnopharmacology*, 122, pp. 76–85.

Zhang, B., Chu, W., Wei, P., Liu, Y. & Wei, T. 2015. Xanthohumol induces generation of reactive oxygen species and triggers apoptosis through inhibition of mitochondrial electron trans-fer chain complex I. *Free Radical Biology and Medicine*, 89, pp. 486–497.

Żołnierczyk, A. K., Baczyńska, D., Potaniec, B., Kozłowska, J., Grabarczyk, M., Woźniak, E. & Anioł, M. 2017. Antiproliferative and antioxidant activity of xanthohu-mol acyl derivatives. *Medicinal Chemistry Research*, 26, pp. 1764–1771.

Index

Note: Page numbers in italic and bold refer to figures and tables, respectively.